Pearson
BTEC National
Construction

Student Book

Simon Topliss
Mike Hurst
Simon Cummings
Sobrab Donyavi
Andrew Buckenham

P Pearson

Published by Pearson Education Limited, 80 Strand, London, WC2R 0RL.

www.pearsonschoolsandfecolleges.co.uk

Copies of official specifications for all Edexcel qualifications may be found on the website: www.edexcel.com

Text © Pearson Education Limited 2017
Edited by Julia Sandford-Cooke
Typeset by Tech-Set Ltd
Original illustrations © Pearson Education Ltd
Illustrated by Tech-Set Ltd
Picture research by Susie Prescott
Cover photo © Anusorn Abthaisong / Shutterstock.com

The rights of Simon Topliss, Mike Hurst, Simon Cummings, Andrew Buckenham and Sohrab Donyavi to be identified as authors of this work have been asserted by them in accordance with the Copyright, Designs and Patents Act 1988.

First published 2017

2023
10 9

British Library Cataloguing in Publication Data
A catalogue record for this book is available from the British Library

ISBN 978 1 292 18404 3

Printed in Great Britain by Ashford Colour Press Ltd

Acknowledgements

We would like to thank Paul Monroe and Tom Watson for their invaluable help in reviewing this book. We would also like to thank Terry Grimwood for his assistance in the development of this book.

The publisher would like to thank the following for their kind permission to reproduce their materials:

p. 7 Table 1.4 taken from Notable updates in V2.0, ICE (Inventory of Carbon & Energy), reprinted with permission of Carbon Solutions: www.carbonsolutions.com/resources/ice%20v2.0%20-%20jan%202011.xls; **p. 12** Table 1.13 reprinted with permission of GreenSpec: www.greenspec.co.uk/building-design/insulation-materials-thermal-properties/; **p. 51** Table 1.36 and **p. 72** Table 1.55 reprinted with permission of The Engineering ToolBox: www.engineeringtoolbox.com/light-level-rooms-d_708.html; **p. 72** Table 1.56 used with permission of CLEAR (Comfortable Low Energy ARchitecture): www.new-learn.info/packages/clear/visual/daylight/analysis/hand/daylight_factor.html; **p. 122** Table 2.7 taken from CIBSE *Guide A: Environmental design*, by permission of the Chartered Institution of Building Services Engineers; **p. 152–3** Table 3.3 reproduced from *Journal of Management in Engineering*, November 1998, Vol 14, Issue 6, with permission from ASCE; **p. 196** Table 4.3 republished by permission of British Standards Institution; p. 197 Figure 4.9 used with permission of David Norbury; **p. 317** illustration reproduced from Fane, B. and Byrnes, D. *AutoCAD 2014 for Dummies*, © 2013 John Wiley & Sons – republished with permission; **p. 336** illustration of the Free University of Berlin republished by permission of Nigel Young / Foster + Partners; sketch republished with permission from Circle Line Art School; **p. 443** Figure 13.1 drawing by Pippa Hodson.

The authors and publisher would like to thank the following individuals and organisations for permission to reproduce photographs:

(Key: b-bottom; c-centre; l-left; r-right; t-top)

1 Shutterstock.com: TFoxFoto; **20 123RF.com:** Montree Laoseeku (2/tr/Rust); **Alamy Stock Photo:** Tihon L1 (5/cr/Steel); **Shutterstock.com:** Tom Gowanlock (7/br/Wet Rot); Stefan Ataman (4/cr/Mortar); Paul Vinten (6/br/Woodworm); Mahey (3/cr/Brickwork); Donikz (1/tr/ASR); **79 Shutterstock.com:** Nadya Eugene; **86 Alamy Stock Photo:** Yon Marsh (bl); **88 Alamy Stock Photo:** Andrew Holt (t); **Shutterstock.com:** Rex / High Level Photography (b); **94 Alamy Stock Photo:** Xiaobo Chen (t); **96 Alamy Stock Photo:** scenicireland.com / Christopher Hill Photographic; **111 Fotolia.com:** Lev; **112 Fotolia.com:** Lev (b); **113 123RF.com:** Bloodua (cr); **120 Alamy Stock Photo:** Athol Pictures (cl); **127 Alamy Stock Photo:** Dave Ellison (bl); **135 Alamy Stock Photo:** Paul Springett A; **141 Shutterstock.com:** Ndoeljindoel (br); **147 Alamy Stock Photo:** Alex Segre (cl); **151 Alamy Stock Photo:** Frances Roberts (cl); **156 Alamy Stock Photo:** Blend Images (cl); **159 Alamy Stock Photo:** Moodboard (cr); **160 Alamy Stock Photo:** Kelly Redinger / Design Pics Inc (cr); **161 Fotolia.com:** Robert Kneschke (br); **177 Shutterstock.com:** benik.at; **181 Shutterstock.com:** Ruslan Sitarchuk (tl); **182 Shutterstock.com:** Dmitry Kalinovsky (tr); **183 Shutterstock.com:** Dmitry Kalinovsky (tl); **185 Alamy Stock Photo:** Michael Doolittle (br); **Shutterstock.com:** SV Production (c); **188 Alamy Stock Photo:** David Wootton (br); **189 Alamy Stock Photo:** Westend61 (cr); David Wootton (tl); **190 Shutterstock.com:** Htneim (cr); **221 Shutterstock.com:** Bikeriderlondon (tr); **222 123RF.com:** Flippo (tl); **224 Shutterstock.com:** Aisyaqilumaranas; **233 Alamy Stock Photo:** Chris Ratcliffe (bl); **236 Getty Images:** DigitalVision / Morsa Images (tl); **238 Shutterstock.com:** Monkey Business Images (t); **240 Alamy Stock Photo:** Hongqi Zhang (tl); **242 Alamy Stock Photo:** CBsigns (br); **255 Shutterstock.com:** SpeedKingz (tr); **256 Pearson Education:** Studio 8 (tl); **257 Shutterstock.com:** Dmitry Kalinovsky; **275 Shutterstock.com:** Vaschenko Roman (cr); **299 Shutterstock.com:** Rob Byron (tr); **300 Pearson Education Ltd:** Gareth Boden (tl); **301 Shutterstock.com:** StockLite; **304 123RF.com:** Decha Anunthanapong (cl); **DK Images:** Steve Gorton (tl); **Shutterstock.com:** Feng Yu (bl); **305 Plantec:** (cl); **Shutterstock.com:** Slavoljub Pantelic (tl); **341 Shutterstock.com:** Wavebreakmedia (tr); **342 Pearson Education Ltd:** Jules Selmes (tl); **343 Shutterstock.com:** Jaka Azman; **348 North Lincolnshire Council;** **365 Shutterstock.com:** Ashwin (tr); **366 Pearson Education:** Jules Selmes (tl); **367 Getty Images:** Sturti / E+; **379 Shutterstock.com:** Imfoto (cr); **400 Shutterstock.com:** Cornfield (tl); **403 Shutterstock.com:** Racorn (tr); **404 Pearson Education:** Gareth Boden (tl); **405 Shutterstock.com:** Lev Kropotov; **416 Shutterstock.com:** Dmitry Naumov (br); **417 Alamy Stock Photo:** MSRF (tr); **437 Shutterstock.com:** Pressmaster (tr); **438 Pearson Education Ltd:** Gareth Boden (tl); **439 Shutterstock.com:** Solis Images; **463 Shutterstock.com:** Wavebreakmedia (tr); **464 Shutterstock.com:** Robert Brown Stock (tl)

Cover images: *Front:* **Shutterstock.com:** Anusorn Abthaisong

All other images © Pearson Education

Websites

Pearson Education Limited is not responsible for the content of any external internet sites. It is essential for tutors to preview each website before using it in class so as to ensure that the URL is still accurate, relevant and appropriate. We suggest that tutors bookmark useful websites and consider enabling students to access them through the school/college intranet.

A note from the publisher

In order to ensure that this resource offers high-quality support for the associated Pearson qualification, it has been through a review process by the awarding body. This process confirms that this resource fully covers the teaching and learning content of the specification or part of a specification at which it is aimed. It also confirms that it demonstrates an appropriate balance between the development of subject skills, knowledge and understanding, in addition to preparation for assessment.

Endorsement does not cover any guidance on assessment activities or processes (e.g. practice questions or advice on how to answer assessment questions), included in the resource nor does it prescribe any particular approach to the teaching or delivery of a related course.

While the publishers have made every attempt to ensure that advice on the qualification and its assessment is accurate, the official specification and associated assessment guidance materials are the only authoritative source of information and should always be referred to for definitive guidance.

Pearson examiners have not contributed to any sections in this resource relevant to examination papers for which they have responsibility.

Examiners will not use endorsed resources as a source of material for any assessment set by Pearson.

Endorsement of a resource does not mean that the resource is required to achieve this Pearson qualification, nor does it mean that it is the only suitable material available to support the qualification, and any resource lists produced by the awarding body shall include this and other appropriate resources.

Contents

How to use this book

Welcome to your BTEC National Construction course!

Choosing to study a BTEC National in Construction is a great decision for a number of reasons. The construction industry is a growing sector with many different specialisms. Studying this qualification will give you clear path of progression towards eventually working in this industry.

You may be considering a career working in site management or looking to specialise in building services or civil engineering. In all these cases, this course will help you to develop the underlying knowledge you will need to head into a career in construction.

How your BTEC is structured

Your BTEC National is divided into **mandatory units** (the ones you must do) and **optional units** (the ones you can choose to do). The number of mandatory and optional units will vary depending on the type of BTEC National you are doing.

This book supports all the mandatory units on the Construction and the Built Environment pathways of the qualification. Many of these mandatory units are also part of the Building Services and Civil Engineering pathways of these qualifications at Diploma and Extended Diploma level. This book also contains some popular optional units.

Your complete BTEC National will be made up of a range of mandatory and optional units. This book focuses on supporting the mandatory units that you will have to understand in order to complete the qualification.

Your learning experience

You may not realise it but you are always learning. Your educational and life experiences are constantly shaping your ideas and thinking, and how you view and engage with the world around you.

You are the person most responsible for your own learning experience so you must understand what you are learning, why you are learning it and why it is important both to your course and to your personal development. Your learning can be seen as a journey with four phases.

Phase 1	Phase 2	Phase 3	Phase 4
You are introduced to a topic or concept and you start to develop an awareness of what learning is required.	You explore the topic or concept through different methods (e.g. research, questioning, analysis, deep thinking, critical evaluation) and form your own understanding.	You apply your knowledge and skills to a task designed to test your understanding.	You reflect on your learning, evaluate your efforts, identify gaps in your knowledge and look for ways to improve.

During each phase, you will use different learning strategies to secure the core knowledge and skills you need. This student book has been written using similar learning principles, strategies and tools. It has been designed to support your learning journey, to give you control over your own learning, and to equip you with the knowledge, understanding and tools you need to be successful in your future studies or career.

Features of this book

This student book contains many different features. They are there to help you learn about key topics in different ways and understand them from multiple perspectives. Together, these features:

▶ explain what your learning is about
▶ help you to build your knowledge
▶ help you to understand how to succeed in your assessment
▶ help you to reflect on and evaluate your learning
▶ help you to link your learning to the workplace.

Each individual feature has a specific purpose, designed to support important learning strategies. For example, some features will:

▶ encourage you to question assumptions about what you are learning
▶ help you to think beyond what you are reading about
▶ help you to make connections between different areas of your learning and across units
▶ draw comparisons between your own learning and real-world workplace environments
▶ help you to develop some of the important skills you will need for the workplace, including teamwork, effective communication and problem solving.

Features that explain what your learning is about

Getting to know your unit

This section introduces the unit and explains how you will be assessed. It gives an overview of what will be covered and will help you to understand why you are doing the things you are asked to do in this unit.

Getting started

This is designed to get you thinking about the unit and what it involves. This feature will also help you to identify what you may already know about some of the topics in the unit and act as a starting point for understanding the skills and knowledge you will need to develop to complete the unit.

Features that help you to build your knowledge

Research

This asks you to research a topic in greater depth. These features will help to expand your understanding of a topic and develop your research and investigation skills. All of this will be invaluable for your future progression, both professionally and academically.

Worked example

Worked examples show the process you need to follow to solve a problem, such as a maths or science equation, or the process for writing a letter or memo. They will help you to develop your understanding and your numeracy and literacy skills.

Theory into practice

In this feature you are asked to consider the workplace or industry implications of a topic or concept from the unit. This will help you to understand the close links between what you are learning in the classroom and the effects it will have on a future career in your chosen sector.

Discussion

Discussion features encourage you to talk to other students about a topic, working together to increase your understanding of the topic and to understand other people's perspectives on an issue. These features will also help to build your teamworking skills, which will be invaluable in your future professional and academic career.

Safety tip

This provides advice around health and safety when working on the unit. It will help build your knowledge about best practice in the workplace, as well as make sure that you stay safe.

Key terms

Concise and simple definitions are provided for key words, phrases and concepts, giving you, at a glance, a clear understanding of the key ideas in each unit.

Link

Link features show any links between units or within the same unit, helping you to identify knowledge you have learned elsewhere that will help you to achieve the requirements of the unit. Remember, although your BTEC National is made up of several units, there are common themes that are explored from different perspectives across the whole of your course.

Further reading and resources

This contains a list of other resources – such as books, journals, articles or websites – you can use to expand your knowledge of the unit content. This is a good opportunity for you to take responsibility for your own learning, as well as preparing you for research tasks you may need to do academically or professionally.

Features connected to your assessment

Your course is made up of mandatory and optional units. There are two different types of mandatory unit:

▶ externally assessed
▶ internally assessed.

The features that support you in preparing for assessment are below. But first, what is the difference between these two different types of unit?

Externally assessed units

These units will give you the opportunity to demonstrate your knowledge and understanding, or your skills, in a direct way. For these units you will complete a task, set directly by Pearson, in controlled conditions. This could take the form of an exam or it could be another type of task. You may have the opportunity to prepare in advance, to research and make notes about a topic, which can be used when completing the assessment.

Internally assessed units

Most of your units will be internally assessed and will involve you completing a series of assignments, set and marked by your tutor. The assignments you complete will allow you to demonstrate your learning in a number of different ways, from a written report to a presentation to a video recording and observation statements of you completing a practical task. Whatever the method, you will need to make sure you have clear evidence of what you have achieved and how you did it.

Assessment practice

These features give you the opportunity to practise some of the skills you will need during the unit assessment. They do not fully reflect the actual assessment tasks but will help you to prepare for them.

Plan – Do – Review

You will also find handy advice on how to plan, complete and evaluate your work. This is designed to get you thinking about the best way to complete your work and to build your skills and experience before doing the actual assessment. These questions will prompt you to think about the way you work and why particular tasks are relevant.

Getting ready for assessment

For internally assessed units, this is a case study of a BTEC National student, talking about how they planned and carried out their assignment work and what they would do differently if they were to do it again. It will give you advice on preparing for your internal assessments, including Think about it points for you to consider for your own development.

Getting ready for assessment

This section will help you to prepare for external assessment. It gives practical advice on preparing for and sitting exams or a set task. It provides a series of sample answers for the types of question you will need to answer in your external assessment, including guidance on the good points of these answers and ways in which they could be improved.

Features to help you reflect on and evaluate your learning

⏸ PAUSE POINT

Pause Points appear regularly throughout the book and provide opportunities to review and reflect on your learning. The ability to reflect on your own performance is a key skill you will need to develop and use throughout your life, and will be essential whatever your future plans are.

Hint
Extend

These sections give you suggestions to help cement your knowledge and indicate other areas you can look at to expand it.

Features that link your learning with the workplace

Case study

Case studies are used throughout the book to allow you to apply the learning and knowledge from the unit to a scenario from the workplace or the industry. Case studies include questions to help you consider the wider context of a topic. This is an opportunity to see how the unit's content is reflected in the real world, and for you to build familiarity with issues you may find in a real-world workplace.

THINK ▶FUTURE

This is a case study in which someone working in the industry talks about their job role and the skills they need. The *Focusing your skills* section suggests ways for you to develop the employability skills and experiences you will need to be successful in a career in your chosen sector. This will help you to identify what you could do, inside and outside your BTEC National studies, to build up your employability skills.

Construction Principles 1

Getting to know your unit

Assessment
This unit is assessed by an examination that is set by Pearson.

The construction industry requires the application of knowledge and understanding related to the design of structures and infrastructure, the selection and use of construction materials and the provision of human comfort in buildings. In this unit, you will develop the skills needed to solve a variety of practical construction problems by applying scientific knowledge and carrying out mathematical and statistical techniques. You will learn about the science underpinning manufacture, properties and degradation of construction materials and will apply mathematical principles and techniques to linked calculations relating to being fit for purpose. You will apply scientific principles to heat loss, sound reduction and lighting levels in order to maximise the comfort of people during design, build and refurbishment.

How you will be assessed

This unit is assessed by a written examination set and marked by Pearson. The examination will have both short-answer and long-answer questions. During the supervised assessment period, you will be assessed on your knowledge of:

▶ construction materials and their properties

▶ application of mathematics in construction contexts

▶ provision of human comfort in buildings.

Throughout this unit, you will find assessment practice activities to help you prepare for the exam. Completing each of these activities will give you an insight into the types of question that will be asked and, importantly, how to answer them.

Unit 1 has four assessment outcomes (AO) which will be included in the external examination:

▶ **AO1** demonstrate knowledge of construction terms, standards, concepts, methods and processes; command words: calculate, describe, explain, identify, state/give

▶ **AO2** demonstrate understanding of construction standards, concepts, methods and processes in context, in order to find solutions to real-life construction problems; command words: calculate, describe, discuss, draw, explain, find

▶ **AO3** analyse and evaluate information in order to recommend and justify the use of technologies and methodologies to solve construction problems in context; command words: analyse, discuss, evaluate

▶ **AO4** make connections between information, technologies and methodologies to resolve construction problems; command words: analyse, discuss, evaluate.

Getting started

Look around the room you are in at the moment and write down the physical properties that are making the room comfortable. How could you measure and quantify these properties? If you were going to renovate the room, how would you work out the amount of paint or floor tiles you would need? How could you estimate possible wastage or breakage of tiles to take into account for your calculations?

 Construction materials

A1 Properties of materials

All materials have a series of **material properties**. These properties define how materials can be used, how they will respond to different situations and loads and how they could degrade during use.

Mass, volume and density

Table 1.1 describes the difference between mass, volume and density.

▶ **Table 1.1:** Mass, volume and density

Property	Description	Unit
Mass (m)	A measure of the amount of matter contained within an object.	Kilogramme (kg)
Volume (V)	A measure of the amount of space occupied by an object.	Cubic metre (m³)
Density (ρ)	Describes the compactness of a material by measuring the amount of matter or mass (m) that is contained within an object's volume (V).	Kilogramme per cubic metre (kg/m³)

Stress and strain

Materials used in construction must often support large forces to ensure the safe function of buildings and other structures such as bridges. These forces might be due to the weight of building materials themselves or to loads applied to the structure, e.g. from moving traffic, strong winds or moving water.

▶ Stress (σ) is a measure of the force distribution inside a material under load and is expressed as the force acting per unit area of material. Stress shares the same units as pressure (pressure also describes a force per unit area) and N/m^2 or Pascal (Pa). In construction, forces tend to be large and so often stresses are given in MPa (Megapascal or 10^6Pa)

▶ Strain (ε) occurs when a material is subjected to an externally applied stress. When this happens, it has a tendency to change shape. Strain measures the **deformation** of a body as a proportion of its original length. As strain is the ratio of two lengths, it has no units.

Strength

Tensile and shear strength

Tensile and shear strength are material-specific properties specifying the maximum tensile and shear stresses that can be applied to a material. If the tensile or shear strength of a material is exceeded, it will rupture.

> **Key terms**
>
> **Material property** – factors that are independent of the size and shape of an object that is made from a particular material. For example, density is a material property that does not depend on the size of the object. However, the mass and volume of an object depend on its size and so are not properties of the material itself.
>
> **Deformation** – the change in the shape of a material caused by the application of a force.

Table 1.2 shows these types of force or applied load.

▶ **Table 1.2:** Types of applied loads

Type of applied load	Description	Illustration
Tensile	A pulling, stretching force.	
Compressive	A squeezing force – pushing together or crushing.	
Shear	A cutting force.	

Bending strength

When a plain, square beam is **simply supported** and loaded vertically at its centre, it will tend to bend. This puts the upper surface of the beam into compression and the bottom surface into tension. As compressive strength tends to be greater than tensile strength, the bending strength of the beam is closely related to, and limited by, its tensile strength.

In practice, the deflection and maximum bending strength of beams is quite complex as it depends not only on the properties of the material used but also on the shape of the beam being used. For instance, an engineered timber joist has equivalent strength to a conventional solid timber rectangular joist of similar overall dimensions despite using around 50 per cent less material (see engineered timber in section A2/A3).

Hardness

Surface hardness is a measure of a material's ability to resist permanent **plastic deformation** of its surface and the formation of scratches, indentations or cuts. High hardness provides good wear resistance.

For example, glass exhibits high hardness and is resistant to scratches and wear. However, hard materials also tend to be brittle, prone to sudden fracture under load and have poor impact resistance.

Fracture toughness

Fracture toughness is a measure of a material's ability to resist fracture under **shock loading** or impact. It's usually expressed as the energy absorbed by a sample of material during an impact failure. Failure under shock loading tends to happen when there are pre-existing defects such as surface cracks or internal material faults such as **voids**.

Malleability

Malleability is the ability of a material to undergo permanent plastic deformation when being squeezed by compressive forces. For example, clay can be shaped easily by squeezing or compressing it into a mould.

Key terms

Simply supported – the free ends of a beam supported at either end.

Plastic deformation – permanent deformation of a material which, unlike elastic deformation, will not spring back into shape when stress is removed.

Shock loading – rapidly increasing dynamic loading such as that caused by an impact, for instance a vehicle colliding with a bridge support.

Void – an empty space or gap in a structure.

In metals, malleability tends to increase at high temperatures, which is why forging often involves heating a metal before it is hammered or squeezed into shape.

Workability

Workability is the ease with which a material can be mixed, handled, shaped, smoothed and finished. This usually happens before **curing** and solidifying into a permanent shape. For instance, the ratio of water to cement has a significant effect on concrete's workability. Water allows concrete to flow freely for a longer time before it cures and hardens into a permanent shape. However, extending workability can cause less desirable effects; for example, increasing its water content decreases the final compressive strength of the concrete.

Stiffness

Stiffness describes the elastic behaviour of engineering and construction materials. A material behaves elastically when a load or force can temporarily deform or deflect it until the load is removed and the material returns to its original shape. This only happens when the applied load is less than the elastic limit of the material, above which the material is unable to return to its original shape once the load is removed and stays permanently deformed.

Fatigue

Fatigue failure is a phenomenon encountered in components that are subject to **cyclic loading** in the form of:

▶ full stress reversal, such as experienced by vehicle suspension springs

▶ random loading, such as experienced by a chain holding a vessel at anchor

▶ vibration, such as experienced by an engine mounting.

In all these instances fatigue can cause premature failure at loads considerably below the normal tensile strength of the material. Although eventual failure can come with little warning, fatigue cracking takes some time to develop and spread through the material until the remaining intact cross section can no longer support peak loading and the material fractures.

In **ferrous metals**, such as steels, there are stress levels below which cyclic loading will not produce fatigue fracturing. This is known as the fatigue limit.

Creep

Creep is encountered in a range of materials, predominantly metals and polymers, where the application of constant tensile stress produces plastic deformation and elongation over time. Creep generally occurs in three phases:

1 Primary creep – the rate is initially high as the movement of dislocations causes straining. This soon slows and then stops as work hardening in the material takes effect. Primary creep usually occurs soon after initial loading.

2 Secondary (or steady-state) creep – a slower process and occurs at a steady rate over a considerable period of time. There are numerous mechanisms that contribute to secondary creep.

3 Tertiary creep – at this stage the strain rate increases rapidly. The internal structure of the material will have sustained considerable damage. Voids and cracks will be present at grain boundaries and around inclusions, and **necking** occurs. When the remaining intact grain structures become overloaded, they fail and the material fractures.

The effects of creep are accelerated at high temperatures but, for metals with low melting points, such as lead, creep can be observed at room temperature. Increasing stress also has an accelerating effect on all the mechanisms involved in creep.

However, for a material to exhibit creep, it must be loaded above its limiting creep stress at any given temperature. Below the limiting creep stress, little or no creep will be observed.

> **Key terms**
>
> **Curing** – the process that causes workable substances, such as wet concrete, paint or glue, to permanently set and harden as a result of chemical reactions.
>
> **Cyclic loading** – constantly varying dynamic loads that are applied in repeated cycles.
>
> **Ferrous metals** – metal alloys containing iron, such as cast iron, steel and stainless steel.
>
> **Necking** – a phenomenon in ductile materials once they exceed their elastic limit and undergo plastic deformation as they approach tensile failure. As the material is stretched it can form a narrow neck with a reduced cross-sectional area at the point of eventual fracture.

Fire resistance

Due to the seriousness and life-threatening risk of fires in buildings, the use of fire-resistant building materials is often an important consideration. Some materials, such as timber and plastics, are flammable and allow fires to become established and spread. Others, such as structural steel beams, can be severely weakened at the high temperatures present in fires. This can lead to the failure of the beam and structural collapse.

Fire resistance of building materials can be increased by using fire-retardant additives in polymers, applying chemical treatments to timber and adding fire-retardant insulation coatings on steelwork.

Electrical conductivity and conductance

It is helpful to be aware of the difference between electrical conductivity and electrical conductance.

▶ Electrical conductivity (σ) describes a material's ability to conduct electricity or the ease with which electrons are able to flow through it. As a material property, it is independent of the physical size of an object or component made from the material. It is measured in Siemens per metre (S/m).

▶ Electrical conductance (G) is a measure of the ease with which electricity is able to pass through an object or component made from a material with known electrical conductivity, length and cross-sectional area. A component with high conductance exhibits little resistance to the flow of electricity.

Copper in electrical cables has high electrical conductivity. To provide the electrical conductance required to carry different electric currents over varying lengths of cable, it is available in a range of standard sizes. In any application, the maximum current and length of a cable can be used to calculate the minimum cross-sectional area of copper required.

Thermal conductivity, conductance and resistance

Due to the role of free electrons in conducting heat energy through a material, most good electrical conductors are also good thermal conductors and vice versa. Table 1.3 describes some thermal properties.

> **Link**
>
> For more information about thermal resistance, see learning aims B and C in this unit.

A range of insulating materials with low thermal conductivity are used in construction to prevent heat loss from buildings and reduce energy wastage. To provide low thermal conductance, thermal insulating materials are available in a range of standard thicknesses. Increasing the thickness reduces the thermal conductance of the insulating layer and reduces the flow of heat energy through it, so that less heat escapes and is wasted.

> **Key term**
>
> **Kelvin (K)** – a unit of temperature. In many scientific calculations temperature must be stated on the absolute, or kelvin (K), scale. 0° Celsius (C) corresponds to 273 K, 20°C corresponds to 293 K and minus 20°C corresponds to 253 K.

▶ **Table 1.3:** Thermal conductivity, conductance and resistance

Property	Description	Unit of measurement
Thermal conductivity (λ or k-value)	A material's ability to conduct heat. Materials with high thermal conductivity are effective conductors of heat energy. This is measured via the heat flow rate through a unit area of a material, induced by a temperature difference of 1K in a perpendicular direction.	Watts per metre per **kelvin** (W/mK)
Thermal conductance (C-value)	The ease with which heat passes from one side of a material to the other. A wall with low thermal conductance will resist the free flow of heat energy through it and act as an insulator. It is defined as the heat flow rate through a unit area of material with known thickness and thermal conductivity, induced by a temperature difference of 1K.	Watts per square metre per kelvin (W/m²K)
Thermal resistance (R-value)	The difficulty with which heat passes from one side of a material to the other. A wall with high thermal resistance will resist the free flow of heat energy through it and act as an insulator. It is defined as the temperature difference required across a unit area of material with known thickness and thermal conductivity that will induce a unit heat flow rate, and is linked to thermal conductance.	m²K/W

Resistance to moisture and vapour penetration and degradation

Water and vapour penetration cause damp and this can lead to many problems in buildings.

▶ Moisture resistance – this is a measure of the degree to which a material resists water penetration. Water must be controlled to prevent damp. Damp contributes to, and encourages, many degradation processes.

▶ Vapour resistance – this is a measure of the degree to which a material resists water vapour penetration. Water vapour helps the formation of condensation, which must be controlled to avoid damp.

Of the wide range of materials used in construction, most are prone to one or more forms of material degradation. However, once environmental conditions are taken into account, materials can be specified that can resist likely forms of degradation. Alternatively, additives and/or protective coatings can further reduce material degradation.

Embedded (or embodied) energy

To make construction projects more sustainable it is necessary to minimise the embedded energy present in building materials. Embedded energy measures all the energy used for the extraction, processing, manufacture and transport of the materials. This allows informed choices to be made when selecting sustainable materials.

▶ **Table 1.4:** Embedded energy of a range of building materials

Material	Embedded energy MJ/kg	Material	Embedded energy MJ/kg
Aluminium	218	Sawn softwood	7.4
Polystyrene	86.4	Plasterboard	6.75
Polyethylene	83.1	Portland cement	4.6
PVC general	77.2	Autoclaved aerated block	3.5
Stainless steel	56.7	Bricks	3
Steel	35.3	Asphalt (4% bitumen)	2.86
Aluminium (recycled)	29	Mortar 1:3 (cement: sand)	1.4
Glass (toughened)	23.5	Concrete 1:1:2 (cement: sand: aggregate)	1.39
Glass (primary)	15	Concrete 1:2:4	0.95
Plywood	15	Concrete (28/35MPa)	0.82
MDF	11	Concrete block (10MPa)	0.67
Sawn hardwood	7.8	Aggregate (gravel or crushed rock)	0.083
Steel (recycled)	9.5	Sand	0.081

Recycling potential

Another method of increasing the sustainability of construction projects is to consider what will happen to the materials at the end of their useful life when the building needs replacing. It is best to specify recyclable materials and use construction techniques to allow them to be easily separated during disassembly and demolition as this will increase the recycling potential of the project and so increase its sustainability.

Ⅱ PAUSE POINT   Investigate the effects of material properties on the behaviour and use of materials.

Hint   Put together your own list of definitions of the different properties. Include an example material for each property.

Extend   Collect a range of different materials used in construction. What properties do each of these materials have? How do these affect their use in construction?

A2/A3 Properties of construction materials and their manufacture and processing

This section will look at the common construction materials you will encounter and the particular properties these materials have. This section will also explain how these properties affect their use and performance and how different specifications and scenarios require different types of materials.

Bricks

Bricks of dried mud or fired clay have been used in construction for thousands of years. Walk down any street and you will see buildings and structures made of bricks. Above ground, these are likely to be facing bricks but there are several types of bricks with different properties suited to a range of applications, as shown in Table 1.5.

▶ **Table 1.5:** Types of bricks and their characteristics and use

Type	Characteristics	Use
Facing	• Selected for their aesthetic appearance • Weather resistant • Durable • Aesthetically appealing	• Above ground • External walls of brick buildings • Popular in residential applications for their aesthetic appeal
Class A and B engineering	Selected for their physical properties: • high compressive strength • high toughness • low water absorption • high water resistance • frost resistant • acid resistant Class B have lower compressive strength than Class A	• Underground structures • Manholes • Sewers
Commons	• Low compressive strength • Less consistent appearance • Generally lower quality	• Internal walls • Filling

In the UK, the principal ingredient of bricks is clay, which varies in colour. Clay is excavated using a shave cutter and transported by conveyor belt to the process plant where it is processed so it can be moulded or extruded. **Extruded** bricks are cut by wires from a continuous length of extruded clay.

After processing, bricks are air-dried and stacked ready for firing in a kiln. This process hardens the brick and gives it its unique colour. Sand can be added to the surface faces of the brick to produce different colours. Wire-cut bricks and moulded bricks produce different textured faces. After this, bricks are stacked in packs, evenly distributed by colour, and shrink-wrapped for transportation.

Concrete

Concrete is soft and malleable when first mixed and strong and durable when hardened. This means it can be formed into a variety of shapes and is used for building everything from bridges to office blocks and can even be used underwater. Its structural strength is suitable for beams and columns and its hard-wearing properties are suitable for floors and roadways.

Concrete is a mix of **fine** and **coarse aggregate** coated in a paste made from Portland cement and water. The proportions of the mix must ensure the cement/water paste evenly coats the aggregate and fills any spaces between grains. Too little paste and the aggregate will not be fully coated: the mix will be stiff, difficult to lay and the hardened concrete will have a rough, porous structure. Too much paste and the mix will crack, lacking the reinforcement provided by the aggregate. It will also be unnecessarily expensive as the cement is by far the most costly ingredient. In both cases the strength and durability of the concrete is reduced.

Other key factors affecting strength of concrete are the quality of the cement and the

Key terms

Extruded – method of production where soft and malleable materials are pushed through a hole in a die plate to form long lengths of shaped material with uniform cross section.

Fine aggregate – sand with a grain size of less than 5 mm.

Coarse aggregate – gravel, crushed stone, recycled crushed concrete or blast furnace slag with a grain size greater than 5 mm.

proportion of water used in the cement/water paste. The water causes it to harden over time in a chemical reaction called **hydration**. A high cement/water ratio, often referred to as a strong mix, provides the best results.

In practice, the minimum amount of water is used to make concrete sufficiently workable, ensure thorough mixing and allow effective consolidation to remove voids and trapped air after laying. Excessive water weakens the mix and has a similar effect on the hardened concrete, reducing its final compressive strength.

The type, size, regularity and range of aggregates are also important factors that must be carefully controlled in order to ensure that the quantity of water/cement paste required to fully coat and fill the spaces between them can be calculated accurately.

Finally, a wide range of additives can be used to enhance some of the characteristics of concrete, such as its colour, water resistance and workability.

Although concrete will set after a few hours, and will harden sufficiently to perhaps walk on if laid as flooring, the hydration reaction will continue for several weeks. Concrete generally does not achieve its full strength and hardness until 28 days after being laid.

Table 1.6 shows the difference between common concrete mix ratios.

> **Key term**
>
> **Hydration** – a chemical reaction in which a substance combines with water.

▶ **Table 1.6:** Typical concrete mix proportions and uses

Nominal proportions (by volume) (Cement: sand: coarse aggregate)	Usage
1:1:2	High-strength concrete often used to support high structural loads.
1:1.5:3	Often used in floor slab, column and other load-bearing structures.
1:2:4	Often used in the construction of buildings under three storeys.
1:3:6	Low-strength concrete often used as non-structural fill.
1:3:0	Screed mix, containing no coarse aggregate, often used as a final smooth and level layer on a concrete floor base.

There are several ways of specifying the required concrete mix proportions when ordering bulk delivery of premixed concrete or for mixing smaller batches as required on site. Table 1.7 describes some common specifications.

▶ **Table 1.7:** Specification of concrete mixes

Type of mix	Specification
Prescribed mixes	Ordered by specifying the exact proportions of the cement, sand and coarse aggregates.
Design mixes	Ordered by specifying the required performance of the mix but leaving the exact proportions of the cement, sand and aggregates to the supplier.
Mixes by ratio of volume	On-site volume is easy to measure using a container or shovel to add roughly equal proportions into a mixer. Water is often added until the mix attains the required workability. Measuring the mix ratio by volume requires little specialist equipment but can lack accuracy and vary between batches.
Mixes by ratio of weight	Bulk delivery concrete is mixed in more controlled conditions using specialist weighing equipment to ensure that accurate proportions of sand, coarse aggregates of differing sizes, cement, water and additives are used consistently in every batch.

Screed mixes

Screed mixes generally contain no large aggregates and are used to smooth and level a concrete floor base (sometimes encapsulating pipework for underfloor heating). The screed itself is generally unsuitable for use as a final finish as the lack of larger aggregates gives it poor wear resistance. However, it does provide an excellent surface to take a final hard-wearing flooring material such as ceramic tiles.

Cements

Portland cement is the key ingredient in concrete, mortar and screed mixes. When mixed with water it forms a paste which hardens to bind the other ingredients used in the mix.

Cement itself can be manufactured from a range of quarried or industrial waste materials, with the exact mix often depending on local material availability. Common ingredients include limestone, chalk, clay, silica sand, iron ore, blast furnace slag and fly ash.

Quarried raw materials such as limestone first have to be crushed in a series of powerful crushers and mills until all the material is reduced to grains that are no bigger than 75 mm.

At this stage, other ingredients such as clay and iron ore are added, followed by further grinding and mixing before being fed into a kiln heated to around 1450°C.

The rotary kilns used to manufacture cement are enormous, around 100 m in length and 3 m in diameter. They are set at an angle so that gravity helps to feed the mix of raw ingredients from the top to the bottom as the kiln rotates. The kilns are usually heated using fossil fuels such as gas or coal burning at high temperatures in a forced draft of air.

As the raw materials pass through the kiln, the extreme heat causes a series of chemical changes to take place to form cement clinker which looks like small grey pebbles. These emerge from the kiln and are cooled before being mixed with small amounts of other minerals such as limestone and fed into grinding machines.

Once it has been ground into an extremely fine powder, the finished Portland cement is stored in silos or packaged into bags ready for use.

Sulphate-resisting cement is a blended cement designed to improve performance where there is a risk of sulphate attack.

As a consequence of the large amounts of energy consumed in cement production, cement-based materials contain high levels of embedded energy so more sustainable alternatives should be considered before specifying cement-based products.

Concrete blocks

Concrete blocks are an alternative to traditional bricks with a range of applications. They are inexpensive, quick and easy to lay but lack the aesthetic appeal of facing bricks as a final finish for building exteriors. Table 1.8 describes some common types of concrete block and their uses.

▶ **Table 1.8:** Types, characteristics and uses of concrete blocks

Type	Manufactured from	Characteristics	Use
Aerated	Cement, lime, sand, pulverised fuel ash (PFA), aluminium sulphate	• Lightweight • Prone to impact damage • Good insulating properties • Medium strength	• Limited to low-rise load-bearing walls • Partitions
High density	Cement/sand/aggregates	• Durable • Tough • High strength • Poor insulator	• Load-bearing walls
Insulated concrete form (ICF)	Cement/recycled wood chip	• Interlocking to create formwork • Poured around steelwork in block cavities to stabilise wall structure • Integrated mineral wool insulation to enhance thermal performance	• Single skin load-bearing walls

Mortar mixes

Mortar is a cement-based gap-filling adhesive used in the construction of brick and blockwork. It fills and seals the irregular gaps between blocks, sticking them together to provide resistance to side loading and allow the effective transmission of vertical loads through the walls it is used to construct. As Table 1.9 shows, a mortar mix should be selected to match the characteristics of the building materials being used.

▶ **Table 1.9:** Types, characteristics and uses of mortar

Type	Characteristics	Use
Cement mortar 3:1 ratio sand: cement	Very strong, very hard, water resistant, quick setting.	Often used to join engineering bricks (which are similarly hard and water resistant) in applications where little structural movement or settling is anticipated which might otherwise cause cracking.
Lime mortar 3:1 ratio sand: hydraulic lime	Soft, porous, slow setting.	Widespread use before development of cement mortar. Now used primarily to repair or restore old buildings. Matches the characteristics of natural stone building materials, allows the movement of moisture to the surface of stonework where it evaporates and accommodates a certain amount of movement or settling, common in older structures, without cracking.
Cement lime mortar 1:1:6 ratio cement: hydrated lime: sand	Less strong, hard, water resistant and quick setting than cement mortar.	Often referred to as 'compo', this substitutes some of the cement content of cement mortar for hydrated lime. Hydrated lime (not to be confused with hydraulic lime) has no binding characteristics and is used solely as a filler material which reduces the strength, hardness and water-resistant properties of the mortar. Cement lime mortar is able to accommodate some movement and settling without cracking and allows some movement of moisture, making it suitable for a wider range of applications of bricks and blocks.
Coloured mortar	As above but containing a coloured additive.	Used where colour matching to existing stone or brickwork is desirable.

Sand

Sand is a vital ingredient in concrete, screed and mortar providing the small aggregate component of the mix. Table 1.10 describes three common types used in construction.

▶ **Table 1.10:** Types, characteristics and uses of sand

Type	Characteristics	Use
Building	Fine sand containing grains in the range 125–250 μm.	Mortar in brick and blockwork, etc.
Sharp	Coarse sand containing larger grains in the range 0.5–1 mm.	Concrete in flooring slabs, structural applications, etc.
Silver	Very fine silica sand containing grains in the range 63.5–125 μm.	Grout in the joints between floor tiles and paving slabs.

Plasterboard

Plasterboard is an easy to install, quick way of lining interior walls and ceilings ready for the application of a final finishing coat of plaster. It is supplied in sheets composed of a gypsum core sandwiched between layers of lining paper. It generally comes in 1200 mm-wide sheets to suit standard 600 mm wall stud spacing and can be quickly nailed or screwed in place. Joints are then taped and a skim of finishing plaster applied.

The gypsum used in plasterboard gives the material some useful properties: fire resistance, sound and thermal insulation and the ability to store moisture and so help to regulate humidity levels.

Glass

The transparent properties of glass have been used for centuries to allow natural light into buildings while preventing the entry of wind and rain. Modern manufacturing techniques, surface treatments and coatings mean that more glass is used in construction than ever before and in a wider range of applications. Table 1.11 describes three common types of modern glass, while Table 1.12 describes some common finishes.

▶ **Table 1.11:** Types, characteristics and uses of glass

Type	Characteristics	Use
Float	Describes general purpose glass sheet. The name comes from its production, which floats molten glass on a bed of molten metal, usually tin, to create extremely flat, smooth glass with uniform thickness.	General purpose, windows, double-glazed units.
Laminated	A type of safety glass consisting of alternating layers of glass bonded to a shatterproof polymer. Even when the glass is broken the pieces remain bonded together and the sheet retains some strength.	Architectural applications, shop windows and secure areas.
Toughened (or tempered)	A type of safety glass that has undergone a process to increase its strength and impact resistance. It forms small granular pieces when broken, which are less likely to cause injury.	Glass panels in showers, shelves, interior doors.

▶ **Table 1.12:** Types, characteristics and uses of glass finishes

Type	Characteristics	Use
Clear	Used in most conventional applications, principally to allow natural light into buildings and for occupants to look out.	Commonly encountered in windows, conservatories, etc.
Obscured	Still allows passage of natural light but obscures vision through it to maintain privacy. It uses raised patterns, sandblasting or stencilled sand blasted patterns on its surface.	Bathroom windows, front doors, offices, shops.
Smart	Switches between clear, obscured or opaque light transmission characteristics by applying an electrical current to a specially developed coating on the glass.	Architectural use, meeting rooms, regulating inflow of natural light and heat (solar gain) in intelligent buildings.

Insulation materials

Insulation materials are designed primarily to prevent heat loss from buildings through walls, lofts and other unheated spaces. Many also provide sound insulation. Table 1.13 describes some common types of insulation.

▶ **Table 1.13:** Types, description and uses of insulation materials (Source: **www.greenspec.co.uk/building-design/insulation-materials-thermal-properties**)

Type	Description/Available forms	Thermal conductivity (typical) W/mK	Usage
Fibre glass wool	• Fibres spun from molten glass. • Up to 30% recycled content. • Available in: rolls, batts.	0.035	Lofts, walls, cavities.
Expanded polystyrene	• Expanded polystyrene foam. • Available in: boards, loose fill.	0.034	Walls, cavities, floors.
Celotex (phenolic foam)	• Expanded phenolic resin foam. • Available in: boards.	0.020	Walls, cavities, floors.
Mineral wool	• Fibres spun from molten rock. • Available in: boards, batts, rolls.	0.032	Lofts, walls, cavities, boilers, pipework.
Cellulose	• Made from shredded newspaper (treated to impart resistance to fire, mould, insects). • Available in: loose fill.	0.035	Lofts, walls, cavities.
Straw	• Agricultural by-product from cereal production. • Available in: bales.	0.080	Bales can be used in load-bearing construction.
Polyurethane	• Expanded polyurethane foam. • Available in: spray-on coatings.	0.023	Sprayed directly onto roof tiles, concrete slabs or wall cavities. Allows retrofit of wall cavity insulation by spraying through holes made in an existing cavity wall.

Construction plastics

Polymer materials, or plastics, are used to manufacture countless consumer products. In construction their use is limited by their low compressive and tensile strength. However, they still have important non-structural applications in products such as waterproof barriers, pipework and window frames, as shown in Table 1.14.

▶ **Table 1.14:** Types, characteristics and uses of polymers (or plastic) materials

Type	Characteristics	Use
Polyethylene (polythene)	Can be manufactured into membranes of various thicknesses, waterproof, flexible (even at low temperatures), resistant to being torn, chemical degradation and rot.	Damp-proof membranes (DPM) used under concrete floors. Damp-proof courses (DPC) used just above ground level in brick and blockwork.
PVC (plasticised)	Can be manufactured into sheets of various thicknesses and surface textures, strong, waterproof, flexible, durable, can be coloured, easy to clean, easy to recycle.	Vinyl floor coverings, inflatable structures.
uPVC (unplasticised)	Can be extruded into complex mouldings or pipes, stiff, strong, lightweight, durable, resistant to impact, chemical degradation and rot.	Doors and window frames, soffits, bargeboards, fascias, guttering.

Timber

Hardwood tree species are generally slow growing and deciduous, shedding their leaves and becoming dormant over the winter months. As Table 1.15 shows, hardwood timber tends to have higher density, strength and durability than softwoods but it is usually more expensive.

▶ **Table 1.15:** Types, characteristics and uses of hardwoods

Type	Characteristics	Use
Oak	Hard, tough, strong	Furniture, interior woodwork, structural beams, flooring.
Beech	Hard, tough, strong, has a tendency to warp	Furniture, tool handles, chopping boards.
Ash	Tough, flexible	Furniture, tool handles, oars, ladders.

Softwood tree species are generally fast growing and evergreen with needles instead of broad leaves, which they retain throughout the year. Softwood timber tends to have lower density and lower durability than hardwoods but is usually significantly less expensive. About 80 per cent of all timber worldwide is softwood.

▶ **Table 1.16:** Types, characteristics and uses of softwoods

Type	Characteristics	Use
Redwood (Scots pine or European redwood)	Has greater hardness and bending strength than whitewood species.	Load-bearing beams, joists and rafters.
Whitewood (any species of spruce, pine or larch which produces timber with similar properties)	Suitable for interior structural applications. Low surface hardness. Low density.	Load-bearing beams, joists and rafters.
Cedar	Durable, water-, rot- and fungi-resistant.	Exterior cladding.

Manufactured boards

Manufactured composite boards have some general advantages over natural timber products: they are available in large sheet sizes, make use of waste from timber processing, are dimensionally stable, easy to machine, less likely to bow or warp and free from naturally occurring defects such as knots or cracks. Table 1.17 describes some common types of manufactured boards.

▶ **Table 1.17:** Types, characteristics and uses of manufactured boards

Type	Characteristics	Use
Plywood	Thin veneers or plies of wood glued together with alternate plies rotated by 90°. Cross graining in this way gives increased tensile strength in all directions. Available in various types and grades: softwood, hardwood, exterior, marine, flexible.	Roofing, floors, walls, general internal and external joinery, furniture and structural applications.
Chipboard/ particle board	Compressed wood chip or wood shaving waste from timber processing held together by a polymer resin. Denser but less strong than natural timber or plywood with poor aesthetics. Prone to water damage and for interior use only.	Flooring, flat pack furniture (when suitable veneer is applied to visible surfaces).
MDF (medium density fibreboard)	Compressed wood fibre and fine sawdust from timber processing held together by a polymer resin.	General internal joinery, furniture.

Roofing materials

The primary role of roofing materials is to prevent the ingress of rain into the structure, where it will cause damage and enable degradation processes such as rot to become established. Roofs need to maintain their waterproof characteristics over extended periods of time in extreme conditions, as they will be exposed to wind, rain, UV and high temperatures in summer and frost and freezing temperatures in winter.

Table 1.18 describes some common types of roofing materials.

▶ **Table 1.18:** Types and characteristics of roofing materials

Type	Characteristics
Slate	A sedimentary rock quarried in large slabs. It is split along its natural grain into thin sheets which can be trimmed into standard-sized individual slates for use. The ease with which slate is manufactured means it has the lowest embedded energy of any roofing material so is more sustainable. It is highly water resistant, durable and has good aesthetic properties, making it ideal as a roof covering.
Concrete tiles	Available in a wide range of colours, forms and styles as a less expensive and more uniform alternative to natural or more traditional roof tile materials.
Pantile	S-shaped interlocking tiles made from fired clay, once used extensively in areas of the UK where clay was readily available. They are still used where a traditional finish is preferred, have good aesthetic properties and are more sustainable than concrete alternatives.
Roofing felt	Made from polyester matting saturated with bitumen containing additives to improve its flexibility. This makes it tough, flexible and waterproof. Light duty grades of roofing felt used as sarking (or underlay) beneath slates or tiles. Thicker and more durable grades are applied in layers to flat roofs as the principal roof covering, glued in place and joined using hot liquid bitumen. Often a capping layer of felt with a mineralised surface is used to provide protection from direct sunlight which can cause the bitumen to degrade over time, becoming brittle and prone to cracking.
Thatch	Rarely, if ever, applied to new-build properties, uses bunches of long wheat straw (difficult to obtain since introduction of modern short-stemmed wheat in the UK) called yelms, attached to the roof with twisted hazel sticks called spars. A thick layer of thatch makes an effective barrier against the elements but requires regular maintenance and highly skilled thatchers to carry out repairs. Its use declined once Welsh slate was made available across the country during the Industrial Revolution.
Ridge	Used as cappings on the ridges and hips of pitched roofs where tiled surfaces meet to cover and waterproof the join. Usually manufactured from fired clay or concrete.
Lead flashing	Lead is extremely malleable and sheet lead is used as flashing around chimneys and other features protruding through roof coverings such as roof lights. Lead can be shaped or dressed to cover joins between the roof covering and chimney brickwork or roof light to ensure it remains watertight.

Engineered timber

Structural elements made from engineered timber allow a reduction in the use of raw materials and are manufactured with consistent quality and performance characteristics in any shape or size, overcoming many of the limitations of natural timber products.

Table 1.19 describes two common types of engineered timber.

▶ **Table 1.19:** Types, characteristics and uses of engineered timber

Type	Characteristics	Use
Glulam beams	Glulam is short for glued laminate and reflects the manufacturing technique of gluing together timber laminations to make up a structural beam. The technique allows the manufacture of large load-bearing structural beams in a variety of complex shapes, curves and arches which would not be possible using conventional timber construction. Glulam beams are often left exposed as they provide an interesting architectural aesthetic to an interior.	Structural columns, arches and shaped beams.
Engineered joists	Engineered joists can achieve equivalent and often better performance than a traditional timber joist of similar dimensions, are much lighter and available in extended lengths of up to 14 m (limited only by the practicality of transporting and fitting them). They are manufactured with an I-shaped cross section with flanges top and bottom connected by a web. The elements of the joist are made from manufactured board and the finished beam uses up to 50% less raw material than an equivalent sawn timber product.	In the place of conventional timber joists in floors and roofs.

Steel

Steel is used in structural applications where high tensile, compressive and/or bending strength is required. Skyscrapers use steel frameworks to support their enormous weight and the open framework of struts and ties visible in the structure of electricity pylons relies on the high strength of steel to provide support for electricity cables. Embedded steel bars improve the otherwise poor tensile strength of concrete in reinforced concrete structures. In addition, countless steel screws, nails and fixings hold together the fabric of all types of buildings.

Table 1.20 describes the three main types of steel used in construction.

▶ **Table 1.20:** Types, characteristics and uses of steel

Type	Characteristics	Use
Mild	Mild steel is a form of plain carbon steel containing iron alloyed with between 0.15 and 0.35% carbon. Most structural steels fall into this category. For example, Grade S275JR is a typical hot-rolled steel used in structural applications. It contains a maximum of 0.25% carbon and has tensile strength of 380–540MPa and yield strength of 275MPa.	Welded, bolted and riveted structural applications.
High strength	Steels with carbon content of between 0.30% and 1.7% have greater tensile strength and hardness than lower carbon steels. These can also be heat treated to significantly increase tensile strength and hardness. However, they tend to be brittle with little or no plastic deformation to warn of overloading prior to failure. For example, Grade O1 is a typical hardenable high-carbon tool steel. It contains 0.95% carbon and, when heat treated, has a tensile strength of 1990MPa and yield strength of 1650MPa.	Tools, springs, high strength wires and cables.
Stainless	Stainless steels are alloyed with chromium and nickel to significantly increase resistance to corrosion. For example, Grade 304 is a typical stainless steel. It contains 0.08% Carbon, 18% chromium and 8% nickel and has a tensile strength of 515–600MPa and yield strength of 205–310MPa.	Fasteners, refrigeration equipment, sinks, fittings.

Steel is generally manufactured from molten iron direct from a blast furnace (used to extract iron from iron ore), scrap steel or a mixture of both.

In basic oxygen steelmaking, a mixture of molten iron and scrap is loaded into a vessel called a converter. Inside the converter, oxygen is forced through the molten metal at high pressure. This combines with carbon and other impurities which are driven off as gases such as carbon monoxide. Lime is also added which helps drive out other impurities which then float to the surface as slag.

Steel is formed once the carbon content of the iron has been reduced to less than 1 per cent. The exact proportion of carbon required in the steel will be dictated by the type of steel being produced. A typical structural steel used in construction contains approximately 0.2 per cent carbon.

Once the required carbon content is achieved then the molten steel is tapped off and mixed with any other metals required in the alloy being produced.

Scrap steel can also be reprocessed using an electric arc furnace. This passes an extremely high electrical current through a charge of scrap steel which is melted by an intensely hot electric arc. The molten steel is then treated in much the same way as in basic oxygen steelmaking, using oxygen and other additives to drive out impurities and alter the carbon content of the steel. The electric arc furnace is preferred when manufacturing high-quality steel where the increased process control allows precise material compositions to be achieved.

Aluminium alloys

Aluminium in its pure form is soft, ductile and malleable with relatively low tensile strength. However, when alloyed with small quantities of copper, silicon, magnesium or manganese, its properties are greatly enhanced.

Aluminium alloys are widely used in the production of rolled plates or sections and extrusions such as door and window-frame components. Casting alloys contain silicon to improve the flow of the molten material and are used to make fittings, casings and machine parts.

Aluminium alloys are also available in forms which can be heat treated to further enhance their mechanical strength and hardness.

PAUSE POINT   Look around the room. Identify all the construction materials you can see and suggest why they were chosen.

Hint   In schools and industrial buildings more of the internal fabric/structure of the building is left visible, making this task easier.

Extend   Look at the exterior of the building. Identify all the materials you can see and suggest why they were chosen.

A4 Degradation of construction materials

All building materials can suffer from decay and degradation over time if they are not protected. This section will explain some of the common scenarios that can lead to decay of materials and the methods that can be used to help reduce the impact of these.

Sources of degradation and their cause

Natural agents

Some of the common causes of degradation come from natural processes:
▶ Ageing – the gradual deterioration of materials over an extended period of time. These are slow acting and so in the short term their effects are minor. However, over time, perhaps tens or hundreds of years, minor damage caused by wind, rain, frost, pollution or wear and tear accumulates and can seriously weaken timber, brick or stonework.

▶ Ultraviolet (UV) – radiation from exposure to sunlight can cause deterioration in a range of polymer materials such as acrylic, which is often used in signage and secondary glazing applications. UV penetrates the material's surface and breaks apart the chemical bonds in the polymer, weakening the structure. Discoloration and fading are early signs of UV degradation in polymers. This is followed by brittleness, cracking and the eventual breakdown and disintegration of the material.

▶ Timber decay and infestation – the gradual decay of dead and fallen trees by a wide range of fungi and insects is a vital process in the woodland eco system. However, this can wreak havoc on timber structures and the wooden beams and joists commonly used to support floors, ceiling and roofs. These are all encouraged and accelerated by the presence of damp and elevated moisture content in the material.

▶ Insect attack – in the UK the only significant insect attack on timber is woodworm; this is the wood-boring larvae of one of three species of beetle, most often the common furniture beetle. Wood-boring beetle larvae spend a large part of their life cycle inside timber feeding on the starches and sugars in the wood. When they emerge after several years, these larvae leave small holes on the surface of infested timber. A further sign of active woodworm is the presence of frass, powdery wood dust, beneath suspected sites of infestation. Internal damage to timber can be severe, causing weakening, crumbling and disintegration.

▶ Wet rot/fungal attack – the natural decay of damp timber causes naturally occurring species of fungus to break down dead wood. In buildings, any damp timber can be affected and broken down. Problems with wet rot are often linked to chronic damp, caused by leaking pipes or roofs or from high levels of condensation.

▶ Dry rot/fungal attack – caused by an infestation of a particular species of fungus; this attacks cellulose in wood fibres which provide most of the timber's strength and stiffness. It still requires relatively high moisture levels to become established although often there is no obvious sign of chronic damp.

▶ Lichens and mosses – by retaining moisture and preventing evaporation and drying out of the materials on which they grow, these make stone, tiles and brickwork more prone to frost damage and timber more prone to rot and fungal attack.

Moisture movement

Capillary action describes the mechanism where moisture is drawn up into narrow gaps. This can include the hollow wood fibres in timber, the pores in materials like plaster or brickwork and micro-cracks in the surface of stonework and concrete. Capillary action is the principal reason why water moves up through walls that are standing in wet foundations and up wooden fence posts set into damp ground. Excessive moisture contributes to several processes of degradation in building materials and must be controlled.

Shrinkage

Trees use water to transport nutrients throughout their structure as they grow. When first felled, some species will contain, by weight, more water than wood fibre. If green freshly felled timber is used straight away then the shrinkage and movement as it dries will be considerable and is likely to cause unacceptable gaps or cracks in the timber itself.

On the other hand, if kiln-dried timber with very low moisture content is used in an area with high humidity then it will absorb some of that moisture, swell the wood fibres and expand – causing bowing or distortion in its surroundings.

When timber is delivered to site it is best, where practical, to allow it time to acclimatise to the conditions in which it will be installed, expanding or contracting to match ambient humidity levels before being cut and fitted.

Exposure conditions

There are several sources of degradation linked to exposure to the environment:

▶ Weathering – the combined action of sunlight, rain, wind and frost and any other environmental factors, such as high salt levels in coastal locations or traffic pollution in cities.

▶ Freeze-thaw – water has the unusual thermal property that it expands when cooled below 4°C and solidifies into ice. This means small cracks or pores in the surface of brittle materials, such as concrete or brick, that contain moisture will be forced apart when moisture freezes in cold weather. With every freeze-thaw cycle, these cracks will grow until they consolidate into larger cracks and the surface crumbles way.

▶ Thermal ageing – many of the chemical processes involved in material degradation are accelerated at high temperatures. Repeated cycles of heating and cooling and the associated expansion and contraction of the material causes micro-cracking in the surface. These cracks grow and consolidate into larger, structurally significant cracks that can cause material failure.

▶ Creep – see section A1. Creep is accelerated in high temperatures.

▶ Humidity – a measure of the amount of water vapour present in the air. Relative humidity is the amount of water vapour present compared with the maximum possible moisture content at a given temperature. High levels of humidity can prevent evaporation from drying out building materials, increasing the moisture content of timber and leading to condensation.

- Condensation – caused by warm humid air coming into contact with surfaces at significantly lower temperatures. At lower temperatures the amount of water vapour that air retains significantly reduces and water droplets form as condensation. This is commonly seen on bathroom windows and tiled walls and can cause rot in timber window frames and floorboards

Loading

Some forms of material degradation and eventual failure are directly related to the in-service loading conditions experienced, as shown in Table 1.21.

▶ **Table 1.21:** Types and characteristics of loading

Type	Characteristics
Static	Constant loading such as in the structure of an electricity pylon. Large static loads can give rise to creep deformation and subsequent failure in susceptible materials over time.
Cyclic	Constantly varying dynamic loading such as in a flagpole swaying back and forth in the wind. Dynamic loading can cause the formation of fatigue cracks at stress levels below the tensile strength of the material when subjected to static loading.
Shock	Shock loads are often localised highly dynamic loads, usually as a result of impacts, which can cause catastrophic failure in low toughness materials.

Chemical degradation

All materials are susceptible to chemical degradation to some extent, whether it is the action of atmospheric oxygen and water on iron to form rust or the effects of acid rain in dissolving the surface of stonework. Table 1.22 gives some examples.

▶ **Table 1.22:** Types and characteristics of chemical degradation processes

Type	Characteristics
Acid rain	Burning fossil fuels in power stations and vehicles pollutes the atmosphere with sulphur dioxide and nitrogen oxide (as well as releasing carbon dioxide significantly contributing to global warming). When these gases combine with water in the atmosphere they form nitric and sulphuric acid, which then fall as acid rain. These dissolve the calcite (a form of calcium carbonate) present in limestone and marble stonework. Although concrete is vulnerable to attack from strong acids, the levels of acidity found in acid rain leave it structurally unaffected.
Alkalis	A reaction between the highly alkaline cement paste and the silica commonly found in aggregates can cause cracking and severe weakening in affected concrete structures. This is generally referred to as the alkali-silica reaction (ASR) but is sometimes known as 'concrete cancer'.
Sulphates	When water containing dissolved sulphates enters concrete it reacts with the key binding constituents of Portland cement to form minerals called ettringite and, by a different reaction, gypsum. Both minerals have solid volumes larger than their original constituents and so cause high levels of internal stress causing the formation of fractures and eventual material disintegration. Sulphates are present in seawater, formed by bacterial action in sewers and found in chimneys as a product of the combustion of sulphur-bearing fuels such as coal.
Leaching	Leaching is the action of carrying away dissolved minerals in an aqueous solution and is the mechanism by which chemical attack is able to degrade and weaken the inside of some porous materials.

Corrosion in metals

Metal corrosion is a process of chemical degradation that gradually affects its physical appearance and reduces its mechanical properties.

The corrosion mechanism we are most familiar with is the **oxidation** of iron and steel to form rust. The most common corrosion mechanism encountered in construction is electrochemical corrosion, the main mechanism in the oxidation of iron and formation of rust. This cannot proceed without the presence of a liquid electrolyte to allow the movement of electrically charged ions. The most common electrolyte in the environment is, of course, water – for example, it is no coincidence that cars do not tend to rust in hot, dry climates.

Key term

Oxidation – a chemical reaction in which oxygen is gained.

Measures to prevent and reduce degradation

In most circumstances, it is possible to prevent or slow degradation by taking preventative action and performing regular maintenance. For instance, one of the most recognisable structures in Britain, the Forth Rail Bridge near Edinburgh, is made of steel and is prone to rust. In the past, this meant having a full-time maintenance team painting the bridge. It took so long to paint that, when they reached the end, they would have to start again in a continuous cycle. Improvements in paint technology now mean that a single coat of specialist glass flake epoxy paint protects the bridge and can be left for up 25 years without reapplication.

Smaller external steelwork items, such as gates or posts, are often dipped in hot zinc in a process called galvanising. The zinc forms a protective barrier around the steel but also provides electrochemical protection in which the zinc corrodes in preference to the steel, which significantly extends its life.

Table 1.23 describes some measures, special paints and protective coatings that prevent and reduce degradation.

Research

A survey of a house has found high levels of damp in the kitchen walls and floorboards. Research the role that water plays in degradation processes commonly encountered in construction and explain the potential long-term consequences of not taking action to eliminate damp where it has been identified.

▸ **Table 1.23:** Techniques to prevent material degradation

Degradation	Material	Prevention/remediation
Corrosion	Steel	• Painting to exclude moisture. • Galvanising (hot dip zinc coating which both physically seals out moisture and provides additional electrochemical protection). • Consider replacement with stainless steel.
Wet and dry rot	Timber	• Maintain good ventilation. • Paint the timber to exclude moisture (external applications). • Address any signs of chronic damp. • Apply anti-fungal timber treatment. • Remove and replace infected timber.
Woodworm	Timber	• Maintain good ventilation. • Address any signs of chronic damp. • Apply insecticide timber treatment. • Remove and replace severely weakened timber.
Acid rain	Stonework	• Specialist surface treatment. • Remove and replace severely damaged stonework.
Sulphate attack	Concrete	• Use sulphate-resistant cement. • Address any signs of chronic damp. • Use gypsum (calcium sulphate)-free aggregates.
UV	Polymers	• Use polymers with a UV-inhibiting additive.

Material failure

Often it is the failure of one of the constituents of a composite material, such as the hardened cement holding together concrete that has been weakened by a degradation process, that brings about the failure of an entire structure.

Likewise, a structure composed of a range of building materials, such as a house set on foundations with brick and block cavity walls supporting a timber pitched roof, is only as strong as its weakest link. Failure of any one element can be catastrophic for the rest of the structure. For instance, any movement in the foundations caused by subsidence or flooding will cause uneven loading in the walls and lead to the formation of cracks. Large cracks if left unchecked can lead to the partial collapse of a wall. If this is no longer able to bear any load then the remaining structure must provide additional support. If this process continues, and the remaining structure becomes overloaded or unstable, then complete collapse of the entire structure will follow.

Before they fail completely, most construction materials show clear signs of degradation. If these signs are spotted early enough, damaged areas might be repaired and continued degradation managed. However, if these early signs are missed, or simply ignored, complete failure of the material and potentially any structure it supports will follow. Table 1.24 shows common types of material failure.

▶ **Table 1.24:** Common modes of failure in construction materials

Material	Failure	Image
Concrete	• Sulphate attack • Screed or finish coat delamination • Freeze-thaw cracking • Alkali-silica reaction (ASR)	Typical crack pattern associated with ASR degradation in concrete.
Reinforced concrete	• Corrosion of steel reinforcement	Formation of rust on steel reinforcing bar in concrete has caused spalling of the concrete surface.
Brickwork	• Sulphate attack • Crystallisation of salts from bricks • Freeze-thaw • Structural movement and settling	Severe freeze-thaw damage causing the disintegration of brickwork.
Mortar	• Shrinkage of mortar on drying • Freeze-thaw • Structural movement and settling	Freeze-thaw damage and weathering of mortar joints in a chimney. Once mortar deteriorates to the extent that bricks become loose the structure will have to be taken down and rebuilt.
Steel	• Corrosion	Severe corrosion of structural beams can weaken them sufficiently to cause structural collapse.
Timber (internal)	• Wet rot • Dry rot • Woodworm	Floorboards severely damaged by woodworm.
Timber (external)	• Wet rot • Woodworm	Severe damage to timber window frame caused by wet rot.

Identify early signs of material degradation in a building and discuss the long-term consequences if this progresses and causes the material to fail.

> Hint

Look for cracks in concrete, shelling of the surface of bricks, rusting steel structures or signs of wet rot in timber.

> Extend

Explain actions that could be taken to prevent or slow the degradation you have identified.

A5 Effects of temperature changes on construction materials

All construction materials respond in some way to temperature changes in their environment.

Heat energy that causes a change in the temperature of a material is called sensible heat. The amount of sensible heat required to produce a given rise in temperature varies from material to material and even between the solid, liquid and gas phases of the same material.

Energy that does not cause a temperature rise, but is absorbed by the material as it undergoes a phase change, is called latent heat. For a given material, the latent heat of fusion (phase change from solid to liquid) will be different from the latent heat of vaporisation (phase change from liquid to gas).

Effect of temperature change on material properties

Changes of state

If a solid material is subjected to a continuous input of heat energy, its temperature will begin to rise until it reaches the material's **melting point**. During the change of state, continued heating will lead to no further increase in temperature until the solid has completed its change into a liquid. After this, continued heat will produce a rise in temperature, but at a different rate, until the **boiling point** is reached. Again, during the phase change, continued heat will lead to no further increase in temperature until all the liquid has changed into gas. Once the change of state is complete, continued heating will produce a rise in temperature at another rate.

Linear expansivity

A change in the temperature of a material is associated with an expansion or contraction of its size in all directions. The amount by which the size changes for a given change in temperature differs from material to material and is defined by the material's **coefficient of linear expansion**.

Evaporation

Evaporation is the process where a liquid changes to a gas, even at temperatures below its boiling point. Evaporation happens because, although the average **kinetic energy** of all the **molecules** in a liquid is low, individual molecules can have sufficient energy to escape from the surface as a gas. As these high-energy molecules leave the liquid they take their energy with them. The average energy left in the liquid reduces as does its temperature. This is why humans sweat to aid cooling, as the evaporation of water from our skin reduces its temperature.

> **Key terms**
>
> **Melting point** – the temperature at which a phase or state change from solid to liquid will begin.
>
> **Boiling point** – the temperature at which a change of state from liquid to gas will begin.
>
> **Coefficient of linear expansion (α)** – a material property that describes the amount by which a material expands upon heating with each degree rise in temperature. It is measured in inverse kelvin (K^{-1}).
>
> **Kinetic energy** – the energy possessed by an object due to its motion.
>
> **Molecules** – the smallest unit of a chemical compound, made up of groups of atoms.

Evaporation is a vital process in a range of applications associated with construction, from drying out timber, bricks or other porous materials that might otherwise retain moisture to the evaporation of solvents from paints, allowing them to dry.

The rate of evaporation can be increased by:
▶ raising the temperature
▶ increasing the surface area of a liquid from which molecules can escape
▶ reducing the relative humidity of the air immediately above the surface of the water (an effective method is by providing an air current over the surface of the liquid to carry away the gas molecules as they emerge).

A6 Behaviour of structural members under load

Structural members are the elements of a building or structure that help to support load. In construction these are divided into a number of common types that are designed for different uses and loading conditions, as shown in Table 1.25.

▶ **Table 1.25:** Types of structural members

Structural member	Loading	Description
Beams	Bending	A horizontal structural member made of wood, metal, stone or concrete supported in two (or more) places and used to support loads in a building or other structure.
Lintels	Bending	A type of beam providing support across the top of door or window opening made in walls.
Columns	Compression	A vertical structural member made of wood, metal, stone or concrete used to support loads in a building or other structure.
Wall and frames	Compression	Load-bearing structures that form the external walls and internal partitions of a structure.
Struts	Compression	A rod or bar forming part of a framework and designed to resist forces in compression.
Ties	Tension	A rod or bar joining two elements of a framework and designed to resist forces in tension.

Effect of different loading conditions

The choice of materials used in the construction of structural members is important. As beams, lintels, columns, walls, frames, struts and ties are all loaded in different ways, their materials must have the corresponding strength. For example, concrete is weak in tension and so would not be appropriate for use in ties. Table 1.26 shows the effect of different loading conditions on a range of materials.

▶ **Table 1.26:** Relative strengths of a range of materials used in structural members loaded in compression, tension and bending

	Concrete	Reinforced concrete	Steel (structural)	Timber (SPL)	Failure mechanisms
Struts, columns, frames and walls loaded in compression	Medium	Medium	Very high	Low	• **Buckling** • Compressive yielding (ductile) • Compressive failure (brittle)
Ties loaded in tension	Low	High	Very high	Medium	• Tensile yielding (ductile) • Tensile failure (brittle)
Beams and lintels loaded in bending	Medium	High	Very high	Medium	• Shear failure • Bending

Key term

Buckling – a structural failure caused when a member loaded in compression becomes unstable and deflects sideways. It is more common in long, thin structural members and can cause failure well below the material's compressive strength.

Types, configuration and effect of loads

The loads that a structure must be designed to support can be divided into several different types, as shown in Table 1.27. Dead or static loads are straightforward to analyse, as they are constant and relatively easy to deal with. However, the effects of live or dynamic loads can be less easy to predict. Dynamic cyclic loads are associated with fatigue failure and loads which vary at the natural frequency of a structure can cause large deflections and premature failure.

▶ **Table 1.27:** Types and characteristics of loads

Type	Characteristics
Dead	Static loads acting on a structure that do not move or change over time, for example the vertical force exerted by the mass of the structure itself.
Live or imposed	Dynamic loads acting on a structure that continually vary, for example the vertical forces exerted by the mass of people moving around in an office building.
Wind	Dynamic loads acting on a structure caused by the movement of air pressing on its external surfaces, for example a high-rise office building will have substantial horizontal loading in high winds.
Point	A load acting at a single location or point on a structure, for example the load exerted on the horizontal beam of a jib crane acts at the point where the support cable is positioned.
Distributed	A load distributed over the length surface area of a structural member, for example the weight of a beam does not act at a single point but is distributed evenly along its entire length.

Characteristics, properties and use of types of supports

Structural supports and joints are divided into types, depending on the directions in which they are designed to prevent movement. For instance, a door is mounted on pinned or hinged supports that allow free rotation but stop any vertical or horizontal movement. Table 1.28 gives some examples of common supports.

▶ **Table 1.28:** Types, characteristics and uses of supports

Type	Characteristics	Uses
Pinned or hinged	• Allows free rotation. • Prevents vertical and horizontal movement.	Join the struts and ties in the framework of electricity pylons.
Roller	• Allows free rotation. • Allows free horizontal movement. • Prevents vertical movement.	Allows bridge structures to expand and contract without imposing large horizontal forces on their supports.
Fixed	• Prevents rotation. • Prevents vertical and horizontal movement.	Beams supported at one end such as those supporting a balcony use a single fixed support.

❙❙ PAUSE POINT Choose a suitable material for a 4.5 m structural beam used to support the brick frontage above the full-width doors and windows of a shop on the ground floor of a three-storey building.

Hint Consider how the beam will be loaded and choose the material best able to support that type of load.

Extend What would be the consequences of making the wrong choice?

1 A loft space is being insulated using fibreglass wool. Which of the following is the key specification requirement for insulation materials? *(1 mark)*

    A Resistance to vapour penetration

    B Embedded energy

    C Thermal conductance

    D Density

2 A glass panel used in a shower cubicle is made using toughened glass. Explain one reason why toughened glass is used in this application. *(2 marks)*

3 Plywood is a manufactured board used extensively in construction. Explain two advantages of using plywood instead of natural timber. *(4 marks)*

4 Roofs are a vital element of any build and are designed to prevent water penetration into the fabric of a building.

    a. Give a type of covering suitable for a pitched roof. *(1 mark)*

    b. Describe the use of lead flashing in roof construction. *(2 marks)*

    c. Give the key material property of lead that makes it suitable for use as flashing. *(1 mark)*

5 The lifespan of any building is limited by the ability of its constituent materials to resist degradation and decay.

    a. Describe the freeze-thaw cycle and how its action damages stonework. *(4 marks)*

    b. State two causes of degradation in exterior timber. *(2 marks)*

    c. State one method used to prevent degradation in exterior timber. *(1 mark)*

6 Lintels are common structural elements used in the construction of buildings.

    a. Explain how lintels are used in a structure. *(2 marks)*

    b. State the material most commonly used for lintels in high-load applications. *(1 mark)*

Plan
- Have I read and fully understood the questions?
- How will I approach the task?

Do
- Have I recognised the command word used in the question (give, state, explain, describe, etc.)?
- Do I understand my thought process and can I explain why I have chosen to answer the questions in a particular way?

Review
- Have I answered the questions fully?
- Can I identify which elements I found most difficult and where I need to review my understanding of a topic?

B   Solving practical construction problems

B1 Application of mathematical and statistical methods and techniques used in practical construction contexts

Algebraic techniques

These are used to represent things like volume, load, time or temperature in the form of equations.

Linear equations

Linear equations take the form $y = mx + c$

where the pattern of numbers increases or decreases by the same amount, enabling a straight line to be drawn on a graph. In this case, m is the slope, and c is where the line intercepts the y-axis.

Link

See Graphical techniques later in this section for more information about equations of straight-line graphs.

Addition and subtraction

Only 'like' terms can be added or subtracted.
For example, the expression

$$30d + 20w - 10d + 30w$$

relates to the number of doors (d) and windows (w) on a particular housing development. Because doors and windows are different, they have to be treated separately, thus:

Total $= 20d + 50w$

Therefore, on the housing development there are 20 doors and 50 windows.

Multiplication and division

As with addition and subtraction, you can multiply or divide 'like' algebraic terms, just as you would numbers. For example:

$b \times b = b^2$ and

$$\frac{k}{\cancel{k}} \times \cancel{k} = k$$

Brackets and factors

Brackets are generally used when multiplying two terms together, and to avoid writing a multiplication sign ('×') that could be confused with an algebraic term 'x', which represents a physical number or property.

Brackets are used to simplify expressions, or to change an expression which has terms added together to one with terms multiplied together. The same can be done with algebraic expressions. For example, the expression $6x + 2y$ can be written as $2(3x + y)$ because 2 is a common factor of both $6x$ and $2y$ and can be divided out of the expression, leaving $3x + y$ in brackets. Therefore:

$$6x + 2y = 2(3x + y)$$

where 2 and $3x + y$ are factors of $6x + 2y$.

Note that sometimes both factors are in brackets. To multiply them out, we can use the 'FOIL' rule:

▶ **F**irst terms (the terms at the start of each bracket) multiplied
▶ **O**utside terms (the terms at either end of the pair of brackets) multiplied
▶ **I**nside terms (the terms in the middle of the pair of brackets) multiplied
▶ **L**ast terms (the terms at the end of each bracket) multiplied.

Worked example

Multiply out the following brackets using the FOIL rule:

$(x + 2)(2x + 3)$

First $= x(2x) = 2x^2$

Outside $= x(3) = 3x$

Inside $= 2(2x) = 4x$

Last $= 2(3) = 6$

Therefore, putting them all together:

$(x + 2)(2x + 3) = 2x^2 + 3x + 4x + 6$

$$= 2x^2 + 7x + 6$$

Quadratic expressions are ones where the highest exponent (power or index) of a variable is a square (^{2}). For example: $a^2 + 2ab + b^2$

Link

For more about quadratic expressions, see Graphical techniques later in this section.

Theory into practice

Multiply out the brackets and then write down the following expressions in their simplest form:

$5x(3x - 8) - 2x^2$

$2(x + 3y - z) - 3(x + y - z) + 4(x - y)$

$2x(y + 3z) - 2y(3z + x) - 3z(2x - 2y)$

$(2x - 1)(3x^2 - x - 3)$

$3(y^4 - y^2 - 2) - 2y(y^3 + 2y - 1) - y^2(y^2 + 1)$

Exponents (powers or indices) and roots

This is where a base number is multiplied by itself a number of times. For example:

$3^1 = 3$

$3^2 = 3 \times 3 = 9$

$3^3 = 3 \times 3 \times 3 = 27$

$3^4 = 3 \times 3 \times 3 \times 3 = 81$

The number being multiplied (or 'raised to a power'), in this case '3', is called the base and the exponent is the raised number, 1 to 4 as shown. If the exponent is a fractional value, then this represents a root of the base number. For example the exponent $\frac{1}{2}$ represents the square root and $\frac{1}{3}$ represents the cube root.

$$16^{\frac{1}{2}} = {}^{2}\sqrt{16}$$

$$27^{\frac{1}{3}} = {}^{3}\sqrt{27}$$

In algebra, exponents can be dealt with using the rules shown in Table 1.29. Note that in the table the '×' has been included for clarity but in practice it is often omitted.

▶ **Table 1.29:** Exponent rules

Rule name	Rule	Example
Product rules	$a^n \times a^m = a^{n+m}$	$3^2 \times 3^3 = 3^{2+3} = 3^5 = 243$
	$a^n \times b^n = (a \times b)^n$	$2^2 \times 3^2 = (2 \times 3)^2 = 6^2 = 36$
Division rules	$a^n \div a^m = a^{n-m}$	$3^5 \div 3^3 = 3^{5-3} = 3^2 = 9$
	$a^n \div b^n = (a \div b)^n$	$4^3 \div 2^3 = (4 \div 2)^3 = 2^3 = 8$
Indices rules	$b^{n\,m} = b^{n \times m}$	$(2^3)^2 = 2^{3 \times 2} = 2^6 = 64$
	$^m\sqrt{b^n} = b^{n/m}$	$^2\sqrt{3^6} = 3^{6/2} = 3^3 = 27$
	$b^{1/n} = {}^n\sqrt{b}$	$8^{1/3} = {}^3\sqrt{8} = 2$

Formulae and equations

Formulae and equations are used to calculate physical quantities. One of the most common is the formula used for the area of a triangle, where A, b and h represent the area, base length and height of any triangle.

Area = ½ × base × height, or $A = \frac{1}{2} bh$

With the equation in the form above, the values of the base length and height must be known to find the triangle's actual area. It is also important that the units of both b and h are the same, for example both are in metres (m). This will give the value of the area (A) in square metres (m²). Both sides of an equation must always balance. The equals sign shows that the left-hand side (LHS) of the equation balances the right-hand side (RHS).

Rearranging formulae

Worked example

If we know the area and the base length of a triangle we can find the height (h). We need to transpose or rearrange the formula so the value we need to find is by itself on the left-hand side (LHS) of the formula. We want to get h on its own on the LHS. To do this, first switch both sides over – the equation remains balanced and the signs do not change.

$$\frac{1}{2} \times b \times h = A$$

We now want to remove the b which is multiplying the h-value from the LHS. So we divide both sides by b.

$$\frac{1 \times b \times h}{2 \times b} = \frac{A}{b}$$

Next, we can cancel the 'b's on the LHS because b divided by b equals 1. Therefore, we have:

$$\frac{h}{2} = \frac{A}{b}$$

Now, we want to remove the 2 from the LHS. As it is dividing into h we need to multiply both sides by 2 to balance the equation.

$$\frac{2 \times h}{2} = \frac{2 \times A}{b}$$

Finally, 2 divided by 2 equals 1, so it cancels, leaving the formula that we want:

$$h = \frac{2 \times A}{b} = \frac{2A}{b}$$

With addition or subtraction, when a number or a term needs to be moved to the other side of the equals sign, we must do the same thing to both sides to make sure that the equation remains balanced. For example:

$$25 + 10 = 41 - 6$$
$$35 = 35 \checkmark$$

Adding 6 to both sides keeps the equation balanced:

$$25 + 10 + 6 = 41 - 6 + 6$$
$$25 + 10 + 6 = 41$$
$$41 = 41 \checkmark$$

The 6 appears to have changed its sign when it moved to the other side of the equals sign. The same is also true for terms involving letters, for example:

$$R + S = T + U$$
$$R = T + U - S$$

The equation is still balanced because we subtracted S from both sides (or changed the sign from '+S' to '−S' as it moved from one side to the other).

Theory into practice

Rearrange the following equations to change the subject as indicated:

$v = u + at$	Make a the subject.
$a^2 = b^2 + c^2$	Make b the subject.
$C = \dfrac{N - n}{2\pi}$	Make n the subject.
$V = \pi r^2 h$	Make r the subject.
$T = 2\pi \sqrt{\dfrac{L}{g}}$	Make g the subject.
$a = b + \sqrt{b^2 + c^2}$	Make c the subject.

Substituting values into and evaluating formulae

Formulae allow us to find out physical information about construction quantities and factors.

Worked example

A rectangular swimming pool, shown below, is s metres wide and t metres long, and is surrounded by some timber decking, which is s metres wide. Find a formula in its simplest form for the area of the decking in terms of s and t.

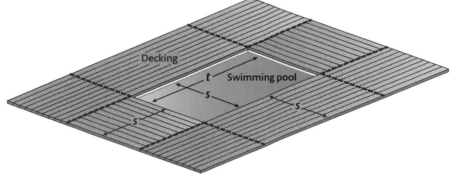

▶ **Figure 1.1:** Finding an equation for the area of decking

The decking is made of a series of rectangles and squares. At the corners of the decking there are four squares, each of s by s m². Next to the longest sides of the pool, the decking is *t* metres long by *s* metres wide, and the shortest sides have decking next to them of *s* metres long by *s* metres wide. Therefore, the formula for the area can be constructed:

Total area of decking $= 4(s \times s) + 2(s \times t) + 2(s \times s)$
$$= 4s^2 + 2st + 2s^2$$
$$= 6s^2 + 2st$$

This formula can now be used to calculate the area of decking for any values of *s* and *t*, by substituting in the known values for *s* and *t* to find the area.

If, say, we know that *s* = 5.5 m and *t* = 12.5 m, we can substitute these values into the formula to work out the area. Therefore, the total area of decking would be calculated as follows:

Total area of decking $= 6s^2 + 2st$
$$= 6(5.5\,m)^2 + 2 \times 5.5\,m \times 12.5\,m$$
$$= 181.5\,m^2 + 137.5\,m^2$$
$$= 319\,m^2$$

Simultaneous equations

To solve equations with two unknowns we need to have two equations, known as simultaneous equations.

Worked example

Find the values of *x* and *y* to solve the following simultaneous equations.

$x + 2y = 8$ [equation 1]

$5x - 3y = 1$ [equation 2]

Step 1: Rearrange both equations to get *x* on its own.

$x = 8 - 2y$ [equation 1a]

$x = \dfrac{1 + 3y}{5}$ [equation 2a]

Step 2: As the expressions on the RHS of both equations equal *x*, they must equal each other:

$8 - 2y = \dfrac{1 + 3y}{5}$

Step 3: We now have one equation in one unknown (*y*), and we can solve this by transposing and simplifying its terms. First, bring the denominator 5 to the top of the LHS by multiplying both sides by 5, then multiply out the brackets, simplify and collect up all the *y* terms and all the other numbers:

$5(8 - 2y) = 1 + 3y$ [multiply both sides by 5]

$40 - 10y = 1 + 3y$ [multiply out the brackets]

$40 - 1 = 3y + 10y$ [add 10y to both sides and subtract 1 from both sides]

$39 = 13y$ [simplify]

$y = \dfrac{39}{13} = 3$ [divide both sides by 13]

Cont...

Step 4: Put the value that you have found for y back into either equation 1 or 2 to find the other 'unknown', x.

Using equation 1:

$x + 2y = 8$
$x + 2(3) = 8$
$x + 6 = 8$
$x = 8 - 6$
$x = 2$

Step 5: Finally, put both values into equation 2 to check your answers!
$$\text{LHS} = 5x - 3y = 5(2) - 3(3) = 10 - 9 = 1 = \text{RHS} ✓$$

Quadratic equations

Quadratic equations are used in more complex calculations, often where maximum or minimum values are needed. These equations take the form:
$$ax^2 + bx + c = 0$$

where a, b and c are numbers, and x can usually have two values. Quadratic equations can be solved in the following ways. They can also be solved using graphical techniques (see page 38).

Solving quadratic equations: Factors

Some quadratic expressions can be factorised.

Factorising means rewriting the expression as the product of two brackets.

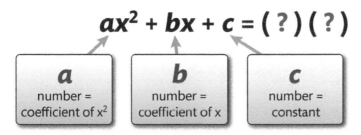

$$ax^2 + bx + c = (\ ?\ )(\ ?\ )$$

a	**b**	**c**
number = coefficient of x^2	number = coefficient of x	number = constant

We can use factorisation to help us solve quadratic equations for x provided we always rearrange them to equal zero.

Consider the following quadratic equation that needs to be factorised:
$$x^2 - 2x - 3 = 0$$

Start by identifying the values of a, b and c. In this example, they are $a = 1$, $b = -2$ and $c = -3$. The method for factorising is then as follows:

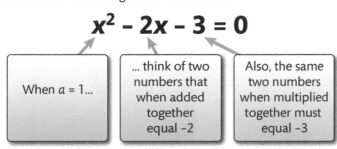

$$x^2 - 2x - 3 = 0$$

When $a = 1$...	... think of two numbers that when added together equal −2	Also, the same two numbers when multiplied together must equal −3

$x^2 - 2x - 3 = (x + ?)(x + ?)$

Now, −3 and +1 seem to fit because:
$$-3 + 1 = -2 \text{ and } -3 \times 1 = -3$$

So let's try -3 and $+1$ to see if we get what we want when we multiply out the brackets:

$$(x - 3)(x + 1) = x^2 = x - 3x$$
$$= x^2 - 2x - 3$$

which is the quadratic expression that we need! Therefore, we can write:

$$x^2 - 2x - 3 = (x - 3)(x + 1) = 0$$

So $(x - 3)(x + 1) = 0$

and the factors are $(x - 3)$ and $(x + 1)$.

Now, if the expression $(x - 3)(x + 1)$ has the value zero, it must be true that either:

$$(x - 3) = 0 \text{ or } (x + 1) = 0$$

This means that either:

$$x = 3 \text{ or } x = -1$$

To check if this is right, substitute these answers for x in the original quadratic equation. Let's first try $x = 3$ to see if the answer we get is zero:

$$x^2 - 2x - 3 = (3)^2 - 2(3) - 3$$
$$= 9 - 6 - 3$$
$$= 0 \checkmark$$

Let's now try $x = -1$:

$$x^2 - 2x - 3 = (-1)^2 - 2(-1) - 3$$
$$= 1 + 2 - 3$$
$$= 0 \checkmark$$

Both of these make the quadratic equation balance, so the solution to the quadratic equation

$x^2 - 2x - 3 = 0$ is $x = 3, x = -1$

Solving quadratic equations: Completing the square

The method converts a quadratic equation to two equal factors or sides of a 'square'. We've learned that the standard format for a quadratic equation is $ax^2 + bx + c = 0$. To find the values of x, we first move the numerical constant c to the RHS of the equation. For example, to find the values of x that solve the equation $2x^2 + 8x - 10 = 0$, first move 10 to the RHS of the equation.

To find the values of x that solve the equation $2x^2 + 8x - 10 = 0$, first move the numerical constant c to the RHS of the equation.

$$2x^2 + 8x = 10$$

Make $a = 1$ by dividing through both sides by a. In this example, divide by 2:

$$x^2 + 4x = 5$$

Now, add $\left(\frac{b}{2}\right)^2$ to both sides of the equation, where b in this example is 4.

$$x^2 + 4x + \left(\frac{4}{2}\right)^2 = 5 + \left(\frac{4}{2}\right)^2$$
$$= 5 + \frac{16}{4} = 9$$

Therefore:

$$x^2 + 4x + \left(\frac{4}{2}\right)^2 = 9$$
$$x^2 + 4x + 2^2 = 9$$

Next, 'complete the square' by factorising the LHS.

$$x^2 + 4x + 2^2 = 9$$
$$(x + 2)(x + 2) = 9$$
$$(x + 2)^2 = 9$$

This can be pictured as a square with sides $x + 2$, as illustrated.

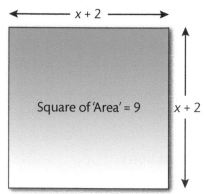

Finally, rearrange and solve the linear equation for x by finding the square root of both sides of the equation.

$$(x + 2)^2 = 9$$
$$x + 2 = \sqrt{9}$$
$$= \pm 3$$

Therefore, $x = 3 - 2 = 1$ or $x = -3 - 2 = -5$

These values can be substituted into the original equation to check that your answers are correct.

For $x = 1$:

$$2x^2 + 8x - 10 = 2(1)^2 + 8(1) - 10 = 2 + 8 - 10 = 0 \checkmark$$

And for $x = -5$:

$$2x^2 + 8x - 10 = 2(-5)^2 + 8(-5) - 10 = 2(25) - 40 - 10$$
$$= 50 - 40 - 10 = 0 \checkmark$$

Solving quadratic equations: Formula

The solution to the quadratic equation $ax^2 + bx + c = 0$ can be found by substituting the values into the following formula:

$$x = \frac{-b \pm \sqrt{b^2 - 4ac}}{2a}$$

So using the example that we used above for solving by factors:

$$x^2 - 2x - 3 = 0$$

we see that $a = 1, b = -2, c = -3$.

Substituting these into the formula, we get:

$$x = \frac{-(-2) \pm \sqrt{(-2)^2 - 4(1)(-3)}}{2(1)}$$
$$= \frac{2 \pm \sqrt{4 + 12}}{2} = \frac{2 \pm \sqrt{16}}{2} = \frac{2 \pm 4}{2}$$

Therefore, $x = \frac{6}{2} = 3$ or $x = \frac{-2}{2} = -1$

which are the same answers as we had before.

ⅠⅠ PAUSE POINT   State the most common rules for calculating with exponents (indices).

Hint   Look back over this section of the unit.

Extend   Simplify the following expressions involving exponents:
$3^n \times 3^m$ and $3^n \times 4^n$

Accuracy of calculations

For calculations to be performed accurately they must be done in the correct order. The simple acronym – BIDMAS – can help you remember the order in which to do the separate parts of the calculation:
Bracketed calculations are done first.
Indices or powers are calculated next (e.g. 4^2, where 2 is the power; $4^2 = 4 \times 4 = 16$)
Division and **M**ultiplication take equal priority and are done next.
Addition and **S**ubtraction are done last and are of equal priority.

Significant figures (s.f.)

There may be times when quoting the exact answer is unnecessary. For example, if the tolerance of a new garage door is to the nearest 10 mm, we could 'round' the answer to a more convenient value, for example 5610 mm. To help round numbers to convenient and usable values there are some basic rules:

▶ Figures ending in 5 and above are rounded up and the next figure to the left increases by 1.
▶ Figures ending in 4 and below are rounded down and the next figure to the left remains the same.

For example:
$14.6539 \approx 14.654$ (rounded to 5 s.f.)
≈ 14.65 (rounded to 4 s.f.)
≈ 14.7 (rounded to 3 s.f.)
≈ 15 (rounded to 2 s.f.)

Using significant figures is good for rough checks and estimations. However, select the most appropriate significant figure to round to. This depends upon the level of accuracy you need.

In some situations, it is more sensible to round down, even when the answer has a remainder of 0.5. For example, how many 2-metre lengths can be cut from a 7-metre long copper pipe? The answer is 7 m ÷ 2 m = 3.5. However, you can only cut three 2-metre lengths; the rest is wasted. So you should round down.

Decimal places

Decimal places (d.p.) are the numbers appearing to the right of the decimal point. So, in the above example:
$14.6539 \approx 14.654$ (rounded to 3 d.p.)
≈ 14.65 (rounded to 2 d.p.)
≈ 14.7 (rounded to 1 d.p.)
Decimal places are useful when dealing with distance. In construction, we measure lengths and distances in metres (m) or millimetres (mm), where 1000 mm = 1 m. This often requires measurements in metres to three decimal places so the accuracy is to the nearest 1 mm.

Standard form

In technical calculations you sometimes have to deal with very large or very small numbers. Standard form is a convenient way of writing and using these numbers. It is used by scientific calculators to display them, and it is how you need to input them into your calculator when doing construction calculations.

Most scientific calculators allow you to toggle through standard form in powers of multiples of 3. This is often called engineering standard form because we quote physical properties in kilos, where 'kilo' means 1000 or 10^3. In standard form a number is split into two parts: a decimal number (N) multiplied by the number 10 raised to a power (n) that can be a positive or negative whole number. Standard form works like an equation, where the left-hand side (LHS) is equal to the right-hand side (RHS), $N \times 10^n$.

For example: $78{,}531 = 7.8531 \times 10^4$

where:

▶ $10^4 = 10 \times 10 \times 10 \times 10 = 10{,}000$
▶ and $7.8531 \times 10{,}000 = 78{,}531$ (the original number).

A simple way to remember this is that for big numbers (i.e. much greater than 1) the decimal point moves to the left and the power to which 10 is raised is always positive, e.g. when the decimal point moves 11 places to the left:

$235\,000\,000\,000.0 = 2.35 \times 10^{11}$ Positive power of 11

Standard form also works with small numbers because negative powers of 10 mean dividing by 10.

For example, the very small number 0.0000000541 is written 5.41×10^{-8} in standard form, where the decimal point moves 8 places to the right.

Use of approximation to check a calculation

It's a good idea to do a mental check on the calculation to make sure the answer is what you were expecting. To do this, you need to know your multiplication tables – reciting and remembering these basic number relationships will help you with your mental arithmetic.

Suppose you need to calculate:

$52.3 \times \dfrac{27.4}{91.8}$

Your mental check could be:

$\dfrac{50 \times 30}{100} = 15$

The answer on your calculator (rounded to 2 decimal places) would be:

52.3 × 27.4 ÷ 91.8 = 15.61

Both answers are similar, so you can be sure that your calculated answer is right.

Effects of rounding errors

Rounding errors are the difference between a rounded numerical value and the actual value. A mathematical miscalculation caused by rounding a number, using significant figures or decimal place criteria, might be small, but it can have a cumulative effect in calculations using the rounded value. For example, a square, turfed lawn with a side length of 12.56 m if rounded to 2 s.f. would be 13 m. This difference of 440 mm might seem small, but, if applied to calculating the area of the turf, would give a rounding error of 7 per cent. Therefore, you should always consider the impact a rounding error could have on your final calculation.

 **PAUSE POINT**
Find the value of the following and carry out a mental check to confirm your answer:

$$\sqrt{\frac{55 + 31.7 - 7.8}{2}}$$

Hint
As the square root applies to the whole expression, remember to put it in brackets.

Extend
Using your calculator, practise and check expressions of your own making.

Geometric techniques

Geometry involves the study of lines, angles, curves and planes. This section will introduce you to some useful techniques to enable you to deal with a variety of practical geometric calculations.

Pythagoras' theorem

Pythagoras' theorem relates the lengths of the three sides of a right-angled triangle, a, b and c as shown in Figure 1.2. The hypotenuse is the side opposite the right angle and is labelled length 'c'. The two shorter sides forming the triangle are 'b' and 'a'. Pythagoras' rule links these three sides by the equation $c^2 = a^2 + b^2$.

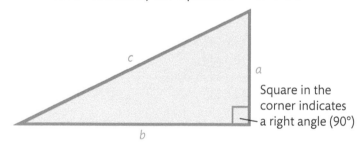

Square in the corner indicates a right angle (90°)

▶ **Figure 1.2:** An example demonstrating Pythagoras' theorem

This formula is used regularly by surveyors when setting out walls and other features that need to be built at right angles to one another. It can be rearranged to calculate the length of side a, b or c. For example:

$c = \sqrt{a^2 + b^2}$ or

$a = \sqrt{c^2 - b^2}$

Trigonometric ratios

The study of trigonometry looks more closely at how angles in triangles affect the lengths of their sides. In a right-angled triangle (Figure 1.3) the sine (sin), cosine (cos) and tangent (tan) of the angle are θ defined by the lengths of the sides of the triangle such that:

$$\sin \theta = \frac{\text{opposite}}{\text{hypotenuse}} \qquad \cos \theta = \frac{\text{adjacent}}{\text{hypotenuse}} \qquad \tan \theta = \frac{\text{opposite}}{\text{adjacent}}$$

These are all known as trigonometric ratios or 'trig ratios' for short.

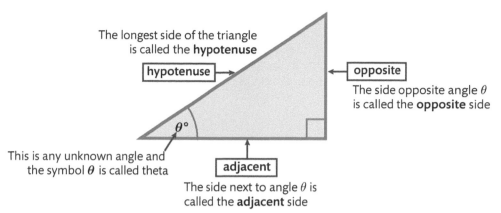

The longest side of the triangle is called the **hypotenuse**

| hypotenuse |

| opposite |

The side opposite angle θ is called the **opposite** side

This is any unknown angle and the symbol θ is called theta

| adjacent |

The side next to angle θ is called the **adjacent** side

▸ **Figure 1.3:** Trigonometric ratios

Worked example

The cross section through a triangular, cantilevered canopy for a proposed warehouse is shown in Figure 1.4. Using trigonometric ratios, work out:

- the height of the canopy strut BC
- the width of the cantilevered overhang AC
- the angle that the rear stay makes with the warehouse roof at D
- the length of the rear stay BD.

Now undertake a suitable alternative check calculation for BD.

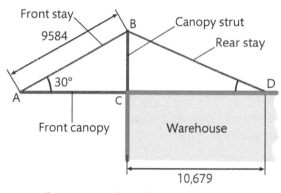

Cross-sectional elevation
(All dimensions in mm)

▸ **Figure 1.4:** Cross section through a canopy for a proposed warehouse

In Figure 1.4, there are two right-angled triangles. In the left-hand side one, labelled ABC, we are given two pieces of information so we should be able to work out the rest of the dimensions and angles for that triangle. Once this is done, we should be able to move on to triangle BCD using the height BC.

1 To find the height of the canopy strut BC, we can look closely at triangle ABC and identify the sides in relation to the angle; that is, find the hypotenuse, opposite and adjacent sides to the angle.

Step 1: It is always a good idea to label them on a separate sketch collating all the information, as shown in Figure 1.5.

▸ **Figure 1.5:** Finding the height of the canopy strut BC

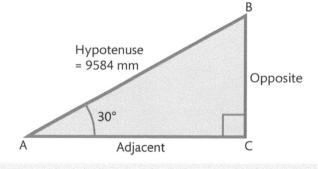

Hypotenuse = 9584 mm

Opposite

30°

A   Adjacent   C

Step 2: Apply the trig ratio. If we want to find BC, the 'opposite' side to the angle, we need to use a ratio which has all the known values and the 'opposite' specified. This can only be the sine ratio because:

$$\sin \theta = \frac{\text{opposite}}{\text{hypotenuse}}$$

and we know the length of the hypotenuse is 9584 mm, and the size of the angle θ is 30°.

Step 3: Substitute in the known values:

$$\sin 30° = \frac{\text{opposite}}{9584\,\text{mm}} = \frac{BC}{9584\,\text{mm}}$$

Step 4: Rearrange the equation to make 'opposite' the subject of the formula and use your calculator to find the value of sin 30°:

Opposite $= 9584\text{mm} \times \sin 30°$
$= 4792\text{mm} =$ length of strut BC

The answer will be in mm units because the distance we put into the formula was in mm.

2 To find the width of the cantilevered overhang AC, using trig ratios again, we can see that AC is the 'adjacent' side to the 30° angle. So we need to use the cosine ratio:

$$\cos \theta = \frac{\text{adjacent}}{\text{hypotenuse}}$$

$$\cos 30° = \frac{\text{adjacent}}{9584\text{mm}} = \frac{AC}{9584\text{mm}}$$

AC $= 9584\text{mm} \times \cos 30° = 8300\text{mm} =$ width of canopy AC

3 To find the angle that the rear stay makes with the warehouse roof at D, let's look at the other triangle that is part of the canopy structure, the right triangle BCD.

Step 1: Draw out the triangle and jot down all the values that we know as well as the unknown angle that we are trying to find. Figure 1.6 shows an example.

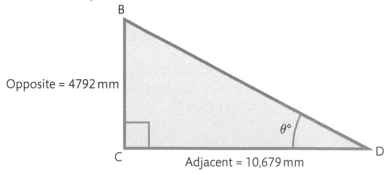

▶ **Figure 1.6:** Drawing out the triangle

Step 2: We want to find the angle $\theta°$, so we need to use a trig ratio which has 'opposite', 'adjacent' and the angle specified.

$$\tan \theta = \frac{\text{opposite}}{\text{adjacent}}$$

Step 3: Substitute in the known values and work out the ratio.

$$\tan \theta = \frac{4792\,\text{mm}}{10{,}679\,\text{mm}} = 0.44873$$

Cont...

Step 4: Convert the tangent ratio calculated into an angle value that we can recognise. To do this, we need to identify the 'inverse trigonometric ratios' on our scientific calculators. Find the button marked with the symbol tan⁻¹ (inverse tangent ratio). Therefore:

$$\tan^{-1}(0.44873) = 24.1672 \text{ decimal degrees}$$
$$= 24° \, 10'$$

4 There are two ways to find the length of the rear stay BD: trigonometric ratios or Pythagoras' theorem. Let's use trig ratios.

Step 1: Collate all the information in a sketch, as in Figure 1.7.

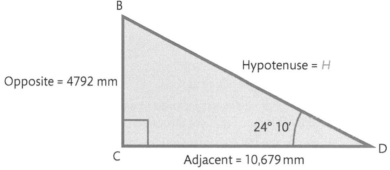

B

Hypotenuse = H

Opposite = 4792 mm

24° 10'

C

D

Adjacent = 10,679 mm

▶ **Figure 1.7:** Collate all the information in a sketch

Step 2: Apply the trig ratio. We can use either the sine ratio or the cosine ratio to find the hypotenuse H. Let's do it using the sine ratio, where:

$$\sin \theta = \frac{\text{opposite}}{\text{hypotenuse}}$$

Step 3: Substitute in the known values and rearrange to get the unknown on the LHS on the top.

$$\sin 24° \, 10' = \frac{4792 \text{mm}}{H}$$
$$\therefore H = \frac{4792 \text{mm}}{\sin 24° \, 10'}$$
$$H = 11{,}705 \text{mm} = 11.705 \text{m}$$

5 To undertake a suitable alternative check calculation for BD, we can either apply the cosine ratio or use Pythagoras' theorem. Using Pythagoras' theorem:

$$BD = \sqrt{(10{,}679 \text{mm})^2 + (4792 \text{mm})^2}$$
$$= 11{,}705 \text{mm} = 11.705 \text{m}$$

Circular measures

Circles have many important geometric properties and use terms which occur regularly in construction problems. Figure 1.8 identifies the main components in circular and curved geometry, where r is the radius of the circle. The basic elements are:

▶ circumference of circle $= 2\pi r$

▶ length of arc $(s) = \dfrac{\theta°}{360°} \times 2\pi r$

▶ area of circle $= \pi r^2$

▶ area of sector $= \dfrac{\theta°}{360°} \times \pi r^2$

Theory into practice

A lean-to conservatory has a base which is 7.5 m long by 4 m wide. The glazed roof makes an angle of 36° to the horizontal, across the 4 m width. Calculate:

- the length of the sloping roof in m to 1 d.p.
- the area of glazed roof in m² to 1 d.p.

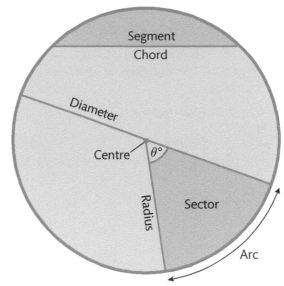

▶ **Figure 1.8:** Parts of a circle

The size of an angle is measured in either degrees or radians. Degrees are normally used by surveyors and construction professionals.

Angles in the UK are measured using the sexagesimal system, where there are 360 degrees (360°) in a whole circle. The sexagesimal system splits up parts of a degree into fractions of a degree as follows:

▶ 1 whole circle = 360 degrees, written as 360°
▶ 1 degree = 60 minutes, written as 60′
▶ 1 minute = 60 seconds written as 60″

So, for example, an angle could be written as 46° 30′ 45″.

In simple calculations, where high accuracy is not required, angles can be quoted in decimal degrees to one or two decimal places. For example 72.5° is equivalent to 72° 30′ because 30′ is half a degree.

For more complex calculations used in civil engineering, a radian angle is defined. This is the angle made at the centre of a circle when a length, equal to the radius, is traced out around its circumference (see Figure 1.9).

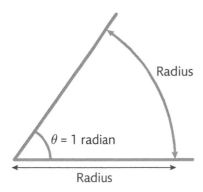

▶ **Figure 1.9:** Radian measurement

We know that the circumference of a circle = $2\pi r$. Therefore the number of radians contained within one whole circle of 360° is $\frac{2\pi r}{r}$

Therefore:

▶ $360° = 2\pi$ radians
▶ $180° = \pi$ radians

which means:

▶ 1 radian $= \frac{180°}{\pi} \approx 57°$

▶ 1 degree $= \frac{\pi}{180°} \approx 0.01745$ radians

From this definition, the following two properties of a circle can be found:

▶ Length of arc (s) = radius × angle in radians subtended at the centre of the circle ($s = r\theta$)

▶ Area of sector = ½ × (radius)² × angle in radians subtended at the centre of the circle ($A = \frac{1}{2} r^2 \theta$)

Properties of lines and angles

A line is a straight distance, usually joining two points together. Lines can be parallel, if they have the same slope or gradient, or they can cross at an angle to form an intersection. An angle forms where two lines intersect. Examples of different types of angles can be seen in Figure 1.10. Table 1.30 lists the different categories of angles.

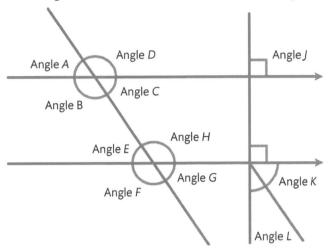

▶ **Figure 1.10:** Different angles at an intersection

▶ **Table 1.30:** Types of angle

Example from Figure 1.10	Type of angle	Size in degrees
Angle A	Acute	Less than 90°
Angle J	Right	Exactly 90°
Angle D	Obtuse	Between 90° and 180°
Angle B + Angle C + Angle D	Reflex	Between 180° and 360°
Angles K and L	Complementary	Sum to 90°
Angles A and B	Supplementary	Sum to 180°
Angles C and E	Alternate or 'Z'	Equal
Angles E and G	Vertically opposite	Equal
Angles D and H	Corresponding or 'F'	Equal

Ⅱ PAUSE POINT Can you explain the difference between the following types of angle and sketch some simple examples?

Acute, reflex and obtuse.

Hint Consider the angle made between two lines.

Extend Draw an isosceles triangle and show how this can be split into two right-angled triangles (see page 38).

Properties of triangles

Similar triangles

Similar triangles contain the same internal angles but their sides are different lengths. Therefore, their sides are in the same ratio. The use of similar triangles is very common in construction calculations. It involves applying ratios to find unknown lengths. The ratio of their sides is such that:

$$\frac{AB}{BC} = \frac{AD}{DE}$$

Worked example

A monopitch roof is a triangle. The height of the roof BC = 2.8 m and the span of the roof AB is 5 m. Calculate the height of the vertical post DE, if the distance from the eaves AD is 2 m.

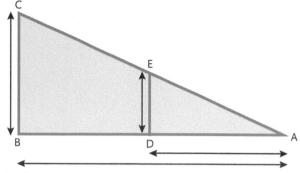

▶ **Figure 1.11:** Monopitch roof

By similar triangles:

$$\frac{AB}{BC} = \frac{AD}{DE}$$

$$\frac{5.0\,m}{2.8\,m} = \frac{2.0\,m}{DE}$$

Rearranging the formula by multiplying both sides by DE and by 2.8 m and, then dividing both sides by 5.0 m, we get:

$$DE = \frac{2.0\,m}{5.0\,m} \times 2.8\,m$$

$$= 1.12\,m$$

Graphical techniques

Cartesian co-ordinates

Most graphs are based on the Cartesian co-ordinate system. This system fixes a point by two numbers within a grid. These numbers are called the *x–co-ordinate* and the *y–co-ordinate* of the point and written are (x, y). To define the co-ordinates, two perpendicular directed lines (the *x*-axis and the *y*-axis) are set up on the grid and scales are marked off on the two axes. The point where the axes cross is called the origin, O.

The Cartesian co-ordinates of points that fall on a line or curve on the grid will fit the algebraic equation of that line or curve. For example, the circle of radius 3 units and centre the origin, may be described by the equation $x^2 + y^2 = 9$; when $x = 0$, $y = 3$ and when $y = 0$, $x = 3$, as can be seen in Figure 1.12.

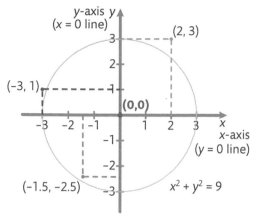

▶ **Figure 1.12:** Cartesian co-ordinate system

Polar co-ordinates

Polar co-ordinates fix the position of a point relative to a given direction and an origin. The point is located according to the angle it makes with this reference direction and the direct distance between the origin and the point. This form of point fixing is used in surveying, with the reference direction being 'north' and the relative angle known as the bearing angle or whole circle bearing (WCB) measured clockwise from north.

Intersections of graph lines with axes

Graphs are most useful at showing how one measurable quality varies against another. Figure 1.13 shows a variety of graphs that display how one variable impacts on others.

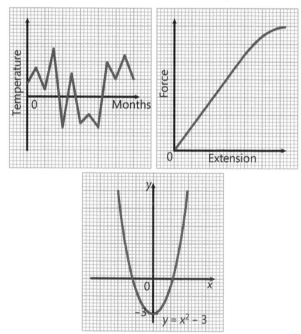

▶ **Figure 1.13:** Graphs showing how one variable has an effect on others

The first graph shows how temperature varies with time. You can clearly see where the temperature drops below freezing if the *x*-axis represents zero degrees on the temperature scale.

The second graph indicates how much a material stretches when it is pulled. Clearly, if there is no 'pull' then the material will not be stretched; hence the red line starts at (0, 0).

The relationship between the x- and y-values that satisfy the equation $y = x^2 - 3$ is such that when $x = 0$ the y-value will be –3, as the curve cuts the y-axis at –3.

Gradients of straight line graphs

The gradient is the slope of a graph and can be negative or positive depending on its shape. A straight line has a constant gradient. The gradient is usually given using the symbol 'm'. A line sloping up from left to right has positive gradient, while a line sloping down from left to right has negative gradient. The value of the gradient can be found by dividing the vertical distance by the horizontal distance:

$$m = \frac{\text{change in } y}{\text{change in } x}$$

Equations of graphs

An example of an equation that links the x- and y-variables is $y = mx + c$ where:

▶ m is the gradient of the line
▶ c is the point where the line cuts the y-axis, called the intercept.

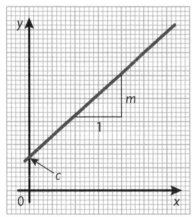

▶ **Figure 1.14:** Straight line graph

In a similar way, more complex simultaneous and quadratic equations can be solved graphically.

Simultaneous equations involve finding the values of x and y that uniquely solve two separate equations, for example:

$x + y = 10$
$y - 2x = 1$

The equations can be solved by rearranging them and using substitution, but often an easier method is by using the straight line graph method as demonstrated here.

Worked example

Find the values of x and y which solve the following pair of simultaneous equations by plotting a graph:
$x + y = 10$
$y - 2x = 1$

Step 1: Rearrange each equation into the standard straight line form of $y = mx + c$:
$y = 10 - x$
$y = 1 + 2x$

Step 2: Plot the graphs by choosing suitable values for x (see Figure 1.15 below). This may take some time by trial and error to get both lines to cross.

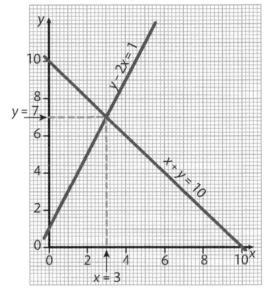

▶ **Figure 1.15:** The solution to the pair of simultaneous equations is where both lines cross

Step 3: Find where the two lines cross then read off the x- and y-co-ordinates of this point. This gives the solution of $x = 3$ and $y = 7$.

Step 4: As a check, substitute the figures back into the original equations to see if both equations balance:
$x + y = 3 + 7 = 10$ ✓

Also:
$y - 2x = 7 - 2(3) = 1$ ✓

Quadratic equations may have terms in x^2 as well as x which means that there are often two sets of solutions. The standard form of a quadratic equation is $ax^2 + bx + c = 0$, where a, b and c are numbers. The x^2 term will always give rise to a particular type of parabolic curve. If the number 'a' is positive this curve will appear v-shaped. If it is n-shaped, the number 'a' is negative.

Worked example

Find the solution to the quadratic equation $x^2 - x - 2 = 0$ by plotting the graph of this equation.

Step 1: Rearrange the equation, put in terms of y as the subject of the equation:
$y = x^2 - x - 2$

Step 2: Work out the y-co-ordinates for a range of selected x-co-ordinates by substituting them into the equation.

x	-3	-2	-1	0	1	2	3	4	
$+x^2$	9	4	1	0	1	4	9	16	
$-x$	3	2	1	0	−1	−2	−3	−4	ADD
-2	−2	−2	−2	−2	−2	−2	−2	−2	
y	10	4	0	−2	−2	0	4	10	

Step 3: Plot the graph (see Figure 1.16) and see where the line $y = 0$ cuts the curve $y = x^2 - x - 2$ to find the solutions to the equation.
The two required solutions are $x = -1$ and $x = 2$, when $y = 0$

Step 4: Check your answers by substituting back into the original equation.

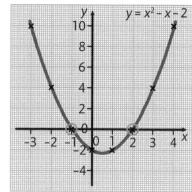

▶ **Figure 1.16:** The solutions to the quadratic equations are where line y cuts the curve
When $x = -1$:
$x^2 - x - 2 = (-1)^2 - (-1) - 2 = 1 + 1 - 2 = 0$ ✓
When $x = 2$:
$x^2 - x - 2 = (2)^2 - (2) - 2 = 4 - 2 - 2 = 0$ ✓
Therefore, both values of x are correct.

Interpolation and extrapolation

Construction materials and their properties are often tested to investigate and prove basic concepts. A linear relationship between two variables (for example if the x-value is doubled, the y-value trebles) can be shown by plotting a best-fit straight line between the experimental co-ordinates and then working out the equation of the straight line produced.

Estimating the best-fit line through a pattern of plotted points is called interpolation. This allows intermediate values to be estimated based on the interpolated line that has been selected.

Extrapolation is another technique used to anticipate trends beyond the data set. Graphical extrapolation means extending the best-fit line beyond the plotted points to a smaller or larger value along the x-axis.

The data from an experiment is shown in Figure 1.17, where the crosses represent the measured values recorded. The red line between points A and B is interpolated to enable intermediate values to be estimated, e.g. a weight of 35 N generates a frictional force of 7.5 N. The line is extrapolated back to $x = 0$, giving a value of 0 N, and also to $x = 70$, giving a value of frictional force of 15 N. These fractional force values have been estimated based on the available data.

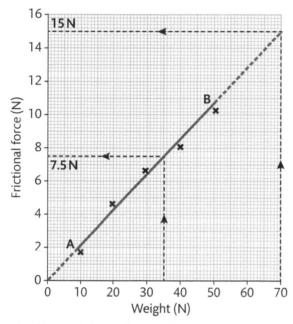

▶ **Figure 1.17:** Example of interpolation and extrapolation

Areas under graphs

Finding the area under a graph line is often useful. Where both axes are in units of distance, such as surveying co-ordinates, the area under the graph represents the plan area of land. Where the graph shows velocity plotted against time, the area under the line represents the distance travelled.

The area under a straight line graph can be calculated by breaking it down into simple shapes such as rectangles and triangles.

PAUSE POINT   Can you explain the equation for a straight line?

    Hint   Consider the gradient of the line and where it cuts the y-axis.

    Extend   Plot the following two lines on a graph and determine which values of x and y satisfy both:
    $y = x + 3$ and $y = 5 - x$

Mensuration techniques for quantity surveying and buying

These techniques rely on calculating perimeter lengths, centre line lengths, and cross-sectional and surface areas, so you need to know some basic formulae and how to apply them. The most common formulae for finding the area and perimeter of rectangles, squares, triangles, trapeziums and circles are shown in Table 1.31. It is good to note that, in some complicated situations, a perimeter or area can be broken down into a combination of simpler shapes. Similarly, for calculations regarding three-dimensional objects, such as prisms, spheres, pyramids and cones, you should refer to the standard formulae in Table 1.32.

▶ **Table 1.31:** Area and perimeter formulae

Shape	Area	Perimeter	Shape	Area	Perimeter
Square L	$L \times L = L^2$	$4L$	Trapezium a h b	$\dfrac{h(a + b)}{2}$	
Rectangle a b	$a \times b = ab$	$2a + 2b = 2(a + b)$	Parallelogram h b	$b \times h = bh$	
Triangle – standard formula c h a b Note: h is the perpendicular height	$\frac{1}{2} \times b \times h = \frac{1}{2}bh$	$a + b + c$	Circle r r = radius d = diameter	πr^2 or $\pi d^2 \div 4$	$2\pi r$
Triangle – half perimeter formula c a b	$\sqrt{s(s - a)(s - b)(s - c)}$ $s = \dfrac{a + b + c}{2}$	$a + b + c$	Ellipse a b	πab	$2\pi\left(\dfrac{a^2 + b^2}{2}\right)$ (approx.)

▶ **Table 1.32:** Volumes and surface area formulae for 3D objects

Object	Cross-sectional area (CA)	Volume	Surface area (SA)	Object	Area of base	Volume	Surface area (SA)
Rectangular prism Length L Height h Width w	wh	whL	$2(hL + wL + hw)$	Square oblique or right pyramid h = height a a	a^2	$\frac{1}{3}a^2h$	$a^2 + 2a\sqrt{(a/2)^2 + h^2}$

▶ **Table 1.32:** *Continued...*

Object	Cross-sectional area (CA)	Volume	Surface area (SA)	Object	Area of base	Volume	Surface area (SA)
Triangular prism	½ *bh*	½ *bhL*	$L(a + b + c) +$ 2(CA)	Oblique or right cone	πr^2	$\frac{1}{3}\pi r^2 h$	$\pi r^2 + \pi r \sqrt{r^2 + h^2}$
Cylinder	πr^2	$\pi r^2 L$	$2\pi r L + 2(CA)$	Sphere		$\frac{4}{3}\pi r^3$	$4\pi r^2$

Theory into practice

1 Calculate the area of turf required for the following lawns, in square metres:
   - square with a side length of 12.75 m
   - rectangular with sides of length 7.6 m and 3600 mm.
2 Calculate the areas of the following triangular walls in m² to 3 d.p. where:
   - base = 12.56 m, height = 5.39 m
   - base length is 2855 mm and height is 3870 mm
   - the lengths of the sides are 21.30 m, 14.95 m and 18.45 m.
3 Calculate the area of a circular duct, with a diameter of 375 mm, to be cut through a steel beam.
4 Calculate the cross-sectional area of an elliptical steel duct that fits snugly through a rectangular service hole measuring 0.6 m × 0.4 mm. Give your answer to the nearest 10 mm².

Centre line calculation

The centre line is commonly used by quantity surveyors when calculating either the length or volume of materials required. In order to work out the **centre line** perimeter of a wall (width *w*), given the external dimensions of the wall:
1 Calculate the external wall perimeter length using standard formulae.
2 For every external corner of the wall, deduct *w*.
3 For every internal corner of the wall, add *w*.
4 Sum up for the length of the centre line perimeter.

Compound and irregular shapes and objects

The areas and volumes of these are determined by breaking down the irregular shapes and objects into regular shapes and objects whose area and volume formulae are known. The following example shows how this can be done.

Key term

Centre line – the length of a line that runs horizontally through the centre of a construction element, such as a foundation or cavity wall, when viewed from above.

Worked example

The layout of a proposed modern stained-glass window, shown in Figure 1.18, is made up from a semicircle and a triangle. In order to price the job the glazier needs to calculate both the length of the lead cames (slender, grooved bars that hold the panes together) and the area of glass. By applying the basic formulae provided above, calculate this information. Give the area of glazing in mm², and allow 5% wastage for the length of lead came.

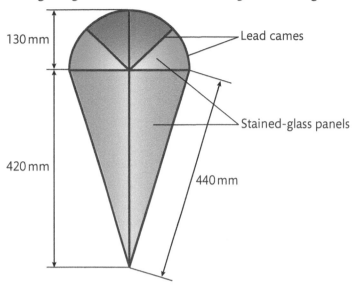

▶ **Figure 1.18:** Calculate the length of the cames and the area of glass in this window

First, calculate the length of cames required. It may at first appear we have not got enough information about the size of the panel. However, the top part of the panel is a semicircle and the radius of the circle is 130 mm. This allows us to add up the separate cames that radiate out from the centre of the semicircle as the length of each of these is the same as the radius.

Length of internal cames in the semicircle
$= 5 \times 130\,mm = 650\,mm$

Length of internal came in the lower triangle
$= 420\,mm$

Length of external cames for lower triangle
$= 2 \times 440\,mm = 880\,mm$

Length of curved came for semicircle
$= \left(\frac{1}{2}\right) 2\pi r$
$= \frac{1}{2} \times 2 \times \pi \times 130\,mm = 408\,mm$

Adding up length of lead cames $= 2358$ mm

Allowing for 5% wastage:

$2358 \times \dfrac{5}{100} = 118\,mm$

Total length of lead came
$= 2358 + 118$
$= 2476\,mm$

The calculation of the whole area can be worked out by splitting up the window into a semicircle and a triangle. The area of a semicircle is half the area of a circle, and the area of the lower triangle can be found by either the standard formula of ½ base multiplied by height or the half-perimeter rule. We shall do both.

The area of the semicircle
$= \pi r^2 \times \frac{1}{2} = \pi \times 130^2 \times \frac{1}{2} = 26{,}546\,mm^2$

cont...

The area of the triangle by standard formula
$$= \frac{1}{2}bh = \frac{1}{2} \times 260 \times 420 = 54{,}600\,mm^2$$
Total area = 26,546 + 54,600 = 81,146 mm²

Alternatively, using the half-perimeter formula:

The area of the triangle
$$= \sqrt{s(s-a)(s-b)(s-c)}$$

where
$$s = \frac{a+b+c}{2}$$
$$s = \frac{440 + 440 + 260}{2} = 570\,mm$$

Hence, area, $A = \sqrt{570(570-440)(570-440)(570-260)}$

$$= \sqrt{570(130)(130)(310)}$$

$$= \sqrt{2.986 \times 10^9}$$

$$= 54{,}646.4\,mm^2 \approx 54{,}600\,mm^2$$

The small difference between the two answers occurs because of the tiny error within the original triangle lengths – the true length of the external came is actually 439.659 mm, not 440 mm!

Statistical techniques

Statistical information may be displayed in tabular or visual form, as shown in Table 1.33. Diagrams are used to show the general pattern of data, including the maximum or minimum values, and spread of data across different categories.

▸ **Table 1.33:** Methods of presenting data

Name	Description
Histograms	Show distributions of variables, with ranges of the quantitative data grouped into 'bins' or intervals. When the column tops are joined together with straight lines they form a 'frequency polygon'.
Bar charts	Compare variables using categorical data.
Line graph	Plot data points which are joined by lines to indicate possible patterns or links between them.
Venn diagrams	Show all possible logical relations between a finite collection of different sets of data. Depict elements as points in the plane, and sets as regions inside circles.
Pie charts	A circle subdivided into sectors, each representing a category of the data. The areas of the sectors are proportional to the sizes of the angles at the centre of the pie chart and are determined as a fraction of a whole circle.
Scatter diagrams	A diagram using Cartesian co-ordinates to display values for typically two variables for a set of data. Data is displayed as a collection of points with positions on the horizontal and vertical axes.
Distribution curves	Show all possible values of the data and how often they occur (frequency).
Tables	Organise and group data that may have elements in common but may not be linked visually.

Data can be collected as several different types:
- ▸ Discrete data – only whole number values are used for the data items and intermediate values do not exist, for example the point scores in a test or the number of people who do a particular job.
- ▸ Continuous data – all values are possible for the data items, for example when looking at the time taken to perform a given construction task in work study analysis.

- Grouped and ungrouped data – data that has been grouped in categories, for example a series of prices of goods may be grouped in price brackets (£0–£5, £5–£10, £10–£15, etc.). Histograms and graphs often use grouped data. Ungrouped data is data that has not been grouped together.

Another example of presenting data is via a climate map. This is a depiction of prevailing weather patterns in a given area. It usually consists of a conventional map overlaid with colours representing climate zones, and analysing climate maps is a good way to develop your knowledge and understanding of the suitability of specifications, materials and details in specific locations.

The UK Meteorological Office produces a range of climate data maps to present its forecasting predictions, which are used to assist in the design of facilities and the planning of construction operations. The most common types of data used for construction are:

- wind speed (miles per hour)
- wind direction (16 compass points, N, NNE, NE, etc.)
- temperature (degrees Celsius, °C)
- visibility (descriptive text: good, moderate, etc.)
- rainfall (mm)
- snow accumulation (cm)
- sunshine (hours per day)
- humidity (percentage, %)

Processing large data groups

When processing large data groups there are several important definitions to know:

- **Mean** – the total of the numerical values divided by the total number of values. It is expressed by the notation $\bar{x}$. For example, if a company is currently running five building contracts for a total sum of £3.5 million, the mean value of contract $\bar{x}$ is £3.5 million divided by 5 = £0.7 million.
- **Mode** (or modal average) – the data value with the highest frequency, e.g. the most commonly occurring result from a given set of options or values.
- **Median** – the value found in the middle when all the data values are arranged in ascending order. It is often quoted as the data value corresponding to 50 per cent of the cumulative frequency (see Figure 1.19).
- **Cumulative frequency** – used to estimate the median average and the spread of values about the median. Table 1.34 shows the test results of 30 applicants for a senior construction management post, with the third column showing the cumulative frequency.

A cumulative frequency curve can be plotted from the information in the table, as seen in Figure 1.19. From the curve we see that the median score was 29.

- **Table 1.34:** An example of cumulative frequency

Test score mark	Frequency	Cumulative frequency
0–10	1	1
11–20	4	5
21–30	12	17
31–40	8	25
41–50	5	30

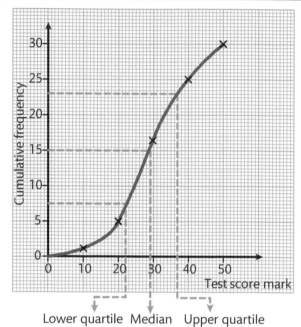

- **Figure 1.19:** A cumulative frequency curve

Application of mathematical techniques used in structural analysis

Force and mass were defined earlier in this unit. Remember: mass is the basic unit for the amount of substance but it is not the same as weight or force. At sea level, gravitational force has a constant value (g) of 9.81 newtons per kilogram. All objects are 'weighed down' with this force, which has a measurable size of 9.81 times the mass in kilograms. Therefore, the units of weight can be expressed as:

Weight (force) of an object = mass in kilograms × 9.81 N/kg

The gravitational constant (g) is defined as the acceleration caused by the Earth's mass towards its centre. At the Earth's surface its value is 9.81 metres per second per second (m/s2). However, in structural design we tend to use *newtons* per kilogram (N/kg) as this is more relevant to building weights.

The terms 'weight' and 'load' are interchangeable.

Coplanar, concurrent and non-concurrent forces

Coplanar forces exert their force in one plane; that is, they act within an invisible layer as indicated in Figure 1.20, where the forces upwards act in the same plane as the forces downwards. An example of coplanar forces in action is those causing the movement of balls on a snooker table.

All the forces on this beam are working in one plane as indicated by the dotted line

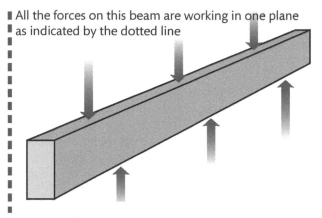

▶ **Figure 1.20:** Coplanar forces

Forces can be concurrent or non-concurrent. Concurrent forces are a collection of two or more forces that join at a common point of intersection. They can either pull away from the common point or point towards it, as Figure 1.21 illustrates.

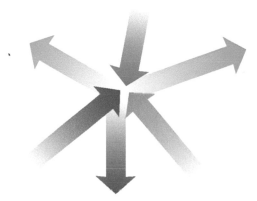

▶ **Figure 1.21:** Concurrent forces

Each arrow represents a force. Non-concurrent forces are simply a collection of forces that do not intersect at a common point, as shown in Figure 1.22. They consist of a number of forces or vectors whose magnitude would involve great effort to try to calculate in structural design.

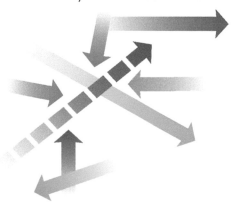

▶ **Figure 1.22:** Non-concurrent forces

Compression, tension, bending and shear

Compression happens when a vertical force pushes down on a structure; for example, a foundation base.

The material within the foundation has to push back with an equal force to resist being crushed and/or deforming under the load.

Tension happens when a force pulls on a structure; for example, a steel tie bar. The material within the bar has to pull back with an equal force to resist being stretched due to the load.

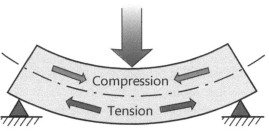

▶ **Figure 1.23:** The effects of compression and tension

Bending occurs in beams. A beam bends when it is loaded. The beam tries to resist this bending by creating stresses inside it. Compressive stresses occur at the top of the beam, above its centre line, while tensile stresses occur below it.

Shear occurs where two parallel but opposing forces attempt to cut a structural element. The shear stress set up within the element tends to resist the cutting action.

Stress, strain and modulus of elasticity

When forces act upon a construction element at right angles to its cross-sectional area, tensile or compressive stresses are formed depending on whether the force is pulling or pushing. When they act parallel to the cross-sectional area then shear stresses are created.

The value of stress can be calculated by dividing the force by the area over which it acts. Thus:

$$\text{Stress} = \frac{Force}{Area}$$

It is usually calculated using the units N/mm^2 or kN/m^2 where:

N = newtons

kN = kilonewtons.

Worked example

A steel tie with a cross section measuring 25 mm by 25 mm is subjected to an axial pull of 100 kN. Calculate the tensile stress in N/mm^2.

$$\text{Stress} = \frac{Force}{Area}$$

$$= \frac{100,000 \, n}{25\,mm \times 25\,mm} = 160 \, N/mm^2$$

Strain

An object placed under stress from a force is in a state of strain, i.e is made longer or shorter. Strain is generally defined as a ratio of the change in length divided by the original length. Thus:

$$\text{Strain} = \frac{\text{Extension or reduction in length}}{\text{Original length}}$$

Strain is just a ratio so does not have any units of measurement.

Worked example

A steel bar 1.25 m long was subjected to a tensile force of 150 kN and extended by 1.5 mm. Calculate the strain that resulted.

$$\text{Strain} = \frac{\text{Extension or reduction in length}}{\text{Original length}}$$
$$= \frac{1.5\,\text{mm}}{1250\,\text{mm}} = 0.0012 = 0.12\%$$

Elasticity and Hooke's law

Robert Hooke (1660) discovered that the extension of a steel wire was directly proportional to the load applied, (i.e. if the load is doubled, the extension will also double). This is known as elasticity. The recovery is complete for most materials (the material will return to normal after the load is removed), as long as the load has not exceeded the material's elastic limit. If this limit is exceeded, the material permanently deforms even when the load is removed, as shown in Figure 1.24.

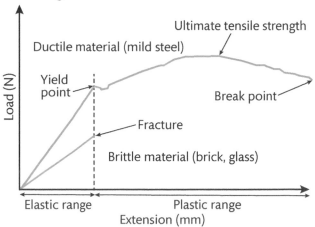

▶ **Figure 1.24:** Graph showing change in length with applied load

Hooke's law states that load is proportional to extension where the extension can be negative or positive, depending on whether the material is pushed or pulled.

Load = constant × extension

This may be written as $F = -kx$ for compressive loads or $F = kx$ for tensile loads, where:

▶ F = load
▶ k = constant
▶ x = extension.

If we divide load by the area over which the force applied (stress), and also divide the extension by the original length (strain) we find:

$$\frac{\text{Stress}}{\text{Strain}} = \text{constant} = E$$

The constant is known as Modulus of Elasticity or Young's Modulus and given the symbol E with units the same as stress: N/mm^2. It can also be shown that this is proportional to the gradient of the slope of the above graph within the elastic limit. It represents the stiffness of materials and the greater the value of E, the stronger (or more resistant) they are to deformation.

Worked example

In a tensile test on a 9 mm diameter steel bar, a load of 20 kN caused an extension of 0.072 mm on a 50 mm length of bar. Assuming that the limit of elasticity had not been reached, calculate the value of Young's Modulus to 3 s.f.

$$\text{Stress} = \frac{\text{Force}}{\text{Area}}$$
$$= \frac{20\,\text{kN}}{\pi r^2} = \frac{20\,\text{kN}}{\pi \times 4.5^2} = 0.314\ \text{kN/mm}^2$$

$$\text{Strain} = \frac{\text{Extension or reduction in length}}{\text{Original length}}$$
$$= \frac{0.072\,\text{mm}}{50\,\text{mm}} = 0.00144$$

$$\text{Young's Modulus } (E) = \frac{\text{Stress}}{\text{Strain}}$$
$$= \frac{0.314\,\text{kN/mm}^2}{0.00144}$$
$$= 128\ \text{N/mm}^2$$

Equilibrium conditions to ensure stability of a beam

For a simply supported beam to remain stable, all the forces acting on the beam must balance, and to do this two basic equations of equilibrium must be satisfied

▶ All the loads acting down on the beam must equal all the supporting reactions acting upwards on it, where the reaction is the force offered to the beam by the supporting wall:

Loads = Reactions

▶ All the clockwise turning actions on the beam must be balanced by all the anticlockwise turning actions on the beam. The turning actions are known as 'moments':

Clockwise moments = Anticlockwise moments

The turning action of a force is called a moment around any given point, which acts like a pivot. The value of the moment can be calculated by multiplying the value of the force by the shortest distance to the chosen point, which is at right angles to the line of action of the force. That is:

Moment = Force × shortest distance to pivot

The unit of moment is kNm as it is force multiplied by distance.

Determination of support reactions for simply supported beams with point and distributed loads

In the beam in Figure 1.25, a simply supported beam carries two point loads of 50 kN and 100 kN as shown. Calculate the left-hand and right-hand support reactions R_L and R_R.

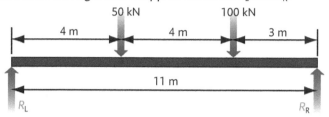

R_L = Reaction left
R_R = Reaction right

▶ **Figure 1.25:** A simply supported beam

We apply the two basic equations of equilibrium:

Reactions = Loads

$R_L + R_R = 100\,kN + 50\,kN = 150\,kN$

Also the turning moments at any point on the beam:

Clockwise moments = Anticlockwise moments

Taking moments about the end of the beam R_L we have:

$(50\,kN \times 4\,m) + (100\,kN \times 8\,m) = 11\,m \times R_R$

$\therefore R_R = \dfrac{200\,kNm + 800\,kNm}{11\,m} = 90.91\,kN$

Also:

$R_L + R_R = 150\,kN$

$\therefore R_L = 150\,kN - R_R = 150\,kN - 90.91\,kN = 59.09\,kN$

We can check this by taking moments about the other end of the beam R_R.

$\therefore R_L = \dfrac{(50\,kN \times 7\,m) + (100\,kN \times 3\,m)}{11\,m} = 59.09\,kN\ ✓$

Another type of load is a uniformly distributed load (UDL), measured in kN per metre length. To work out the reactions for these, the two equations of equilibrium are applied, as with point loads, except that for an UDL the total load is taken to act through the centre of the UDL. In the beam in Figure 1.26, a simply supported beam carries a UDL of 10 kN/m. We need to calculate the left-hand and right-hand support reactions R_L and R_R.

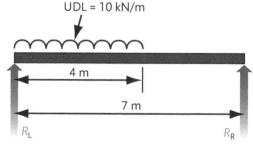

▶ **Figure 1.26:** A uniformly distributed load

We apply the two basic equations of equilibrium:

Reactions = Loads

$R_L + R_R = 10\,kN/m \times 4\,m = 40\,kN$

Also the turning moments at any point on the beam:

Clockwise moments = Anticlockwise moments

Taking moments about the end of the beam R_L and remembering that the total UDL acts through its own centre, we have:

$(10\,kN/m \times 4\,m) \times 2\,m = 7\,m \times R_R$

(Note that 2 m is the distance between the centre of the UDL and R_L)

$\therefore R_R = \dfrac{40\,kN \times 2\,m}{7\,m} = 11.43\,kN$

Also:

$R_L + R_R = 40\,kN$

$\therefore R_L = 40\,kN - R_R = 40\,kN - 11.43\,kN = 28.57\,kN$

We can check this by taking moments about the other end of the beam R_R.

$\therefore R_L = \dfrac{(10\,kN/m \times 4\,m) \times 5\,m}{7\,m} = 28.57\,kN\ ✓$

If a beam has both point loads and uniformly distributed loads, the reactions can be calculated in the same way as described above.

Shear force and bending moment for a simply supported beam with point and distributed loads

The shear force is the net vertical force attempting to cut the beam, and the bending moment is the net moment attempting to bend the beam. Once the support reactions R_L and R_R are known for the beam it is possible to calculate the shear force and bending moment at any point on the beam. Knowing the size and variation of these two quantities along the length of the beam assists in its design and specification in terms of its size and material properties.

The shear force at any point X on the beam is determined from the net vertical force on the beam to ONE side of point X.

Shear force at X = Reaction − ∑ Loads
(taken to one side of point X)

The units of shear force are N or kN.

The bending moment at point X is determined by applying the following relations to the reaction moment and load moments to ONE side of the point X.

Bending moment at X =
Reaction moment − ∑ Load moments
(all taken to one side of point X)

The units of bending moment are Nm or kNm.

For Figure 1.25 involving point loads:

Shear force at centre of beam
= Reaction − $\sum$ Loads
= 59.09 kN − 50 kN = 9.09 kN

Bending moment at centre of beam
= Reaction moment − $\sum$ Load moments
= (59.09 kN × 5.5 m) − 50 kN × 1.5 m = 250 kNm

(Note that 1.5 m is the distance from the 50 kN load to the centre of the beam).

For Figure 1.26 involving uniformly distributed loads:

Shear force at centre of beam
= Reaction − $\sum$ Loads
= 28.57 kN − (3.5 m × 10 kN/m) = −6.43 kN

Note the negative value indicates the shear force is acting downwards.

Bending moment at centre of beam
= Reaction moment − $\sum$ Load moments
= 28.57 kN × 3.5 m − (3.5 m × 10 kN/m) × $\frac{3.5m}{2}$
= 38.75 kNm

Note the $\frac{3.5m}{2}$ is the distance from the centre of the 3.5 m long UDL to the centre of the beam.

The relationship between shear force and bending moment and their effect on a beam cross section

Shear force and bending moment are very important when designing the cross-sectional shape, depth and material properties of structural beams.

In simply supported beams, like the example in Figure 1.25, we can see the following:

▸ Maximum shear occurs near the ends of the beam. This is where the beam is most likely to be cut due to loading. This means the beam has to be made stiffer near the supports to prevent this.
▸ The maximum value of bending moment occurs where zero shear occurs (in Figure 1.25 you can see where the 12 kN point load should be). This is the point where the bending stresses, in tension in the bottom and compression in the top of the beam, are at their greatest. This means that the beam has to have the greatest resistance to bending at this location which, in some materials, means that this is the deepest part of the beam.
▸ The positive bending values in a simply supported beam indicate 'sagging', which is the shape of the defective beam under load. Negative values occur with cantilevered ends and in this case the bending is called 'hogging'.

Application of mathematical techniques involving the human comfort effect of temperature on construction materials

Effects of temperature change on materials and coefficients of thermal expansion

Almost all materials expand as they get hotter and contract as they cool. When designing and specifying building components we need to ensure that this change in shape does not seriously affect the stability of the component so that it cannot function properly.

The physical property that governs the rate and amount a material expands and contracts is its coefficient of linear expansion. This is defined as the change in length of a specimen one unit long when its temperature is changed by one degree.

The formula which enables the calculation of the effects of temperature on materials due to expansion or contraction is as follows:

$$\Delta L = L \times \alpha \times \Delta \theta$$

where:

▸ ΔL = change in length (expansion is '+' while contraction is '−')
▸ L = original length
▸ α = coefficient of linear expansion (per °C)
▸ $\Delta \theta$ = change in temperature.

Typical values for α are given in Table 1.35, in descending order.

▸ **Table 1.35:** Typical coefficients of linear expansion

Coefficients of linear expansion for some common construction materials	
Material	**α (per °C)**
Lead	0.000029
Aluminium	0.000023
Copper	0.000015
Steel	0.000012
Concrete	0.000012
Cast iron	0.000011
Invar (nickel steel)	0.00000017

There are many examples of thermal expansion in construction. Lead has by far the greatest potential for expansion and contraction so, when it is used for roofs and flashings, movement joints should be used. It is also important to note that steel and concrete have the same value because, if they did not expand and contract by the same amount, reinforced concrete would not be fit for purpose as the bond between the two would break.

Calculation of *U*-values

Thermal conductivity (*k*) is an important concept when considering the temperature of a room. The *U*-value, or thermal transmittance, is the amount of heat loss in one second through 1 square metre of 1-metre thick material, with a 1-degree temperature difference between the faces. The units are W/m °C (watts per metre per degree Celsius). The thermal conductivity values for a range of construction materials are shown in Table 1.36.

▶ **Table 1.36:** *U*-values for different materials (Adapted from the CIBSE Guide A: *Environmental design*)

Material	U-value (W/m °C)	Material	U-value (W/m °C)
Concrete, dense (2100 kg/m³)	1.4	Concrete, lightweight (1200 kg/m³)	0.57
Stone, sandstone	1.7	Concrete block, lightweight (600 kg/m³)	0.19
Glass	1.022	Plaster, lightweight	0.16
Brickwork, exposed (1700 kg/m³)	0.84	Plasterboard	0.16
Brickwork, internal (1700 kg/m³)	0.84	Timber, softwood	0.13
Concrete block, medium weight	0.51	Timber, hardwood	0.18
Asphalt roofing (1700 kg/m³)	0.75	Fibre insulating board	0.1
Bitumen felt layers (1700 kg/m³)	0.5	Glass wool, mat or fibre	0.04
Plaster (dense)	0.5	Mineral wool	0.039
Rendering, external	0.7	Polyurethane (foamed) board	0.025
Screed (1200 kg/m³)	0.41	Expanded polystyrene (EPS)	0.034

Thermal resistivity (*r*) is sometimes more convenient to use than thermal conductivity (*k*) of materials. It is the reciprocal of thermal conductivity, that is:

$$r = \frac{1}{k}$$

Thermal resistance (*R*) of a construction material is its thermal resistance multiplied by its thickness (*t*):

$$R = \frac{t}{k}$$

To calculate the *U*–value of a construction element, such as a wall, we need to establish the thermal resistances of the materials used to build it and add to this the internal and outside surface resistances, which are due to the thin insulating layer of air that 'sticks' to these faces as a result of capillary action. The *U*-value is then calculated.

$$U\text{-value} = \frac{1}{R_T}$$

where R_T is the total of all the thermal resistances that make up the construction element. The thermal resistance of the inside face of the element is abbreviated to r_{si} and the outside face to r_{so}. If the wall has a cavity, this too has to be considered as having a surface thermal resistance. All these are constant values, as shown in Table 1.37, and can be included when calculating R_T.

▶ **Table 1.37:** Typical thermal resistances

Typical internal surface resistance (R_{si})	0.13 m² °C/W
Typical external surface resistance (R_{so})	0.05 m² °C/W
Typical cavity resistance, 5 mm − 19 mm cavity (Rsc)	0.11 m² °C/W
Typical cavity resistance, 20 mm + (R_{sc})	0.13 m² °C/W

Thus the total thermal resistance to the passage of heat is calculated from:

$$R_T = R_{si} + \sum \frac{t}{k} + R_{sc} + R_{so}$$

where:

▶ R_{si} = inside surface thermal resistance in m² °C/W
▶ t = thickness of the construction material element in m
▶ k = thermal conductivity of the material in W/m °C
▶ R_{so} = outside surface thermal resistance in m² °C/W
▶ R_{sc} = cavity surface thermal resistance in m² °C/W.

Then the *U*-value can be calculated from:

$$U\text{-value} = \frac{1}{R_T}$$

Worked example

Calculate the *U*-value for the following cavity wall:

- 15 mm dense plaster finish to inside walls
- 100 mm concrete blockwork (medium weight)
- 100 mm wide uninsulated cavity
- 102.5 mm thick facing brickwork.

The thermal resistance of a slab of building material can be calculated in two ways. In both instances, you need to know the thickness of the material in metres.

$$R_T = R_{si} + R_{plaster} + R_{blockwork} + R_{sc} + R_{brickwork} + R_{so}$$
$$= 0.13 + \frac{0.015}{0.50} + \frac{0.100}{0.51} + 0.13 + \frac{0.1025}{0.84} + 0.05$$
$$= 0.658 \, \text{m}^2 \, °C/W$$

Therefore *U*-value $= \frac{1}{0.658} = 1.52$ W/m² °C

Worked example

In most modern forms of wall or roof construction, the insulation is the component that is added to the element to improve its energy efficiency. In the above worked example a U-value of 1.52 W/m² °C is high and would not meet current Building Regulation approval under part L. The cavity was uninsulated so we now will calculate the thickness of polyurethane board to be placed in the cavity to improve the U-value to less than 0.3 W/m² °C.

$$R_T = \frac{1}{U\text{-value}} = \frac{1}{0.3} = 3.333 \text{ m}^2 \text{ °C/W}$$

Therefore

$$3.333 = 0.13 + \frac{0.015}{0.50} + \frac{0.100}{0.51} + 0.13 + R_{insulation} + \frac{0.1025}{0.84} + 0.05$$

$$= R_{insulation} + 0.658$$

therefore $R_{insulation} = 3.333 - 0.658 = 2.68 \text{ m}^2 \text{ °C/W}$

But thermal resistance of the component is

$$R = \frac{t}{k}$$

Therefore the thickness of polyurethane insulation board needed can be calculated:

$$t = k \times R = 0.025 \text{ W/m°C} \times 2.68 \text{ m}^2 \text{ °C/W} = 0.067 \text{ m or } 67 \text{ mm, say } 70 \text{ mm}$$

Therefore to improve the insulation of the wall to meet the minimum standard of U-value 0.3 W/m²°C, there must be 70 mm thick polyurethane insulation board included within the cavity of the wall.

Calculation of structural temperature profiles

Structural temperatures represent the temperature of the actual components within a construction element, such as an insulated brick wall. The temperature drop across a particular component can be obtained from the following formula:

$$\Delta\theta = \frac{R}{R_T} \times \theta_T$$

where:

▶ $\Delta\theta$ = temperature difference across a particular layer
▶ R = thermal resistance of that layer
▶ θ_T = total temperature difference across the element
▶ R_T = total thermal resistance of the element.

Using this, a profile of how the temperatures vary through the element can be calculated and plotted.

Calculation of dew point temperature profiles

The **dew point** is the temperature at which a sample of air becomes fully saturated with water vapour.

Vapour resistance (R_v) describes the resistance of specific thicknesses of material to the passage of water vapour and is calculated from the following formula:

$$R_v = r_v \times L$$

where:

▶ R_v = vapour resistance of that material (GN s/kg; giganewton seconds per kilogram)
▶ L = thickness of material (m)
▶ r_v = vapour resistivity of the material (GN s/kg m).

Using a similar method to the structural temperature profile, dew point temperatures can be calculated and plotted for the construction element. The two temperature profiles can then be combined to find if condensation within the element will occur where the structural temperature falls below the dew point temperature.

> **Key term**
>
> **Dew point** – the temperature to which humid air must be cooled to reach saturation (the maximum amount of water vapour that can be contained within the air). Further cooling forces water vapour to condense as water droplets.

The vapour pressure drop across a particular component can be obtained from the following formula:

$$\Delta P = \frac{R_v}{R_{vT}} \times P_T$$

where:

▸ ΔP = vapour pressure drop across a particular layer (measured in pascals (Pa))
▸ R_v = vapour resistance of that layer
▸ P_T = total vapour pressure drop across the element
▸ R_{vT} = total vapour resistance of the element.

Worked example

Determine the structural and dew point temperature profiles of the following wall construction and whether condensation will occur.

The external wall is constructed with 12 mm plaster coat, then 25 mm of expanded polystyrene (EPS) insulation then 125 mm dense concrete. The inside temperature is 22°C and the outside temperature is 1°C.

Take the vapour resistivity of plasterboard as 50 GN s/kg m, EPS as 100 GN s/kg m and concrete as 30 GN s/kg m. Assume the vapour pressure on the inside is 1400 Pa and the outside is 600 Pa.

Calculate the structural temperature profile using a tabular format such as the one in Table 1.38.

▸ **Table 1.38:** Calculating the structural temperature profile

Layer	Thermal conductivity, k (W/m °C)	Thermal resistivity, r (m °C/W)	Thickness, t (m)	Thermal resistance, R (m²°C/W)	Temperature drop, $\Delta\theta$ (°C)	Boundary temperature, θ (°C)
	–	$r = \dfrac{1}{k}$	–	$R = r \times t$	$\Delta\theta = \dfrac{R \times \theta_T}{R_T}$	
Inside	–		–	–	–	22
Internal air surface	–	–	–	0.130	$\dfrac{0.13 \times (22 - 1)}{1.079} = 2.5$	
Boundary	–	–	–	–	–	19.5
Plaster	0.16	6.25	0.012	0.075	$\dfrac{0.075 \times (22 - 1)}{1.079} = 1.5$	
Boundary	–	–	–	–	–	18.0
EPS	0.034	29.41	0.025	0.735	$\dfrac{0.735 \times (22 - 1)}{1.079} = 14.3$	
Boundary	–	–	–	–	–	3.7
Concrete	1.4	0.71	0.125	0.089	$\dfrac{0.089 \times (22 - 1)}{1.079} = 1.7$	
Boundary	–	–	–	–	–	2.0
External air surface	–	–	–	0.050	$\dfrac{0.05 \times (22 - 1)}{1.079} = 1.0$	
Outside	–	–	–	–	–	1.0
				$R_T = 1.079$		

Step 1: Enter in the values for thermal conductivity (k) or thermal resistivity (r).

Step 2: Enter in the thickness of the material.

Step 3: Calculate the thermal resistance, $R = r \times t$.

cont...

Step 4: Total up thermal resistances R to obtain total thermal resistance R_T.

Step 5: Calculate the thermal temperature drop across the layer.

$$\Delta\theta = \frac{R \times \theta_T}{R_T}$$

Step 6: Calculate the boundary temperature.

Step 7: Plot the structural temperature profile through the wall as a graph, with the temperature on the y-axis and the thicknesses of the individual elements on the x-axis.

Calculate the dew point temperature profile using a tabular format such as the one in Table 1.39.

▶ **Table 1.39:** Calculating the dew point temperature

Layer	Thickness, t (m)	Vapour resistivity, r_v (GN s/kg m)	Vapour resistance, R_v (GN s/kg)	Vapour pressure drop, ΔP (Pa)	Vapour pressure at boundary (Pa)	Dew point temperature at boundary, θ (°C)
			$R_v = r_v \times t$	$\Delta P = \dfrac{R_v \times P_T}{R_{vT}}$		
Internal air surface	–	–	negligible	–		
Boundary	–	–	–	–	1400	12
Plaster	0.012	50	0.6	$\dfrac{0.6 \times (1400 - 600)}{6.85} = 70$		
Boundary	–	–	–	–	1330	11.3
EPS	0.025	100	2.5	$\dfrac{2.5 \times (1400 - 600)}{6.85} = 292$		
Boundary	–	–	–	–	1038	7.3
Concrete	0.125	30	3.75	$\dfrac{3.75 \times (1400 - 600)}{6.85} = 438$		
Boundary	–	–	–	–	600	0
External air surface	–	–	negligible	–		
			$R_{VT} = \textbf{6.85}$			

Step 1: Enter in the thickness of the material (t).

Step 2: Enter in the values for vapour resistivity (r_v).

Step 3: Calculate the vapour resistance ($R_v = r_v \times t$).

Step 4: Calculate the vapour pressure drop across the layer.

$$\Delta P = \frac{R_v \times P_T}{R_{vT}}$$

Step 5: Calculate the vapour pressure at the boundary by consecutively subtracting the vapour pressure drop, working from the inside to the outside, e.g. vapour pressure between plaster and EPS is $1400 - 70 = 1330$ Pa.

Step 6: Look up the corresponding dew point pressure using a standard psychrometric chart (you will easily find one online); remember dew point temperature occurs at 100% relative humidity, and 100 pascals = 1 millibar of pressure.

Step 7: Plot the dew point temperature profile on the same graph as the structural temperature profile using the same scales, as shown in Figure 1.27. The profiles show there is a region within the wall where structural temperature falls below the dew point temperature, and here condensation will occur. External insulation needs to be installed to prevent this.

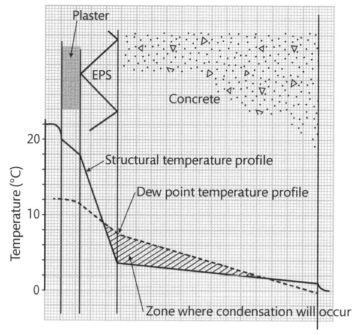

▶ **Figure 1.27:** Structural and dew point temperature profiles

Calculation of sound absorption coefficients, reverberation, actual and optimum reverberation times

Sound is measured in decibels (dB). Sound fades as sound energy is absorbed by surfaces. There is more information about reverberation in section C2 below.

▶ Standard reverberation time is the time taken for sound to die away to a level 60 decibels below its original level.

▶ The optimum reverberation time for a room depends upon its use. For a medium-sized, general purpose auditorium used for music, around two seconds is desirable. The optimum for a lecture room is 0.7–1.0 seconds.

Reverberation time is strongly influenced by the absorption coefficients of the room's surfaces. Highly reflective surfaces lengthen reverberation time while more absorbent surfaces reduce it. Typical absorption coefficients for sound at 500Hz are shown in Table 1.40 below.

▶ **Table 1.40:** Typical sound absorption coefficients

Surface finish	Absorption coefficient
Concrete (poured)	0.02
Vinyl tile on concrete	0.03
Gypsum plasterboard, 12.5 mm	0.05
Concrete block (painted)	0.06
Heavy carpet on concrete	0.15
Timber floor (suspended), ordinary window glass	0.2
Curtains, acoustical plaster	0.5
Acoustic tile (suspended), upholstered seating	0.6
Upholstered seating, occupied per person	0.8

The actual reverberation time (t) is calculated using Sabine's Formula as follows:

$$t = 0.16 \times \frac{V}{A}$$

where

- t = reverberation time
- V = volume of the room
- A = absorption in sabins (m²)
  = $\sum$ area × absorption coefficient for all finishes in room.

Worked example

A lecture hall is 20 m long × 10 m wide and is 5 m high and is used for teaching. It has 6 windows, each 1.2 m × 2.2 m. It seats 50 people on upholstered chairs. The walls are concrete block, the floor is carpeted and the ceiling is plasterboard.

Calculate the reverberation time and state whether it is acceptable for a lecture room.

Table 1.41 shows one way of doing this.

▶ **Table 1.41:** Calculating reverberation time

Location	Area (m²)	Absorption coefficient	A (sabins)
Windows	15.84	0.20	3.17
Floor	200.00	0.15	30.00
Walls	300 − 15.85 = 284.15	0.06	17.05
Ceiling	200.00	0.05	10.00
Occupied chairs	50 No.	0.80	40.00
		Total	100.22

Therefore:

$$t = \frac{0.16\,(20\,m \times 10\,m \times 5\,m)}{100.22} = 1.6\ \text{seconds}$$

This reverberation time is too long for a lecture room, so additional sound insulation board materials need to be provided within the room.

Determining lighting requirements

A light source's intensity is measured in units known as candelas (cd). The light emitted from a source spreads out the further it travels, and the intensity of the light passing through unit area decreases. The illuminance on a surface is measured in the unit known as lux (lx), where one lux is equivalent to 1 cd per metre squared.

The formula that links light intensity at the source to illuminance on any given surface that the light falls is:

$$E = \frac{I}{d^2}$$

This is known as the inverse square law of illumination where:

- I = intensity of a point source (cd)
- d = distance between source and surface (m)
- E = illuminance on that surface (lx).

When light hits a surface not located at right angles to it, then the light is spread over a larger area, and we have to adjust the inverse square law to take this into account.

As shown in Figure 1.28, the illumination at point A can be found from adapting the inverse square law by applying a reduction factor of cos θ. This is known as the cosine law of illumination.

$$E = \frac{I}{d^2} \cos\theta$$

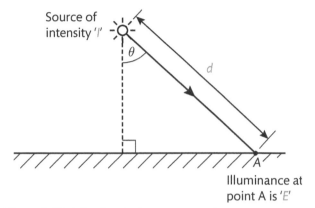

▶ **Figure 1.28:** Adjusting the inverse square law

Worked examples

1 A street lamp suspended 3 m high has a luminous intensity of 1500 cd acting as a point source. Calculate the luminance produced on the ground directly below in lux.

$$E = \frac{I}{d^2}$$
$$= \frac{1500}{3^2} = 166.7\ lx$$

2 A snooker table has one 1000 cd light suspended over its centre point at a height of 1.3 m above the table. If the snooker table is 3.6 m long and 1.8 m wide what is the illumination on one of the side pockets?

$$d = \sqrt{1.3^2 + 0.9^2} = 1.58\,\text{m by Pythagoras' theorem}$$

Also $\tan\theta = \dfrac{\text{opposite}}{\text{adjacent}} \therefore \theta = \tan^{-1}\left(\dfrac{0.9}{1.3}\right) = 34.7°$

∴ Illuminance at the side pocket:

$$E = \frac{I}{d^2} \cos\theta = \frac{1000}{1.58^2} \cos 34.7 = 329\ lx.$$

Ⅱ PAUSE POINT   Can you calculate the illuminance in the previous worked example for one of the corner pockets of the snooker table?

Hint   Use Pythagoras' theorem to work out half the diagonal distance across the table when finding d and θ.

Extend   At what point on the snooker table is the illuminance at its maximum and what is the angle θ?

Lumen method of design

The rate of flow of light energy is known as luminous flux and is measured in lumens (lm). The lumen method of lighting system design takes into account a number of factors such as efficiency of the luminaires (lamps), the reflectance from surfaces in the room and the general loss of output due to deterioration and dirt.

The number of luminaires required is given by the following formula:

$$N = \frac{E \times A}{F \times UF \times LLF}$$

where:

▶ N = number of lamps required
▶ E = illuminance level required at the working plane (lx)
▶ A = area at working plane height (m²)
▶ F = initial luminous flux output of each lamp
▶ UF = utilisation factor
▶ LLF = light loss factor.

Two other factors you need to be familiar with are:

▶ Utilisation factor (UF) – the ratio of the total flux reaching the working plane compared to the total flux output of the lamps. It takes into account surface reflectance of ceilings and walls and the shape of the room in the form of the room index (RI). From this information the utilisation factor (UF) can be obtained from manufacturers' data sheets.
▶ Light loss factor (LLF) – the ratio comparing the illuminance provided when the lamps are new to when they have been in use. It depends on a number of factors such as decline of output, dirt on reflectors and dust accumulated on room surfaces reducing reflection. Typical values for LLF after 12 months' use are 0.95 for a clean air-conditioned environment and 0.7 for a dirty industrial area.

Using the lumen design method, the spacing between the array of lamps needs to meet the following minimum requirement:

▶ $S_{max} = 1.5 \times H_m$ for fluorescent tubes in diffusing luminaires and
  $S_{max} = 1.0 \times H_m$ for direct lamps.

Worked example

An office space measuring 30 m by 15 m and 3.5 m in height requires an illuminance of 400 lx at desk height, which is 1 m above the floor. Fluorescent tubes have been selected which provide a luminous flux of 5000 lm. Assuming that the utilisation factor (UF) is 0.6 and the light loss factor (LLF) is 0.95, calculate the number of lamps required and their spacings.

$$N = \frac{E \times A}{F \times UF \times LLF}$$

$$= \frac{400\,lx \times (30\,m \times 15\,m)}{5000\,lm \times 0.6 \times 0.95} = 63.15$$

cont...

The minimum number of lamps required is 64, so the suggested layout would be five rows of 13 lamps with spacings of 15 m/6 = 2.5 m and 30 m/14 = 2.14 m for the room.

Check spacing using $\quad S_{max} = 1.5 \times H_m$

$$= 1.5 \times (3.5\,m - 1\,m) = 3.75\,m$$

So the suggested layout is satisfactory as the distance between the lamps is not greater than 3.75 m.

Application of daylight factor using a simplified desktop method

The daylight factor (DF) is the ratio of the interior natural illuminance received at a point (E_i) of a reference plane to the simultaneous external illuminance (E_o) of a horizontal surface in an unobstructed site under a standard overcast sky. The higher the specified daylight factor the higher is the requirement for natural light in the design.

$$DF = \frac{E_i}{E_o} \times 100$$

The average daylight factor can be used at an early design stage to estimate the total glazed area required. It is calculated from the following formula:

$$A_g = \frac{\overline{DF} \times A(1 - R^2)}{\theta \times T}$$

where:

▶ A_g = glazed area of window required
▶ $\overline{DF}$ = average daylight factor
▶ θ = angle of visible sky, measured in section from the centre of the window opening in the plane of the inside window wall
▶ T = transmittance of glazing to diffuse light, including the effect of dirt. Typical values are 0.7 for clear double glazing and 0.65 for low emissivity glazing.
▶ A = total area of enclosing room surfaces (ceiling + floor + walls including windows)
▶ R = mean reflectance of enclosing room surfaces. Typical values are 0.5 for a room with a white ceiling and light-coloured walls, and 0.3 for a normal office or living room.

Assessment practice 1.2

1 A swimming pool with a cross section as shown in Figure 1.29 is to be built. The external width of the pool is 5 m. Some preliminary costings are required so you will need to find out the following for the proposed work:

   a. Calculate the angle that the slope of the pool bottom makes with the horizontal to the nearest degree, and hence the length of the sloping portion of the bottom of the pool.

                                       (5 marks)

   b. Find the minimum amount of soil to be removed from the site to the nearest 0.1 m³.   (4 marks)

   c. Calculate the volume of concrete given that the thickness of the pool walls is 200 mm and assuming 7% of the volume of the pool walls is the reinforcing steel. State your answer to the nearest 0.5 m³ of concrete and the assumptions that you made.   (8 marks)

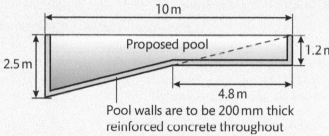

10 m

Proposed pool

2.5 m   1.2 m

4.8 m

Pool walls are to be 200 mm thick reinforced concrete throughout

▶ **Figure 1.29:** Swimming pool cross section

2 A cavity wall is constructed from 15 mm dense plaster, 100 mm inner skin concrete blockwork, 90 mm cavity, 100 mm outer skin concrete blockwork and 25 mm external render.

   a. Calculate the U–value for the wall in W/m² °C.   (5 marks)

   b. Find the thickness of polyurethane board that will be needed to insulate the wall to reduce the U–value to below 0.3 W/m² °C.   (5 marks)

3 A simply supported structural beam spans 4.7 m and supports a uniformly distributed load of 2 kN/m along the whole of its length. It also supports two point loads, the first at 2 m from its left-hand end of 5kN and the second at 1 m from its right-hand end of 4 kN.

   a. Draw the loading diagram and calculate the support reactions, R_R and R_L.   (5 marks)

   b. Draw the shear force diagram for the beam noting the size and location of the maximum shear force.   (5 marks)

   c. Draw the bending moment diagram for the beam noting the size and location of the maximum bending moment.   (6 marks)

4 A rectangular concrete block has plan dimensions of x and y, and a height of z, all measured in metres.

   a. Write down an expression for the total surface area (A) of the beam in terms of x, y and z.

   b. The total surface area of the block is 7.5 m² and x is 1 m and y is 1.2, transpose the formula for its surface area (A) to make z the subject and calculate the height of the block in metres to 2 d.p.   (6 marks)

Plan
- Have I read and fully understood the questions?
- How will I approach the task?

Do
- Do I understand my thought process and can I explain why I have chosen to approach the task in a particular way?

Review
- Have I answered the question fully?
- Can I identify which elements I found most difficult?
- Did I use the correct units to measure the physical quantities required?

C Human comfort

Discussion

What might the effects be on the occupants of a building with insufficient heating during the winter, or which gets too hot in the summer?

Describe how comfortable you would feel outside on a still, dry day in temperatures of 5°C, 15°C, 25°C and 35°C.

The comfort levels of the eventual occupants of a building need to be considered throughout its design and construction. The principal factors that affect human comfort in the built environment are heat (hot or cold), sound (acoustics) and lighting.

C1 Heat

Managing heat is an import consideration in controlling thermal comfort and providing comfortable living and working environments.

Scientific principles and their application in the built environment

There are several key parameters you must be familiar with in order to understand the behaviour of heat in the built environment. These are described in Table 1.42.

▶ **Table 1.42:** Environmental parameters affecting the behaviour of heat in the built environment

Parameter	Description	Units	Measured with
Air temperature	Inside buildings the air temperature is controlled to help maintain appropriate levels of comfort for its occupants.		
Mean radiant temperature	A measure of the average temperature of all the surfaces surrounding a body or object, with which it will exchange thermal radiation. Radiant exchange is an important factor affecting thermal comfort levels and has a greater influence on how we lose and gain temperature than simple air temperature. Sources of significant radiant heat in a working environment might include electric heaters, ovens, machinery or hot surfaces.	°C or K	Globe thermometer
Relative humidity	The amount of water vapour present in air compared with the maximum possible moisture content at a given temperature. Humidity affects the rate at which sweat evaporates so the human body is less able to cool itself effectively in an environment with high humidity.	%	Hygrometer
Air movement	Some air movement inside a building is essential to maintain air quality and avoid still or stagnant areas where occupants might feel 'stuffy'. It also helps disperse and dilute pollutants and remove water vapour from areas of high humidity to reduce condensation and damp.		
Air velocity	The speed and direction of air movement. This affects the rate of heat transfer between air and the surfaces it moves over. High air velocity can be a good thing, for example an integral fan increases the dispersion of heat by moving air over the element in an electric heater, or a desk fan blows cooling air over the skin. Air movement over the surface of a liquid also increases the rate of evaporation and helps eliminate damp by promoting drying. It can also be a bad thing, for example movement of cool air might be perceived as an unpleasant draught and have a negative impact on thermal comfort.	m/s	Anemometer
Dry bulb temperature	The temperature recorded by a conventional thermometer directly measures air temperature and is independent of humidity. To record an accurate dry bulb temperature the thermometer should be kept dry and be shaded from sources of radiant heat such as direct sunlight.	°C or K	Thermometer
Wet bulb temperature	The temperature recorded by a thermometer with a damp cloth moistened with distilled water wrapped around its bulb. The temperature recorded is reduced by the cooling effect of water evaporating from the cloth. The rate of evaporation and the size of the cooling effect depends on the relative humidity in the air around the bulb. At low humidity, evaporation will be rapid and a significant cooling effect observed. At high humidity, evaporation is slow and its cooling effects will be reduced. Wet bulb temperature is useful in measuring the apparent temperature felt by sweating skin and so is more useful than dry bulb air temperature when gauging thermal comfort.	°C or K	Wet bulb thermometer

Mechanisms of heat transfer

Conduction

Conduction in solids and liquids involves the transmission of heat energy from one atom to another through physical contact. Atoms with high heat energy will pass some of this energy on to adjacent low-energy atoms, establishing a flow of heat energy through the material. A secondary heat transfer process contributes to conduction, where free electrons move through the material, transferring heat energy. This helps to explain why good electrical conductors tend also to be good thermal conductors.

Conduction is more difficult in gases because the molecules or atoms are not in permanent contact. Heat energy can only be transferred during collisions between high-energy molecules or atoms and those with low heat energy. This helps to explain why gases are generally poor thermal conductors.

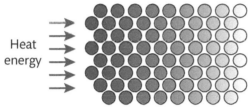

▶ **Figure 1.30:** Heat transfer by conduction occurs when heat energy is passed from one atom to another through physical contact

Convection

Heat transfer by convection only occurs in liquids and gases where molecules are free to move. Any local heating in one part of such a material (caused by conduction or radiation) will cause localised expansion and a reduction in density. Fluid with lower density than its surroundings tends to rise, carrying its heat energy with it. Low-temperature fluid then flows in to replace the risen fluid and it, in turn, is warmed, expands and rises away. This establishes a convection current.

Forced convection occurs when a fluid is forced to flow over a heat source to distribute heat energy to its surroundings. This is the principle employed in a hair dryer – cool ambient air is blown over an electric heating element and emerges at a much higher temperature.

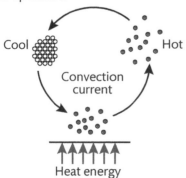

▶ **Figure 1.31:** Heat transfer by convection

Radiation

Radiation is the transfer of heat energy without physical contact. Instead, energy is transmitted in the form of electromagnetic waves (similar to those carrying light or radio signals). This explains why we can feel the warmth of the sun, despite it being millions of miles away, through the vacuum of space.

> **Discussion**
>
> Describe how a hot radiator is able to raise the air temperature throughout a room.

Measurement instruments

Some of the key measurement instruments and their uses have already been described, including thermometers, globe thermometers, hygrometers and anemometers (see Table 1.42). Some of the other types of environmental measurement and control equipment you will encounter in the built environment are shown in Table 1.43.

▶ **Table 1.43:** Heating control and monitoring equipment

Instrument	Description
Electronic control systems	These process information on environmental conditions within a building that is monitored by a series of sensors. These take actions to ensure comfort levels are maintained.
	For example, on a bright sunny day the solar gain through large windows can cause the temperature inside to rise. When this is detected, automatic reactions address it, such as increasing air flow using ventilation fans to distribute excess heat, opening windows or roof lights to vent warm air or activating smart window glass or moving shutters or blinds to reduce solar gain.
Thermostats	These monitor and/or respond to temperature change. In domestic central heating systems, electronic thermostats monitor room temperatures and send information to heating control systems.
	In less complex systems, purely mechanical thermostatic valves are used on radiators to limit or cut the flow of hot water in the radiator once a preset temperature has been reached.
Remote monitoring systems	Modern telecommunications technology has made it possible to integrate monitoring and control of intelligent building environmental management systems with mobile and/or remote devices.
	Increasingly homeowners use smartphone apps to control the heating in their homes, allowing real-time monitoring of temperatures and control of domestic heating from anywhere at any time. These systems give consumers better control of their energy usage and can lead to fuel efficiency savings.

Acceptable thermal comfort parameters

Thermal comfort is a measure of whether an occupant of a building is neither too hot nor too cold and so feels comfortable. As feelings of thermal comfort are wholly subjective and vary from person to person it can be hard to define. It is influenced by air temperature, radiant temperature, air velocity and humidity but also depends on a range of personal factors. In practice, thermal comfort levels tend to be measured by the number of employees or occupants of a building who complain of discomfort.

Current building regulations

The Domestic Building Services Compliance Guide (Approved Document L: Conservation of fuel and power) contains guidance on using heating and cooling systems efficiently to help manage thermal comfort. Approved document F: Ventilation specifies minimum air change rates and other ventilation guidance which will have an impact on maintaining thermal comfort.

Additionally, health and safety regulations in the UK govern the temperatures that are acceptable in the workplace. Remember, however, this is not the only factor that needs to be considered when assessing thermal comfort.

The Workplace (Health, Safety and Welfare) Regulations 1992 oblige employers to consider the temperatures their employees work in and provide a 'reasonable' temperature in the workplace. The related Approved Code of Practice issued by the Health and Safety Executive (HSE) suggests a minimum workplace temperature of 16°C, lowered to 13°C when employees are carrying out strenuous physical activity.

Personal factors and thermal comfort requirements

There is no maximum working temperature in law, although employers are required to ensure the health and safety of their employees in the workplace. As part of this, employers need to consider maintaining thermal comfort and the potential effects of discomfort in the workplace. Some of the personal factors that influence thermal comfort are shown in Table 1.44.

▶ **Table 1.44:** Personal factors influencing thermal comfort

Clothing	Used to insulate the body against loss of heat to its surroundings. The amount of insulation required to maintain thermal comfort is altered by simply adding or removing layers. Clothing prevents effective evaporation of sweat from the skin, so hinders the body's natural ability to cool itself. This can be a problem where a uniform or protective clothing must be worn in the workplace and causes overheating and discomfort.
Level of activity	High levels of physical activity generate metabolic heat and someone performing physical work or exercising will tend to feel warmer than someone sitting at a desk using a computer.

▶ **Table 1.44:** *Continued ...*

Metabolic rate	Metabolic rate differs between genders (with women having a metabolic rate some 5–10% less than men) and reduces with age. Those with low metabolic rates tend to have increased sensitivity to cold and difficulty in staying warm.
State of health	Illness can affect the body's ability to regulate and maintain body temperature and changes our perceptions of thermal comfort.
Age	With advancing age there is a decline in both metabolic rate and level of activity, and elderly people are more sensitive to the cold and have difficulty in keeping warm.
Gender	See metabolic rate. Pregnant and menopausal women often feel hotter than other people.

Principles of heat losses and gains in buildings and controlling them

To maintain thermal comfort levels in a building it is necessary to maintain a constant comfortable internal temperature. This is not always easy if the temperature outside is significantly warmer or colder than the required internal temperature. There are also different types of heat loss, as described in Table 1.45.

> **Link**
>
> Thermal conductivity and resistance is important for understanding the principles of heat losses and gains from buildings. See learning aim B for more information.

▶ **Table 1.45:** Types of heat loss in buildings

Type of heat loss	Explanation
Fabric heat loss	Loss of heat by conduction through the materials used in walls, floors, roofs and windows.
Ventilation heat loss	Loss of heat by the flow of warm air escaping through gaps in the structure of the building, such as around doors or through cracks or small openings.
Thermal (or cold) bridges	Areas in the structure of a building with significantly higher thermal conductivity than their surroundings so heat loss in these areas is also significantly higher. For example, in steel frame construction thermal bridging occurs when continuous steel beams penetrate thermal insulation.
Contribution of air change to heat loss	Air change describes the rate at which the air inside a building is replaced. It is typically expressed in air changes per hour where one air change is equivalent to all the air in the building being replaced during one hour. Where air change rates are high, generally in old buildings that do not comply with the air tightness levels required in new builds, ventilation heat loss will also be high.

A wide range of factors contribute to heat loss from buildings, as Table 1.46 shows.

▶ **Table 1.46:** Factors contributing to heat gains and losses

Contributing factor	Explanation
Insulation of building	Insulation provides a layer of material with a very low thermal conductivity between the inside and outside of a building. This helps to reduce heat transfer through floors, walls and ceilings.
Surface area of the external shell	Surface area is directly proportional to the rate of heat transfer by conduction. The larger the surface area the greater the rate of heat loss.
Exposure and impact of local climatic conditions	Heat loss will increase in exposed positions where a structure might be subjected to rain, snow and high winds.
Temperature difference between inside and outside	Temperature difference is directly proportional to the rate of heat transfer by conduction. The larger the temperature difference, the greater the rate of heat transfer.
Building use	Industrial processes carried out in factories or other commercial premises that generate waste heat can have a significant warming effect on the inside of buildings. Even the metabolic heat generated by workers in an open office space can impact internal temperatures. However, large numbers of employees or visitors regularly coming and going from a building means entrances and exits are frequently opened, which can significantly increase ventilation losses.

Significance of the insulating material

A material with a low thermal conductivity will make an effective insulating material by resisting the flow of heat energy. The flow of heat energy through an insulating layer of material is proportional to thermal conductivity, surface area and temperature difference and is inversely proportional to thickness. This means that, as thickness increases, the rate of flow of heat energy decreases.

Determining fabric heat loss

Fabric heat losses through any given surface in a building is calculated using the formula:

Fabric heat loss $= U\,A\,\Delta T$

where:

▶ U = U-value (reciprocal of thermal resistance or R-value) W/m^2K
▶ A = area of the surface in m^2
▶ ΔT = temperature difference between inside and outside in K.

Fabric heat loss is therefore the sum of individual losses through floors, walls, roofs, windows and doors which can be calculated using the summation formula:

Total fabric heat loss $= \sum (U\,A\,\Delta T)$

Determining ventilation heat losses

Ventilation heat loss due to air change in a building is calculated using the formula:

Ventilation heat loss $= \rho\,v\,C\,\Delta T$

where:

▶ ρ = density (kg/m^3)
▶ v = ventilation rate (m^3/s)
▶ C = specific heat capacity (J/kgK)
▶ ΔT = temperature difference between inside and outside (K).

This can be expressed in terms of air change rate and the internal volume of the building using the formula:

Ventilation heat loss $= 0.33NV\Delta T$

where:

▶ N = air change rate (1/h)
▶ V = internal volume of the building (m^3)
▶ ΔT = temperature difference between inside and outside (K).

Link

For more information on calculating required insulation thickness and U-values look back to section B1 earlier in this unit.

And it is assumed that:

▶ Density of air $= 1.2\,kg/m^3$
▶ Specific heat capacity of air $= 1000\,J/kgK$.

Methods for controlling heat loss from buildings

A range of methods is employed by architects and designers to reduce heat loss from buildings. Table 1.47 gives some examples.

▶ **Table 1.47:** Methods of controlling heat loss from buildings

Method	Explanation
Roof, wall and floor insulation	Conventional building materials such as bricks, blockwork and timber are relatively good conductors of heat energy. To help prevent excessive losses these materials are insulated with a non-structural layer of insulating material (see above for more information on insulating materials).
Double/triple glazing	These windows have gaps between each pane of glass containing a partial vacuum. These gaps significantly reduce heat loss through the window unit by minimising heat transfer by conduction or convection.
Low emissivity glass	This has a coating applied during manufacturing that allows the passage of visible light but reflects the infrared frequencies responsible for transmitting radiant heat. In this way, radiant heat from inside a building is reflected back inside, thus reducing heat loss. Double glazing which also incorporates low emissivity glass reduces heat loss by conduction, convection and radiation.
Secondary glazing	Secondary glazing installed just behind existing windows is a less expensive option than changing over to double-glazed units. The gap maintained between the existing and secondary windows isolates the exterior glass from convention currents inside the building.
Draught reduction	Low air tightness in draughty old buildings increases the exchange of interior warm air and exterior cold air which is a significant source of heat loss.
Insulated building materials	Some building materials have better insulation properties than others and their use can reduce reliance on secondary insulation.
Location and type of heating installations in a building	Traditionally, radiators are positioned at floor level on exterior walls, often beneath windows where cold air sinks towards the floor. The radiator rewarms this air and establishes a convection current that helps to warm the space and eliminate cold draughts. Modern, well-insulated homes with double (or tripled) glazing often use underfloor heating. This ensures even heat distribution throughout the space.

> **Ⅱ PAUSE POINT**
>
> What materials and techniques have been employed to reduce heat loss in the building you are in at the moment?
>
> **Hint**   Start by looking at the doors and windows. Are they double-glazed? Are door seals fitted?
>
> **Extend**   What additional measures could be taken to further reduce heat loss from the building?

Condensation

Condensation occurs when humid air with high water vapour content is cooled to below its dew point temperature. This often happens when humid air comes into contact with cold surfaces and leads to water droplets forming on those surfaces. Excessive condensation can lead to problems associated with chronic damp such as rot and mould growth.

Sources of water vapour in buildings that can contribute to problems with condensation include:

▶ the activities of the building's occupants, e.g. respiration, perspiration, bathing, showering, cooking, washing and drying clothes and dishes
▶ use of gas appliances as water vapour is one of the products of combustion
▶ house plants, e.g. watering, transpiration
▶ drying out of construction materials in new builds, concrete, plaster and timber
▶ rain penetration.

Impact of structural temperature profiles

Link

More information on calculating structural temperature profiles and dew point temperature profiles can be found in learning aim B of this unit.

A structural temperature profile illustrates the variation in material temperature through a cross section of an insulated wall or other structural element. Generally, walls are constructed of load-bearing and structural elements with insulation layers and sometimes voids or gaps between them. Heat will be conducted at different rates through each of these layers and so there will be variation in temperature through them.

Impact of dew point temperature profiles

Most building materials are porous and allow the passage of water vapour whenever there is a humidity difference between two surfaces. A dew point temperature profile illustrates the variation in dew point (the temperature at which water vapour will condense) through a cross section of an insulated wall or other structure.

Figure 1.32 shows a cavity wall with complete fill insulation. In this example water vapour tends to move through the structure from the high humidity inside to the low humidity outside. The material temperature profile is shown in red. The dew point temperature profile is shown in blue.

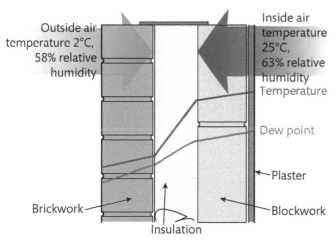

▶ **Figure 1.32:** Movement of water vapour through a cavity wall with complete fill insulation

Prediction and prevention of condensation

Condensation can be predicted using material temperature and dew point profiles and will occur at any point when material temperature reaches or falls below the dew point. Figure 1.33 shows the structure of cavity wall with partial fill insulation. The exterior is substantially colder and more humid than the interior of the building and water vapour will tend to move from the outside in. This situation might occur in rainy winter weather in a centrally heated building.

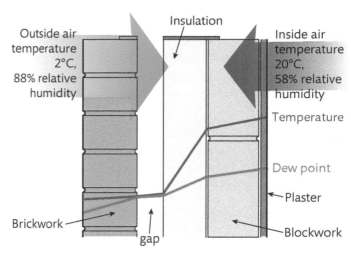

▶ **Figure 1.33:** Movement of water vapour through a cavity wall with partial fill insulation

Condensation occurs where the material temperature and dew point profiles meet. In Figure 1.33 this happens on the inside surface of the outer leaf of bricks. In this situation, applying an external layer of insulating render to the bricks to increase the material temperature inside the structure, or increasing ventilation in the cavity to reduce humidity, will help prevent the formation of condensation.

Figure 1.34 shows the structure of a cavity wall with complete fill insulation. This time the exterior is substantially colder and less humid than the interior of the building and so water vapour will tend to move from the inside out. This situation might occur in dry winter weather in the bathroom of a centrally heated building.

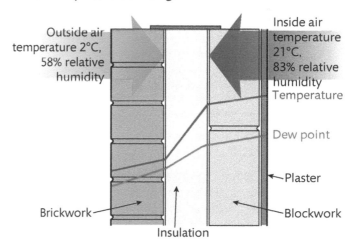

▶ **Figure 1.34:** Structure of a cavity wall with complete fill insulation

Condensation will occur on the inside surface of the outer leaf of bricks. Here, applying a layer of insulating render or increasing ventilation on the inside of the building to reduce the humidity will help prevent this.

Interstitial condensation

Condensation that forms inside the building fabric, such as cavity walls, is described as interstitial condensation. It can lead to chronic damp hidden inside the building fabric which will damage insulation materials and encourage mould growth and rot in timber structures.

Methods for controlling condensation in buildings

As well as adjusting the insulation, condensation can be controlled via internal processes. Table 1.48 suggests some of these.

▶ **Table 1.48:** Methods for controlling insulation in buildings

Method	Explanation
Air conditioning	Air conditioning systems not only control the air temperature but also regulate humidity levels. Although their primary use is to help maintain thermal comfort levels for the occupants of a building, managing temperature and humidity can also help prevent condensation.
Heating and ventilation	Keeping the interior of a building heated to above the dew point temperature ensures that water vapour remains airborne. Ventilation can be used to move this humid air out of the building before it can cool and contribute to condensation.
Dehumidification	Dehumidifiers extract water vapour directly from the air inside buildings. They are often employed on a temporary basis when building materials are drying out. Lowering the humidity not only speeds up drying processes by increasing the rate of evaporation but also prevents the moisture coming out of drying materials from reappearing elsewhere in the structure as condensation.
Extractor fans	Extractor fans are commonly used in domestic premises to ventilate kitchens and bathrooms when cooking or showering. This moves the extremely warm, humid air created by these processes out of the building before it has chance to cool and contribute to condensation.

C2 Acoustics

The difference between sound and noise is subjective and depends on the likes, dislikes and opinions of the listener. Some sounds are considered as undesirable noise such as traffic or passing aeroplanes. Others divide opinion. Music played at high volume will be considered sound and enjoyed by those playing and listening to it; however it may be undesirable noise to neighbours or others nearby.

Sound is a pressure wave travelling through a medium, such as air or water, initiated by some initial displacement or vibration. Sound pressure waves propagate away from the source of the sound in all directions like the ripples made when a pebble is dropped into a pond.

The speed of sound waves depends on the molecular structure of the material they are travelling through and

its temperature. In general, sound travels slowly in gases, quickly in liquids and quicker still in solids.

A pressure wave in a fluid (liquid or gas) is made up of alternating areas of high pressure called compressions and areas of low pressure called rarefactions. Figure 1.35 shows a graph of pressure versus time to illustrate the nature of sound as a more conventional-looking wave.

Sound waves can be described as having three important characteristics:

▶ wavelength – the length of a complete wave or cycle measured (in metres) as the distance between successive compressions (or rarefactions)

▶ frequency – the number of full waves or cycles to pass a certain point in a second, expressed as Hertz (Hz)

▶ amplitude – the magnitude of the wave, expressed as Pa (for a pressure wave).

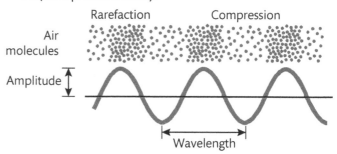

▶ **Figure 1.35:** The compression and rarefactions of a pressure wave and the corresponding graph of pressure vs time

Sound frequency

The human ear is able to detect sound with a frequency range from as low as 20Hz to as high as 20kHz but is most sensitive between 1kHz and 4kHz. The human ear is 'tuned' to these frequencies as most of our important interactions are carried out in this range. As a result, a low intensity sound at 3kHz can be detected far more easily than a sound with the same intensity at 15kHz.

Standard units

The loudness of a sound is measured by the maximum pressure difference in a pressure wave and can be expressed in Pascal (Pa) or converted in a sound pressure level expressed in decibels (dB).

When we hear sound our ears are responding to the varying pressure differences in the pressure waves reaching us. Inside our ears, these pressure waves cause the mechanical movement of the eardrum and a series of tiny bones. This movement is in turn converted into nerve signals which pass information to the brain which we interpret as sound.

The human ear is incredibly sensitive and can detect sound pressures as small as 0.00002Pa. That makes us capable of being able to literally hear a pin drop on a hard floor some distance way. Sound pressures of 200Pa and above, such as those we would experience standing close to a jet engine, are beyond the pain threshold for human hearing and will cause discomfort to the listener. Prolonged exposure to sound at these levels can cause physical damage to the internal structures of the ear and permanent hearing loss.

Because of the huge range of pressure levels audible to the human ear (from 0.000020Pa to beyond 20Pa) it is useful to express sound pressure on a more manageable scale as a sound pressure level (SPL) in decibels (dB). The dB scale is logarithmic and compares sound pressure to a given reference level, in this case the human hearing threshold pressure level of 2.0×10^{-5} Pa (the way that 0.00002Pa is usually expressed in mathematical notation).

The following formulae are used to convert between linear sound pressure in Pa and the logarithmic sound pressure level in dB:

$$SPL(\text{dB}) = 20 \log_{10} \left(\frac{p(\text{Pa})}{2.0 \times 10^{-5}} \right)$$

$$p(\text{Pa}) = 2.0 \times 10^{-5} .10^{\left(\frac{SPL(\text{dB})}{20} \right)}$$

where:

▶ $SPL(\text{dB})$ is the sound pressure level in dB
▶ $p(\text{Pa})$ is the sound pressure in Pa.

Table 1.49 shows common sounds, from quiet to noisy, in terms of their sound pressure and sound pressure level.

▶ **Table 1.49:** Some common sounds and their corresponding sound pressure and SPL

Source	Sound pressure (Pa)	SPL (dB)
Sound level inaudible to the human ear	0.000010	–6
Lower limit of human hearing	0.000020	0
Leaf rustling	0.000056	9
Whisper	0.0002	20
Talking	0.011	55
TV/radio	0.020	60
Passing car	0.10	74
Noisy workplace	0.32	84
Chainsaw	3.56	105
Jet engine (from 100 m)	63.25	130
Pain threshold for human hearing	200	140

Reminder on logarithms

Logarithmic functions are the opposite or inverse of exponent functions.

Exponent function	Logarithmic function
$10^3 = 1000$	$\log_{10} 1000 = 3$
$10^2 = 100$	$\log_{10} 100 = 2$
$10^0 = 1$	$\log_{10} 1 = 0$
$10^{-1} = 0.1$	$\log_{10} 0.1 = -1$
$10^{-2} = 0.01$	$\log_{10} 0.01 = -2$

Addition and averaging of sound pressure levels (dB)

The decibel scale is useful in providing an easy-to-understand scale on which sound levels can be measured but there are complications when adding and averaging sound levels measured on a logarithmic scale.

Logarithmic dB values of SPL must first be converted back to values of pressure on the linear pressure scale measured in Pa.

Worked example

Addition

Two machines next to each other in a workshop have an SPL of 85dB and 76dB respectively. Calculate the sound level when the machines are both running together.

First convert both SPL(dB) back into sound pressures (Pa):

$$p_1 \text{ (Pa)} = 10^{\left(\frac{p1(\text{dB})}{4.0 \times 10-4} \right)} = 10^{\left(\frac{85}{4.0 \times 10-4} \right)} = 0.355 \ldots$$

$$p_2(\text{Pa}) = 10^{\left(\frac{p2(\text{dB})}{4.0 \times 10-4} \right)} = 10^{\left(\frac{76}{4.0 \times 10-4} \right)} = 0.126 \ldots$$

Add the individual sound pressures (Pa):
$$0.355\ldots + 0.126\ldots = 0.481\ldots \text{ Pa}$$
Convert the sound pressure (Pa) back into SPL(dB):

$$SPL(\text{dB}) = 20 \log_{10} \left(\frac{0.481 \ldots}{2.0 \times 10^{-5}} \right) = 87.64 dB$$

The sound level is 87.64dB.

Averaging

The sound levels in an area of a factory are monitored by taking five readings over the space of an hour. The readings taken are shown in the table below:

Reading	1	2	3	4	5
SPL (dB)	68	72	67	64	61

Calculate the average sound level during that hour.

In a similar way to the previous worked example, convert each of the dB readings back into linear sound pressure. When dealing with multiple values it is useful to use a table to organise your work.

Reading	1	2	3	4	5
SPL (dB)	68	72	67	64	61
Calculated sound pressure (Pa)	0.050...	0.079...	0.044...	0.031...	0.022...

Sum of calculated sound pressures = 0.226...Pa

cont...

Average sound pressure $= 0.228.../5 = 0.045...$ Pa

In a similar way to the previous worked example, convert the average sound pressure (Pa) back into SPL(dB):

$$SPL(dB) = 20 \log_{10}\left(\frac{0.045...}{2.0 \times 10^{-5}}\right) = 67.19 dB$$

The average sound level is 67.19 dB.

Sound reduction indices

Sound reduction indices are used to measure the sound insulation properties of the different elements of a structure such as a wall, door or window. Both are measured in decibels (dB); see Table 1.50.

▶ **Table 1.50:** Sound reduction indices

Index	Description
Sound reduction index (R)	A measured quantity which characterises the airborne sound insulating properties of a material or element of a building in a particular sound frequency band. Calculated using the formula: $R = L1 - L2 + 10 \log (S/A)$ where: • L1 is the average SPL in the source room (dB) • L2 is the average SPL in the receiving room (dB) • S is the area of the test specimen (m^2) • A is the equivalent sound absorption area of the receiving room.
Weighted sound reduction index (R_w)	A simplified single-number quantity which characterises the airborne sound insulating properties of a material or element of a building over a range of sound frequencies.

Reverberation time

In an open area, sound travels away from a source in all directions and, from a listener's point of view, will dissipate quickly. In enclosed spaces, such as concert halls or theatres, sound will be reflected from walls, floors and ceilings back into the space, perhaps several times, before it dissipates. This extends the time that a sound persists and can be heard by a listener, a phenomenon called reverberation.

Reverberation time (RT_{60}) is the time taken for the SPL of a sound wave emitted in an enclosed space to dissipate by 60dB. It is dependent on the enclosed volume in which the sound is made and the equivalent surface area of sound absorbing material (from which no sound is reflected) used on its internal walls, floors and ceilings. It is calculated with the formula:

$$RT_{60} = 0.161\frac{V}{A}$$

where:

▶ V is the enclosed volume (m^3)
▶ A is the equivalent sound absorption area (m^2).

Acceptable acoustic comfort parameters

The UK Building Regulations 2010 Part E: Resistance to the Passage of Sound specifies minimum standards of sound insulation between adjacent dwellings as well as acoustic requirements for large communal spaces within residential developments and schools. The regulations are split into four parts, described in Table 1.51.

▶ **Table 1.51:** Summary of the four parts of the Building Regulations 2010 Part E: Resistance to the Passage of Sound

Part	Title	Description
E1	Protection against sound from other parts of the building and adjoining buildings	Provides guidance on the construction of walls and floors to provide minimum insulation requirements against the passage of airborne and impact sound from other parts of the same building or adjoining properties.
E2	Protection against sound within a dwelling-house, etc	Provides guidance on the construction of walls and floors to provide minimum insulation requirements against the passage of airborne and impact sound through internal walls and floors in the same property (does not apply to internal walls containing doors).
E3	Reverberation in the common internal parts of buildings containing flats or rooms for residential purposes	Sets limits on reverberation in the stairways, corridors and entrance halls of residential buildings.
E4	Acoustic conditions in schools	Sets requirements for protection from noise disturbance in a school setting.

The Control of Noise at Work Regulations 2005 must also be taken into consideration when designing factories and commercial premises. These oblige employers to ensure that employees are protected from excessive noise in the workplace that could cause permanent hearing loss.

A range of personal factors can affect the perceived level of acoustic comfort. These can include age, previous exposure to noise, state of health and the activity being undertaken at the time.

Noise criteria indices

The noise criteria index is a single number used to rate the sound quality of a room based on a distribution of SPL over a range of frequencies. The noise criteria (NC) bands are illustrated in Figure 1.36.

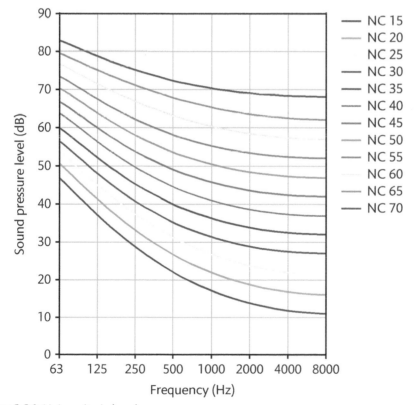

▶ **Figure 1.36:** Noise criteria bands

You will notice that the NC bands are curved to reflect the fact that the sensitivity of human hearing varies with sound frequency. Sensitivity is low at frequencies below 125Hz and highest in the range 1000–4000Hz. As a consequence, a sound at 125Hz and 40dB and a sound at 4000Hz and only 15dB are actually in the same NC band.

The NC rating applied to a particular room corresponds with the highest NC band reached by the measurements of sound level taken in the room over a range of different frequencies. These ratings can be compared with recommended NC levels for given room types and uses; see Table 1.52.

▶ **Table 1.52:** Examples of maximum recommended levels for NC

Type of room	Maximum recommended NC
House	25–30
Classroom	25–30
Factory	40–65
Office	30–35

Acoustic comfort can also be influenced by a range of personal factors; see Table 1.53.

▶ **Table 1.53:** Personal factors influencing acoustic comfort

Age	As we get older, the range of frequencies of sound we detect reduces. For example, young people are far more sensitive to higher frequency sounds than adults.
Previous exposure to noise	Exposure to excessively high levels of noise over prolonged periods can damage hearing and reduce the range of sound frequencies we can detect.
Activity	Any sound can be a distraction when you are trying to concentrate on a task or trying to get to sleep. On the other hand, sound levels when dancing in a nightclub are usually far higher than would be considered comfortable in other circumstances.
State of health	Hospitals manage noise levels to maintain patient comfort as illness can increase sensitivity to sound.

Measurement of sound levels

Sound level measuring equipment works by converting the electrical signals from one or more microphones into a sound pressure level (dB).

The requirements for the function and operational accuracy of sound level meters is laid out by IEC standard 61672. According to this standard, Class 1 devices need to have the precision necessary for laboratory testing and scientific experiments and Class 2 devices must be suffciently precise for general purpose sound level measurement and monitoring.

Hand-held Class 2 sound level meters are widely used in industrial and commercial applications for monitoring sound levels to ensure that they comply with the exposure limits set in health and safety regulations.

Sound insulation and sound absorption

Sound insulation reduces the transmission of sound from one room or enclosed space into another. Sound absorption manages the acoustic behaviour of a space by absorbing sound incident on floors, walls or other surfaces. This reduces the amount of sound reflected back into the room (see reverberation).

Airborne and impact sound

Airborne and impact sound are slightly different in the way that they are caused.
▶ Airborne sound is produced directly in air by, for example, a voice or loudspeaker. These sound waves cause vibrations in solid objects they encounter such as walls or floors.
▶ Impact sound is caused by the vibrations in solid objects that have been struck by other objects, for example footsteps striking a tile floor. The vibrations spread throughout the objects that have been struck and in turn transmit vibrations into the air surrounding them, creating audible sound waves.

Flanking transmission

Flanking sound is not transmitted directly through walls or floors but works its way around them through other pathways available in the structure of a building. This might include sound transmission through ventilation ducts, ceiling voids, recessed lighting fixtures, windows, etc.

Excellent sound insulation in walls and ceilings is no guarantee of effective sound isolation, unless flanking transmission has been considered at the design stage.

Why sound insulation and reduction are required

The primary reason why sound insulation and reduction are required in buildings is to reduce or eliminate unwanted noise. In a residential setting, even relatively low levels of persistent noise can severely impact quality of life and can even cause long-term health problems relating to stress and sleep disturbance. In a work setting, exposure to high noise levels can cause permanent hearing loss.

Table 1.54 shows a range of approaches that can be applied in order to ensure effective sound insulation.

▶ **Table 1.54:** Approaches to sound insulation

Approach	Description
Source-path-receiver approach	An approach to address the problem at one or more of the three stages of nuisance sound transmission. 1 Analyse the source – can the source of the noise be changed to reduce its noise output? 2 Analyse the path – can the noise transmission pathway be blocked or disrupted to prevent noise from reaching the receiver?

▶ **Table 1.54:** *Continued ...*

Approach	Description
Source-path-receiver approach	**3** Consider the receiver – can the receiver (person hearing the sound) be moved away from the source of noise or be provided with ear protection?
Improving structural elements	Tackling the main means of transmission by improving the sound insulation and/or absorption properties of walls, floors and ceilings. This is best achieved in the early stages when a building is being designed and constructed.
Controlling flanking sound	Identifying and then altering or disrupting pathways for flanking sound. For example, a sound pathway via the windows in adjacent rooms can be significantly reduced by installing double or secondary glazing.
Use of appropriate materials to reduce sound	Materials used in construction have a significant effect on noise transmission. A heavy solid concrete floor will reduce airborne sound transmission because of its high mass. Covering this floor with underlay and carpeting will reduce any impact sound at source.

Ⅱ PAUSE POINT   Explain the difference between sound insulation and sound absorption.

Hint   Consider the two from the point of view of being inside a room containing a source of sound and being outside that room.

Extend   Explain the impact sound absorption materials have on reverberation times.

C3 Lighting

All buildings should ensure that there is enough light, both allowed into the building from outside (natural light), and from light fittings in the building (artificial light), to provide comfort and to make the building easy to use for its intended purpose.

The principal form of natural light is sunlight. The sun illuminates our surroundings in daylight hours with a duration and intensity that vary according to season and geographical position. The light from the sun consists of electromagnetic waves spanning a continuous spectrum of frequencies. This includes the visible spectrum, those wavelengths of light our eyes have evolved to detect and our brains are able to interpret. The human eye can detect wavelengths of light between 400 and 700 **nanometres (nm)** and our brains interpret different ranges of these wavelengths as different colours.

Our brains interpret normal sunlight as white (or colourless) because it contains a continuous distribution of wavelengths across the whole visible spectrum. However, when sunlight passes through water droplets held in the atmosphere it splits into bands of differing wavelengths that we see as individual colours. The red, orange, yellow, green, blue and violet bands in a rainbow display the full range of the visible spectrum.

When we do not have access to natural light, artificial light is required. For centuries, we relied on open flames to provide light so the invention of the electric lamp was revolutionary. These developed from the filament lamps to fluorescent and discharge lighting and now modern high output, high efficiency light-emitting diodes.

Artificial light sources tend not to radiate wavelengths over the complete visible spectrum. A typical incandescent bulb emits most of its light in the red/orange/yellow part of the spectrum. Cool-white fluorescent lamps emit light in the yellow/green part of the spectrum. For this reason, plants, which rely on the full spectrum of wavelengths in natural sunlight to carry out photosynthesis, do not grow well under artificial light.

Illuminance levels

Illuminance levels define the amount of visible light energy on a surface per unit area (see luminous flux). In practice this indicates the amount of light available in a

Discussion

What is providing the light that enables you to read this sentence?

Key term

nanometre (nm) – the unit of measurement for the tiny wavelengths of visible light, where 1nm = 0.0000000001m.

particular area and is an important consideration when designing the lighting for areas in which different tasks need to be carried out. Some examples of typical illuminance levels that are found outdoors are given in Table 1.55.

▶ **Table 1.55:** Typical outdoor illuminance levels

Outdoor conditions	Typical illuminance (lux)
Day (bright sunlight)	107000
Day (shade)	10700
Day (heavy cloud)	1070
Twilight	10
Night (full moon)	0.1
Night (heavy cloud)	0.0001

Daylight factor

The use of natural daylight in buildings reduces the need for artificial lighting during the day and reduces energy costs and their associated impact on the environment.

Daylight factor quantifies the amount of useful natural light entering a room. It is a ratio of the illuminance available inside a room and that available outside at the same time under an overcast sky. It is calculated by using the formula:

$DF = 100 \times E_{in} / E_{ext}$

where:

▶ E_{in} is the illuminance at a point inside the room
▶ E_{ext} is the illuminance at a point outside the room under an overcast sky.

Table 1.56 gives examples of the appearance of a range of average daylight factors and their implications in terms of energy used on artificial lighting.

▶ **Table 1.56:** Average daylight factor

Average daylight factor (DF)	Appearance	Energy implications
<2%	Gloomy	Inadequate for most tasks. Artificial lighting required throughout the day.
2 to 5%	Reasonably bright	Adequate for basic tasks. Some supplementary artificial light required for part of the day.
>5%	Very bright	Adequate for most tasks. Artificial lighting not required during the day. However, high brightness might lead to issues with glare and solar gain (the warming effect of sunlight) and cause discomfort for the occupants.

Discussion

Explain the energy implications if insufficient natural daylight is available inside a building.

Glare indices

Glare describes the discomfort caused by the presence of high contrast areas between light and dark in a person's field of vision.

▶ Disability glare is the disabling effect of being dazzled and unable to see properly, for example when caught in the headlights of an approaching car on a dark night.
▶ Discomfort glare is the discomfort or annoyance caused by a bright spot in the field of vision, for example a direct line of sight to a lamp in an overhead luminaire. It is often quite subjective and people have a range of tolerances and responses, from being totally unaware and unaffected to suffering headaches or fatigue.

Glare indices have been developed in an attempt to quantify and measure glare. One of the design criteria for new lighting installations in commercial buildings and offices is to minimise this glare index. This is calculated from complex mathematical analysis of photometric data provided by lighting designers. Alternatively, approximations of glare index can be made by using design data tables of standardised values.

Direct and reflected light

Direct light reaches the observer from the light source by line of sight. A proportion of the light on a surface will be reflected, depending on the reflectance of the surface. Mirrors are highly efficient reflectors and even white painted walls reflect around 80 per cent of incident light.

The colour of an object is defined by the particular wavelengths in the visible spectrum that it reflects. For instance, red objects appear red when viewed in natural light because they reflect only the red wavelengths of incident light while absorbing those in the rest of the visible spectrum.

Luminous intensity

Luminous intensity refers to the power of the light source. For centuries, candlepower, literally the brightness of a standard candle, was used as the unit of measure for the luminous intensity or brightness of a light source. The modern SI unit of luminous intensity, the candela (cd), is based on a scientifically well-defined light source. The candela is equal in magnitude to the old unit of candlepower which remains in widespread use, especially in the USA. When defining luminous intensity, the power or brightness of a light source, it is useful to relate it to luminous flux.

A light source with a luminous intensity of 1 candela emits a luminous flux of 1 lumen per **steradian**.

Key term

Steradian – a solid angle defining the size of a three-dimensional cone, used in the measurement of light.

Luminous flux

Luminous flux measures the flow of visible light energy emitted by a light source. It is a measure of the total flow of light energy emitted by a source of visible light.

A uniform point source, which emits an equal number of lumens in all directions, with a luminous intensity of 1 candela (or 1 lumen per steradian) will emit a total of 4π (or 12.56) lumens of luminous flux.

Illuminance

Illuminance is used to measure the concentration of luminous flux on a work plane or surface per unit area.

Illuminance is an important factor in lighting design and determines the light available on the work plane of a particular area such a desk or workbench.

Standard units of light measurement

Table 1.57 further explains the standard units of light measurement.

▶ **Table 1.57:** Standard units of measure for lighting quantities

Quantity	Unit of measure	Definition of unit of measure
Luminous intensity (I)	candela (cd)	1 cd is defined as the luminous intensity (power) of a light source of wavelength 555.016 nm which has a radiant intensity of 1/683 watts per steradian.
Luminous flux (F)	Lumen (lm)	1 lm is defined as the luminous flux (flow of light energy) emitted by a uniform point light source with a luminous intensity of 1 candela in a three-dimensional cone defined by a solid angle of 1 steradian.
Illuminance (E)	Lux (lx)	1 lx is defined as a luminous flux of 1 lumen incident on a surface with area 1 m².

Acceptable illuminance levels for different activities and building use

There are established guidelines for the illuminance required to enable different tasks to be carried out comfortably (see Table 1.58). Architects and lighting designers need to ensure adequate light levels are provided in a building. For a given application, the working plane could be the floor, a table or a workbench.

▶ **Table 1.58:** Typical indoor minimum illuminance levels recommended for different uses

Indoor usage	Typical illuminance (lux)
Emergency evacuation routes	1

▶ **Table 1.58:** Continued ...

Indoor usage	Typical illuminance (lux)
Working area where tasks are only occasionally carried out	100
Homes, warehouses, storage areas	150
Classroom	250
Office, library	500
Supermarket, mechanical workshop	750
Operating theatre	1000
Detailed drawing/illustrating, watchmaker's workshop	2000

Variation of daylight factors in a room

Daylight factor calculations are generally performed on a grid pattern across working planes or surfaces where people will carry out tasks. When displayed graphically as a map, the daylight factor distribution across the working plane can be analysed. A great deal of repetitive mathematics is required to create daylight factor distributions and they are often generated by a computer using specialist simulation software. This allows multiple simulated distributions to be calculated quickly to see the impact of changes to window size and position, internal decoration and workspace layouts to find the optimum solution for a given space.

Principal components of daylight factor

Daylight factor generally has three components, as described in Table 1.59.

▶ **Table 1.59:** Principal components of daylight factor

Component	Description
Sky component (SC)	Light entering through windows or skylights directly from the area of sky visible at the point in the room being assessed. SC is increased by positioning windows/roof lights such that there is a greater proportion of sky directly visible. In built-up areas this needs to take into account the size and position of adjacent buildings.
Externally reflected component (ERC)	Reflected light from surrounding buildings and architecture entering through windows or skylights. Although often not within the control of architects and designers, increasing reflectivity of surrounding buildings, perhaps by painting them white, can increase ERC.
Internally reflected component (IRC)	Light that enters as direct or reflected daylight but only reaches the point in the room being assessed after being reflected from its internal surfaces. Using light colours on interior walls and ceilings helps to increase IRC.

Artificial light sources

There are many different types of artificial light source that can be used in buildings. Often the main difference is the type of bulb inserted into light fittings in buildings. Table 1.60 shows the most common types of artificial light sources.

▶ **Table 1.60:** Types, characteristics and uses of artificial light sources

Technology	Lamp type	Typical electrical power (W)	Typical efficacy (lm/W)	CRI	Typical service life (hr)	Notes
Incandescent filament	Tungsten filament (GLS)	60–100	7–14	100	1000	Now banned in the EU due to their extreme inefficiency.
	Tungsten halogen (Halogen GLS)	28–105	13–18	100	2000	Deposits evaporated tungsten back onto the filament allowing longer life and higher operating temperatures. Widely used as a replacement for GLS lamps.
Fluorescent lighting	Compact fluorescent (T3 Spiral)	11–24	56–66	85	8000	Widely used as a replacement for conventional GLS lamps.
	Compact fluorescent (PLL)	18–55	67–82	85	10000	Widely used in office and retail lighting.
	Fluorescent tubes (T8 Triphosphor)	15–70	64–90	85	15000	
Discharge lighting*	High pressure sodium (HPS) (High output HPS (SON HO) elliptical)	70–400	90–133	20	28000	Commonly used in street lighting, industrial lighting and warehousing. They take several minutes to reach full brightness and produce most light in the yellow/orange part of the visible spectrum.
	Low-pressure sodium LPS (SOX)	18–180	100–177	–40	16000	Highest lumen output per watt of any discharge lamp. Often used in low power exterior lighting. Take several minutes to reach full brightness. Light produced is almost entirely in the yellow part of the spectrum. Poor CRI means it is extremely difficult to distinguish colours under LPS lamps.
Solid state light-emitting diodes	Light-emitting diodes (LEDs) (LED GLS Thermal plastic)	6.5–17.5	72–87	80	15000	Revolutionising lighting with their small size, high efficacy, excellent colour rendering and extremely long service life.

*Discharge lighting works by passing an electrical current through an ionised gas which causes the emission of light energy. All discharge lamps require an electromagnetic ballast to stabilise the current in the lamp and provide the high voltages necessary to initiate and maintain the ionisation of the gas in the lamp.

Fluorescent lighting is a special form of discharge lighting. Ultraviolet light is emitted inside the lamp which then excites a fluorescent coating on the interior lamp walls which, in turn, emits visible light.

 **PAUSE POINT** Describe the differences between natural daylight and that produced by a low-pressure sodium lamp.

Hint Consider the range of wavelengths present in the light in each case.

Extend Despite the quality of the light, explain why low-pressure sodium lamps are in widespread use for exterior lighting.

Assessment practice 1.3

1 A survey is carried out to assess the working environment in an office building after complaints by several employees about their thermal comfort.

   a. During the survey, readings are taken using a wet bulb thermometer. State what is being measured and explain why this instrument is used instead of a dry bulb thermometer.
   (3 marks)

   b. Explain three personal factors that affect the perception of thermal comfort.
   (6 marks)

   c. Explain one method by which temperature and humidity can be controlled inside a building.
   (2 marks)

2 You are a building services engineer preparing a presentation on how window design can have a significant impact on heat loss. This will to be delivered to a client with little knowledge of the subject.

   a. Explain the three principal methods of heat transfer. (6 marks)

   b. Explain how windows can be designed to reduce heat transfer through them by considering each of the three principal methods.
   (6 marks)

3 Reducing sound transmission between adjacent properties must be factored into the design of new buildings.

   a. Explain the difference between airborne and impact sound.
   (4 marks)

   b. Describe what is meant by flanking transmission.   (2 marks)

   c. Explain two ways in which the use of appropriate construction materials can help increase the sound insulation properties of a suspended floor.
   (4 marks)

4 Two industrial washing machines running side by side each have a sound pressure level of 75dB. Calculate the combined sound pressure level of the two machines.
   (3 marks)

5 Fluorescent lighting is used widely in office and retail applications. Explain two factors which make fluorescent lighting suitable for these applications.
   (4 marks)

6 Daylight factor is used as a measurement of the amount of external natural daylight that enters a room.

   a. Explain two of the three principal components included in daylight factor calculations.
   (4 marks)

   b. Explain the implications for energy usage where rooms have a daylight factor less than 2%.
   (2 marks)

Plan
- Have I read and fully understood the questions?
- How will I approach the task?

Do
- Have I recognised the command words used in the question (give, state, explain, describe, etc.)?
- Do I understand my thought process and can I explain why I have chosen to approach the task in a particular way?

Review
- Have I answered the question fully?
- Can I identify which elements I found most difficult and where I need to review my understanding of a topic?

Getting ready for assessment

About the test

This unit will be assessed by a written examination, set and marked by Pearson. During this supervised examination period, you will be assessed on:

- your knowledge of construction materials and their properties
- your application of mathematics in a construction context
- your understanding of the provision of human comfort in buildings.

Remember all the questions are compulsory and you should attempt to answer each one. For the assessment test you will need a non-programmable calculator, and a ruler and pencil for sketching.

The questions could be a combination of:

- multiple choice
- short answer questions worth 1–5 marks
- questions that involve mathematical calculations, based on supplied information
- longer answer questions.

Sample answers

For some of the questions you will be given some background information on which the questions are based. Look at the sample questions which follow and our tips on how to answer these well.

Answering short answer questions

1. Reducing sound transmission between adjacent properties must be factored into the design of new building.

a) Explain the difference between airborne and impact sound. (4 marks)

b) Explain what is meant by flanking transmission. (2 marks)

c) Explain two ways in which the use of appropriate construction materials can help increase the sound insulation properties of a suspended floor. (4 marks)

Part (a) requires a written explanation and is worth 4 marks. The answer here is split into two short paragraphs. The first explains what impact sound is, how it works and gives an example for 2 marks. The second goes on to explain how this differs from airborne sound by explaining what airborne sound is, how it works and gives an example for a further 2 marks.

Answer

a) Impact sound is caused by the impact of an object on a solid surface. Vibrations caused by the impact are transmitted to the surrounding air as sound on both sides of the surface. That is why banging on a shared wall in one property can be clearly heard in an adjacent property.

On the other hand, airborne sound is generated directly in the air by a source like a loudspeaker or voice. It is more difficult for airborne sound to start the vibrations in solid surfaces, like walls, which can transmit sound between adjacent properties.

Part b) requires a written description and is worth 2 marks. The answer here includes two characteristics of flanking transmission for 2 marks.

Answer

b) Flanking transmission is where sound does not pass directly through a wall or floor from one room to the next, but finds a path around or over them like through a ceiling void or through pipes or ducts.

Part c) requires a written explanation and is worth 4 marks. The answer here is split into two short paragraphs. The first explains the characteristics of a flooring material used to insulate against airborne sound and provides an example for 2 marks. The second explains the characteristics of a floor covering used to reduce the generation of impact sound for a further 2 marks.

Answer

c) A large mass of material is harder for airborne sound to set vibrating and so a heavy oak or concrete floor will help insulate against airborne sound.

Absorbing the energy of impacts on the floor will help reduce the generation of impact sound and so a thick carpet and underlay to cushion impacts, like heavy footsteps, will help increase sound insulation.

2. The lifespan of any building is limited by the ability of its constituent materials to resist degradation and decay.

a) Describe the freeze-thaw cycle and how its action damages stonework.

(4 marks)

b) State two causes of degradation in exterior timber. (2 marks)

c) Explain one method used to prevent degradation in exterior timber. (2 marks)

Part a) requires a written description and is worth 4 marks. The answer here uses appropriate technical terms, such as capillary action, to explain how water penetrates the surface of the stone. It also includes the fact that water expands as ice is formed which is the key factor that drives freeze-thaw degradation. The explanation of the process is clear and the answer is structured logically.

Answer

a) Water is drawn into the tiny natural flaws, cracks and pores in the surface of stonework by capillary action. In extremely cold weather this water freezes. As water freezes and forms ice, it expands and causes stresses inside the flaws in the stonework which causes cracking.

When the ice thaws again liquid water is drawn deeper into the newly enlarged cracks. Extreme cold causes the water to freeze and expand as ice is formed once more, causing more stress and the further enlargement of the cracks. After many freeze-thaw cycles, small cracks start to join together to form larger cracks that can cause the surface of the stonework to disintegrate.

Part b) requires you to state two causes of degradation in exterior timber and is worth 2 marks. Here the answer simply requires two relevant degradation processes worth a single mark each. No further explanation is asked for and will not be credited with extra marks if it's provided.

Answer

b) Wet rot, insect attack

Part c) requires a written explanation and is worth 2 marks. The answer here gives a method of preventing degradation and briefly explains how this works.

Answer

c) Using a suitable exterior grade of paint on woodwork will provide a physical barrier to prevent water penetration into the timber. The dry timber under the paint will be less prone to wet rot which requires high moisture levels to become established.

Construction
Design 2

Getting to know your unit

Assessment
You will be assessed using an externally set written task.

In this unit, you will learn the principles and practice involved in the design and construction of low- and medium-rise buildings and structures, and will gain an understanding of how design is influenced by client requirements and external constraints.

You will consider the stages involved in the design and construction process. You will also look at how sustainable methods and techniques can be used to protect the environment and reduce the carbon footprint of the building. The design of buildings will be supported by sketching and computer-aided design (CAD) to provide efficient methods of designing, constructing and maintaining structures over their life cycle.

This unit is externally assessed and it is a mandatory unit, as it introduces many of the design skills and processes required in the industry.

How you will be assessed

This unit is assessed under supervised conditions. You will be given a scenario a set period of time before a supervised assessment period so that you can carry out research.

During the supervised assessment period, you will be given a set task that will assess your ability to produce designs to meet client requirements. Pearson sets and marks the task.

The assessment outcomes for this unit are:

▶ **AO1** demonstrate knowledge and understanding of construction design and build concepts and processes

▶ **AO2** apply knowledge and understanding of construction design and build concepts and processes to design a building to meet an initial project brief

▶ **AO3** analyse site, client and construction information to make decisions in order to produce a building design to meet an initial brief

▶ **AO4** be able to develop a reasoned design solution for a building to meet an initial project brief.

As the guidelines for assessment can change, you should refer to the official assessment guidance on the Pearson Qualifications website for the latest definitive guidance.

This unit contains a number of key command terms you will need to be familiar with
for your assessment.

Command or term	Definition
Borehole report	A report that provides information on the soil types and depths within the various strata underneath the surface of the site.
Client details	Information about the client and their requirements.
External envelope	The walls and roof forming the external surfaces of a building, including features such as the windows and external doors.
Ground conditions	Soil type, composition, contamination, level of compaction, water table level, level of saturation.
Ground water table	The depth below ground level of water contained in the ground.
Initial project brief	A document providing information relating to the spatial requirements, desired project outcomes, context of the site and budget.
Internal views	3D internal views of the building.
Medium rise	A building of three to eight storeys in height.
Sketch	A freehand drawing with annotations, using pens and pencils.
Specification	Details of the building fabric that will achieve the required outcomes.
Sub-soil	The soil below the topsoil.
Virtual model	A 3D computer-generated image of a CAD design that can be rotated and viewed from any angle and can be used to generate rendered images of a project.

Getting started

The design and construction of buildings can be a complex process and requires careful consideration. List the stages needed to design and build a new house. Consider each stage and try to identify who might be involved and what factors may influence the process.

A The construction design process

A1 Stages and tasks involved in the design process

Most buildings are complex structures. They represent many things – they may be landmarks, a statement by the architects and designers who want to leave a legacy of their work for the communities using the buildings. They have many different functions, such as homes, schools, hospitals, airports, railway stations, factories or shopping centres. Through their use of materials, style and appearance, buildings represent our cultural values. Even a simple house has many components, is designed to meet everyday needs and provide a safe shelter. Buildings are expensive to construct and over their lifetime may need to be modified or adapted. They also need significant amounts of maintenance to prevent them becoming damaged and ineffective.

A lot of work goes into deciding whether a building project can actually be built. This is known as assessing the **feasibility**.

When a client asks an architect to design a building, they need to complete an appraisal of the criteria for success to determine whether the building can go ahead. Planning for a co-ordinated and structured approach to building is necessary so that all the key players – architects, architectural technologists, structural engineers, services engineers and facilities managers – are aware of the design development as it proceeds.

For example, on a large building project, such as an office block at the edge of a river, the architect would work with the client to create an initial design. Data from the site investigation will provide information on the ground conditions of the site, which will enable the architect and structural engineers to amend the plans. Each of the key players must understand the impact of design changes from the early stages.

The Royal Institute of British Architects (RIBA) Plan of Work 2013

The Royal Institute of British Architects is the professional body for architects in the UK. Its purpose is to promote and develop architecture and share knowledge within the profession.

Most complex building projects are based on the RIBA Plan of Work. This was established in 1963 and has been updated several times to keep pace with changes in the construction industry. The RIBA Plan of Work 2013 framework has the advantage of being clear enough for all the key personnel to understand their roles within it.

There are eight stages of the RIBA Plan of Work 2013. For the purposes of this unit, we will apply Stages 0–4 to the tasks associated with the design of low- and medium-rise domestic, commercial and industrial buildings. Stages 5–7 cover construction, handover and use of the building.

Key term

Feasibility – deciding whether the building is either practicable or will proceed.

▶ **Table 2.1:** Stages 0–4 of the RIBA Plan of Work 2013

Stage	Description
Stage 0: Strategic definition	The architect works closely with the client to determine and prepare the project's long-term objective. At this stage, architects provide clients with an initial assessment of the project to allow the client to decide if it is technically and financially feasible before a detailed brief is created. This would include reviewing potential sites.
Stage 1: Preparation and brief	The **initial project brief** and feasibility studies are completed. The sustainability and environmental impact of the project need to be considered. If the project involves refurbishing an existing building this may include building surveys to create plans and elevations and to determine the condition of the building. The project team are assembled and their roles and responsibilities defined. The use of **building information modelling (BIM)** at this stage will improve the design process. BIM uses digital processes to produce a solution that is supported by data.
Stage 2: Concept design	The architect will usually have produced **sketch plans** of the layout, design and construction, which will allow the client to understand and approve the progression to outline planning proposals. Outline specifications and costs are established and a final project brief produced.
Stage 3: Developed design	The architect obtains final decisions from the client on the design, specification, construction and cost. Final planning approval is applied for.
Stage 4: Technical design	The main design team members complete their technical work and provide the detailed information required for construction. The design work is then completed, although they may need to clarify issues arising during the construction work on site (Stage 5: Construction). When final planning approval is granted, the plans need to be submitted to Building Control for approval, either by a Building Notice or Full Plans submission.

> **Key terms**
>
> **Initial project brief** – the brief prepared by the architect and client to establish the project objectives.
>
> **Building information modelling (BIM)** – a process of creating and managing digital information about the physical appearance and properties of materials used to design and construct a building.
>
> **Sketch plans** – rough plans made by an architect or designer according to the client's requirements.

⏸ **PAUSE POINT**   Do you understand the differences between (a) outline and full planning permission applications made to the local authorities and (b) Building Notice and Full Plans submission to Building Control?

     Hint   Think about the planning process involved in designing and constructing a new house.

     Extend   Can you explain the difference between planning permission approval and Building Control approval?

A2 Factors that influence the design process

Client requirements for the project outcomes

It is important to establish the requirements of the proposed building during the initial project briefing stage.

▶ **Building use** – what the client is going to use the building for, e.g. residential, business or industrial purposes. The Town and Country Planning Order 1987 puts different uses of land and buildings into a range of categories known as use classes, for example:
  - A1 shops
  - B1 businesses

- B2 general industrial
- C1 hotels and hostels
- C3 dwelling houses
- D2 assembly and leisure.

The planning authorities will consider the use of a building carefully and look how this may impact the area. The planned development must operate within its defined use.

▶ **Project spatial requirements** – Each project will have specific requirements that impact on the design process. It is crucial to know the project's proposed size and layout, circulation space and the number of floors and rooms and their use. This information is vital to establish initial cost analysis and to identify suitable sites which could accommodate the project.

▶ **Flexibility and remodelling potential** – The client may have their own ideas and vision of how the building should look. Architects need to be able to work with the client to meet their needs while ensuring they maintain the integrity of the building.

▶ **Future extension potential** – The design should consider the future use of the building and any plans for meeting residential needs or business expansion. This will require investigation into the potential site and the surrounding area. For example, if the client is planning a retail park, how many shops do they intend to build, is there room for more shops in the future and is there space for sufficient parking?

▶ **External and internal aesthetics** – The client's requirements for the **aesthetics** of the inside and outside of the building will have a significant influence on the design. The client may want a modern open-plan style or a more traditional style. The materials the client wants to use are also important to determine, as each will have its own properties and characteristics which can impact on the design.

> **Key term**
>
> **Aesthetics** – the expression and perception of beauty; in construction, how attractive a building or structure looks.

▶ **Sustainability and energy efficiency** – Among the most important considerations for clients are the energy costs and the possible incorporation of energy-efficient processes within the building. Lifetime energy costs are a major financial factor for a building so clients may wish to incorporate new and innovative techniques to reduce these costs. Examples include solar panels, wind turbines, biomass boilers, ground source heat pumps and rainwater harvesting.

▶ **Age demographic of the building user(s)** – It is important to consider the end user of the building, i.e. the target market. The requirements for student accommodation at a university will be different from the requirements of a retirement home, as older people are more likely to have physical disabilities and to spend more time in their rooms.

▶ **Target market sector** – When a client is considering investing in a project they need to know that there will be demand for its use, for example, by providing office space, a new city-centre hotel or student accommodation close to a university. If they are building residential accommodation, they need to know which type of accommodation will be in demand, for example detached, town houses, or apartments. With a good understanding of the market, the client can decide what is most likely to be in demand and thus maximise their profits.

▶ **Needs of different building users** – Legislation such as the Disability Discrimination Act 1995 and the Building Regulations ensure that buildings are accessible to all users, including wheelchair users and others with mobility issues. Designers must also consider the implications of the building being used by specific groups, such as children, who not only need the facilities to be on a smaller scale but also must be kept safe and not harmed by anything relating to the design of the building.

▶ **Security requirements** – The design and construction of secure and safe buildings is a key objective of any design team. Four key areas are:
- fire protection
- health and safety of the users (such as air quality in the building)
- protection against natural hazards such as high winds and flooding
- security of the building users and assets.

Secure building design incorporates methods to deter, delay and respond to attacks or break-ins so that the users of the building, assets and the building structure are protected. Security requirements will depend on the building's use and any security risks which can be anticipated. For example, a bank will require a higher level of security within its design than a domestic house. Related to this, safety features should be designed into the building so that those constructing, using, maintaining and eventually demolishing it can do so safely. The Construction (Design and Management) Regulations 2015 provide guidance for this (see later in this unit).

▶ **Corporate image and branding requirements** – A business's brand symbolises the way it wants its customers to experience the company. It represents the strengths and values of the company such as quality, value for money or customer service. For example, the business may want to emphasise its commitment to sustainability, as in the case study below. Tech companies may wish to have modern, cutting-edge offices that reflect their outlook. Therefore, it is important that the designers know the image the client wants to portray and must design accordingly.

Discussion

Think about the different materials that could be used to construct a domestic building. Perhaps there are new houses being built in your local area – if so, what do they look like? What are the advantages and disadvantages of the different materials that are used?

❚❚ PAUSE POINT   Why is it important to consider all the eventual uses of a completed building during the planning stage?

Hint   Think about why you would plan any project before starting work on it.

Extend   What additional security measures would be used in the design of a bank?

Site information and constraints

One of the first objectives of the design process is to locate suitable sites to build on.

Site features

The features of the site selected for the project will have a major influence on the design. Factors could include:

▶ The **size and location** of the site, for example whether it is in an urban or rural setting. A city-centre project has different considerations to one in a remote field.

▶ The **configuration** of the site (how its different elements fit together and how easy they are to access).

▶ The **orientation** of the building, for example reducing potential heating costs by using solar gain as passive heating. South-facing windows can absorb the sun's heat to help to heat the building in winter. To avoid overheating and glare in the summer, buildings may rely on some form of shading.

▶ **Access** to and around a site, for example the ability of existing roads to deal with increased traffic. It is also important to establish if any **rights of way** exist on the land.

▶ The **topography** of the site (its shape and features such as slopes, hills etc.). If a project is in a rural setting it is important to locate any trees or rivers within the site.

Key terms

Right of way – a privilege allowing someone to pass over land belonging to someone else.

Topography – the surface features of an area.

Theory into practice

Consider possible locations in your local area for residential development. Investigate potential site issues at each and how these will impact on the design process. Share your thoughts with the rest of the group as a short presentation.

Geotechnical investigation including borehole reports

Geotechnical investigations are carried out by geotechnical engineers to provide information on the ground conditions and physical properties of the soil of a proposed site. This enables the design team to determine the type of foundation required to support the building. The geotechnical investigation can also check for ground contamination.

Any geotechnical investigation will include taking soil samples using sampling techniques such as boreholes. A borehole is a narrow shaft drilled into the ground using percussion drilling, hand-auger drilling or rotary percussion drilling. It can determine the different layers of soil within the ground and also be used to take water and soil samples, which can then be tested in the laboratory.

Soil samples from different depths are taken to a laboratory and tested to find out their properties and behaviour, such as strength and compressibility. A good geotechnical report should contain the following elements:

▸ An introduction to the project, its location and a brief outline of the range of the investigation.

▸ On-site investigation and testing, including the equipment and procedures used to bore and test the soils.

▸ Site conditions, including the topographic features of the area, such as general geological history of the area, water features, trees etc.

▸ A general area map, showing geology and faults, and site maps with borehole locations.

▸ Borehole logs for the work carried out on the site, including groundwater levels, soil classifications and sample locations.

▸ Results from the tests on the soil samples, such as sieve analysis, classification tests, strength tests, detection of contaminants and any other tests which were requested.

▸ The results of the investigation and testing should enable the geotechnical engineer to recommend solutions to the substructure requirements of the project.

Ground contamination

It is important to establish if the proposed site is **greenfield** or **brownfield**. If it is a brownfield site, the location and use of existing or previous buildings need to be determined, as there may be associated health risks from the previous use that will have to be eliminated or neutralised. Previous industrial processes might have used a number of potentially harmful substances, such as asbestos, acids, fuel and lead. Tests should be carried out to identify any contamination from such harmful substances, which may be in the water or in the soil.

Building services availability

Whatever the project is, and wherever it is to be situated, the availability and location of building services such as water, gas and electricity must be established. Part of the initial site study will determine what services already exist and if any need to be brought to the site. It is also important to identify existing underground services as these could cause a safety risk during any site investigation or excavation for foundations.

Existing buildings and structures

It is important to identify the existing buildings on a site. This can provide additional information on the ground conditions and type of foundations previously used. The client will then need to decide whether to incorporate the existing building in the new project or if the building is to be demolished. If the existing building is in a conservation area or is a listed building then it will have to be incorporated in the new project unless it is structurally unsafe.

Neighbouring structures and temporary/permanent support

Investigation of the foundations of the neighbouring structures may indicate whether they will be disturbed by vibrations associated with construction work, which could increase the risk of the building collapsing. These buildings may require temporary support during the construction of the project. If the risk is significant, permanent support, such as underpinning the foundations, is required.

Key terms

Greenfield – sites which have not previously been built on.

Brownfield – sites which have been previously built on and may now be disused or derelict.

▸ When might it be necessary to underpin the foundations of a house?

Existing underground services

Construction projects usually require excavation, which presents the risk of striking underground services such as sewers, electricity cables or gas mains. This is not only dangerous; it could also be expensive and disruptive trying to restore these services. Health and Safety Executive (HSE) guidelines and the Construction (Design and Management) Regulations 2015 require designers to obtain records or plans of underground services before work starts. A simple walk-over survey will provide clues to the types of underground services on the site, such as manhole covers and gas main markers.

Trees

It is necessary to establish the location of any trees on the site. The trees may be protected under Tree Preservation Orders (TPO) and so cannot be removed. The new building design must consider how this impacts on the project. If a tree is very close to a building, then its roots could damage the structure. More likely is the effect of soil shrinkage when the roots absorb the moisture in certain soil types, causing the soil to shrink which can then lead to **subsidence** (different from **settlement**).

Rights of way

Public rights of way are paths which the public has legally protected rights to access, to enable them to enjoy walking in the countryside. The local authority is responsible for their upkeep. You may be able to access private land but only if the owner has given permission. If a right of way is identified on a potential site, the client must protect the route or provide a suitable alternative. The right of way must be maintained during and after the project. The character of the route or the alternative must also be maintained.

Underground transport

When considering building above underground transport systems, the design must ensure that the foundations do not impact on the structure of the network of tunnels and services, and that the new structure is not affected by vibrations from the transport system.

Planning constraints

The planning constraints on a project are a major factor influencing the project as whole. The purpose of the planning process is to make sure the right buildings are built in the right location, such as houses, shops, factories, leisure centres or hotels. The planning system can also be used to change and regenerate the character and feel of areas.

Planning consent

Planning permission means submitting a request to carry out building work. Planning and listed building consent applications are made to the local planning authority (LPA). They will either grant (possibly subject to certain conditions) or refuse permission to build. If the design does not comply with planning requirements, the designers will be required to make the necessary changes. If the changes are not made, it could lead to significant costs and even demolition.

Most appeals are made because the LPA has refused permission or consent. A small number of objections, especially those linked with larger proposed developments, are heard by public inquiry. Experts may be required to provide their opinion on the objections to help local authorities consider the overall impact. The final decision in these cases will be made by the secretary of state (the cabinet minister who is head of the government department for Communities and Local Government). If the application conflicts with national policy, the secretary of state can intervene in ('call-in') a planning application and overturn the LPA's decision.

> **Key terms**
>
> **Subsidence** – the downward movement of the ground on the site that is not related to the weight of the building.
>
> **Settlement** – consolidation or decrease in the volume of the soil due to the weight of the building.

Queen Elizabeth Olympic Park

The 560-acre Queen Elizabeth Olympic Park was built for the London Olympics in 2012. This was an opportunity to regenerate a deprived part of East London. The Lower Lea Valley in East London was one of the most deprived communities in the country. Unemployment was high and the public health record was poor. The area also suffered from a lack of infrastructure. The construction process for the project started in 2005.

Many of the key facilities, including the Olympic Village, the main stadium, media centre, hockey complex and warm-up tracks, were built in the area. The Olympic Village provided accommodation and amenities for 17,000 athletes and officials.

1 Why had this area of East London become run down?

2 How do think the building of the Olympic Park in this area helped to regenerate it?

3 Is there an area near your local town or city which could be regenerated? What could be built there to attract business and development?

▶ What impact did the London 2012 Olympics have on the Lower Lea Valley?

Local plan requirements

Local area plans (also known as local plans or area plans) identify how land is to be used with designated or zoned areas such as residential, commercial industrial, **conservation areas**, **greenbelts**, **Areas of Outstanding Natural Beauty (AONB)** and **Sites of Special Scientific Interest (SSSI)**.

Other areas may be identified by the Environment Agency as flood risk areas. Flooding can be caused by the sea and rivers, reservoirs, groundwater and surface flooding. Any planning applications within these areas must be shown to the Environment Agency.

Design sympathetic to local environment

Involving local communities in the planning process can help residents to shape the future of their area. This involvement in the planning process will protect the local environment and ensure any designs are appropriate. For example, replacing multi-paned Georgian windows of an existing Victorian design with large uPVC framed double glazed windows would not be in keeping with the traditions of the area. Local planning must also be in line with the National Planning Policy Framework.

When planners receive an application for planning permission they must weigh up all the positive and negative impacts the project will have on the area. For example, the impact of a modern building on a conservation area will be much more significant than a similar building in an industrial zone. Planners must look at the larger picture.

Planning objections and pressure groups

Individuals and pressure groups may lodge objections to a planned project for several reasons. It may be contrary to local planning policy or it may have a negative impact on the existing character of the area. Pressure groups often have a particular interest in protecting local heritage, trees or the environment.

Clients or companies who are building new projects can address any concerns from the local community by communicating with them. This could mean holding open sessions that enable the local community to hear about the project and ask questions. They could provide information sessions in schools. They could also provide financial support for local community groups. They can highlight the benefits of the project to the community such as increasing jobs and financial investment in the area or improved infrastructure.

The final decision, however, is made by the local authority and not by the community. It is the quality, not the quantity, of any objection(s) that can influence the outcome of any planning decision.

Listed building consent

A listed building is protected under the planning system to retain its architectural and historic importance for future generations. Permission is required to make changes to the building. There are three different grades of listed buildings:

▶ **Grade I buildings** are of exceptional interest (2.5 per cent of listed buildings).

▶ **Grade II* buildings** are of more than special interest (5.5 per cent of listed buildings).

▶ **Grade II buildings** are of special interest (92 per cent of listed buildings).

More than 500,000 listed buildings are identified on the National Heritage List for England (NHLE).

Section 7 of the Planning (Listed Building and Conservation Areas) Act 1990 provides that no person shall execute, or cause to be executed, any works for the demolition of a listed building or for its alteration or extension in any manner that would affect its character as a building of special architectural or historic interest, unless the works are authorised. Section 9 of the 1990 Act makes it an offence to contravene Section 7.

Tree preservation orders (TPO)

These are made by the LPA to prevent cutting down certain trees or woodlands if their removal could have a significant impact on the local environment. Trees have an important role to play in maintaining and protecting the environment and they may be of historical importance. They can also provide shelter for wildlife, protect and screen developments, define the character of an area and contribute to health.

Link

See Unit 5 for more about CDM 2015.

Link

See Unit 8 for more about Building Regulations and Building Control.

Statutory constraints

Legal constraints on the building process must be complied with to ensure a building is constructed safely and to suitable standards. Several pieces of **legislation** need to be taken into account.

Theory into practice

Research your local district, city or borough council website for information on planning. Find the strategic plan for the area. The council provides a list of planning applications that are being considered. Select a planning application and in groups discuss possible constraints which may influence the project.

Construction (Design and Management) Regulations 2015 (CDM 2015)

These regulations are the main set of regulations for managing health, safety and welfare on construction sites. They can be used to support clients who may not have the knowledge or expertise to carry out a construction project. The regulations ensure the client selects the right team and helps them to work together to manage the health and safety of the project. If there will be more than one contractor involved in the project, then the client must appoint a **principal designer** and a **principal contractor**.

A project is more likely to be completed on time and without any accidents if it is properly planned and managed. Construction projects can often be complex and involve high-risk work such as working at height, excavations, electricity, using heavy plant and machinery, creating dust and removing hazardous materials such as asbestos.

The principal designer and contractor must work together, allow enough time for the high-risk activities to be planned and managed, and ensure that the information is clearly communicated to everyone involved in the construction process.

Building Regulations approval

Building regulations aim to ensure the safety, quality and security of houses and commercial buildings. The Building Regulations 2010 are a set of **Approved Documents** setting out the requirements for a building under 14 separate sections that cover critical areas such as structure, fire safety, ventilation, conservation of fuel and power and electrical safety.

Note that the Building Regulations in England and Wales are different from the Building Regulations in Northern Ireland, and in Scotland they are called the Building Standards. They all have the same purpose, which is to provide technical guidance on how to keep a building safe, effective and sustainable.

Most building work in the UK requires approval from local authority Building Control. Approval can be sought by **Full Plans approval** or **Building Notice approval**. With a full application, Building Control will inspect your drawings and spot any issues with the regulations before you build anything. With a Building Notice you can start the work immediately but if there are any problems then Building Control may enforce that the work must be taken down and fixed.

The local authority could demand that a building without approval must be pulled down, or that faulty work is corrected. During Full Plans approval, possible faults or defects should have been identified from the plans at the design stage. Without the correct approval, selling the building may be very difficult.

Party Wall etc. Act 1996

This legislation tries to prevent disputes with neighbouring properties over **party walls** and boundaries. When a project that falls under this Act is to be constructed, the owners of the adjoining building or wall must be informed. The adjoining owners can review the plans and, if they raise an objection, the Act has systems in place to try to resolve the dispute.

The best way of settling any differences with neighbours on party walls is by discussion. Any agreement should be documented. If a suitable agreement cannot be achieved, then the Act allows for the appointment of an **agreed surveyor** to draw up an award, which is a report including a description of the work to be carried out and the duties and rights of the two owners. The surveyor must be agreed upon by both the owner and neighbour.

Alternatively, each owner can appoint a surveyor to draw up the award together. If the two appointed surveyors cannot agree, or either of the owners still do not agree, then a third surveyor may be asked to decide on the award. The responsibility of the surveyors is to resolve the dispute in a fair and practical way. The award is final and binding unless it is overturned in court.

Such disputes can add additional time and costs to the project and, therefore, any possible disputes must be considered by the client as part of the overall project.

Disability Discrimination Acts (1995, 2005) and Equality Act 2010

The Disability Discrimination Acts give disabled people important rights of access to everyday services such as local councils, doctors' surgeries, shops, hotels, banks, churches and schools. These service providers are required to make reasonable adjustments to buildings to enable disabled people to approach and enter the buildings and use the facilities, such as providing disabled car parking bays, sufficiently wide doors, access ramps and contrasting colours.

The Equality Act 2010 simplifies previous equality laws in a single piece of legislation. This law protects people from discrimination by service providers due to a number of factors, including disability. The Act works in conjunction with the Building Regulations as it requires 'reasonable adjustments' to be made to ensure accessibility to buildings and services.

Landlord and Tenant Act 1985

This legislation sets out the responsibilities of landlords and tenants. The landlord or owner of the premises must keep the building up to a suitable standard. The owner is responsible for maintaining the internal service supplies for water, gas and electricity and for maintaining external structural elements such as the roof, walls, windows and doors. The tenant also has responsibilities, such as avoiding damage to the property.

Restrictive covenants on land and property

Restrictive covenants are limitations put on land or property that prevent certain activities from taking place. This is often to prevent building on a section of land or to prevent a specific type of business using the land. If you breach this legally binding agreement, you can be fined or have the agreement terminated.

Examples include ensuring that all buildings in a development maintain a similar appearance to each other, not being able to build additional structures on the land, or not removing internal walls.

> **Key terms**
>
> **Party wall** – a wall that is located on the land of two separate owners. It may form part of the building or be a garden wall.
>
> **Agreed surveyor** – a surveyor who is appointed by the building owner or adjoining owner to resolve a party wall dispute.
>
> **Restrictive covenant** – a legal agreement or clause associated with particular land or property which may limit the use of the building. It is usually a promise to do or not to do something on the land.

Ⅱ PAUSE POINT Which of the statutory constraints would have the biggest impact on constructing a new housing development on a greenfield site?

Hint List as many of the statutory constraints as you can remember.

Extend Explain how these constraints would impact on the project.

Environmental constraints

The construction industry is responsible for producing significant quantities of pollution. This includes air pollution from dust and CO_2 emissions. Every year, several water pollution incidents and thousands of noise complaints are associated with construction projects. It is essential to consider the potential impact of a construction project on the environment before any work is undertaken.

Avoidance of air, water and noise pollution

Air pollution consists of particles introduced into the air that have a harmful effect on health. Construction projects produce high levels of dust from everyday materials such as concrete, cement and timber. Other air pollutants come from the plant and machinery used on site. Demolition of older buildings can also disturb harmful materials such asbestos, which can lead to lung disease such as asbestosis.

Water sprays and sprinklers are used on construction sites to keep dust levels down during activities such as filling skips, breaking concrete and using disk cutters or grinders. Washing the wheels of vehicles leaving the site, if they are carrying mud or waste, will also help to keep air pollution down.

Water pollution can be caused by sources such as diesel, oil, paint and other dangerous chemicals. Polluted surface run-off can affect marine life such as fish and animals that drink the water. Chemicals may seep into the groundwater, which may also be a source of drinking water. Polluted groundwater is difficult to treat.

Steps can be taken to minimise this on construction sites, such as using non-toxic paints, solvents and other hazardous materials wherever possible.

Wastewater that is generated from site activities can be collected in settlement tanks, screened and then the clean water discharged with the remaining sludge disposed of according to environmental regulations.

Construction work such as demolition, construction and repair work can produce significant **noise**. A lot of this noise is hard to avoid, so it is important to strike a balance between the need for the work and the impact on the local residents and environment.

Project planners should be aware of the population density of the proposed area and consider the impact the construction noise may create. Heavy plant and machinery, and daily lorry deliveries can be noisy. Exposure to high noise levels can lead to temporary and, in severe cases, permanent loss of hearing. High noise levels not only disturb the residents but can also affect the wildlife in the area.

The authorities responsible for controlling the noise levels can use several methods to manage the impact, including restricting hours of work, setting minimum noise levels and requiring that everything **reasonably practicable** is carried out to limit the noise levels.

National Planning Policy Framework (NPPF) 2012

The purpose of planning is to provide **sustainable development**. The National Planning Policy Framework (NPPF) lays out the planning policies of the English government. The NPPF has an underlying presumption in favour of sustainable development. It provides a framework within which local councils can monitor and control their own areas. This allows them to keep to the overall plan while still being able to accommodate their own specific needs and requirements.

The plan for delivering sustainable development is broken into thirteen parts. Several are relevant to environment constraints on a building project.

▶ **Part 6 Delivering a wide choice of quality homes** – There is a need to increase the housing stock across England, so local councils must plan to ensure sufficient affordable housing will be available in their area. Local planning authorities have to identify specific sites that can be used for residential development over the next five to 15 years in order to meet the expected growth of the area. In rural areas, local planners must reflect the needs of the local community to prevent inappropriate developments, and look to develop areas of housing rather than promote new isolated homes.

Key terms

Reasonably practicable – a phrase often used in health and safety legislation to mean that reasonable steps should be taken to meet the requirements.

Sustainable development – growth that meets the needs of today's society without impacting on future generations.

▶ **Part 7 Requiring good design** – Good design is a positive way to promote sustainable development. Local planners should encourage designs which reflect the local character of the area.

▶ **Part 9 Protecting green belt land** – A green belt around a town or city prevents urban sprawl and protects the countryside. It will also encourage planners to look at brownfield sites to regenerate.

▶ **Part 10 Meeting the challenge of climate change, flooding and coastal change** – Planners have a major role to play in promoting reductions in greenhouse gases and reducing the impact of climate change. Planners can support improvements to buildings that will improve their energy efficiency. Applications to build in coastal change and flood risk areas should receive careful consideration.

▶ **Part 11 Conserving and enhancing the natural environment** – Planners must support development where the main objective is to protect the environment. Planners must be aware of risks from air pollution, contamination and noise pollution produced from any development.

▶ **Part 12 Conserving and enhancing the historic environment** – Local planners need to prepare a strategic plan to protect the historic environment. Any plans which may lead to the loss of a major heritage site should be refused unless significant benefits outweigh the harm.

Part 1 of the Wildlife and Countryside Act 1981

Under this Act it is illegal to kill or injure wild birds and animals, and to disturb any area used for their shelter or protection. The Act also lists plants it is illegal to pick, uproot or destroy. The Act has a number of schedules which identify the birds, animals and plants that are protected. These restrictions can have an impact on any plans for building refurbishments, for example bats in an old church tower which is due to be refurbished.

Environmental Impact Assessments (EIAs)

An **Environmental Impact Assessment** provides the planning authorities with information about the environmental effects of a project. In England, the Town and Country Planning (Environmental Impact Assessment) Regulations 2011 provide the legal framework for implementing an EIA. The planning authorities are responsible for determining whether an EIA is necessary for any proposed project.

Two lists of types of building projects are identified within the EIA regulations:

▶ Schedule 1 identifies the projects for which an EIA is compulsory, e.g. power plants, chemical plants and metal production plants.

▶ Schedule 2 identifies projects where an EIA may be required. It specifies criteria, such as floor space, under which a project needs to be considered. The limitations set out in schedule 2 will determine whether an EIA is necessary.

The key question is how significant an effect the project will have on the environment and the surrounding area. The EIA should identify the environmental, social and economic impact of a project before any decisions on the project are made. It should enable planners to consider all the positive and negative consequences of a project before deciding if it will be granted planning permission. Designers may need to incorporate elements into their design that reduce the environmental impact.

> **Key term**
>
> **Environmental Impact Assessment (EIA)** – the process of assessing the environmental effects of a development project, implemented under the Town and Country Planning (Environmental Impact Assessment) Regulations 2011.

 **PAUSE POINT**   What environmental constraints apply to a construction project?

         Hint   Consider environmental constraints that may apply in your local area.

         Extend   Which environmental constraints do you feel are the most important?

Social constraints

A major constraint for any project is the impact it may have on the social and personal lives of people and communities in the area where the development is due to take place. These social needs can be defined as acceptance, belonging, safety and security.

An example of a major construction project that had an impact on society was the Three Gorges Dam Project in China, which was completed in 2012. The benefits of the project for the Chinese people are flood control, power generation, navigation and tourism. However, the social impact was the destruction of thousands of towns and villages and approximately 1.5 million people forced to move from their homes.

▶ The Three Gorges Dam Project

Neighbours' rights

Neighbour disputes can be difficult. If neighbours are objecting to a construction project they should make sure their objections are not unreasonable. An example of a reasonable objection may be to loss of light or the privacy of their property due to the planned construction project. They should seek advice from a competent solicitor with specialist knowledge of property and planning law.

Local community objections

Local communities may object to plans for building projects which they feel will have a detrimental impact on the area. It is important that the objections are legitimate such as:

▶ Detrimental impact on **local amenities**, e.g. the location of a fast food restaurant could produce litter, increase traffic and produce unwanted smells.

▶ Traffic congestion– increased traffic in an area creates more noise and air pollution and may increase accident rates.

▶ Loss of privacy, such as an office block overlooking the gardens of a housing estate.

▶ Negative impact on the local environment. A new building may not fit in with the style of the surrounding buildings and detract from the beauty of the locality.

▶ Loss of light – natural light inside buildings is very important for comfort. The law has to balance the need for light for existing buildings with the need for new buildings. A multi-storey office block close to a residential area could reduce the illumination from the windows in the houses. Therefore, an objection needs to show how the project would cause a significant reduction in illumination on the existing buildings.

- Inadequate parking or access – a commercial or industrial development will require sufficient parking to prevent users of the facilities clogging local streets and parking areas, and causing annoyance and congestion. A sports stadium could attract 50,000 people on match days so there needs to be sufficient parking or public transport.
- Impact on drainage – if not designed correctly, new developments can increase the risk of flooding due to poor drainage. New projects must use drainage systems that do not increase the risk of flooding or pollution. Drainage impact assessments are often carried out, along with flood risk assessments. Sustainable drainage systems (SUDS) reduce the risk of surface flooding. Hard surfaces such as concrete and bitumen used for roads and car parking prevent natural soak away of water. Building Regulations Part H deals with drainage from roofs and paved areas.
- Not complying with local and government planning policies.

In planning law, 200 people objecting to the same planning issue carries less weight than three people objecting to different valid planning issues. Everyone in a community could oppose a project but it could still go ahead if there are not enough valid reasons, or those reasons could easily be overcome.

Green space requirements

Green spaces include parks, playing fields, riverside walks and allotments and are recognised as improving healthy living and meeting the recreation needs of the community. Local authorities should prepare a green-space strategy to protect and maintain them for the benefit of their community. A strategy can also limit possible sites for building projects or instruct that suitable green space is incorporated within their design.

Mixed and balanced development

New developments such as housing should be safe, secure and affordable. The homes should meet the needs of all the users, young and old, and must not have a detrimental impact on the environment. With the increase in the population of the area it is vital there are sufficient local amenities such as schools, doctors' surgeries and shops. These community facilities need to be included to promote and maintain the sense of belonging and culture of the area.

There should be a mix of residential, commercial and local amenities in a new development, e.g. a new school if a housing estate is going to increase the population significantly, and enough potential employment opportunities for new residents.

Project budget and economic constraints

The overall cost of a project or the budget that a client has to spend is one of the most important constraints. Once the design team know what the budget is they have to work with then they can start to plan accordingly.

Cost planning

All construction projects have constraints, mainly cost, time and scope.

- **Cost** – all projects will have a budget and the client wants value for money. It is important that a proper cost plan is prepared and that the costs are managed throughout a project.
- **Time** – the client will have a deadline by when they want the project completed.
- **Scope** – if the brief is not well defined, or changes, this can increase the cost and time for a project.

Available funds and other sources of additional funding

Available funds are the amount of money an investor can spend on a project. This may not be sufficient to finance a project and so the investor or client must seek other sources of funding to finance the project, such as grants, e.g. for the maintenance or repair of historic buildings.

Government incentives might be available, for example for first-time buyers of homes and shared-ownership schemes, or for incorporating renewable energies into the building, such as solar panels.

European funding has been provided in accordance with strict rules and guidelines to ensure the money is spent correctly and in compliance with the requirements of the funding programme used. EU funding is complex, with many different programmes available and a significant amount of paperwork associated with the monitoring of the spending.

▶ The Peace Bridge in Northern Ireland benefited from EU funding

Local land prices

Land values vary due to demand for land in different regions. The location of the property will have a major impact, such as whether it has a sea view, is in the centre of a town, close to a park or near local schools. If there is limited space for building and high demand for accommodation, land prices are at a premium, such as in central London. Table 2.2 indicates that the average house price in London is significantly greater than any other region in the UK.

▶ **Table 2.2:** Average house price by country and government office region July 2016 Source: **www.gov.uk**

Country/region	Price	Country/region	Price
England	£232,885	London	£484,716
Northern Ireland	£123,241	North east	£129,750
Scotland	£143,711	North west	£150,082
Wales	£144,828	South east	£313,315
East Midlands	£173,783	South west	£176,598
East of England	£273,806	West Midlands	£151,581
		Yorkshire and the Humber	£151,581

Find out the average house price in your area. How does it compare with those in Table 2.2? Have house prices gone up or down in your area recently? What factors influence house prices in your local area?

First-time buyer residential accommodation

Residential developments are commercial enterprises – the developer intends to make a profit from the sales. Encouraging people to buy their first home is key as, once they are on the property ladder, it is easier to buy and sell houses in the future.

However, first-time buyers are not usually particularly wealthy and may need support to be able to afford a mortgage. The government offers several incentive schemes to improve first-time buyers' borrowing potential, such as shared-ownership schemes (where the buyer buys a percentage of the house and pays rent on the rest until they can afford to buy more) and Help to Buy ISAs (where the government contributes an additional 25 per cent to savings in a special account). It also gives developers financial incentives (such as grants, loans and tax relief) to encourage first-time buyers to buy on new housing estates.

Life-cycle costs

Life-cycle costs are the costs of the entire project, including design, construction, maintenance, operation and demolition of the building. They should enable the client to understand the financial investment that will be required and help decide if the project is worth undertaking. Almost three-quarters of the total costs are the maintenance and operation costs. For example, the whole life costs of a shopping mall need to considered against the projected incomes to determine if it is a financially viable project.

Case study

Wembley Stadium

The demolition of the old Wembley Stadium and dismantling of the famous twin towers started in 2002. The new Wembley Stadium, a 90,000 all-seater stadium, eventually opened in 2007; it is the home of the England national side and hosts major sporting events. The final cost was estimated to be £750 million, making it the most expensive sports stadium in the world.

The cost of the land to build the stadium was £120 million. Major delays during the construction process significantly increased the costs. The original building company was removed, due to delays and disputes. The company then sued the engineering consultants who claimed their services were unsatisfactory. Health and safety issues also had an impact on the project: a worker was killed and another injured when a scaffolding platform apparently collapsed on top of them. Fees to bankers and lawyers are reported to be £82 million.

Check your knowledge

Investigate the cost of other major sports stadiums. How does the cost of Wembley compare?

PAUSE POINT From your investigation identify the issues that may occur which can cause the cost of the project to increase. What is a budget and why is it important for a construction project?

Hint Consider the range of economic constraints which may apply to a construction project.

Extend How can the cost impact on the quality of the project?

You are working as an assistant to the project manager of a multi-disciplinary engineering company. The company has recently been awarded a contract to design and build a new 250-bedroom budget hotel. The hotel will be located on an inner-city brownfield site and can be up to five storeys high.

The client's vision

The client is a major hotel chain that wishes to provide great value city-centre accommodation, within walking distance of the city-centre amenities, to meet the needs of a diverse range of travellers. It wants to create a safe and secure environment with sufficient communal space for relaxing and socialising, and comfortable bedrooms. It wants to impress guests by providing spacious, bright and affordable accommodation that is finished to a high specification.

Task

Research the design issues of buildings for hotel accommodation, including:

Reception areas	Parking
Bedrooms	External envelope
Forms of construction	Restaurant/bar areas
Meeting/conference areas	Typical layouts
Furnishings	Spatial requirements
Cost planning	Circulation space
Hard and soft landscaping	Aesthetics and style
Bathroom facilities	Design constraints

Plan
- What is my research being asked to address?
- How will I find the information I need to complete the task?

Do
- I can identify the range of uses for buildings and how these may affect the client's plans and the design process.

Review
- I can explain what the task was and how I approached development of my research.
- I can explain how I would approach the more difficult elements differently next time.

B Project information and building design production

B1 Project information

In order to begin designing a building, several key pieces of information are required.

Client requirements

As we have seen, many different client requirements can influence the design of a building. The client's requirements should be established at Stage 0 Strategic Definition of the RIBA Plan of Work 2013. The **strategic brief** should identify the core project objectives. These could include:

▶ sustainability and materials used
▶ building use and the style and aesthetics
▶ accommodation (size, type and number of rooms)
▶ budget and timescales
▶ site (area, location, access and services)
▶ the impact of planning and building regulations.

> **Key term**
>
> **Strategic brief** – initial objectives of the project are established and the client may consider different sites or choose between a refurbishment and new build.

> **Research**
>
> Many programmes on television, such as *Grand Designs* and *Location, Location, Location* feature clients planning new homes and stating their requirements. Watch some examples from shows like this on the internet. Are there any common themes? What sort of issues do the clients focus on?

Case study 1

A couple have recently had a third child and want to move to a bigger house. They are currently living in a two-bedroom flat in the centre of a busy town. They want to move out of town to a detached house with a garden. They would like to have separate rooms for the children and at least two rooms for entertaining and study. They also want to be within the catchment area of the local schools.

Check your knowledge

4 What other information would you require from the couple to help you prepare a client brief? (Look at learning aim A for ideas.)

5 Are there other possible options that you might suggest to the couple?

6 Which do you think will be the most important factors for these clients?

Case study 2

A budget hotel chain is looking for a site to build one of its standard hotels in a seaside town with a population of about 100,000. It has estimated that approximately 200 bedrooms would be feasible. The site should be located centrally in the town, have good access roads and be close to public transport connections. It should have 200 parking spaces. The chain needs to open the hotel within two years.

Check your knowledge

1 What other information would you require from the

client to help you prepare a client brief? (Look at learning aim A for ideas.)

2 What other options might you suggest?

3 Which do you think will be the most important factors for this client?

We will return to these case studies throughout this section to demonstrate the impact of different constraints when planning a building.

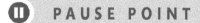 **PAUSE POINT**   For a project to be successful, the client's requirements must be clearly identified and recorded early on. How do you think the clients' requirements might differ in the two case studies?

   Hint   What would the most important requirement for each client be?

Extend   Why would the requirements for each client be different?

Site constraints

The site constraints for any construction project will be unique so they must be established at an early stage. This will enable the design team and the client to quickly decide if the location is worth considering.

For example, the family in the case study may be looking for a newly built house in a housing development outside the town centre with four bedrooms, two reception rooms and a garden. Alternatively, they might look for a greenfield site – a plot of land in a more rural location on which they can build their own home.

On the other hand, the budget hotel chain is unlikely to find a greenfield site in a suitable location in the centre of a town. It is more likely to be a brownfield site. This could mean any existing buildings on the brownfield site will have to be refurbished, extended or demolished. A new building may have to be designed which meets the client's needs. In addition, the brownfield site may be in an area of the town that has become derelict and needs regeneration.

Planning constraints

The client and the design team need to be aware of the planning constraints or restrictions which may apply to a location. Planning constraints include flood risk areas, contaminated land, listed buildings, conservation areas and assets of community value. These planning constraints may prevent planning permission being granted or may require the design to be adapted to meet the requirements of the planners.

As part of the **feasibility study**, it is critical that the planning constraints of any site are investigated. The client and design team would need to:

▶ Identify any zoned areas from the local area plan.
▶ Understand the requirements for the different stages of the planning process (outline planning permission and full planning permission).
▶ Apply for outline planning permission at an early stage of the design process to determine whether a proposal is likely to be approved by the planning authority. This should prevent any significant costs being incurred on preparing detailed designs which may never be approved.

Two possible examples of planning constraints from the case studies in the section above include the following:

▶ If the family is looking at a greenfield site in a rural area then they would need to establish if they were in an AONB or SSSI. Also, there may be a neighbourhood plan that only permits affordable housing. If they are designing a new house it would need to be in keeping with the local environment.
▶ The budget hotel chain would need to look at the local area plan to identify areas zoned for commercial development. They would also need to establish if the existing buildings were listed or within a conservation area.

Key term

Feasibility study – an assessment of the likely success of a project by looking at the advantages and disadvantages of the proposed project at an early stage. This will generally relate to the cost of the project and the benefits.

Theory into practice

Research the local area plan for your district, borough or city council. List the different types of zoned areas in your location. Consider possible locations in your local area that could be used for commercial development such as a hotel. Investigate the advantages and disadvantages of the site. Share your thoughts with the rest of the group as a short presentation.

Statutory constraints

Any new building, renovations or alterations to existing buildings must comply with current legislation. The design team needs to be aware of those that are relevant to the project. For example, in the case of the city-centre hotel, they should consider the following points:

▶ Construction (Design and Management) Regulations 2015 (CDM): designers must ensure that health and safety is considered from the planning and design of the building onwards.
▶ Building regulations: the building should conform with relevant requirements for structural elements, fire safety, noise resistance and energy use.
▶ Party Wall etc. Act 1996: this will only be relevant if the hotel shares a wall with another land owner but this may be the case if, for example, it is on a brownfield site close to existing buildings.
▶ Disability Discrimination Acts 1995 and 2005: these acts make it unlawful to discriminate against disabled people, for example by failing to ensure access to a public building. The hotel should be accessible to all, for example by including ramps and lifts, accessible toilets and ground-floor bedrooms spacious enough for wheelchair users.
▶ Equality Act 2010: this gives legal protection from discrimination in the workplace and in wider society on the grounds of race, disability and gender. The impact on designing a hotel is again mainly on making reasonable adjustments to ensure accessibility.
▶ Landlord and Tenant Act 1985: Section 11 covers the obligation of the landlord to repair and maintain any building they rent out. This is most relevant to domestic settings but may also apply if the hotel building is leased to a third party.

Environmental constraints

The completion of an Environmental Impact Assessment can help clarify the impact of a project and assist with selecting the most appropriate solution for the client. As you read in section A2, the National Planning Policy Framework 2012 sets out the requirements a project must meet in terms of impacting on the natural and built environments.

For Case study 1, a family dwelling in the countryside could complement the existing style of buildings in the area and incorporate many environmentally friendly technologies to minimise the energy use, for example:

▶ maximise solar gain (large south-facing windows to use the warmth of the sun)
▶ eco-friendly, natural materials, cheap to heat
▶ unobtrusive in the landscape
▶ light streams in throughout the day
▶ easy to manage and natural garden
▶ sheltered from the prevailing winds (for example, by trees).

For Case study 2, a 200-bedroom town-centre hotel will have a more significant impact on the environment than a single dwelling, for example significant energy use for lighting and heating, increased traffic creating more noise and CO_2 emissions. But if it is built on a derelict existing site then there are many benefits to regenerating an area to balance against the environmental impact. It will also create employment, increase tourism and revenue in the area.

Social constraints

There is demand for all sorts of development, such as schools, hospitals, roads, power stations and prisons. Everyone accepts these types of projects are required (although some people do not want the buildings near where they live) but they feel the construction work or finished structure will have a negative impact on their lifestyle. Residents could raise objections to the projects because of noise, dust, contamination, increased traffic or damage to the local environment.

> **Discussion**
>
> Watch an episode of *Grand Designs*. Note the clients' requirements, the difficulties they faced and how they decided on a solution. How do the results compare with the original requirements? Why were changes made?

> **Theory into practice**
>
> In September 2016, a deal was signed for the construction of the £18bn Hinkley Point power station in Somerset. The construction project is estimated to last for 10 years and create 25,000 jobs. The capacity of the power station is predicted to be 3.2GW and should produce enough energy to power 5.8 million homes.
>
> How will the construction project impact on the local communities? Think about the size of the project and how long it will last.
>
> What are the advantages of the project to the local community?

Economic constraints

Planning the costs of a construction project is vital yet the construction industry has a poor reputation for delivering projects on time and within budget. The government's National Audit Office (NAO) has investigated the problem of construction project cost overruns many times and is continually looking at ways to improve processes.

The available funds for a project depend on the type of project. Government grants or funding may be available for large projects. All possible sources of additional funding need to be investigated.

▸ In Case study 1 (earlier in B1), a family looking to design and build a detached four-bedroom house will be limited by the amount of money they have to put down as a deposit and also by the amount they can borrow from a bank or building society. The size of their mortgage will depend on their ability to pay from their monthly income. The budget for the design and construction of the house will be limited and they will not really be able to manage many significant unforeseen cost increases. They may need to compromise on their desired requirements, for example, if a four-bedroom house in the countryside is too expensive, they have to decide whether they want the larger house in a more urban area or remain in the countryside but in a smaller house.

▸ In Case study 2, the budget hotel chain will have significant financial resources and may also be able to draw funding from grants if they meet certain criteria. The eventual cost may not be as important to them as the date of completion as, once they are open, they will start to generate income.

❚❚ PAUSE POINT   Research your local property guides and find a selection of new and old four-bedroom houses.

       Hint   For each example list the constraints that may have applied during their construction.

       Extend   Review the difference in the constraints between the older houses and newly built houses.

B2 Initial project brief

The initial project brief is completed during RIBA Plan of Work 2013 Stage 1: Preparation and Brief. It should record the outcome of discussions between the client and the design team and clearly identify the client's objectives.

The initial project brief is one of the most important documents to be agreed with the client. If the project objectives are clearly defined from the start, there is a much greater chance that the project will be completed successfully.

The success of any project can be measured in terms of the time taken to complete it and the final cost. A good initial project brief is the starting point as it summarises the client's requirements and specifies clear objectives. This allows everyone working on the project to understand the client's required outcomes and avoid delays and disagreements. The size of a project will dictate the detail and accuracy required for the brief.

▶ Case study 1: a family is looking for a four-bedroom, two-reception detached house in the country, with good access to schools. Their budget is approximately £250,000. Not much detail is needed at the initial stage.

▶ Case study 2: a budget hotel chain is looking to build a 200-bedroom hotel in a town centre. A significant amount of detail is required for the initial project brief, for example the size of the hotel, the size of each room, and any minimum standards expected by the brand. A specialist consultant may be called in to prepare a suitable brief.

To have a clear and successful initial project brief there are certain basic requirements that must be identified and clearly defined.

Desired project outcomes

The initial project brief should be a clear statement of the client's key requirements, resulting from discussions between the client and the design team and taking all the constraints such as planning, environmental issues and budget into consideration. This should enable a project to remain focused and on track.

Spatial requirements

The space required depends on the use of the building – residential, commercial or industrial. This can be produced as a schedule of accommodation, detailing the number of rooms required, the size of each room (the floor area in m²), the function of the room and quality of finish, along with an approximate number of people who may use the space. Functions may include bedrooms, kitchens, dining areas, office space and storage space. The number of floors required would also be detailed at this stage.

The client's preferences for the layout of the building will be specified, such as an open-plan style with lots of windows to maximise the natural light. Human comfort factors should also be considered in relation to the size of the spaces, including temperature and acoustics, which will be influenced by function and the number of end users expected for each room. Access and egress from the building and circulation flows must be identified. Parking requirements are another essential consideration.

Defining the spatial requirements at this stage can prevent problems later in the design process, e.g. thermal requirements for walls and floors possibly impacting on room size and minimum height.

Site or context

The size and budget of the project will dictate the amount of information that can be obtained through site investigations. A site survey may identify any existing buildings on the site, the topography, the geology and geotechnical conditions, availability of building services, possible contamination, traffic and transport links and site boundaries. It may also identify any access, planning or other legal constraints.

The team should also consider any building surveys completed on existing buildings that record the condition of the property and identify faults or defects, such as the presence of asbestos, which may mean additional expenditure.

Budget requirements

The amount of money available for a project is critical and must be established as early as possible. The budget is basically how much the client is willing or able to spend. Knowing the amount will allow the design team and client to know what is possible within the financial constraints. By looking at similar projects, the design team can inform the client of the likely costs and help them adjust their expectations.

The budget should include a breakdown of expected costs such as:
▶ land purchase
▶ design and approval fees
▶ surveys, e.g. site investigations or building surveys
▶ planning costs and consultant fees
▶ construction costs and insurance.

It should also allow for inflation and have a suitable amount set aside for **contingencies**.

> **Key term**
>
> **Contingency** – a future event or circumstance that may occur but was not planned, which may impact the schedule or budget of a construction project.

Use of an initial project brief to generate and develop design ideas and specifications

A good initial project design brief will enable the design team to create possible solutions for the client. It must have:

1 Objectives and goals of the new design
2 Budget
3 Time scale
4 Scope of the project
5 Overall style/appearance
6 Any definite 'do nots'.

Case study 1: Possible options a design team might offer a family looking for a four-bedroom/two-reception detached house, might be:

▶ a bungalow or chalet bungalow
▶ two- or three-storey house
▶ ultra-modern design
▶ refurbished existing dwelling.

The style of the house could be modern or traditional. A wide range of materials could be used for the finish of the house. This will allow the client to refine their choices and preferences.

Completing the initial project brief

The initial project brief should be written so that the client and the design team can clearly understand the objectives of the project. It should state the client's wishes clearly and concisely. As we have seen, the client could range from an individual to a multinational business. Therefore, it is important that the design team have a clear understanding for whom the brief is being written.

The style of the initial project brief should reflect how the client thinks and feels about the project. It should represent the essence of what the client wants from the building, e.g. value for money, safe, secure etc.

The language used in the brief must be suitable for the client. Any technical terms used should be explained otherwise this may lead to disputes or complications at a later stage.

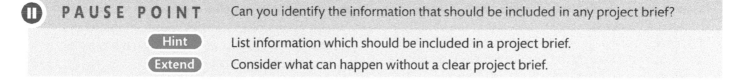

PAUSE POINT   Can you identify the information that should be included in any project brief?

Hint   List information which should be included in a project brief.

Extend   Consider what can happen without a clear project brief.

B3 Design production

This section explains how creative and innovative designs can be developed to meet the requirements of the initial project brief requirements. Designs could cover a wide range of buildings, most often low- and medium-rise domestic, commercial and industrial buildings.

The design team must present their ideas in such a way that the client can visualise and get a feel for how the building will look and perform. A skill of the designer is to interpret the client's wishes through discussion and investigation to produce an accurate and engaging initial project brief.

A designer may create solutions for a range of different types of buildings and constructions. Although for each of these there will be different focuses, the fundamental process the designer follows will be similar.

> **Theory into practice**
>
> Research residential properties suitable for a family of five in central London (for example, within five miles of the postcode E1) which could be bought for a budget of £500,000. What do you think are the factors that contribute to differences in the cost of the houses?

Outlining a solution with 2D and 3D sketches of initial ideas

When working with client briefs, the first step for any designer is to outline a 'solution' to the client's brief. This will take all the information the client has supplied about what they desire from the final product, and demonstrate how they will use the space available to meet those needs and what the designer suggests should be constructed to do this. This will often involve creating sketches of the final design.

Types of sketches

Sketch designs are a series of sketches, often three dimensional, that illustrate the designer's interpretation of the initial brief. The client may choose their preferences from several sketches and the ideas are combined into a final sketch design for approval. These sketches could also include two-dimensional sketches of the front elevation of the building, a site plan of the area and the building orientation and floor plans showing possible layouts of the rooms.

There are several types of drawing that you may encounter and these may include the following:

▸ Freehand sketch – these will be used for developing initial concepts, and are often quick and simple, not drawn to exact scale or to be used for detailed planning. Instead they are used to form ideas and are then developed into more detailed drawings.
▸ Single-point perspective – these are drawings that show an item from one position. In these drawings items get smaller the further away they are, all converging on a single 'vanishing point' on the horizon. It allows buildings to be drawn in a 2D way, while appearing 3D.
▸ Two-point perspective – similar to single-point drawings, but with two vanishing points. For construction drawings, the corner of a room and building will face towards the viewer, with each of the walls connected to it shrinking to 'vanishing points' either side.
▸ Planometric drawings – these are drawings of items from above, with the item placed at a slight angle – for example looking down on a room plan from a 45° angle.
▸ Isometric views – these show three sides of a an object, but none of them are a true shape with 90 degree corners. The vertical lines are drawn vertically but horizontal lines are drawn at 30 degrees to a base line. This creates a 3D image in the sketch.

3D effects can also be created in sketching by increasing the thickness of lines around one or two sides of an object. This creates a visual illusion of depth on one side, making it appear that you are looking at it from a slight angle. Using shading around edges or projecting from the object can also help to create a 3D effect in a sketch (to do this you will need to decide what direction the 'light' in the picture is coming from so that it is consistent across the whole picture). Freehand rendering (adding colour, shading and texture to an image) can also help to make it look more realistic, engaging and give a client a better idea of what the final build will look like.

Clear communication

It is important that your drawings are clear and easy to understand and that they communicate all the information the client or contractor will need in order to plan and cost the work to be carried out.

All drawings should have a suitable level of technical annotation, describing key elements or individual components, or identifying particular components or materials. It is important that this annotation is clear and concise and that is it presented clearly and in a manner that is easy to read and understand.

As part of this, you will need to ensure that your drawings clearly identify the key features of the structure you have planned. This may include:

▸ The external fabric of the building – e.g. what materials is this made from? What technical specification is it designed to meet?
▸ Roof type – e.g. what is the design of this roof? What materials have been used to construct it? What is its likely load?
▸ Service access – e.g. where will this be positioned for each key service (gas, electricity, water etc.)?
▸ Location of windows and doors.

B4 Computer-aided design

Computer-aided design (CAD) allows everyone, including the client, to visualise what a proposed project will look like. There are two methods of producing CAD drawings. The choice depends on the nature of the project.

▸ **2D digital project information** – small-scale domestic projects, such as house extensions and alterations, may use 2D CAD, which mirrors the process of hand-drafted drawings.
▸ **3D digital project information** – this uses CAD to produce a set of virtual architectural objects which combine to form a 3D model of the project. Advantages over the 2D CAD system include:
  • flagging up clashes such as a load-bearing column clashing with the main sewage pipe
  • embedding information about individual objects (which links to BIM)
  • improving design productivity and efficiency
  • high-quality rendered presentation images.

As well as producing 3D models, CAD can also produce 2D project information to help with detailed planning of the building or structure. 2D drawings will lack the details and manipulation possibilities of 3D models, but allow the user to show different information around the project. These can include the following:

▶ Elevation – this shows a front-on view of an object as if you were standing looking at it, and could be used to show a contractor where finishes are to be made or the precise location of objects.

▶ Plan view – this shows the appearance of an object if you looked at it from above. The floor plan will have full details of measurements and be vital for planning placement of objects and spacing solutions.

▶ Section – this shows a view of a structure if it has been 'cut' in half – so it might show the floors and walls in a building if you cut the front half of the building away.

The pace of software development is such that even for small projects architects are using packages to produce 3D rendered images so that clients can understand what the project will look like when it is built. Virtual reality (VR) presentations are also now used by architects to enable clients to roam around the virtual space of the proposed project. This is the best available way of bringing the project to life for the client. They can adjust the design to meet their personal requirements.

Setting up a CAD project

There are different elements required in setting up a CAD project. This may involve:

▶ Number of floors – the floor number can be stated on each drawing starting with ground level.

▶ Floor levels – a datum level can be placed on the structural floor level to indicate its height from the site datum or temporary bench mark provided by the designer.

▶ Linking elements with anchors – anchors are a set-up in CAD that tells the software that certain elements are connected (for example a window would be anchored to the wall it sits in).

▶ Building footprint – the building and its external works can be shown on one drawing. The building footprint will be illustrated on each floor plan.

▶ Component libraries – these are a library of standard components that many manufacturers provide to make it quick and easy to add common elements to a drawing.

▶ Saving in an appropriate format – the most common CAD file format is DWG, but this can be converted to PDF or Tiff files to print or attach to emails so that people without CAD software can see the plans.

Use of basic CAD methodologies

Dimensional control, sizing and scale

Dimensional control (capturing accurate dimensions and measurements) is normally established in relation to the centre lines and grid of the central supporting structure. This is used as a bench mark for the rest of the structure, for example a steel frame. The column centres would be used to form a grid pattern in CAD. All the other elements in the construction would align to this grid, for example, beams would be centred on the grid.

CAD drawings are always drawn to full-sized dimensions. You can alter the scale used in the template to metres or millimetres for dimension input. You can use an imperial units template (in feet and inches) if you are working on American projects. With the full-sized dimensions you can then set a viewport to select the drawing paper size that you will print to. The viewport can be used to scale a drawing within its boundary.

Detail levels

The level of detail within a drawing will depend upon the intended audience and the final stakeholders. For example, a fire evacuation plan only needs to show the exit routes and signage details in colour. It does not need construction details as these are not relevant to the general audience that needs to know about the fire escape routes.

The level of detail will depend upon the audience and the purpose of the drawing.

This detail could be fine, medium or coarse. Fine work will involve high-dimensional control (very accurate plotting) and may include the finishing stages of a project, where all details need to be completely accurate and to scale. Coarse detail is used for rougher, or early stage, work; for example the outline excavation of a project where dimensions do not need to be accurate.

Drawing elements in CAD

Table 2.3 demonstrates some of the elements that may have to be formed within a CAD drawing and which may be subjected to a detailed drawing.

▶ **Table 2.3:** Examples of drawing elements in CAD software

Element	Level of detail
Hidden element features	Particularly used in engineering, a hidden feature, for example the depth of a bolt hole, is essential information. Hidden features can be shown on floor plans where openings are positioned across walls which do not have doors.
Composites	A composite is made up of several different elements. This may be a material or installation upon site, for example a precast concrete floor sitting on a structural steel frame. The level of detail between each, in terms of structural integrity, needs to be drawn. Similarly with a roof and wall panel, standard CAD drawing composite details can be provided by a manufacturer.
Openings (placement and positioning)	Doors, windows and shutters, along with other openings in buildings, will need to be detailed so that weather protection measures are co-ordinated along with the structural fixings for each unit. For example, a standard door set has to be fixed into an opening leaving a tolerance for the frame to fit. The gap left will then have to be sealed. The drawing also needs to indicate the gauge of the brick used, while dimensions need to course in, both in terms of height and length.
Fixtures, furniture and fittings	This covers stairs, fitted units, furniture, kitchens, etc. Standard elements can be again used to represent many of these features. Standard sanitary symbols represent bath, shower tray and WC. Similarly, electrical symbols are used for light fittings, sockets and smoke detectors.

Use and manipulation of CAD software to produce virtual models

CAD provides the facility to draw in three dimensions: x (vertically), y (horizontally) and z (depth, 3D). The third dimension produces a virtual model that can be manipulated and rendered with the actual appearance of the materials it will be manufactured from, for example bricks, roof tiles, window glass etc., allowing the user to really get a sense of what the final product will look like.

3D models can be viewed in several different ways on screen. You can rotate or re-orient the image to look at it from different angles, as well as zoom in or out of features. The difference between 2D and 3D is reflected in Table 2.4.

▶ **Table 2.4:** Comparison of 2D and 3D

2D	3D
Elevations can be drawn	Model can be rotated and viewed from any direction
Ground and floor plan views	Surface details can be added to produce a realistic model
Detailed sections through elements	3D perspectives can be introduced with a number of vanishing points

Rendered images

Rendering is a design tool used to add colour, texture and lighting to a 2D or 3D drawing to create realistic images of structures. Natural textures, colour and shading give the impression of real materials.

CAD can be used to allow several different 'camera views' when looking at an object. You can change the camera angle, its position in relation to the object and introduce shadows and similar effects to the object to get a better idea of what it might look like when completed. The lighting effects that you can add to an object could be with external or internal lighting, depending on the requirements of your drawing. You can also add seasonal and weather effects to the drawing.

CAD software is becoming an increasingly vital tool for architects and designers when working on buildings or installations. Get access to a CAD software program and try creating a model of a room that is familiar to you. Then use the program to redevelop the room in some way. What information do you need to do this? How can CAD help your planning?

PAUSE POINT   How can the design team communicate their ideas and concepts for a project to the client?

> Hint   List the different methods the design team can use to present their ideas and concepts to the client.

> Extend   Think about why it is important for the design team to produce a range of options for the client.

C Construction methods and techniques

Each building has its own challenges when planning and carrying out its construction. This section will introduce you to some key construction methods and techniques for designing and building structures.

> **Link**
>
> Many parts of this learning aim link with Unit 4: Construction Technology. Where parts of this unit's content will be covered in more depth later in this book, this has been clearly signposted.

C1 Forms of low- and medium-rise structures

Functional requirements of key primary and secondary elements

The past hundred years have seen more change, in terms of prosperity, welfare and in society generally, than any other century. There have also been huge developments in science and technology. These changes have also impacted the construction industry, revolutionising the ways buildings are designed and built.

Buildings are now much more than just shelter; they need to provide suitable sanitation and clean water, prevent damp, and allow sufficient natural light and natural ventilation. Another major change is that the need to conserve energy has become a major driving force in building design and construction methods in recent years.

A lot of legislation has been introduced to improve the quality of construction work and to standardise processes and procedures. These include Building Regulations, planning laws and public health laws. British and European Codes of Practice cover the quality and sizes of materials and are linked to construction methods. Other pressures influencing building design are financial controls and costs in public housing.

The primary functional requirements of any design are the major construction elements: walls, floors and roofs. As illustrated in Figure 2.1, their functions are vital – for example, the function of the roof is to provide thermal insulation and protection from the weather.

> **Discussion**
>
> Prefabrication is becoming more popular. In groups, think about the following:
> - How would you explain prefabrication?
> - Would you use this method to build high-quality homes?
> - Would you consider this to be a sustainable method of construction? Why?

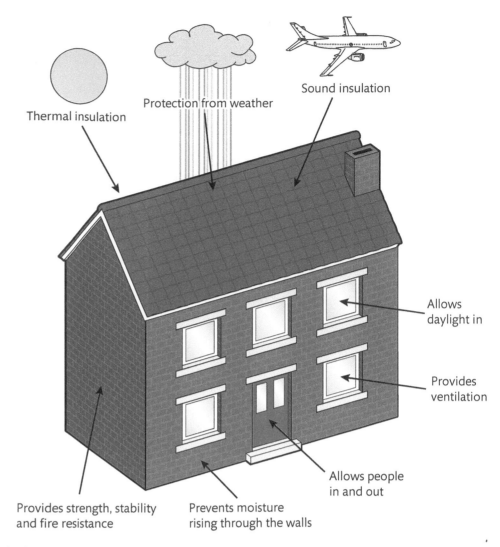

Thermal insulation

Protection from weather

Sound insulation

Allows daylight in

Provides ventilation

Allows people in and out

Provides strength, stability and fire resistance

Prevents moisture rising through the walls

▶ **Figure 2.1:** The primary functions of the external envelope of a house

▶ **Frames** – Many methods and different materials can construct the frame. Composite structures combine two or more materials, each of which retains its own distinctive properties. For example, a tall steel frame structure incorporating concrete floors could be precast or poured in situ. Steel is light, fast to erect and uses less space for the structural elements (beams and columns) than other materials. The concrete adds stability to the structure and has good fire resistance. Composite structures combine the advantages that steel is good in tension and concrete is good in compression.

▶ **Ground floors** – It is important to assess the ground conditions before any ground floor is constructed. As the ground will ultimately support the load of the building, if it moves under loading this will impact the performance and function of the floor. The use of the building may affect the floor type; for example, a factory warehouse must support machinery and stored materials. Other factors, such as what chemicals might be used in any industrial processes, could affect the performance of the floor.

▶ **Walls** – Walls are external or internal, or load bearing or non-load bearing. The main purpose of an external wall is to provide strength and stability, provide protection from the weather and fire and to provide thermal and sound insulation. Wall construction encompasses solid walls made of bricks or stone, **cavity walls** made of an outer leaf of brick, a space or void and an inner leaf of block, and even timber stud walls used for internal partition walls.

Key term

Cavity wall – two walls with a gap in between known as a cavity. The outer leaf is built with bricks and the inner leaf, which is the load-bearing wall, is built of blocks. Insulation is placed in the cavity to minimise heat loss.

▶ **Roofs** – These provide a protective covering for the building. They can generally be classified as pitched (sloped) or flat and can be constructed from many different materials depending on the purpose of the building. For example, domestic houses may have roofs constructed from ceramic to slate tiles or even straw in a thatched roof. Industrial buildings may use metal sheeting, such as copper, galvanised steel or even cement sheeting or corrugated steel.

Secondary elements of a building include windows, doors, internal walls and staircases. Cladding, glazing systems and structural insulated panels (SIPs) can be connected to the main frame to meet the requirements of the client.

Ⅱ PAUSE POINT   Can you identify the primary and secondary elements of a construction project?

　　　Hint　　　List the elements used to construct the main frame of a building.

　　　Extend　　Identify all the elements used to complete a building structure and investigate their function.

Types, characteristics and application of construction techniques

The following sections discuss the types, characteristics and application of construction techniques and methods for different types of structure.

Traditional construction

Old traditional

These houses were built using solid bricks (often made in the local area), on basic brick foundations. The roofs were usually pitched and made with timber, had no underfelt and were completed with local slates and cast iron pipes. Ground floors were a mixture of solid floors under the kitchen area and suspended timber process for other areas and upper floors. Most rooms had open fireplaces as the source of heating.

One of the major points is that they did not contain any damp course. The need to prevent damp problems was initially tackled by the gradual introduction of damp courses in the 1920s. The need to retain heat began to become important in the 1930s so brick external cavity walls were introduced. Roofs still had no insulation.

Prefabricated 1945–50

The destruction of so many buildings during the Second World War created a huge demand for housing, particularly in cities that had been affected by bombing. A new construction method was required to meet this high demand. Factories produced prefabricated materials such as asbestos, steel and concrete, which were used as cladding for houses. This reduced the time and amount of skilled labour required to build the houses.

Changes to traditional approaches 1950 to present

Improvements in transport infrastructure allowed the introduction of new materials rather than the previous dependence on local materials. Using blocks rather than bricks for internal walls and inner leaves of cavity walls became common practice. In the late 1950s, plastic gutters and pipes were increasingly used.

The majority of new housing built today in the UK uses traditional masonry construction. The move to more modern methods of construction is often resisted as the major house builders are reluctant to take the commercial risks on non-proven, more expensive methods.

Research

Consider the house you live in. Find out when it was built. What method of construction was used to build it? What materials were used in the construction process? Research old photographs of the area.

The construction process

The main features of on-site construction are:

▶ excavation and installation of suitable foundations
▶ building of cavity walls made up of an inner and outer leaf
▶ the inner leaf (block work) is load bearing and supports the roof and internal floors
▶ the outer leaf (brickwork) is for aesthetics and provides the first stage of defence against the weather
▶ the outer and inner leaves are joined by wall ties
▶ the cavity is filled with insulation to retain heat and improve the thermal efficiency of the building.

The advantages of traditional construction include:

▶ flexible in design and construction
▶ can be cost effective
▶ common form of construction so there is a large supply of suitable tradespeople
▶ historical record of success, accepted by financial institutions
▶ local materials available and many materials are recyclable
▶ alterations easily accommodated
▶ good sound insulation, fire protection and thermal characteristics.

The disadvantages of on-site construction include:

▶ delays due to weather conditions, lack of labour, loss of materials etc.
▶ room on site required to store materials
▶ labour intensive and longer time needed for completion
▶ lack of quality control
▶ materials causing issues, such as health hazards associated with the disturbance of asbestos.

PAUSE POINT   How would you define traditional construction methods?

 Hint   List the materials that are used for traditional construction methods.

Extend   Identify the advantages and disadvantages of traditional construction methods.

Modern methods of construction (MMC)

Modern methods of construction such as timber framed houses, light steel framed houses and modern concrete framed houses can reduce construction waste, promote sustainable building and shorten the time to build projects.

Government research suggested that, by using MMC rather than traditional methods, up to four times as many houses could be built with the same number of workers on site. This would contribute to reducing the demand for new housing each year and reduce construction time, and performance would match traditional building techniques. (Source: **www.nao.org.uk**)

Timber frame construction

This is a proven and flexible MMC using standard, prefabricated timber sheets for walls and floors.

The modern timber framed house uses breathable membranes, high-quality insulation and vapour controls to ensure the durability of the building. This type of building meets the requirements for thermal, acoustic and fire protection set out in the Building Regulations.

Timber frame building is one of the fastest methods of construction available. The house can be weathertight within four or five days once the concrete slab foundation is completed.

Advantages of timber frames include:
▸ can use off-site construction methods
▸ higher speed of erection (as elements prefabricated and then assembled on site)
▸ reduced overall build time and risk of delay and simplified project planning process
▸ predetermined routes for wiring and plumbing reducing time and skilled labour requirements
▸ predetermined door and window openings with the outcome of improved accuracy
▸ minimal site storage requirements and reduced trip hazards
▸ reduce site waste due to factory controlled quality assurance
▸ fast heating due to low thermal mass
▸ uses sustainable building materials
▸ high **BREEAM** rating.

The disadvantages of using timber frames include:
▸ additional design and engineering time requirements
▸ lack of skilled builders and erection teams
▸ higher transportation costs due to larger prefabricated pieces being delivered to site
▸ possible loss of quality control after prefabrication due to all materials being assembled on site
▸ adverse weather could damage the structure before it is weathertight
▸ work is needed to meet the requirements for fire rating of individual walls etc.
▸ significant exposure to moisture can lead to timber decay
▸ may be damage to integrity of the building from follow-on trades such as plasterers and joiners.

> **Key term**
>
> **BREEAM** (Building Research Establishment Environmental Assessment Method) – an assessment process that measures the design, construction and operation of a building against targets based on performance bench marks.

▸ What are the advantages and disadvantages of timber frame construction?

> **Discussion**
>
> Compare the advantages and disadvantages of traditional construction methods and timber frame construction. Which factors are most influential? Choose which method of construction you would use for your house and justify your decision.

Steel frame construction

Steel frames are often used in small light commercial buildings. Their advantages include:
▸ the strength properties of the steel allow for large internal open-plan areas
▸ being fast to erect as they are prefabricated and quick and easy to connect on site
▸ good fire proofing due to boxing the frames in with insulation or painting them with fire resistant paint (intumescent paint)
▸ being manufactured to a high quality in factories
▸ durability and the ability to withstand a significant amount of wear and tear
▸ steel can be recycled
▸ very little waste as the sections are prefabricated

- the ability to adapt steel to many different shapes and forms allows architects to be creative with space without compromising on strength and safety of the building
- savings to be made not only in cost but also in time to complete the project.

The disadvantages of steel frames include:

- inferior acoustic performance
- ability to conduct heat makes it susceptible to **thermal bridging** – this can lead to significant heat loss, build-up of condensation and the formation of mould
- ongoing maintenance costs as the steel may be exposed to the weather and prone to corrosion.

Steel **portal** frame buildings are suitable for construction of large span commercial and storage buildings. This form of modern construction takes advantage of the strength and durability of steel.

Light steel frame construction

Light steel frames are made from C- or Z-shaped sections of galvanised steel and tend to be used for residential and smaller commercial buildings. It is similar to timber framed construction in technique.

Advantages of light steel frame construction are:

- during construction, it can be easily lifted by operatives or delivery personnel and transported so there is no need for any lifting equipment on site
- the walls are straight and the corners are square which may not always be the case with other materials
- no drying out time is required even in damp conditions
- little waste and the material can be recycled
- a strong structure with strong connections held together by rivets and screws
- steel is not vulnerable to rot or fungi in the same way as timber
- provides good fire safety.

Disadvantages of light steel frame construction include:

- steel is a poor insulator thus heat loss can be an issue/thermal bridging
- tradespeople are not familiar with working with this material on site
- susceptible to rust/corrosion in damp moist conditions.

▶ What are the advantages of steel frame construction?

Concrete frame construction

Concrete frame structures are another common modern method of construction and this type of construction generally refers to reinforced concrete. The building is made up of a frame or skeleton of concrete beams (horizontal elements) and columns (vertical elements). The floors consist of concrete slabs.

The advantages of concrete frames are:

▶ can be prefabricated off site and delivered to the site or it can be cast in situ
▶ often used in heavily populated urban areas as it is quick and easy to erect where space is limited
▶ it is a safe construction material as it provides a stable frame structure
▶ requires no additional fire proofing.

Disadvantages of concrete frames include:

▶ low tensile strength
▶ can vary in quality and requires regular quality control checks
▶ limited spans possible with concrete blocks
▶ dense material and thus creates large loads for the foundations
▶ cracks can appear due to shrinkage/drying out
▶ requires large column sections to support the loads
▶ if cast in situ the formwork can be expensive
▶ speed of construction is restricted by curing of the concrete.

Structural insulated panels (SIPs)

Structural insulated panels (SIPs) generally consist of laminate with a 100–200 mm foam core with a structural facing, such as plasterboard or plywood, on each side. They are manufactured under factory quality control processes and can be made to nearly any size or shape. The SIP construction method is regarded as strong, energy efficient and financially viable.

The advantages of SIPs are:

▶ offers a high degree of insulation, **air tightness** and fire protection
▶ prefabricated and cost effective
▶ high level of quality assurance
▶ quick and easy to assemble and minimal skilled labour required.

The disadvantages of SIPs include:

▶ can be damaged by moisture, pests and insects
▶ modifications to the initial design can be expensive
▶ lack of builders with the necessary knowledge to construct the SIPs
▶ high transportation costs for component parts
▶ requires a mechanical ventilation system due to the high levels of air tightness – it is important to circulate fresh air around the building.

Passivhaus construction

Passivhaus or passive house construction has its origins in Germany in the 1990s and is based on detailed design principles about incorporating high levels of energy efficiency into the building. It works on the principle of super-insulation throughout the whole house, eliminating thermal bridging and reducing the air tightness levels of the building as much as possible. With energy costs likely to increase in the future, this construction method could be an ideal solution.

Advantages of Passivhaus are:

▶ the building uses 90 per cent less energy for heating and cooling than standard UK homes

▶ Why is concrete frame construction often used in heavily populated areas?

Key term

Air tightness – resistance to unwanted air leakage from the building. The lower the leakage, the more efficient the building.

▶ the only significant technology used in the system is a mechanical ventilation and heat-recovery system (MVHR). This is a balanced ventilation system providing fresh air, extracting old air and warming air by recovering heat from the extracted air and transferring it to the incoming air

▶ heat generated within the building from people, cooking, lighting and solar gain is retained due to effective insulation and maximum air tightness.

Disadvantages of Passivhaus include:

▶ standards are considerably stricter than current building regulations

▶ it requires knowledge of the building and how it operates – more a way of life than just a building design

▶ the high energy performance of the building must be considered from the initial stages of the design and requires skilled trades and tight controls of the quality of the building materials and processes

▶ any alterations made to the building after completion, such as bringing in cables, may affect its integrity and reduce its efficiency.

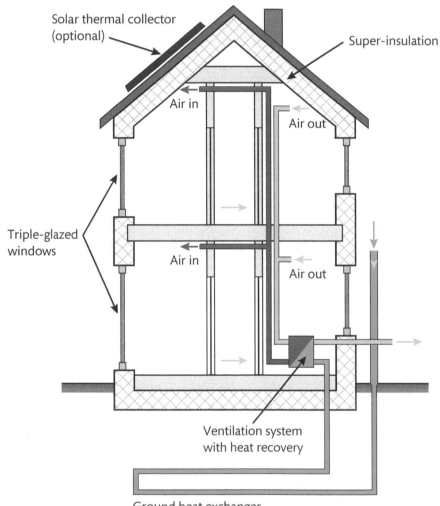

Figure 2.2: Passivhaus principles must be incorporated from the earliest stages of construction to be effective

Link

For more about different forms of low-rise construction, see Unit 4: Construction Technology.

PAUSE POINT What modern methods of construction (MMCs) are available?

 Hint   List the different MMCs.

Extend   Identify the main advantages and disadvantages of each MMC.

C2/C3 Substructure and superstructure construction

The **substructure** is the foundations or supporting part of a structure. It is the building below ground level which transmits the loads of the building to the surrounding ground. All building projects should carry out a detailed site investigation to determine the properties and behaviour of the soil. This may involve carrying out site surveys, either using existing data or by visiting the site, and compiling survey reports identifying any issues. This enables a suitable foundation to avoid any settlement of the building which could affect its structural integrity.

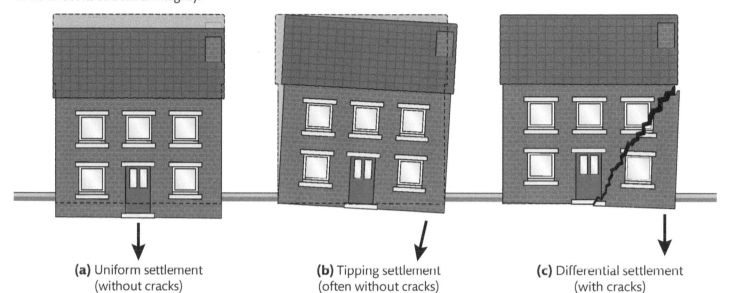

(a) Uniform settlement (without cracks)

(b) Tipping settlement (often without cracks)

(c) Differential settlement (with cracks)

▸ **Figure 2.3:** There are three main types of settlement: (a) uniform, (b) tipping and (c) differential

The **superstructure** refers to the part of the building above ground level. This includes all the primary and secondary internal and external elements of the building such as walls, floors, roofs and stairs, doors and windows.

There are more details about many of the processes and construction types of different buildings in Unit 4: Construction Technology. This unit covers site investigations, soil investigations, groundwater analysis, foundation design and construction methods.

Unit 4 and Unit 10: Building Surveying also cover many of the construction elements of the superstructure.

> **Key term**
>
> **Superstructure** – the part of the building above ground level.
>
> **Substructure** – all work below the ground floor level including damp-proof membrane and foundations.

▸ **Table 2.5:** Elements of the superstructure

External walls	• Solid masonry, cavity walls, curtain walls, infill walling, rain screen, panel, cladding, profiled sheets, rammed earth, straw bale (Unit 10) • Formation of openings, heads, sills, jambs/reveals, thresholds (Unit 4) • Weathertightness, thermal and acoustic insulation and finishes (Unit 4)
Internal walls	• Separating/party, partition/compartment (Unit 4) • Load bearing, non-load bearing (Unit 10) • Finishes (Unit 4)
Structural frames	• Steel, reinforced concrete, timber, structural insulated panels, light gauge steel (Unit 4) • Fire protection (Unit 4)
Ground floors	• Solid and suspended (Unit 10) • In-situ concrete, beam and block, timber (Unit 4) • Thermal insulation, damp proofing and finishes (Unit 10) • Upper floors – composite concrete/profiled steel, precast concrete slabs, in-situ concrete, beam and block, timber/engineered timber (Unit 4) • Fire protection

Roofs	• Flat/pitched forms and terminology (Unit 10) • Traditional, trussed rafter (Unit 10) • Profiled decking, lattice frame, portal frame (Unit 4) • Weather protection, coverings (Unit 10)
Stairs and landings	• Stair and landing terminology/regulations (Unit 4) • Timber, in-situ concrete, precast concrete, steel (Unit 4)
Doors and windows	• Types and construction (Unit 10) • Uses in fire compartmentalisation and escape (Unit 10)

Link

See Unit 4 for more information on site investigation and analysis, as well as factors affecting foundation design. See Unit 10 for more information about foundations.

Key terms

Carbon footprint – sum of all the CO_2 emissions used in a process (usually expressed as tonnes of CO_2 emitted).

Greenhouse gases – gases such as carbon dioxide, methane and water vapour that stop heat escaping from the Earth's atmosphere. Increasing amounts of greenhouse gases can lead to global warming and climate change.

Thermal mass – the property of a material which relates to its ability to absorb and store heat energy.

C4 Sustainability

Sustainable construction techniques are a requirement for modern construction projects. There are ethical, legal and financial reasons to reduce pollution and its impact on the environment. Ethically, the construction industry should not be responsible for destroying the environment. Legislation requires significant improvements in processes to reduce their **carbon footprint**. It is possible to make significant financial savings by reducing energy usage within buildings.

Statistics indicate that in the UK nearly one-third of the waste that goes to landfill is produced from the construction or demolition of buildings and almost half the carbon emissions come from buildings. All these factors have contributed to developments aimed at reducing energy consumption and reducing the impact on the natural environment. (Source: **www.ukgbc.org/resources/key-topics/circular-economy/waste**)

Passive solar gain

Passive solar gain is the process of harnessing the sun's energy to heat and cool living spaces. Using passive solar gain to help heat a house reduces the demand on fossil fuels and, as it does not produce any **greenhouses gases**, helps to protect the environment. To prevent over heating or too much glare, suitable shading should be incorporated in the design.

The process of passive solar gain involves:

▶ gaining the benefits of solar energy through properly oriented, south-facing windows
▶ using building materials with a high **thermal mass** such as concrete and bricks to store energy
▶ transporting stored solar energy through the building by using natural processes of convection and radiation
▶ using high specification windows to maximise the solar heat gain.

Passive stack ventilation

Passive stack ventilation (PSV) systems use air flowing over the roof and the natural buoyancy of warm air to lift the stale, damp air from the kitchen, bathroom and toilet up ducting to the peak of the roof where it escapes into the atmosphere.

By installing trickle vents in the windows, fresh air is drawn into the building. There is no need for any mechanised systems to power the process but the amount of ventilation will depend on the amount of air movement outside the building. A PSV system needs detailed design and installation for it to work effectively. It can be installed as part of a new-build project but it is hard to install in existing buildings.

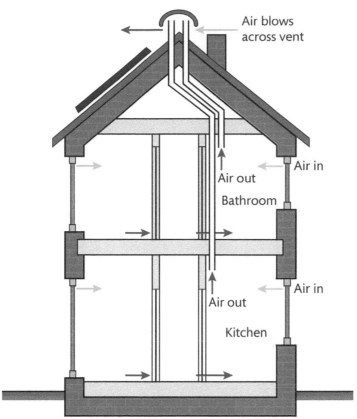

Air blows across vent

Air out

Air in

Bathroom

Air out

Air in

Kitchen

▶ **Figure 2.4:** The effectiveness of passive stack ventilation depends on the air movement outside the building

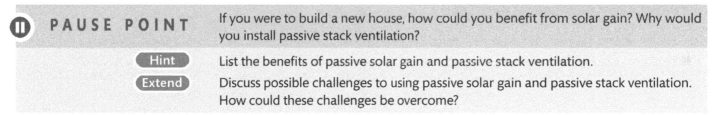

Ⅰ PAUSE POINT   If you were to build a new house, how could you benefit from solar gain? Why would you install passive stack ventilation?

Hint   List the benefits of passive solar gain and passive stack ventilation.

Extend   Discuss possible challenges to using passive solar gain and passive stack ventilation. How could these challenges be overcome?

Water use reduction methods

Almost 97 per cent of the water in the world is too salty to drink. Glaciers and the polar ice caps contain a further 2 per cent, leaving approximately 1 per cent for human consumption, agricultural and industrial use. As the population of the world increases, it is important to protect the existing supplies by being more water efficient. With many water companies raising water charges, it is vital for home owners and businesses to minimise waste to keep costs down. Designers could include a water efficiency strategy in their initial plans and ensure this is developed and implemented throughout the project.

Grey water systems

Grey water is waste water generated from water consumption in houses and offices which has not come into contact with waste from toilets (**black water**) or soiled nappies. Sources of grey water are sinks, showers, baths, washing machines and dishwashers.

Grey water is easier to treat and reuse than black water. Approximately a third of the water that is used in a house flushes toilets. Grey water, when treated, is not safe to drink but can be used to flush toilets and for land irrigation. Using grey water can help significantly reduce water usage and help reduce the water bills in metered properties. Typically, the cost of a grey water system may be £5,000–£6,000 but the system could pay for itself in 3 to 5 years.

Key terms

Grey water – waste water generated from water consumption which has not come into contact with black water.

Black water – water which has come into contact with faeces and may contain bacteria that can cause diseases.

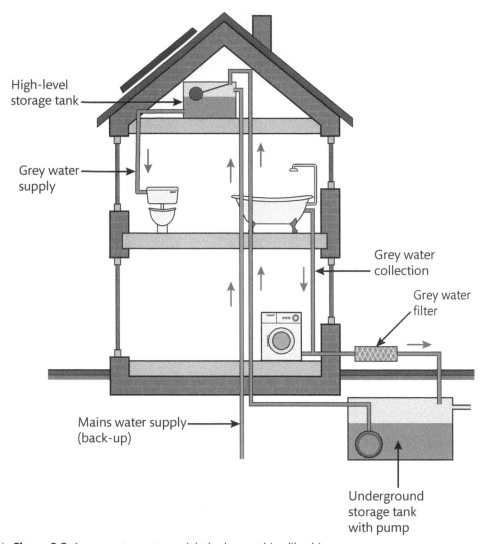

High-level storage tank

Grey water supply

Grey water collection

Grey water filter

Mains water supply (back-up)

Underground storage tank with pump

▶ **Figure 2.5:** A grey water system might look something like this

If grey water is not treated sufficiently it may contaminate the system and lead to a greater risk of infection. There are three ways that grey water can be safely reused.

▶ Untreated grey water can be used directly for watering plants.

▶ Biological systems such as sand filters and wetlands are used to remove organic material from the grey water so that it can be safely reused.

▶ Mechanical filters are used to remove organic material from the grey water so that it can be safely reused.

Rainwater harvesting

This is collecting and storing rainwater for flushing toilets and watering plants to reduce water use by up to 50 per cent, both reducing water bills and protecting the environment. An additional advantage of rainwater is that it is free from contamination and pollutants. Rainwater is usually collected from the roof but may also be collected from paved areas as part of a sustainable urban drainage system (SUDS). If the water is not collected, or harvested, it flows directly into the drainage system.

The main disadvantage of rainwater harvesting is that it is dependent on the amount of rainfall. In some areas it is difficult to predict rainfall. Storage tanks for rainwater can also be large and require space which may not be available in urban areas. They need regular maintenance to prevent bacteria building up or contamination from insects or rodents. Some types of roof coverings can allow chemicals into the water.

Water efficiency measures and fittings

There are many simple ways to reduce water use. Toilets in houses account for approximately 30 per cent of water use in the home while in a business this can be significantly greater. Building regulations require any new toilet to have a maximum flush of 6L and a dual flush system. Low-flush urinals in men's toilets also significantly reduce water usage. Other ways for designers to reduce water inefficiency include:

▶ Specifying water-efficient fittings. For example, taps left running can waste a lot of water so install self-closing or infrared movement detecting taps and hoses.

▶ Specifying landscaping containing drought-resistant plants that require less irrigation.

▶ Installing water metering systems so building users can keep track of their water consumption.

▶ Including leak detection and monitoring systems.

▶ Any leaks or broken fixtures and fittings should be replaced. The use of showers (again with push button controls or proximity sensors) instead of baths also reduces the amount of water used.

Processes that reduce water usage and help protect the environment are looked on favourably by planning authorities. The incorporation of such techniques can help gain planning permission in areas where there can be a risk of flooding and where there can be water shortages and hose pipe bans.

Research

Research and watch videos showing ways of reducing water wastage at home. Make a list of the possible options. Look at the water usage you have at home and calculate the amount of water you use in a week. If you were to design a new house now, what water saving processes would you incorporate?

PAUSE POINT   Do you understand why it is important to reduce the water usage in buildings and do you know the various strategies available to achieve this reduction?

    Hint   List the various strategies available to reduce water consumption.

    Extend   Identify the advantages and disadvantages of each of the different strategies.

Waste reduction measures

Waste reduction is a process of minimising the waste produced. There is a clear hierarchy of waste disposal.

1 **Reduce** – reduce the amount of waste produced
2 **Reuse** – the process of reusing the item or material
3 **Recycle** – using old material for a new process or activity
4 **Energy Recovery** – use the energy contained in materials to generate heat or electricity
5 **Disposal** – when no other option is available the material is disposed of responsibly in landfill.

Segregation of waste

More than 100 million tonnes of waste is produced from construction and demolition work each year in the UK. (Source: **www.ukgbc.org**). The cost of disposing of these materials has risen significantly in recent years due to landfill taxes and the increased cost of recycling packaging.

Reducing construction waste is key in meeting the requirements of sustainable development (along with reducing energy consumption and water use). All construction projects in England are strongly advised to have a site waste management plan (SWMP).

On any site, a positive attitude to reducing waste should be adopted and encouraged, for example by providing rewards for good waste management and issuing fines for poor waste management to contractors on site. Everyone on site needs to be aware of the benefits of reducing waste. There are significant financial savings to be made by reducing and segregating waste. It can also lead to a cleaner and tidier site, reducing the risk of accidents from slips, trips and falls. The waste management policy should be part of the induction process and everyone should be regularly reminded of the processes.

Construction waste can be divided into six main categories: timber, concrete, metals, plastics, packaging and other. Placing skips with a specific colour for each material on site will help encourage contractors and suppliers to segregate their waste.

Recycling

Many materials that are used in the construction process can be recycled or reused. Recycling reduces the amount of material going to landfill and the depletion of natural resources.

Common construction materials that can be recycled are shown in Table 2.6.

▶ **Table 2.6:** Recyclable materials often found on construction sites

Material	Recyclability
Concrete	Recycled concrete can be crushed and used as aggregate. Its most common use is for road sub-base.
Asphalt	Recycled asphalt can be separated into its component parts and reused for new roads.
Metals	Metals used in construction which can be recovered and reused include steel, copper, aluminium and lead.
Glass	Glass is 100% recyclable. Recycled glass (cullet) can be used in the production of many different products such as fibreglass, new bottles or it can be crushed down and used as an alternative to fine gravel.
Plastic	Most plastic can be recycled and used to produce products such as garden furniture, bollards and fencing.
Masonry	Bricks and blocks may be recovered and reused to provide a historic look to buildings or they can be crushed down and used for road sub-base.

⏸ PAUSE POINT What types of materials would be left after a building has been demolished?

 Hint List the types of materials that would be left and how they could be recycled.

Extend Discuss the benefits of recycling construction waste material to the client and the natural environment.

Case study

Olympic sustainability

The London Olympic Games in 2012 aimed to be the most sustainable Olympics ever.

Pre-Games and post-Games sustainability reports recorded and reported on the targets and achievements of the Games in terms of its sustainability. For example:

▶ Buildings – the Olympic Stadium was regarded as the most sustainable stadium ever built at the time. Almost two-thirds of the steel used in the roof of the stadium was recycled. This recycled material significantly reduced the embodied energy of the building. The Velodrome was constructed using sustainably sourced timber, had natural ventilation systems and maximised natural light to reduce energy costs.

▶ Recycling – only 3% of construction waste went to landfill which significantly reduced the impact on the environment and provided savings on landfill tax.

▶ Carbon emissions – the Olympic Delivery Authority (ODA) set a target of a 50% reduction in carbon emissions from the initial plans. This target was almost met (47%) by using energy conservation techniques and green energy suppliers.

▶ Transport – the increased use of buses and trains was very successful and major traffic congestion was avoided.

▶ Water – both grey water recycling and rainwater harvesting were used to flush toilets and for irrigation.

▶ Wildlife – brownfield sites were brought back to life with wildflower meadows being planted to encourage flora and fauna, such as birds and bats.

Check your knowledge

1 Why do you think it was important for London to produce the most sustainable Olympic Games ever?

2 Research the reports on the sustainability of the London Olympics and identify what objectives were set.

3 Discuss the areas that were successful and the areas which could have been improved on.

4 Was there a sustainable strategy for the Rio Olympics?

Alternative energy sources

Alternative energy sources do not use fossil fuels, so are environmentally friendly and produce little, if any, pollution.

Ground source

Ground source or geothermal energy is gathered from beneath the Earth's surface. It is generated by pumping a fluid (mixture of water and antifreeze) through a series of pipes beneath the ground to draw heat from the ground. The ground source pump takes the heated fluid and transfers it to the heating system of the building. The system can be used to heat the building in the winter and cool it in the summer.

Ground source is a reliable renewable energy as it can operate at any time of the day and in any weather. Horizontal systems can be laid in a shallow trench over a large area or vertical systems with pipes in boreholes at depths of 15 metres or more depending on the size of the building. If the building has limited space, then the vertical system is the best option, as the area covered by horizontal collectors cannot be used for anything else.

Air source

An air source heat pump (ASHP) absorbs heat from the external air and uses it to heat a building through radiators or underfloor heating systems. The pump is powered by electricity so it does have some impact on the environment, but much less than fossil fuels.

The system consists of an outdoor heat exchanger coil, which extracts heat from ambient air, and an indoor heat exchanger coil, which transfers the heat to the boiler or heating system.

There are limitations to air source energy such as the heat generated by the heat pump is not as high as that produced by a standard boiler and it should be used as part of an energy package for the building which includes suitable insulation and underfloor heating systems.

The two main types of ASHP are:

▶ Air-to-water system – the heat absorbed from the external air is transferred to the central heating system which is water. This works better with underfloor heating rather than radiators as they work more efficiently at lower temperatures.

▶ Air-to-air system – the heat absorbed from the external air is passed to the internal air which is circulated around the house to provide heating. However, this system will not provide any hot water.

Wind source

Wind power refers to the process of using wind to generate mechanical power or electricity. Modern wind turbines harness the power of wind to generate electricity. Wind power is a renewable energy and reduces the reliance on fossil fuels. Improvements in the technology have brought down the size and price of wind turbines and increased their efficiency, so micro wind turbines can be used for off-grid power generation on domestic buildings or small commercial or industrial structures.

Traditional horizontal turbines, with visible spinning blades, may produce up to 4kW.

Vertical turbines rotate vertically around a shaft, which may be enclosed, and some designs do not have blades, making them safer for human contact. They can capture wind from any direction, and tend to be quieter than horizontal turbines because they rotate more slowly and so create less vibration. They may be as small as 200 mm in diameter so can be used like a battery on small buildings or boats.

The disadvantages of wind turbines are that they are weather-dependent and can only be installed in windy areas.

Solar energy

Solar energy is energy produced by the sun, the ultimate renewable energy. Methods of collecting solar energy are cheap to install and benefit both the user financially and the environment by not depleting the finite fossil fuels. There are several methods of harnessing solar energy:

▶ Solar thermal collectors consist of an insulated black material that absorbs the energy from the sun and transfers the absorbed heat to a fluid (water and antifreeze). This fluid is circulated to areas such as hot water tanks.

▶ Solar concentrators use mirrors to focus the sun's rays onto a specific point to generate very high temperatures which can then be used to create steam to turn a turbine and generate electricity. The mirrors can be turned to follow the sun but this increases the installation costs and can only be used in countries with a lot of sunlight.

▶ Photovoltaic (PV) panels can change sunlight directly into electricity. They do not need direct sunlight, only daylight, so can work in cooler climates such as the UK. PV panels are popular with home owners as their cost has decreased and their efficiency has increased. Government grants and subsidies have also encouraged the installation of PV panels.

All these solar sources of energy are renewable, help protect the environment and can save the owner considerable amounts of money by reducing energy costs and earning money through government schemes.

Research

Investigate the sources of energy at your school or college and determine if it uses any alternative energy sources. Discuss the options and establish the advantages and disadvantages of using the different types for your location.

Energy-efficient electrical and mechanical services installations

Energy-efficient installations within a building relate to its heating and lighting. The renewable processes have been mentioned in the previous sections. Energy-efficient lighting design must encompass lighting as part of the whole sustainable building design and must consider both internal and external lighting.

For indoor lighting design, the quality of the light in an internal space is important and should be linked to the purpose of the space.

▸ **Table 2.7:** Ideal light levels for indoor spaces (Source: *Environmental design CIBSE Guide A* – London: Chartered Institution of Building Services Engineers, 2006)

Activity	Light level (lux)
Home – bedroom	100
Home – bathroom	150
Home – living room	50–300
Home – kitchen	150–300
Business – computer rooms	500
Business – conference rooms	300–500
Business – drawing office	750

For a business, up to 40 per cent of its energy costs can be linked to the lighting costs. This means it is good business practice to look at ways to improve the efficiency of lighting systems. This can be achieved by methods including:

▸ Movement sensors – if there is no one in the room or space the sensors can switch off the lights. Any movement triggers the lights to come back on.

▸ Daylight sensors – only switching on the artificial light when the daylight has dropped below an acceptable level.

▸ Good maintenance planning – regular cleaning of windows will reduce the need for artificial lighting and cleaning and maintenance of light fittings will improve their performance.

Outdoor lighting provides security and enhances the appearance of a building. Modern technology such as motion sensors and photovoltaic panels will reduce the amount of energy used for external lights. Using LED or fluorescent lights can also reduce costs. Timers can be used to turn decorative lighting on for short periods of time.

Good design can also maximise the amount of natural light, which can enhance the ambience of the building and reduce lighting costs. Natural light is vital for wellbeing. Poor lighting has been linked to sick building syndrome (SBS).

> **Key term**
>
> **Embodied energy** – the total energy necessary for the extraction, processing, manufacture and delivery of building materials to the construction site. Embodied energy is used as an indicator of the overall environmental impact of building materials.

 PAUSE POINT Why is it important for a building to consider alternative energy sources?

Hint Think about alternative energy sources that may be available.

Extend Identify the advantages and disadvantages of each alternative energy source.

Sustainable and low-embodied materials

The term '**embodied energy**' refers to the total quantity of energy required to produce the material. The embodied energy of a material includes:

▸ the availability of the raw materials, i.e. the energy needed to collect the raw materials

▸ the manufacturing process, i.e. the energy needed to convert the raw materials into building materials such as timber, concrete or steel, and the energy used to transport the materials to their destination

- the energy needed for construction machinery
- the design life of the building materials
- demolition or renovation.

It is a useful measurement of the effectiveness of energy-saving devices, or the overall environmental impact of creating a new building.

Embodied energy of construction materials depends on the manufacturing process, the availability of raw material in the vicinity, the efficiency of production, and the quantity of material used in actual construction.

Low-embodied energy building materials use less energy in terms of manufacturing process, transportation and construction. The lower the embodied energy, the less the impact on the environment. The main ways to reduce the embodied energy of a building are to:

- reduce waste, e.g. manufacture off site
- recycle material, e.g. existing bricks can be reclaimed and reused
- use low-embodied materials, e.g. straw bales for building blocks, sheep's wool for insulation, timber instead of steel or concrete in some low-rise buildings
- minimise transportation costs, e.g. avoid transporting materials large distances.

> **Research**
>
> Select a construction material and investigate the embodied energy involved in its production, manufacture, transportation and construction.

Insulation methods

Insulating a building helps to maintain comfortable conditions to live in and reduces energy costs. It will not overheat in warm weather and heat created inside the building is retained in colder months. Figure 2.6 shows how heat is lost in an uninsulated house.

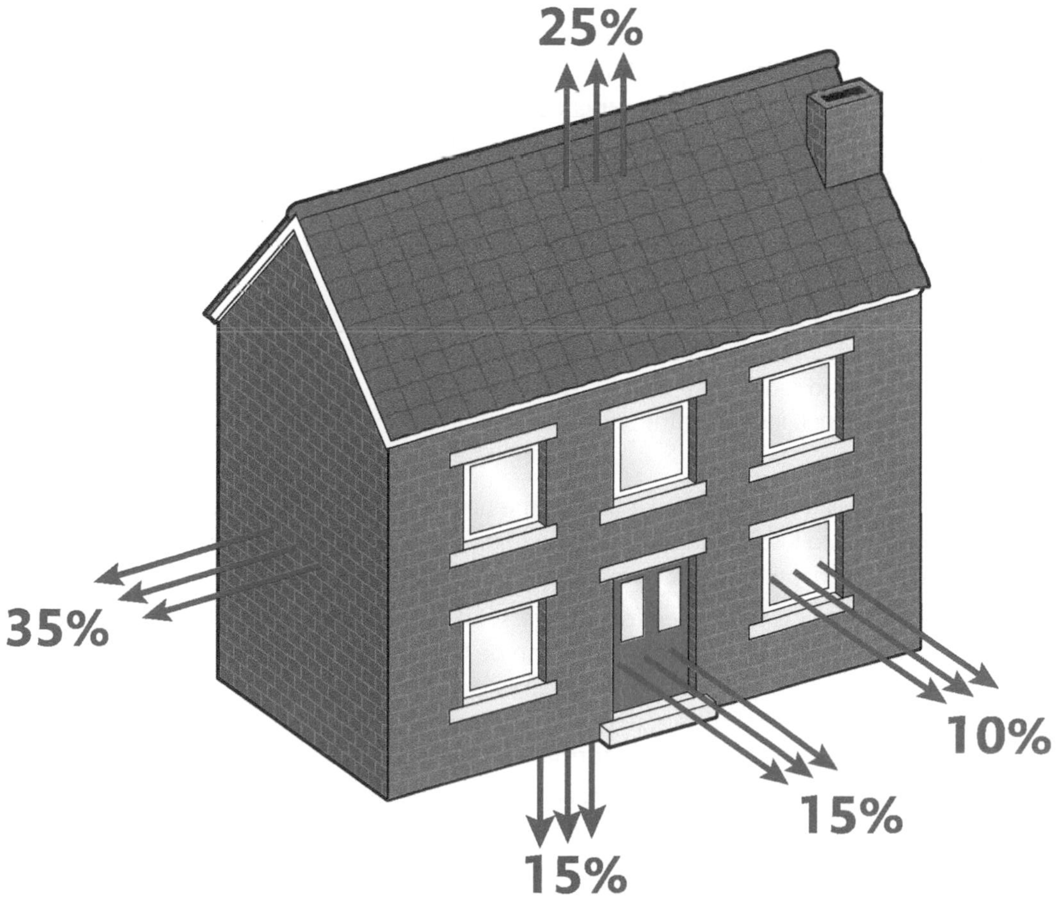

▶ **Figure 2.6:** The main ways in which heat is lost from an uninsulated house

Walls

Figure 2.6 shows that the majority of heat lost from a building is through the walls (almost 35 per cent) so wall insulation should be the first consideration when improving a building's heat retention. Almost a third of the current housing in the UK at present has little or no insulation so there is an urgent need to significantly improve the methods of insulating existing homes.

Houses built before the 1920s are likely to be of solid wall construction, with no cavity. Houses built after this period will have been built using cavity wall construction. The size of the cavity has increased over the years. Cavity sizes adhere to a minimum size, typically 50 mm to 100 mm. The size of the cavity continues to increase to allow for super-insulation but the larger the cavity the less interior floor space is available. New homes can be built using modern materials such as structural insulated panels (SIPs). SIPs can be used for many different elements of a building such as exterior walls, roof and floor constructions. The Building Regulations specify the levels of thermal insulation expressed as U-values. The lower the U-value, the better the insulation.

Types of wall insulation available include:

▶ sprayed foam
▶ SIPs
▶ insulated concrete blocks (aerated concrete blocks)
▶ blanket batts and rolls of mineral wool
▶ polystyrene foam boards
▶ blown-in insulation – polystyrene beads
▶ reflective insulation in the form of foil-faced materials.

Floors

Almost 15 per cent of heat loss from a building is through the floor. The type of floor insulation required will depend on the type of flooring:

▶ Older houses tend to have suspended ground floors. These are above a void or space and are prone to a lot of heat loss. They can be insulated with rolls of mineral wool attached between the joists.
▶ Solid (concrete) floors are found in buildings constructed after the 1930s and are less of a problem than suspended floors but can be insulated by placing a layer of solid insulation on top of the floor.
▶ Modern houses have polystyrene insulation sheets below the concrete floor.

The floors of upstairs rooms are usually suspended and are not insulated if they are above a heated room. If the room is above an unheated space with little insulation such as a garage, then insulating the suspended floor would prevent significant heat loss.

Roofs

In an uninsulated home, 25 per cent of the heat is lost through the roof. There are several ways to insulate the roof of a building depending on the type of roof (flat or pitched), the depth of the rafters and the use of the space.

If the roof space is to be used for storage, then the floor needs to be covered with boards. The space between the joints can be filled with mineral wool rolls but the depth of insulation may not be sufficient to meet the Building Regulations. Additional height may provide sufficient depth for the insulation. If the insulation is compacted, it will lose a lot of its insulation performance properties.

Floorboards can be bought with insulation panels (often polystyrene) pre-bonded to them that are easy to install and meet the U-values required in the Building Regulations for pitched roofs – ceiling level, pitched roofs – rafter level and flat roofs. To achieve U-values of 0.13W/m², a minimum thickness of 300 mm of mineral wool or 180 mm of rigid foam is required.

If the space is to be used for storing decorations and suitcases then it is more economic to insulate the roof at the floor level, while if the space is to be used as another room then it is best to insulate at the ceiling.

Flat roofs that have been added as part of an extension can often suffer from heat loss and leaks. The damp can cause the materials to become damaged and fail to perform. If possible, it is better to insulate a flat roof externally rather than internally. Insulating internally can lead to condensation problems if it is not installed correctly.

⏸ **PAUSE POINT**   Why is it important to insulate a building?

> Hint   Identify the different areas of a building which can be insulated.
>
> Extend   For each of the areas investigate the different methods of providing insulation.

Sustainable urban drainage systems

In the natural environment, rainwater will either drain naturally through the **permeable** ground as groundwater or run off to nearby rivers and streams as surface water. This process of permeating through the ground slows the flow of water and helps to filter it and remove pollutants.

Key term

Permeable – a material that allows water to flow through it.

Construction replaces the natural ground with hard impermeable surfaces, such as roofs, driveways, car parks, roads and footpaths. The natural route for the rainwater to the ground is blocked and much more surface water is created. This means less water is being absorbed into the ground, which can result in a shortage of water available to communities living in the area while also more water to manage through the drainage systems. This increased surface water can overwhelm the drainage systems and overflow the rivers and streams, leading to flooding.

Surface water and groundwater are valuable resources and must be managed correctly, for example by Sustainable Urban Drainage Systems (SUDS). This is a method of managing surface water by simulating natural flow patterns such as permeation. These systems allow the surface water to flow through them and help to control the speed and strength of the water in order to prevent flash flooding and enable pollutants to be removed. The systems' aims are:

▶ attenuation – decreasing the volume of water entering the drainage system and hence delaying the flow and reducing the risk of flooding
▶ retention – storing or holding some of the water for reuse such as irrigation
▶ permeation – allowing water to permeate through the ground
▶ treatment – improving the quality of water by removing pollutants.

Table 2.8 shows how SUDS can benefit the local area.

▶ **Table 2.8:** Features and benefits of SUDS

SUDS structure	Features and benefits
Green roofs and rainwater harvesting	Act as a means of attenuation and storage by helping to slow down the flow of water as well as storing it for later use.
Permeable surfaces	Water soaks through the surface rather than running off it, which controls the flow of water and reduces the risk of flooding.
Swales	Shallow channels to collect and move water away.
Channels	Move water and remove pollutants to improve its quality.
Filter strips	Water is passed over a strip of grass which removes pollutants and improves its quality.
Soakaways	A large void or hole filled with gravel that lets the water soak through a layer of material before reaching the groundwater or another SUDS structure.
Retention and attenuation structures	Retention structures are like ponds that store water while detention structures are for attenuation or slowing down the flow of water. Both reduce the risk of flooding.

Sustainable landscape design

Sustainable landscape design has evolved as a response to the increased pressures modern construction has placed on the environment. These include depleting natural resources, air pollution, water pollution, energy usage and the increase in CO_2 emissions. These changes have damaged natural habitats and affected wildlife and flora and fauna as well as impacting on climate change.

Sustainable landscape design aims to counteract some of these processes and address the damage to the environment. The sustainable landscape is designed to be kind to the environment but also enjoyable for occupants and users. The main aims of a sustainable landscape design are to:

▶ reduce the impact and quantity of surface water such as using green roofs
▶ reduce demand on water by using grey water and rainwater harvesting, and plants that require less irrigation to thrive
▶ use wetlands to biofilter water
▶ create open green spaces which can be enjoyed by the public and wildlife
▶ use recycled products, such as timber, glass and plastic, to build outer decking and furniture
▶ use renewable energy products such as solar-powered landscape lighting
▶ reuse kitchen and garden waste as compost for fertilising the garden
▶ use mulch, a layer of material on the surface of the ground, which can reduce water loss due to evaporation, reduce weeds and reduce soil erosion.

Building Research Establishment Environmental Assessment Method (BREEAM)

Building Research Establishment Environmental Assessment Method (BREEAM) provides sustainable ratings for non-domestic buildings. By applying BREEAM to construction projects, clients can quantify and reduce the environmental impact of the new building. BREEAM assessments produce a rating which evaluates the performance of the building in several areas, as shown in Table 2.9.

▶ **Table 2.9:** BREEAM sections

Environmental section	Explanation	Weighting %
Management	The inclusion and management of sustainable processes in the design, construction and maintenance of the building.	12
Health and wellbeing	Processes which encourage comfort, health and wellbeing within the building such as lighting, ventilation, thermal comfort, landscaping and noise levels.	15
Energy	Specification and design of energy-efficient processes for the building which reduce energy use and CO_2 emissions, for example using high performance glass.	19
Transport	Providing access to sustainable transport for the users of the building (such as being cycle friendly, providing electric car charging points, priority parking for car sharing) and links to public transport such as buses and trains. Reducing the number of car journeys means reducing congestion and CO_2 emissions.	8
Water	Reducing the water usage of the building, for example by recycling of grey water or rain harvesting.	6
Materials	The materials used in the construction and maintenance of the building, looking to use low-embodied energy materials where possible, for example using glulam beams rather than steel or concrete beams in a leisure centre.	12.5
Waste	Processes put in place to manage and reduce the waste during the construction and operation of the building, for example reusing and recycling materials rather than disposing to landfill.	7.5
Land use and ecology	Sustainable land use to protect the landscape and the natural environment within it, for example using brownfield sites and encouraging urban regeneration.	10

▶ **Table 2.9:** *Continued ...*

Environmental section	Explanation	Weighting %
Pollution	Management processes for controlling all forms of pollution such as noise, light, CO_2 emissions, flooding and ground and water contamination.	10
Total		100
Innovation (additional)	The use of innovative ideas and solutions within any of the categories can be awarded additional credits.	10

The BREEAM rating levels allow the sustainability performance of buildings to be measured against other buildings. Typically, less than 1 per cent of UK non-domestic buildings achieve a rating of Outstanding and only 10 per cent achieve a rating of Excellent.

▶ **Table 2.10:** BREEAM ratings

BREEAM Rating	% Score
Outstanding	≥ 85
Excellent	≥ 70
Very Good	≥ 55
Good	≥ 45
Pass	≥ 30
Unclassified	< 30

Case study

BREEAM ratings

▶ One Angel Square, Manchester, has an Outstanding BREEAM rating

An example of an Outstanding BREEAM rated building is One Angel Square, Co-operative Group HQ in Manchester. The open form of this building creates natural heating, cooling and lighting. Other environmentally friendly processes adapted include:

▶ heat recovered from the IT systems being used to assist with heating the building
▶ low energy lighting and IT systems being deployed in the building
▶ grey water and rainwater harvesting systems being used for flushing toilets.

Check your knowledge

1 What are the benefits of achieving an Outstanding BREEAM rating for a building project?

2 Research and identify a range of design features which were incorporated in this building to enable it to achieve an Outstanding BREEAM rating.

Research

Search on **www.greenbooklive.com** to locate buildings in your area which have recently received a certified BREEAM assessment.

Investigate what design features are likely to have achieved credits towards the rating.

PAUSE POINT   What topics are assessed as part of BREEAM?

Hint   List the topics which are assessed under BREEAM.

Extend   Give new examples of how credits can be achieved within each section.

Assessment practice 2.2

The client from Assessment practice 2.1 has now provided you with a more detailed brief for its 250-bedroom budget hotel. It will be built on an inner-city brownfield site and must be safe and secure with sufficient communal space for socialising as well as bedrooms. It should be spacious and bright but affordable to the client.

Site details

- The site is 70 m long and 30 wide and has been levelled.
- Ground conditions are 1 m of fill and then stiff boulder clay.
- Five-storey office blocks are on both sides of the site.
- Main roads are at the front and rear of site.

Spatial requirements

- 200 double rooms each with en-suite bathroom (shower, bath, toilet and wash basin).
- 50 family rooms each with en-suite bathroom.
- One floor to contain conference and meeting rooms.
- Ground floor reception area, bar/restaurant, open-plan relaxation area, laundry room, store area.
- Underground parking.

Activities

1   Communicate to the client the design factors and constraints of the site and the impact they will have on the design.

2   Produce an initial project brief that covers:
- spatial requirements
- desired project outcomes and budget requirements
- site information.

3   Recommend and justify an outline solution to the client that meets their vision in terms of the size of the building and its form and type of construction.

4   Produce annotated sketches of the initial idea:
- 3D external view of the building
- plan view of a bedroom
- 3D internal view of bedroom.

5   Using CAD software, develop a 3D virtual model of the design of the external envelope and the interior of an en-suite bedroom.

Plan
- Do I have enough information to complete the activities?
- Do I have the right skills to produce annotated sketches and CAD models?
- If not, what do I need to learn before I attempt the activities?

Do
- I know how to use the information given to communicate design factors and constraints, produce a project brief, recommend a solution, produce annotated sketches and develop a 3D virtual model.
- I can identify where there are gaps in the information that the client may require.

Review
- I can explain what the activities were and how I approached development of them.
- I can explain how I would approach the more difficult elements differently next time.

Further reading and resources

Anderson, J., Shiers, D. and Steele, K. (2009) *The Green Guide to Specification: BREEAM Specification*, 4th Edition, Bracknell: BRE Press.

Chudley, R. and Fleming, E. (2005) *Construction Technology*, Oxford: Blackwell.

Chudley, R. and Greeno, R. (2016) *Building Construction Handbook*, 11th edition, London: Routledge.

CIBSE (2006) *Environmental Design Guide A*, London: Chartered Institution of Building Services Engineers.

Cooke, R. (2007) *Building in the 21st Century*, Oxford: Blackwell.

Greeno, R. (2005) *Construction Technology*, London: Prentice Hall.

Illingworth, J.R. (2000) *Construction Methods and Planning*, London: Spon Press.

Riley, M. and Cotgrave, A. (2008) *Construction Technology 1*, London: Macmillan.

Websites

Royal Institute of British Architects Plan of Work 2013: **www.ribaplanofwork.com**

National Planning Policy framework 2012: **www.gov.uk/government/ publications/national-planning-policy-framework--2**

Information on the planning process and Building Control: **www.planningportal.co.uk**

British Research Establishment Environmental Assessment Method: **www.breeam.com**

Getting ready for assessment

This section has been written to help you do your best when you take the assessment test. Read through it carefully and ask your tutor if there is anything you are still not sure about.

Sample answers

For your set task you will be provided with some background information on a client's requirement for a building project. You will be required to research the type of project, identify design factors and constraints, and produce an initial brief recommending the building form and layout and type of construction. You will also be required to produce a range of annotated 2D and 3D drawings to present your ideas to the client.

Look at the sample scenarios which follow and our tips on how to answer them well.

Example 1

A college wants to knock down its existing building, which has become too expensive to maintain, and construct a new building to meet the needs of the modern courses. The college needs office space, library, gym and sports hall, training kitchen, beauty salon (hairdressing, make up, nails), joinery and brickwork workshops, teaching rooms with computers, canteen area, toilets, changing facilities, a student recreational area and sufficient parking for staff and students. There is a budget of £15 million.

1 Based on the information provided, identify the design factors and constraints of the project.

2 Produce an initial project brief including:

- spatial requirements
- desired project outcomes
- site information
- budget requirements.

Answer

Many constraints will impact on the design of this project. These include:
- the size of the existing site
- ground conditions
- impact on neighbouring buildings
- planning permission
- environmental impact of the building
- budget of £15 million

Only a few of the design factors and constraints on the project have been identified. The impact and influence of these factors has not been communicated in relation to the design or specification. The use of technical vocabulary is limited.

To develop the answer, you would need to discuss a broad range of relevant factors and constraints. The impact these factors have on the design and specification should be explained. Technical vocabulary should be used accurately throughout the answer.

Answer: Initial project brief

The local college will need a modern fit-for-purpose building that meets the needs of today's learners.

Sports students will need a multipurpose hall that can be used for a range of sports and a fitness suite with up-to-date training equipment. They will also need a separate room for monitoring and assessing clients, with suitable privacy, and showers, toilets and changing facilities for male and female students as well as separate facilities for staff and a store room for equipment. The students will also need access to teaching rooms with computer and internet access.

Beauty students will need a training salon, with space for tables, chairs and ten beds for massage, waxing, and other treatments. They will need changing facilities, toilets, and washing and drying facilities for the towels and sheets. They will need a hair salon, with ten working areas. The students will also need access to teaching rooms with computer and internet access.

Brickwork students require a large open-plan space on the ground floor with sufficient space for an office, large mortar mixing apparatus and a room big enough for 15 students to build. They will require washing troughs for the tools. Ventilation system to remove dust from the workshop is vital. Storage areas are also required for all the materials, bricks, blocks and cement. The students will need access to teaching rooms with computer and internet access.

Joinery is a popular subject and will require two workshops with room for eight work tables (allowing two students at each table). A machine room will be required to house the equipment needed to cut and plane the timber for the classes. Storage will also be required for the timber. The students will need access to teaching rooms with computer and internet access.

Catering students will require two training kitchens with work space for 16 students in each for food preparation. They will also require a restaurant area to present the cooked food to the staff, students and public. Storage, toilets and changing facilitates will be required. The students will also need access to teaching rooms with computer and internet access.

General requirements are parking, reception area, canteen, recreational student area, library, teaching rooms, and office space for teaching and administrative staff.

Spatial requirements

Sports

Sports hall – minimum standard of 34.5 m x 20 m

Fitness suite – 100–200 m^2, ceiling – height 3.5 m

Changing rooms (x2) – min area 25 m^2

Reception area – 20 m^2

Assessment room – 20 m^2

Store room – 20 m^2

The initial project brief contains accurate information but with limited detail. There is some relevant information in the sections but these could be developed further. Think about spatial requirements, desired project outcomes, site information and budget requirements.

This could include a range of possible structural options, consideration of different sustainable construction materials, different types of finishes and environmentally friendly technologies. Is there any flexibility with the budget? How will the money be spent?

Beauty
Salon training room – 100 m² (this should contain 3–4 separate treatment rooms)
Hair salon – 80 m²
Store room – 20 m²
Washing Drying room – 10 m²
Changing facilities – 25 m²

Brickwork
Workshop area – 340 m²
External store area – 20 m²

Joinery
Workshop area 1 – 250 m²
Workshop area 2 – 250 m²
Machine room – 150 m²
Store area – 30 m²

Catering
Training kitchen 1 – 200 m²
Training kitchen 2 – 200 m²
Restaurant – 200 m²
Store – 20 m²
Changing facilities – 20 m²

Other rooms
Library – 300 m² (with computer facilities)
Large open-plan reception – 300 m²
Canteen – 300 m²
Recreational area – 150 m²
40 teaching rooms – 150 m²
Five open-plan offices for administrative and teaching staff – 250 m²
Male and female toilets and disabled toilets
Parking for 300 cars
Building services – Plumbing, heating, lighting, IT facilities, waste collection

Desired outcomes
Modern up-to-date teaching and training facilities which will attract and meet the needs of students in the area. The facilities should enable the students to meet the requirements of the qualifications and prepare them for industry.

Site information
By building a multi-storey structure, more space can be created for the rooms and for suitable parking spaces.

Budget requirements
The budget allowed by the local authority is £15 million.

Example 2

A local council has agreed to allow a derelict former factory in the centre of the city to be demolished and replaced with modern office space but the façade of the factory must be retained. As part of the agreement, the council has specified that the multinational company that is carrying out the work ensures the building has a BREEAM rating of at least Excellent. (The company has a strong record of developing sustainable buildings.) The building needs office spaces, canteen, reception area, a board room, several conference rooms, computer suites, a gym recreational area and sufficient parking for staff. To maximise the clients' profits, they want to have it open within a year and have set a budget of £20 million.

1   Based on the information provided, identify the design factors and constraints of the project.

2   Produce an initial project brief including:
* spatial requirements
* desired project outcomes
* site information
* budget requirements.

Answer: Initial project brief

The client is a multinational company looking for high-quality offices so that they can provide their customers with a fast, technically superior and quality service. They require the office space to be flexible and easy to adapt to meet different needs. They have a sustainability policy and require the construction of the project to be BREEAM –Excellent. They have a standard branding and form to their buildings and want this replicated for the city-centre office.

Time constraints
The client needs to have the building constructed and operational in one year.

Environmental constraints
As the site is a brownfield site, planning will be required for the demolition of the existing building and reuse or recycling of the waste material. There is also the possibility that bats are roosting in parts of the building.

Site investigations will need to be carried out to determine the soil conditions and identify any contamination.

Maximising the energy efficiency of the building and minimising the construction waste are important design factors to the company.

Social constraints
The site is in an area of the city that has become run down and neglected. Work should be carried out to engage the local community and look for other opportunities to develop the area.

Planning constraints

Planning permission will be required for the project and thus plans must be submitted to the local planning authority.

Due to its location, the height of the building is limited to between five and eight storeys.

Planning for the retention of the façade of the building is necessary to protect the historic value of the building.

Financial constraints

The purchase of the old factory site is part of a £20 million investment in the city-centre project by the company. The government is also providing financial support for the urban regeneration and EU funding has also been secured for the project.

A range of design factors and constraints on the project have been identified. The impact and influence of these factors has been communicated in relation to the design or specification. There is evidence of some use of technical vocabulary. To develop the answer, you would need a broader range of relevant factors and constraints. The impact these factors have on the design and specification should be explained in more detail. Technical vocabulary should be used accurately throughout the answer.

Tendering and Estimating

3

Getting to know your unit

Measuring, estimating and tendering are an essential part of a construction company's business activities and begin with the estimator. The estimator is the first key person to become involved in pricing the tender for the client and contractor, using measurements to produce quantities and estimate prices, and following the company's procedure for submitting a tender.

Measurement is the physical act of using a tape measure or scale rule to produce a value that can be used to create a meaningful and accurate quantity. Measurement is primarily used in the quantity surveyor's role in the construction process. This unit looks at the process of estimating associated with measurement, producing an estimated project cost, cost modelling and tendering procedure by considering external factors and risks when developing a tender.

How you will be assessed

This unit is assessed under supervised conditions. You will be given a set task that assesses your ability to produce designs that meet the requirements of a client.

Throughout this unit you will find activities that will help you prepare for your assessment. Completing these activities will not mean that you have achieved a particular grade, but you will have carried out useful research or preparation that will help you when you do your external assessment. As the guidelines for assessment can change, you should refer to the official assessment guidance on the Pearson Qualifications website for the latest definitive guidance. This unit has three assessment outcomes (AOs) that will be included in your external assessment.

The assessment outcomes for the unit are:

▶ **AO1** Apply knowledge and understanding of the tendering and estimating process and techniques to determine estimated costs.

▶ **AO2** Analyse information to determine tendering and estimating outcomes in order to make evaluative judgements and commercial decisions in context.

▶ **AO3** Be able to apply the tendering and estimating process, techniques and outcomes in order to produce a justified tender submission relative to the scenario.

This unit contains a number of key command terms you will need to be familiar with for your assessment.

Borehole report	A report that provides information on the soil types and depths within the various strata underneath the surface of the site.
Build up	An estimating process used to produce the nett cost per unit, including all the components of cost for that item.
Company profile	Information about the company that will be useful when making tendering and estimating decisions.
Complete	Input items necessary to fill in all sections of a document.
Measured work section	The section of the bill of quantities containing measured items, with quantities, into which unit rates are inserted.
Produce	Create required information as a result of own calculations or selecting and inputting from given information.
Tender sum	The final price that the company submits as a bid for the project.
Unit rates	The cost of a measured item per unit of measure, including materials, wastage allowances, labour and sundry plant requirements.

Getting started

The information contained in a tender document is essential for planning building projects. List the different factors and elements that might need to be included in tender documentation. For each of these factors, try to identify how it could impact on a company's tendering strategies and commercial risk.

 # Commercial risk

A1 Action on receipt of tender documentation

All construction companies work on a tendering and estimating basis. **Tendering** is a process where clients invite companies to bid for work that they have planned. All bids must be submitted by a deadline. Clients need to put together a full **specification** and details of the work needed and send it to companies for bidding.

The main purpose for the main contractor and, ultimately, the subcontractors and suppliers, is to win the contracted work on a competitive basis and to ensure that the profit margin within the tender is maintained or exceeded (as this will mean the contractor makes a profit).

Based on the information in the tender documentation, a contractor will make a detailed price offer to the employer to execute and complete the works and to make any repairs required by the contract. In other words, the tender put forward by the contractor is the sum of money, time and other conditions required to complete the specific construction work.

Submitting a tender is competing for work but is also a **commercial risk** because it may be unsuccessful.

Any company considering submitting a tender will need to review their **organisational capacity** for taking on the work connected with the tender. To do this the company will need to review their workload, the size of the business, their resources and the impact of any other work they may have scheduled.

Documents contained within a tender

The method of tendering undertaken in order to select a contractor for a project will depend on some of or all of the following:
▶ size and geographical location of the project
▶ financial stability of the construction company tendering for the work
▶ competency of the contractors with regard to health and safety
▶ available physical resources of labour, plant and facilities
▶ reputation and references of a company.
All tenders will be made up of a series of documents and detailed measurements, taken from the contract drawings and specifications. These are used to identify the work and materials required, and to calculate the quantities required for each item.

▶ Tender documentation indicates the different factors which need to be carefully measured and considered by any company responding to the tender. All contractors, subcontractors and suppliers must extract the key information to help them understand the work required.

▶ Any measurements should contain all the necessary information for an estimator to calculate a price. Based on this initial information, the tender documentation is prepared and both the client and contractor will calculate their budget based on this. This can directly affect the whole tendering strategy and the entire process of the project production.

> ### Key terms
>
> **Tendering process** – a client invites companies to submit a formal, fully costed written proposal for carrying out particular work.
>
> **Specification** – a description of the materials and workmanship required for a specific construction project.
>
> **Commercial risk** – when a company invests money in doing something that does not have a guaranteed profit return.
>
> **Organisational capacity** – the ability of the business to take on a certain amount of work, for example a small independent builder would not have the organisational capacity to build a hospital development.

Bills of quantities or specifications

Bills of quantities are included within the tender documents to cover items that have not been measured. The bill of quantity is prepared by the quantity surveyor, who is a cost consultant. It provides the quantities of the items in the project (for example, in terms of numbers, area, width, length, volume, weight or time), each of which is identified in the drawing and specification found in the tender documentation.

The whole project is measured and unit quantities are placed within the bill of quantities. The exceptions are:

- contingencies – money allocated for unforeseen items
- dayworks – money for time-related charges levied by the contractor for variations that cannot be measured
- provisional sums – money included for works not fully designed or specified.

A bill of quantities will have section totals which are carried to collections that summarise each section. These are totalled to form a final summary which adds up to the tender price.

The specification and design need to be completed in order to prepare a bill of quantities. This helps tenderers to estimate the cost of construction fairly and accurately.

Tender drawings

The set of drawings received with the tender should provide sufficient information for the estimator to price the unit rates within a bill of quantities. If the tender only contains drawings and a specification it should provide information for **taking off** of quantities. It is advisable to stamp the drawings with the wording 'tender drawing' and a date. This provides a reference set of the drawings that accompanied the tender on which the contractor's price was based. These can be checked against any revised drawings during the final account stage of the project in order to highlight changes and any contract variations.

Schedules

Schedules are often prepared for certain material elements and may be issued with the tender documentation. A typical schedule would include estimated quantities of:

- windows and doors
- ironmongery and joinery
- internal finishes
- concrete reinforcements.

Schedules are an easy way of counting similar specified items. A single document is easier to look through than several drawings, as the window schedule in Figure 3.1 illustrates.

Description	W1	W2	W3	W4	W5	W6
1200 x 1200 uPVC window in white with trickle vents to top, including uPVC window board and cill, Pilkington K glass double-glazed units with 20 mm air gap.	✓					✓
2400 x 1200 uPVC window in white with trickle vents to top, with two side opening lights 600 mm wide and 1200 mm high, including uPVC window board and cill, Pilkington K glass double-glazed units with 20 mm air gap.		✓		✓		
900 x 1200 uPVC window in white with trickle vents to top, including uPVC window board and cill, Pilkington K glass double-glazed units with 20 mm air gap.			✓		✓	

▶ **Figure 3.1:** Window schedule

Tender information form and submission documentation

The tender information form provides all the data that has been extracted from the tender documents. It is where all the information that is used during the tender adjudication process is placed.

An invitation to tender might include: a letter of invitation to tender, the form of tender, the form of contract, specification, design drawings, a drawing schedule and tender pricing documentation.

The tender form is prepared by the client or their consultants and the tenderer needs to complete and sign it for the client. It is a formal acknowledgement that demonstrates the tenderer fully understands and accepts the terms and conditions of all the tender documentations and any other requirements.

The information contained includes: the deadline for returning documents, a tender reference number, the price for which they are supposed to do the works and any adjustments that have been agreed within the scope of work, the period of time change, a completion date, the date for which the price remains valid, any qualifications that apply, confirmation of who will accept the cost of preparing the tender and a confirmation of what laws govern the contract. This information can be filled out by the client or their consultants, and the tenderer just needs to sign it.

	WMCC		£	p
Item	**SUBSTRUCTURE**			
A	Excavate to remove hard standings or turf, and to reduce levels average 250 mm deep, and cart away (approx. 23 m^3)	Item		
B	Excavate to reduced level for foundation trench not exceeding 1.00 m deep, 600 mm wide, and cart away surplus spoil, level and ram bottom to receive concrete, part backfill trench with hardcore internally in 150 layers, and with spoil externally to level of existing ground (12 m^3)	Item		
C	Provide and lay ground floor complete on prepared ground, comprising 150 mm thick consolidated hardcore, well rolled, 25 mm sand blinding, 1200 gauge visqueen damp-proof membrane turned up wall at edges and lapped into dpc, 100 mm thick. A 252 mesh reinforced concrete slab, tamped finish Gen 1 concrete – 20 aggregate, 100 thick Kingspan insulation laid butt jointed, floor overlaid with laminate flooring (approx. 45 m^3)	Item		
D	Provide and lay concrete Gen 1 foundations poured against face of excavation 250 mm deep (approx 8 m^3)	Item		
E	Provide and lay 300 mm wide dense 7 n/mm concrete foundation blocks, in cement mortar (1:3) (approx. 25 m^3)	Item		
F	Provide and lay three course class B engineering bricks 102 mm wide, in sand lime cement mortar (1:1:6) (approx. 4 m^3)	Item		
G	Provide and fix hy-load dpc 100 mm wide, including all necessary laps at joints, steps, etc. (approx. 15 m)	Item		
	To Collection	**£**		
4.3	**Section – Schedule of works**			

▶ **Figure 3.2:** A specification for a tender

PAUSE POINT   Do you understand the different documents that are included in tender documentation?

        Hint   Each document has a clear purpose – what are the important things you would need to know if you are responding to a tender?

        Extend   What could be the impact if a bill of quantities was inaccurate?

Quality and accuracy of information

It is vital that all information should be as accurate as possible. All bills of quantities need to be based on industry standards to avoid any misunderstanding or ambiguities for tendering companies when preparing their quotes. This will help to avoid any disputes caused by different interpretations on the specification. The most common source of standards for general construction work is the New Rules of Measurement (NRM2), a new standard that became operative in January 2013.

It is also vital that all bills of approximate quantities and the **schedule of rates** produced by tendering companies are accurate. This will ensure that the proposal is correct and that the client can make an informed decision based on accurate information. It also ensures the tendering company bases their decision to apply for the contract on correct information and on an estimate they can actually achieve.

Tender period

A client will invite contractors to tender for the work during a fixed timeframe. Companies may be informed of this process through adverts, or may be directly invited to tender for the work. Normally the initial enquiry from the client is a telephone call or email asking if the contractor would like to tender for the work. It is good practice for the contractor to decide quickly so the client has an option to add another contractor to the tender list.

The contractor will arrange quotes for the main construction, supply of equipment, materials, demolition, labour, etc. in response to the client's specifications.

The tenderers will need to submit their prices within a timeframe defined in the specification. The tender period depends on the nature and complexity of the tender and the amount of work needed for a company to prepare for it, which will vary considerably from project to project.

At the end of this period, the client will shortlist the contractors who they think are most appropriate for the project. They will then select the contractor, based on the tender and possibly after interviews with potential contractors.

Completion of internal tender information form and review of tenders

Any company putting together tenders will need a team to collect the information needed to meet the requirements of the client.

The estimating team will use an enquiry form (often called an internal tender information form) to find out details of the tender proposed by the client. They can use this to decide whether to proceed with the tender or not.

An enquiry form records the information sent to a supplier and subcontractor which can be used as evidence in the event of a dispute over prices (see Figure 3.4). The enquiry form will be passed to the person who copies the drawings and pages from the specification which are sent to the subcontractor or supplier.

Different perspectives and expertise are needed within contractor companies to meet the requirements for tender documents. Each member of the team will need to review the tender documents to use their own experience and skills to check it is suitable for the company and meets the needs of the company.

▶ The bid manager will manage the entire process and has the authority to make key decisions.

▶ The estimator will work on outlining the estimated cost of the project, and may also work on health and safety plans and construction programmes as part of this tender. They will interact with specialist subcontractors to make sure they have all the information they need.

▶ The planning engineer will prepare and update the project schedule based on the contract to ensure the plans meet the client's specification.

▶ The quantity surveyor is responsible for managing all the costs of the project from the initial estimates to the final acquisition.

▶ The buyer will be responsible for purchasing (and ensuring the delivery of) materials and plant resources.

▶ The contracts manager will manage the contracts for subcontractors and suppliers and will help to resource the project.

Project Horizon Laboratories		TENDER TIMETABLE	Simtop Construction

Description	Latest date	Apr 11 M	12 T	13 W	14 T	15 F	18 M	19 T	20 W	21 T	22 F	25 M	26 T	27 W	28 T	29 F	May 2 M	3 T	4 W	5 T	6 F	9 M	10 T	11 W	12 T	13 F	16 M	17 T	18 W	19 T	20 F
Project appraisal																															
Check documents		■																													
Tender information sheet		■																													
Tender timetable		■																													
Document production (d&b & drg & spec)																															
(Drawing)																															
(Specification)																															
(Bills of quantities)																															
Code bill items and enter computer bill			■	■																											
Enquiries																															
Abstract, prepare, dispatch – subs	18 Apr		■	■																											
Abstract, prepare, dispatch – mate	20 Apr						■	■	■																						
Date for receipt of quotations – mate	4 May																		■												
Date for receipt of quotations – subs	9 May																					■									
Project appraisal																															
Site visit	19 Apr							■																							
Tender method statements														■																	
Tender programme	12 May																								■						
Pricing																															
Labour and plant												■	■	■	■	■	■														
Materials																					■	■	■								
Subcontractors																								■							
PC and provisional sums																									■						
Project overheads																									■						
Reports																															
Checking procedures and summaries																										■					
Tender																															
Review meeting(s)	17 May																											■			
Submission documents																															
Submission	19 May																													■	

Figure 3.3: An activity schedule for a tender

Tender supplier	Drawing nos	Specification pages	BoQ pages
Aggregates Ltd	2007/01/a	P23 item 1	
Tender subcontractor	**Drawing nos**	**Specification pages**	**BoQ pages**
Topliss Electrical	2007/21/a		P47 items 1–8
	2007/22/b		
	2007/22/c		

Figure 3.4: A subcontractor enquiry form

▶ What is the role of an estimator?

A2 Tendering considerations and strategies

The main aim of tendering is for the client to get the best value for their work, with a good balance between quality and cost. Remember, the cheapest solution may not meet requirements. Both the client and contractor should always consider the total costing, not just the initial outlay. The client also needs to make sure the tendering company does not get squeezed to the limit financially.

The tendering company has to consider all the decisions and dealings and must be clear, transparent and auditable. For an effective and appropriate tendering strategy some approaches will obtain the best results. These include appraising and understanding the whole life and cost of the project, legal issues, terms and conditions of the contract, environmental factors, ethical issues and the final quality of the project.

There are two schools of thought for preparing a strategic tender:

1 Draw up a detailed list of all the requirements, rate them by degree of importance, then re-evaluate the tender based on the importance of each item before a final decision is made.

2 Focus on the fundamental goals of the tender (the 'big picture') and only discuss the detailed requirements in broad terms.

Accepting or declining an invitation to tender

When invited to tender, all companies need to decide whether to proceed with the tendering process. This decision will require the company to review its capabilities and resources and decide if it can commit to delivering the contract, if it is awarded to them.

They must consider all long-term costs and consequences for tendering. It is important to minimise risks by having all members of the team work together to ensure that everyone understands all the technical issues. It is worth spending resources and time to avoid bigger risks from bad decisions.

To make a decision on accepting or declining the invitation, companies need to fully review all the information available in the tender documentation.

If a company declines the invitation to tender, it should send an apology and a full and detailed written explanation to the client to ensure that the tenderer does not lose future opportunities.

If the tender invitation is accepted, the tenderer must ensure that all documentation has been received and that it is repeatedly checked and confirmed correct. Provide copies of all supporting documents.

Serial (regular) tendering

Serial tendering is carried out in response to an offer of a package of works. These works would be of a similar nature, such as a series of schools, community projects or community police stations.

There are several benefits to this type of tendering method. The successful contractor will be able to use the experience gained from the first job on the next one. The client will also be assured of their long-term commitment. However, the tendering company must be certain it can meet this commitment.

Priority tender (fully committed)

Some companies may invite contractors to tender after they have completed a pre-qualification questionnaire (PQQ). A company may advertise this PQQ to encourage contractors to apply. Its main purpose is to enable the client to produce a shortlist of appropriate contractors. This can make the tender process more efficient and productive, by identifying, in advance, the companies best placed to deliver on the tender.

Case study

A new school is going to be built on the site of a derelict factory. The school is a prestige construction project and, when complete, will serve a large area of a major town. There is considerable media attention on the works and pressure from both local and national government for a timely and within-budget outcome for what is seen as a flagship construction project and a key component of the government's new education initiative.

The factory buildings, and the equipment they house, are mostly intact, although they have deteriorated badly and are considered unstable and highly dangerous. The complex was built before the Second World War and was last used by a paint, solvent and wood preservative manufacturer. The manufacturing processes involved toxic chemicals, from an era when hazardous material handling regulations were less stringent than today. A site survey was carried out and asbestos discovered in many of the buildings.

A key condition of planning permission being granted is the modification of the current road system to enable it to cope with the influx of traffic and pedestrians arriving at, and leaving, the school each day, as well as a regular flow of people using its state-of-the-art sports and gym facilities in the evening.

Check your knowledge

Working in teams of four, hold a planning meeting to discuss the issues around tendering for this work. Look at the problems and produce a proposal on how you would deal with each one and what you would need to include in the final estimate for the work.

Tender preparation programme

Any decisions made during the tender will have a huge impact on the final direction of the project. So it is important to ensure all the project requirements are fully understood. Most organisations have a standard approach to tendering: estimating, reviewing estimates and then submitting the tender.

Activity schedules are programmes for visually controlling the tendering and estimating process. They enable the co-ordination of tenders using a master programme, so that the estimator can control the workload. The activity schedule can assist the monitoring process by identifying any items likely to miss milestones. Where more than one tender is being priced at a time, a master activity schedule may have to be produced to control the whole process within the estimating department.

Timeline of tender preparation

An established timetable highlights important dates and deadlines during the tender process. Important stages in this process include:

- Dispatch of enquiries – preparing a list of what is needed for the project. These lists are issued by the tendering company to suppliers and subcontractors and include all project details and documentation.
- Receipt of quotations – written bids, sent to the client's contracting officer in response to a request for a quotation. It is necessary that it is received before the closing date. Quotations must be signed and dated by the contracting officer upon receipt.
- Bills of quantities production – these itemise the work for a project and are usually prepared by a cost consultant or quantity surveyor and are based on the specification and any detailed drawing it contains.
- Visit to the site and locality – the tendering company will send a representative, or a team, to see where the project will take place and identify any issues related to its location. This will help them better understand the project and create more accurate estimates. Aspects of the site will be examined to include any potential contamination, site boundaries, health and safety, transport access and sustainability issues.
- Completion of pricing and measured rates – the completed tender pricing documents are a priced bill of quantities, setting out the rates and costs of the project. This allows the client to compare the tender and the cost plan. A cost consultant can then identify where the value lies between different tenders and see where any savings can be negotiated with the contractors.
- Finalisation of the tender works programme – this is the outline that the tendering company prepares, indicating how they will carry out the work, including suggested deadlines it will aim to meet.
- Internal co-ordination and review meetings – the tendering company will need to ensure they have thoroughly checked and reviewed the tender documentation and are happy with the costs and timelines they are suggesting in the tender.
- Tender settlement meeting – this is the final stage before submitting the tender to the client company. All the estimates are finalised and checked to ensure they meet with the client's requirements. The risks to the tendering company will be reviewed, including the penalties for non-delivery and its confidence in the work proposed and with any assumptions in the estimates. Formal minutes should always be taken.
- Tender adjudication – at this stage the client company will review all the tenders and decide which one to select. A client will review a tender based on:
  - its compliance with the original brief
  - their judgement of the ability of the tendering company to deliver what they have stated in the tender
  - the price quoted in the tender.

Allocating resources

When putting the tender together, a company needs to ensure that its resources meet the needs of the tender, in order to complete the work by the deadline, and that it is has allocated these appropriately.

- Tendering and pre-tender budgets need to be realistic. If either of these are unrealistically low then it may lead to complications with both the tender and final product, such as time delays, sub-standard workmanship, low profit for contractor, poor value for money, high maintenance costs resulting from inferior materials, work practice or the installation of low-quality equipment.
- The tendering company must ensure that the estimating team has the time and resources they need to fully complete the tender in sufficient detail by the deadline.
- Sufficient time is also required for working with the subcontractors and suppliers who will supply quotes for estimating the tender. This is vital, as all materials have to be priced accurately. If requests for estimates and costs on material and labour

are sent late or do not provide enough time for responses, this will lead to serious delays in completing the tender.

▶ A visit to the site of the planned project needs to be completed at an appropriate time, preferably early in the estimating process, so that any issues identified can be planned for when completing the tender.

Ⅱ PAUSE POINT   There are many different stages that need to be considered when planning and completing a tender for a project. Do you understand how this process works?

Hint   How can delays in completing each stage of the tendering process affect the final tender?

Extend   Why is it important to plan the work and resourcing of each member of the team working on the tender?

A3 Contractual arrangements

Contractual arrangements are written agreements between two or more parties, such as a client and contractor, which are enforceable by law. In a contract, all the rights and obligations are stated and agreed between the two parties to the contractual arrangement: the client will set out what they expect the contractor to deliver and the contractor will agree the timelines and costs they will meet to achieve this.

The project's contractual arrangements are designed to ensure a successful delivery of the service or project and the level of control the client will have over the project work. This can also help to minimise the level of risk and improve the level of communication, its details and content.

Project contractual arrangements promote the delivery of products to satisfy both parties and state the expected business benefits and value for the money that the project will produce. Contracts bind organisations together for some time and involve a large degree of dependency, so mutual trust and understanding in an open and constructive environment is vital.

Conventional lump sum form of procurement

A lump sum contract is the most common type of contract in the construction industry. The client agrees to pay a single 'lump sum' price for all the work. From this lump sum the contractor will complete all the work, supplying materials and labour. The money not spent on this is kept by the contractor as profit.

Lump sum contracts give some risk to the contractor as well as the client, as they have promised to deliver something to a specific timeframe and cost, but allow the client to be more certain about the amount they will pay for the work. The prices of lump sum contracts can, however, change in response to changes in the project or if unexpected issues arise during the construction process.

Joint contract tribunal contracts

The joint contract tribunal (JCT) is a standard building contract designed for construction projects. It is divided into two groups: with quantities (elemental and trade bills of quantities) and without quantities.

▶ The JCT standard contract with quantities is usually used for complex and large-scale construction projects where a highly detailed contract is needed. This type of contract is suitable for projects procured via the traditional or conventional method.

▶ The JCT standard contract without quantities is usually used where either the client or their representative:
  • provides drawings and specifications detailed enough to negate the need for the tendering company to draw up a bill of quantities
  • employs a quantity surveyor of their own to draw up the bill of quantities.

Research

Can you find examples of projects where there was a dispute around what was agreed in the contract and what was finally delivered? Why did these disputes happen? How were these issues resolved? What impact might these have had on future contracts?

Design and build contracts

A design and build contract is for projects where the contractors will carry out both the design and the construction work. It is suitable where the project needs to meet a quick deadline or where detailed provisions are needed around what the design must include and the standards the final construction must achieve.

Term contracts

A term contract specifies the details of the agreements within a fixed time. It can be short term or long term, part time or fixed term, depending on the nature of the project.

⏸ PAUSE POINT   There are many reasons for producing a contract for business work. Do you understand why this is?

> **Hint**   What would you want to have a formal agreement on if you were working with someone?

> **Extend**   Try putting together a contract for a simple job with a friend. What terms and conditions do you need to agree? What disagreements are there?

A4 Supply chain

All companies need to work with a chain of suppliers that will provide materials and labour to help them complete a contract. Companies need to work closely with their supply chain as they will rely on these suppliers to provide them with accurate costs and quotes for services and materials. These quotes will be crucial in preparing their final tender submission for their clients.

Case study

A main contractor is tendering for the renovation of a large old building in central London. The client wants the frontage and general style and feeling of the building to be retained. There is a large amount of decorative plasterwork on the ceilings and walls. This will require a specialist company, which is experienced in renovation and restoration work. There are two companies within 100-mile radius that qualify for the work. Their quotes for the work are similar. The contractor's bid manager needs to make a decision.

Check your knowledge

Draw up a list of questions to ask each of the specialist subcontractors before deciding to invite them to tender for the work.

Quotations from suppliers and subcontractors

When preparing their tender, the main contractor relies on subcontractors and supply enquiries to calculate their bid price. The suppliers will provide quotes for the parts of the tender that are relevant to them. The suppliers are effectively 'bidding' for parts of the tender with the main contractor. There are several different types of quotation that can be supplied by subcontractors and suppliers:

▶ **Fixed price** – these quotes offer a single non-negotiable price for the work or goods.
▶ **Fluctuating price** – long-term contracts may experience changes in price for goods and services perhaps because of changes to the economic situation of the country, such as **inflation**. For these contracts, the quote will calculate this possible change in price over the life of the project and include this (for example, if the price of labour is expected to increase over a five-year project, the quote will take this into account).
▶ **Variant bids** – these quotes are more flexible than regular quotes or bids. Companies submitting variant bids will offer two quotes, one that matches the original criteria

Key term

Inflation – an economic situation which leads to an increase in prices and a fall in the value of money.

and one that is a variation (such as different pricing structure or offering a new way of delivering the service) to improve their chances of winning the bid.

▸ **Incomplete quotation and prices** – any incomplete information can, during the tender process, directly affect the estimate and cost time and money. A client could disregard an incomplete tender.

Suppliers of materials

The relationship between a contractor and its suppliers is a very important factor in the tendering and estimating process. Good relationships with suppliers can bring better opportunities for the firm, as suppliers are more likely to offer better terms and prices for work to be carried out. Similarly, experience with suppliers will help companies learn what to expect from working with them.

Negative experiences, such as late delivery or an increase in cost, can often mean a company is less likely to work with that supplier again. Positive experiences, such as the supplier offering additional value for money, will encourage companies to build closer relationships with that supplier and work with them again in the future.

▸ Why is it important to have reliable suppliers?

Named manufacturer

At times, the client may supply a list of preferred (named) suppliers or manufacturers from which the tendering company will, in turn, invite tenders for the supply of services, materials or equipment.

Generic specification

There are projects that consist of a series of similar or identical works. For example, a programme of constructing sewage pumping stations across a particular county, which are all to be built to a specific standard and design. In this case, a generic specification would be drawn up to detail the construction of these buildings.

Generic specifications are also used by organisations to detail particular standards for any construction or building services works they commission. The Ministry of Defence, for example, has a set of generic specifications covering electrical, plumbing and heat, and ventilation works carried out in their premises.

Range of alternative suppliers

There are many potential suppliers available for companies to work with, and the estimator and the quantity surveyor will review all the possible suppliers for different goods and services and work out which one is best placed to supply what is required. As part of this process, estimators and quantity surveyors will review the capacity of the supplier to deliver the goods or services required. For example, a large order will probably require a large firm to successfully deliver it, while a small order may have many smaller companies who can fulfil the order.

Subcontractors

Subcontractors generally work under contract to the principle contractor. The principle contractor has to ensure that subcontractors are competent and are provided with all the relevant information needed for working on the contract. The contractor has to provide **method statements** so that a subcontractor can co-ordinate any work. On a construction project, a subcontractor may carry out a specific part of the job, for example on a building site a subcontractor may be hired to erect scaffolding or to install a gas supply.

A general contractor's ability to select the best subcontractors is a key competitive advantage when tendering and estimating. Subcontractors can transfer their unique skills and talents for completing specialised work.

A shortage of rigorous or skilled subcontractors can lead to serious problems, such as cost and time overruns, disputes, failing to meet building standards and client dissatisfaction, any of which could disqualify the contractor and lead to them losing the whole project.

Many relationships between contractors and subcontractors will develop over time, and are based on previous experiences. This is a reliable way to choose subcontractors as both parties know their capabilities and have developed a level of trust. This relationship can be both professional and based on personal experiences.

The reputation of subcontractors is also important. A subcontractor with several endorsements and positive comments from other businesses or colleagues is more likely to be approached about work.

Subcontractors need to work closely with the main contractors to ensure that their resources and skills are transferred to the project and working towards its success. The skills subcontractors may need to transfer include their **operational resources**, **functional skills** and **managerial skills**.

Several factors can affect how subcontractors are selected and how effectively this relationship works:

▶ Relevant experience of working on similar projects is a prime consideration.
▶ The number of available subcontractors in that specialist area – some tasks may be more specialist than others so only a few subcontractor companies will have the expertise and experience to take on the project, making the choice of firms narrower.
▶ The geographical area of operations – the location where the building project is taking place may also reduce the choice of subcontractors, as some areas will have a smaller range of firms operating in them, or an otherwise ideal subcontractor may be based too far away.
▶ The level of supervision the contractor will wish to have over the project or the subcontractor may need to complete

the job. For example, a high level of supervision would be required if the work is unusual and unlikely to be within the experience of a subcontractor, or if the project is a prestige job where quality of work is of the utmost importance and certain high standards must be maintained.

▶ References and recommendations the subcontractor has had from other companies.
▶ The level of insurance the subcontractor holds.
▶ If the subcontractor holds, or requires for the job, a certain quality assurance (QA) registration. Examples of this are registration with the Construction Industry Scheme (CIS), the Gas Installers Registration Scheme or National Inspection Council for Electrical Installation Contracting (NICEIC).
▶ The health and safety record of the subcontractor and its staff.
▶ Any **collateral warranties** connected to the contract.
▶ Any commitment on the project or with the client to work within considerate contractor policies, encouraging best practice on site for respecting the community, protecting the environment, health and safety and valuing the workforce.

Key terms

Method statement – documents which identify the methods used to price the work items, the plant and the labour required for each activity.

Operational resources – assets to the business that help it to function, including human resources; physical resources such as machinery and industrial equipment; a distribution network; and intangible resources such as brand, licences and reputation.

Functional skills – skills connected to running the business, such as marketing and commercial awareness, project management, logistic and procurement skills (receiving, storing and stock control), financial and operational skills (management of production costs and deadline).

Managerial skills – skills needed by senior staff such as leadership, strategic planning, use of different management and business intelligence tools, ability to allocate resources or refining an existing strategy.

Collateral warranty – a contract where a contractor or subcontractor warrants to a third party (such as the client) that it has complied with all the terms of its contract, possibly including using materials of an agreed quality or carrying out work in a professional manner.

Nominated suppliers and subcontractors

The main contractors have responsibility to nominate, select, organise, co-ordinate and manage the relationship with subcontractors and suppliers. Nominating the right suppliers and subcontractors and setting the correct degree of collaboration and co-ordination between them and all other parties involved can contribute to the success of a project. A company placing a tender must have researched possible suppliers and have a good idea if they can deliver and to what budget.

A main project will be broken down into a series of **packages of work**. Each of these will require a subcontractor to work on them. The work package given to a subcontractor must include any specific design elements to follow as part of their work.

Most subcontractors will have their payments calculated as a prime cost (PC) sum. This will cover the cost of the work or materials supplied by the subcontractor and is exclusive of any profit mark-up that can be introduced by the contractor. Payments are made based on the invoices plus any agreed percentages for overhead costs or profits.

Suppliers and **subcontractors** can be selected (or **nominated**) by the client themselves, if there is a specific material or firm that the client wishes to work with. Nominating the right contractors can help to ensure the success of the project.

⏸ PAUSE POINT　　What factors can affect which suppliers and subcontractors might be invited to work on a project?

> (Hint)　　How do you choose who you would like to work with when you need to work as a team on a job?

> (Extend)　　What can be the impacts of selecting and working with a subcontractor who is inappropriate for the project?

A5 Commercial risk analysis

Risk is an inevitable part of any construction job, from estimating and tendering to completion. As risk can be dangerous for a business, there are many factors which need to be considered when estimating for a tender. These factors can directly or indirectly affect the strategic decision about whether to put forward a tender or not.

Current workload of the company and maintaining turnover

One of the main factors that will decide if a company can commit to a tender is a review of their current workload. If a company has many projects it is already committed to, quality may suffer if it takes on more work.

A company needs to commit to maintaining its **turnover** in order to survive. For example, taking on more work than the company can deliver or can support with its current size and means might mean that turnover is lower than the amount the company is spending on its commitments.

Market conditions and economic climate

When the economy is booming, a company may look to expand and take on more work as they are confident that funding for the project will last. However, in times of economic downturn, the company may look to reduce its workforce, or downsize its commitments to protect its turnover and, therefore, its profits.

Project considerations

A company will need to review the project and decide if it is a good fit for their company. Do the needs of the project match up with the strengths of the business?

Buildability

Buildability is defined by the Construction Industry Institute (CII) as 'the optimum use of construction knowledge and experience in planning, design, procurement, and field operations to achieve overall project objectives'. In other words, it is about using experience and expertise in planning, purchasing materials and working on site to achieve the aims of the project.

Key terms

Package of work – a group of related tasks within a project. These can be tendered for by subcontractors, and allow the main contractor to transfer some risk to other companies.

Nominated subcontractor – a subcontractor who is appointed by the client and has already tendered for the work package for the client. The main contractor is then instructed to appoint this subcontractor and the value is offset against a provisional sum placed within the main tender documents.

Turnover – the amount of money that a business receives within a given time.

Unclear designs, drawings and technical specifications will slow down or prevent construction as they will require continual clarification and updated information. This will have a major impact on the ability of a business to deliver. If the business does not believe the project meets these buildability standards, it may decide it is not worth putting itself forward to take on the job.

Previous experience of construction methods

Contractors can also minimise the risk by tendering only for projects that match their previous experiences. Previous experience of the contractor with the client, of similar projects, and of local custom and environmental characteristics mean that the firm knows what to expect and how to solve typical problems.

Site and location factors

Site and location issues generally divide into two different categories:

▶ issues connected with the site itself, such as the physical surroundings and make-up of the site
▶ issues connected with the location of the site, such as its access to transport and other facilities.

Both need to be monitored as they might create risk for the actual construction work. These logistical issues and requirements have a significant bearing on a contractor's decision whether to tender or not, as seen in Table 3.1.

▶ **Table 3.1:** Issues connected with the site and its location

Issues connected with the site	Issues connected with the geographical location of the site
Ground contamination in the site can lead to complications or extra work.	Availability of tipping facilities – some sites may have poor access to waste disposal facilities.
Soil types and ground conditions will need to be carefully considered and understood before work can begin.	Availability of skilled workforce – if there is a shortage of expertise in the area, a company may need to bring in specialist workers from elsewhere at extra cost.
Existing site features that need to be preserved (or need to be worked around) will need to be taken into consideration when planning work.	Industrial relations and local labour agreements may vary from area to area and could affect the expectations of workers on site.
Existing trees may need to be worked around, especially those with preservation orders.	Temporary roads and services could be needed if a site is in a remote location.
Presence of protected species need to be worked around and their safety ensured. This might mean relocating some species and providing a suitable habitat for them.	Security requirements can vary according to location. They may need to take into account crime statistics or the risk of theft and vandalism in the area. Security may also be needed if the site is close to schools or playgrounds to protect children. This could be affected by the attitudes of the people living in the area towards reporting crimes and the reputation the area has.
Availability of space for contractor's accommodation, storage and distribution of plant, tools and materials will need to be planned carefully around the restrictions on site.	Local climatic conditions could be flooding risks or higher than average rainfall.
	Local pressure groups may be hostile to the proposed project. These groups may work to delay the project or to obstruct it in some way.
	Location of contractor's office and depot – time and costs will be associated with transporting people, tools, plant and materials from the contractor's main depot to the site.

Programme factors

When the client puts together the information needed for their project, they may have certain key considerations that they want all those companies tendering for the project to commit to. These factors will directly and indirectly affect the work and therefore need to be carefully considered within the tendering process.

Commencement date

Some clients may wish the work to start on a particular date. This could affect the project, and the plans of the tendering companies, in different ways:

▶ Weather and climate – if the work needs to start at a time of the year when working on site is difficult (such as winter) this could affect their ability to work successfully, effectively and safely.

▶ Completion of other products – a company may have committed to other projects whose completion date overlaps with the client's desired commencement dates. In these cases, the company may decide not to tender as there are concerns about overstretching its resources.

▶ What external factors could increase the cost of a building project?

Specific contract conditions

The client may also introduce different conditions to the contract that could affect how easy the tendering company finds it to work on site. Their nature depends on the project, but might significantly affect the cost of the project. Possible conditions include:

▶ Methods of working – the client may require the company to work in a specific way that may not be best suited to the tendering company's organisational structure, resources or work practices.

▶ Restrictions on working methods or access to the site – the client may need to restrict certain working methods due to the requirements of the site, or may only be able to allow limited access to it, for example construction of a road project may need more major work done at night rather than during rush hour.

▶ Factors that interrupt the regular flow of trades – problems within the contracting or subcontracting companies may affect their ability to carry out the work assigned to them. For example injury, sick leave or resignation of staff.

▶ Factors that may affect the duration or sequence of project activities – construction projects are planned using project management tools such as Gantt charts. These define critical paths through the project and meeting the milestones along this path are vital to meeting the project's end date. Late delivery of materials, adverse weather conditions, re-work caused by poor workmanship, and changes requested by the customer can all lead to missed deadlines. Provision has to be made for catch-up activities to keep the project on track.

▶ Factors that may increase the cost significantly, or are hard to quantify or calculate in advance – there may often be problems encountered on site that could not have been foreseen. Examples of this are the discovery of hazardous substances, obstacles in the path of pipework or cabling or complaints about noise from the site early in the morning or in the evening, which will restrict working times.

Financial issues

Financial stability is often the major factor for whether companies tender for a project and take on the commitments that will come with winning the tender.

The client's reputation and financial capability, the financial value of the projects, the availability and cost of materials and the stability of the construction industry will control the bidding decisions of contractors. The following factors are all considered as part of any financial considerations:

▶ Cash flow forecasting for the project and own company to predict how much revenue is entering and leaving the business and project. This gives the business an indication of whether it can afford to take on a project and if it will leave enough money for it to survive.

▶ Available finance – this is the money immediately available to the contractor, either from their own funds or a loan.

▶ Contract bonds – issued by an insurance company or bank to guarantee satisfactory completion of the job by the contractor. If the company is worried it will default on the contract, a contract bond will make them liable to reimburse any costs to the client.

▶ Retention percentage – the client will retain a percentage of the payment – usually about 5 per cent – to ensure the contractor properly, and completely, finishes all the duties required by the contract.

▶ Payment period and frequency of payments – the client may wish to make a series of payments through the lifespan of the project, or a payment on completion of each stage, or a single payment on completion. This payment plan will affect the amount of money coming into the contracting company, and therefore affect their decision on whether their cash flow can support working on the project.

▶ Financial checks on the client – the contractor may wish to find out more about the client by checking their public accounts, credit rating and to confirm they have funding in place for the projects. If the tendering company is not confident on the client's finances and funding, it will not wish to take on work with them.

▶ **Table 3.2:** Risk identified factors at pre-construction stage (Adapted from *J. Manage. Eng* (2016) (11))

Type of risk	Risk
Client related	• Client's change/changes in owner's requirements • Client's brief and type of client
Planning and design related	• Extent of completion of pre-contract design and specification • Defective design and specification • Planning requirements or restrictions • Little or no information about mechanical and electrical services • Procurement system • Project scope and legal requirements
Cost related	• Type and quality of cost planning data • Changes in estimating or cost planning data • Underestimation
Market related	• Tender period • Market condition • Project location • Unforeseeable fluctuation in labour and materials prices
Project related	• Complexity of design and construction • Quality of information and flow requirements • Availability of design information • Project team's experience of the construction type • Method of construction and expertise of consultants • Site investigation information (geological/subground conditions) • Site constraints • Type of project and structure • Inadequate cost plan/tender documentation

▷ **Table 3.2:** *Continued ...*

Type of risk	Risk
Bidding requirements related	• Availability and supplies of materials and labour • Type of bidding and contract conditions • Tender inflation and zonal rates
External	• Government legislation/policy

Ⅱ PAUSE POINT   Do you understand how commercial risks can affect the decision of a company to tender for a project?

Hint   What makes you worried about committing your time and money to doing something? How might these same feelings affect a business?

Extend   Research company financing. How do companies structure themselves and plan for their financial future?

A6 Commercial intelligence

To make decisions on whether to tender for a project, a company will need to understand their commercial environment, for example knowing about other companies in their industry, whether they will have competition for the tender and, if so, how many competitors.

Gathering commercial intelligence

Many companies engage in gathering commercial intelligence. This involves investigating the sector of the industry that company exists in, and finding out more about the other companies in that sector and how they operate. This knowledge allows the company to increase their understanding of the sector and can help them to tailor and prepare their tenders more effectively. Table 3.3 shows some of the different tactics that can be used to research and gather commercial intelligence.

▷ **Table 3.3:** Advantages and disadvantages of different methods of gathering commercial intelligence

Method of commercial intelligence gathering	Advantages	Disadvantages
Speaking to suppliers and subcontractors	• First-hand knowledge and experience • Suppliers may be aware of latest innovations and technologies	• Will present themselves advantageously • Will want to sell their own products and services
Networking	• Information from a wide variety of sources • Opens opportunities for all sides	• Opinion not fact • Self-interest colouring information given
Press releases	• Ready packaged information that delivers the facts • Up-to-date information	• Reliability of the source
Using intelligence services or research through journals	• Frees up a company's own resources by buying in information • Journals will reflect current trends and technology	• Cost when using a third party for research • Time spent vetting and obtaining quotes for intelligence gathering • Journals can be costly and time consuming to read

Competitive tendering

All tenders are likely to be competitive, with multiple companies applying. The more information companies can gather about the other organisations likely to tender, and about their background, the more companies can design their tenders to counter the strengths of their rivals. For example, if a rival company has a particular expertise in a certain area, the tendering company will want to demonstrate that they can match it. This can make a company's tender stronger and therefore more likely to be selected.

When researching other companies, a company could also focus on:

▷ Recent tendering success – how many successful tenders has this company made? Why were these tenders successful?

- ▶ Recent price levels – what have they charged on recent tenders? Is it possible to undercut these prices or offer a better deal?
- ▶ Capacity to take on new projects – does the company have the scope to take on new projects without delay? Do they have the resources and materials to do this?
- ▶ Current tendering workloads – how committed is the company to current contracts? How many additional jobs do they have the capacity to take on without affecting the quality of their work?

Assessment practice 3.1

Your company has been invited to tender for six new police stations located throughout the county. They will act as replacements for the existing buildings, and form part of a reorganisation of the county constabulary structure. The locations of these police stations cover a wide geographic area.

Three of them are in the centres of large towns, while others are located in smaller market towns and one is in a village. There are four different types of building, each designed around location and need. One of them is a one-off design and will act as the county headquarters. The contract also requires that the stations are built in two simultaneous lots of three, in order to shorten the life of the overall project.

Your company is a regional contractor with an annual turnover of £20 million. It employs 275 staff and operatives, including bricklayers, carpenters, plasterers and labourers. Other specialist trades are generally subcontracted. The company is currently operating at 55 per cent capacity.

It has had a cautious attitude to commercial risk and has grown steadily to a net value of £8.5 million. It has a philosophy of rapid expansion to increase the worth of its stocks, a commercial risk which has caused concern among some of the company's managers because this is outside the usual risk-averse philosophy of the company.

Write a report covering the following areas:

1. Explain the company's ability and suitability to tender for this work. (2 marks)
2. Suggest which tendering method (serial or priority) is best used for this project. (1 mark)
3. Conclude your report by analysing the commercial risk attached to the following items:
   - current workload
   - need to maintain turnover of work
   - market conditions and economic climate
   - project considerations. (6 marks)

Your report should conclude with a recommendation for the tendering strategy and a justification for this approach.

Plan
- What is the task? What am I being asked to do?
- How can I order the information in order to produce the required document?

Do
- I know what it is I'm doing and what I want to achieve.
- I can identify when I've gone wrong and adjust my thinking/approach to get myself back on course.

Review
- I can explain what the task was and how I approached the task.
- I can explain how I would approach the hard elements differently next time.

B Estimating

Estimating is one of the key stages of preparing a tender, as it involves putting together the initial quotes and prices on which the company will base their tender and that the client will use to decide whether to accept the tender. The estimator will prepare this based on the client's tender documentation which will be either:

- ▶ a set of drawings with a specification
- ▶ a set of drawings and a bill of quantities.

From these, the estimator will prepare a rate against each quantity. If using the drawings with a specification, the estimator will measure, count up and price quantities. Multiplying the rate by the quantity gives the total price for each taken off item. When all these are summarised, the estimate is the **net costs**. The **gross estimate** contains the items that must be added to the **net estimate**. These items are:

- profit
- **overheads**
- risk and uncertainty items.

B1 Materials and subcontract quotations

When working through the tender preparation process, the main contractor will engage with subcontractors and suppliers to gain prices for materials to calculate the final price for the tender. The estimator will need this information to prepare accurate estimates for their tender documentation.

Subcontractors

Subcontractors are mainly selected by the main contractor who will send out additional enquiries for the specialist works that they cannot undertake. Sometimes the contractor has to use subcontractors who are approved by the client and, in these cases, a list accompanies the tender.

Subcontractor prices are influenced by:

- how busy they are
- their historical relationship with the main contractor
- the level of the main contractor's discount - the contractor may offer a discount to the client (often to ensure prompt payment). This is then passed on to the subcontractor who will price their work accordingly
- how long the main contractor takes to pay the subcontractor
- the location of the work
- how specialised the work is.

Suppliers

Suppliers normally only provide materials or a piece of equipment and do not undertake subcontracted installation works. Therefore, they are asked to provide materials quotations from schedules prepared by the estimator and these materials are purchased from the supplier using a purchase order. Suppliers' prices and rates are affected by:

- the payment terms negotiated
- nationally agreed rates from the contractors' buying departments
- the level of discount negotiated
- the location of the supplier
- how quickly the materials are required.

Suppliers can also be 'named' or nominated by the client. In this case, the price is already agreed and the main contractor is paid a percentage profit as stated in the tender document plus attendances.

Factors that can affect quotes

There are several factors that could affect the quotes received from suppliers and subcontractors:

- Lack of information - many enquiries for quotes are based on previous experience rather than more recent information or systematic evidence.
- Errors in information - not giving companies the full information needed to take into account the background and nature of the project can also lead to errors in the future. Simple mistakes in the information supplied can have a serious impact

---

Key terms

Net estimate and net costs – the estimate for the total costs of labour, plant, materials, preliminary items and subcontractors minus the financial benefits (profit) to be gained by the work for the contractor. They do not include overheads and costs incurred by extras or uncertainties (see Gross estimate).

Gross estimate – the net costs plus overheads and profit and risk items.

Overheads – costs that need to be met in order for the company to continue to run, such as staff salaries, cost of premises, price of equipment.

Theory into practice

A main contractor requires quotes for fire alarm installation in a new supermarket to be built in a commercial park on the edge of a large town. The fire alarm system is to be a state-of-the-art system incorporating cutting edge detection and fire extinguishing technologies. The project has already fallen behind, so many contractors are asked to work round the clock over a period of time. A specialist electrical company is required.

In small groups, draw up an invitation to tender, explaining what is required and the issues around the tight timescale.

on the accuracy of the quotes supplied, for example errors relating to timeframes, quality of documents and subcontractor resource level.

▶ Gap analysis outcomes – the process from estimating through to placing materials on site is complex, and during this a business may identify a number of gaps in its operations. In these cases, the business will conduct an analysis of the 'gaps' it has that prevent it from delivering on a tender, and identify how to fill these.

▶ Variant offers – an item offered that is outside the original contract requirement. This could also apply to improvements in technology which enhance the original equipment or services offered.

▶ Abnormal quotes – some quotes will be more or less than normal prices, and will usually be withdrawn from the tender.

Ⅱ PAUSE POINT   Do you understand the relationship between the main contractor, suppliers and subcontractors?

Hint   How close are each of them to the original client?

Extend   Put together a rough structure chart for the team working on a construction project.

▶ A site visit is an excellent opportunity for the potential contractors to ask questions of the client and better establish what they want from the project

Discussion

As a group discuss what factors you would need to find out about when conducting a site visit. Why is it important to gain as much information as possible? What could the impact be on a tender if you do not get all the information you need?

B2 Site visit

The site visit completed during the tender preparation process will help the estimator to become familiar with the area where the project will take place, and to start to form their estimates.

The site visit should happen early to clarify any concerns which the contractor, subcontractor or supplier may have with the tender documents, the scope of the work and any other requirements of the project. Formal site visits are usually planned and carried out for works procurement and more complex goods requirements.

When a site visit is planned, all the details such as date, time and number of visitors must be clear and agreed in advance with the client, especially if it is private land that requires permission to visit, or if there are health and safety issues. During a site visit, different factors and elements need to be carefully considered; these may include the following:

▶ A careful assessment will be made of any structures that may require demolition.

▶ An assessment of any repairs or alterations that may be required and their likely cost.

▶ Any site clearance requirements in place from the client.

▶ Any security issues or site access restrictions that may affect the work to be carried out.

▶ An assessment of any relevant community issues – is there any controversy in the community around the build that will need to be addressed or any community pressure issues that will affect the work?

▶ The amount of space available for site establishment and movement of the workforce.

▶ What methods for distributing materials will be most appropriate on site.

During the site visit the contractors can all ask questions of the client which may help clarify, or alleviate any doubts about, information provided in the documentation. A careful assessment of the site visit creates a much better chance for an accurate estimation.

Site visits should ideally be held before any pre-bid meeting. This allows the bidders to make additional queries to the client and any issues can also be discussed at the pre-bid meeting. It may also identify issues that the client may need to communicate to all bidders.

B3 Completion of the estimate

A client's budget will need to be accurate when the project costs run into millions of pounds and may have to be financed through loans.

The estimation of labour costs is based upon the agreement on pay rates that the company has with the operatives. From this, all the additional costs will need calculating, for example employer's National Insurance, holiday credit, sick pay and any other allowances. This will give a rate that can be used by the estimator to recover all the overheads against the labour in the tender.

One or more typical estimating methods could be applied.

▶ The **unit or number method** estimates cost on a unit basis, for example a seat in a cinema or a bed in a hospital. Very simple calculations can produce a cost for a potential project by using previous contract final accounts and the unit number of occupancy. Obviously, this method may be very inaccurate and does not take into account the complexity of the design.

The cost per unit method is used for estimating costs for budget preparation using a physical unit, for example a hospital bed, a school pupil place or a workstation. The method relies on the fact that there is some relationship between value and the number of units.

Worked example

A 250-bed hospital costs £350 million to construct. This has been taken from historical information on the cost of building a hospital. Estimate the cost of a proposed 300-bed hospital.

£350,000,000 / 250 = £1,400,000 per bed space

Proposed hospital 300-bed space × £1,400,000 = £420,000,000

There can be a vast difference between actual cost and estimated costs and this method does not have a high degree of accuracy. An experienced estimator would need to look at the provisional design against the historical design from which the unit rate was obtained in order to ensure some degree of consistency.

Worked example

A 76,000-seater stadium cost £37 million to construct. Estimate the cost of a proposed 60,000-seater stadium using similar construction methods and design.

£37,000,000 / 76,000 = £486.84 per seat

New stadium = 60,000 × £486.84 per seat = £29,210,400

▶ The **area method** takes previous historical contracts, often by the Royal Institution of Chartered Surveyors, to produce and compile square-metre cost rates and cost estimates. The square-metre cost rate is found by measuring the total floor area of a project and multiplying it by the rate plus

or minus any adjustments. This method produces a fairly accurate estimate for the new project.

Calculating the cost per unit area, for example square metres of gross floor area, involves using historical rates, increased in line with inflation, on a similar project in order to produce a realistic rate for a new building project's floor area. This can be illustrated using the following worked example.

Worked example

Historical cost of building A = £300,000 Floor area = 40 m × 20 m = 800 m^2 = £300,000

Area rate = £300,000 / 800 m^2 = £375 per m^2

Proposed new building B of similar design floor area = 650 m^2

= £375 per m^2 × 650 m^2

= £243,750

▶ The **cubic method** is where the volume of a historical project (its length, width and depth calculation) is measured and the original cost divided by this volume. This produces a rate per cubic metre that can then be applied to a new project in order to produce a cost estimate for the client's budget. This method does not provide the most accurate estimate.

▶ The **approximate quantities method** provides additional detail compared with the other methods described above. It uses the architect's sketch designs to produce some approximate quantities. The estimator's experience needs to be called on here to include the items of work that will be required for the final design. By using current cost rates or **price books**, a realistic estimate can be prepared for the client's budget.

Key term

Price book – a published book that contains current prices and rates for items of work based on NRM2 (Royal Institution of Chartered Surveyors New Rules of Measurement 2).

▶ The **cost of functional element method** involves breaking down the estimate into functional elements, or small packages of work. Unit rates per square metre, metre or number can then be applied to produce a cost estimate. This method relies on the knowledge and experience of the estimator to apply historical element costs from previous work and upgrade these to the new proposal. It allows costs to be taken from different historical works to provide a budget plan from which further work can be developed.

▶ **Elemental estimating** involves breaking down the

proposed design into elements, for example foundations, ground floor construction, first floor construction, structural frame and roof finishes. Cost estimates can then be prepared against each element and a final budget produced.

▶ The **NRM quantities method** is undertaken as part of the preparation for tendering and the procurement of a contractor to carry out the contract work. It is the most accurate of the estimate processes compared to the above methods.

The accuracy of all the estimating methods will vary greatly. Preparing area unit or cubic estimates will depend on the accuracy of the drawn information from the architect and the historical cost information used to produce a unit rate, i.e. the effects of inflation on prices. Once a detailed design is approved, accurate quantities can be taken off and a better cost estimate refined from the initial studies.

Element	Quantity	Element unit rate	Totals
Substructure	150 m²	£50	£7500
Structural frame			£20,000
External walls	2500 m²	£55	£137,500
Roof construction	1500 m²	£45	£67,500
Internal partitions	2000 m²	£28	£56,000
Windows and doors	40	£250	£10,000
Wall finishes	1500 m²	£12	£18,000
Floor finishes	1500 m²	£25	£37,500
Plumbing			£30,000
Electrical			£50,000
External works	2000 m²	£20	£40,000
Drainage	1000 m	£25	£25,000
Total			**£499,000**

▶ **Figure 3.5:** A typical cost estimate based on elements

Demolitions and associated temporary works

Demolition is the process of removing the existing structure from the site in order to redevelop. Often these are structures near the end of their useful life and cannot be adapted or refurbished due to cost constraints.

Demolition may be instantaneous, using explosives, or slow and methodical, using machinery to carefully cut and remove the building. There is a vast amount of planning for the destruction of a building using explosives. The building has to be structurally weakened in preparation for drilling and charging holes with explosives. The police, fire and highway authorities have to be contacted regarding road closures and notices to evacuate residents on the day of demolition.

Similarly, demolition involving the use of machinery has to be planned because of the noise, dust and volume of road traffic. The modern approach to demolition is to recycle parts of the structure. Metals can be reprocessed and brickwork and concrete can be crushed to produce a hardcore for reuse in filling materials.

Pricing alterations, repairs and conventions

On many large contracts there are numerous variations to the original specification. This, in turn, will alter the cost of the work. Provision for these variations, and the likelihood of their being required, need to be included in the tender. Maintenance and repair work may also need to be costed into the estimate.

International regulatory conventions can also control issues around construction, for example human rights protecting workers, and conventions in areas that might contain endangered species.

Building up unit rates

The tender price will be built up from the following costs:

▶ Approximate quantities – prepared by the quantity surveyor, this quote will evolve through the tender's life cycle as more information becomes available. It can sometimes be based on a rough design drawing issued prior to the final version being issued.

▶ Selection of material price to use – appropriate materials selected in terms of quality and price, including any discounts available.

▶ Coverage rates – the amount of materials needed for each unit rate, for example the number of bricks needed per square metre.

▶ Appropriate wastage percentage – allowance for wastage of materials through breakages, incorrect deliveries or over-ordering.

▶ Offloading and storage costs – unloading delivery vehicles and setting up secure storage for materials and equipment must be included. Also, safe storage for hazardous substances such as gas for welding will be required.

▶ 'All-in' labour rates – wages and other costs for labour, such as National Insurance and income tax payments.

▶ Labour 'constants' – calculation of labour output which enables timescales to be calculated. This can sometimes be ascertained from time and motion studies, which research the amount of time taken to complete specific tasks.

▶ Plant requirements – **cost** of plant to be either purchased or hired, which may include maintenance and fuel costs for equipment owned by the contractor, and also training of staff in its use.

▶ Are labour costs classified as direct or indirect costs?

Completing provisional and prime cost (PC) sums

Inclusion of provisional sums

Provisional sums are used for any specific job or service that it is impossible to define a price for at the time the tender is being prepared. An estimator will calculate the most likely price and then include that cost in their final tender. For example, if a contractor is being contracted to knock down a building and replace it with a new construction, they may not be able to fully investigate the ground and soil conditions under the building until it has been demolished. In this case a provisional sum would be quoted for any part of the work that will be affected by this, which can then be revisited once the building has been demolished.

Some provisional sums are 'undefined'. These place greater risk on the clients, as they could add a considerable cost to the project if the work is more extensive and expensive than originally supposed.

Inclusion of prime cost (PC) sums

Prime cost sums are the amount calculated for the supply of all work or materials by a contractor for the client. Prime cost is the direct cost of producing an item. Calculating this cost can help the contractor to decide how much more it needs to charge to make a reasonable profit. It also lets the contractor know that if it charges less than this it will be making a loss on the deal.

Raw materials + direct labour = Prime cost

Businesses will need to factor in several other factors in addition to their prime cost sums when preparing a final tender, in order to reach a final quote. These will include:

- overheads
- profit
- attendance and special attendance
- contingency sums.

For example, if you have been asked to build a wall you will need to know the exact cost of all the bricks and mortar and how much you would need to pay anyone helping you to build the wall. You may also factor in the transport costs (your overheads) to get the bricks and labourers there. As you will want to make some money yourself as well (your profit) you will need to add this on top.

Profits

Construction companies do not consider profit to be measurable and so do not include this within a typical contract. However, businesses have to make money in order to survive, and profit can be included in an estimate by either adding a percentage to each of the rates within the bill of quantities or by adding to each of the items contained within the preliminary section.

The level of profit that is applied to a cost estimate will depend on the following factors:

- the amount of work that the construction company currently has on its order books
- the level of competition for the work
- the amount of risk associated with the project
- the complexity of the work
- the nature of the **procurement route**, for example, **partnering agreements**
- the payment terms of the contract.

There is no set level of profit that can be applied to an estimate. Each tender should undergo an **adjudication process** where the senior managers of the company discuss the factors listed above and arrive at the level of profit that will be applied to the estimate.

Overheads

A company's overheads are those costs that have to be met in order to run the head office, and include:

- departmental costs
- insurances such as public and employer's liability
- company cars
- IT equipment.

A percentage of the overhead is often recovered from additional costs added into the estimate. To calculate this percentage, the total value of the company's overheads per year needs to be assessed. Taking the turnover for the year (the amount the company takes in receipts) and dividing this by the overhead costs and multiplying by 100 gives the

percentage that needs to be applied to future estimates as long as the turnover does not drop below this level.

Overheads / Turnover × 100% = Percentage to add to tender

There are a number of ways of **reconciling** the overhead costs and recovering these against tendered works. Overheads can be costed in several ways:

- by not including them, but using an increased profit margin to cover their cost
- by establishing their cost divided by the total turnover and adding this percentage to tender submissions as noted above
- on larger projects, by moving head office functions onto site and recovering these costs through the preliminaries.

▶ The costs of running an office are part of a company's overheads.

Attendance and special attendance

Attendance is the work carried out by the main contractor for services it has to provide to assist suppliers or subcontractors to deliver their part of the project. This will always have a labour or financial cost connected to it. Possible services the main contractor may supply include handling and storing materials, delivery of goods, waste clearance or carrying out a supporting task such as scaffolding.

- General attendance refers to general support offered by the main contractor to help a subcontractor, such as access to the site, using electricity on the site, use of equipment such as cranes or hoists, waste disposal and welfare and security support.
- Special attendance refers to the specific support a subcontractor may need, such as using specialist

equipment provided by the main contractor or using the main contractor's employees to perform a specialist task.

If the main contractor is taking responsibility for this, they will need to cost for that in their quotation to the client, particularly if they are using a subcontractor nominated by the client.

Inclusion of contingency sums

Contingency sums are built into a quote to provide a reserve amount for risks that may emerge on the project that cannot be foreseen, for example bad weather affecting work on site, or an issue with the site that affects the use of certain plant or equipment. They can also be used as an emergency reserve if the project is running late and needs to be ready for a specific deadline.

A contractor will usually add a small percentage to their quote to use as a contingency.

Inclusion of subcontractor quotations

Each tender will include a number of quotations for work carried out by subcontractors on the project. This is for individual jobs on the project that a subcontractor will complete for the main contractor completing the tender. Just like the tender the contractor is preparing for the client, these quotes from the subcontractor will need to take into account several factors for the subcontractor including unit rates, lump sums, pricing for attendance, profit and overheads.

⏸ PAUSE POINT   Do you understand what factors need to be built into a calculation of prime cost and provisional sums?

          Hint   Why is it important for a company to make sure that it is covering all its overheads on a project?

          Extend   Investigate a building project in your local area – what would the possible contingency and attendance issues be there?

Pricing dayworks

Every year, or following price increases, the BCIS (Building Cost Information Service) publishes daywork rates which are hourly rates for various specialist trades, for example bricklayers or plumbers. The bill of quantities will have these daywork costs inserted by the client. Sometimes a rate for variations cannot be used from the bill of quantities and instead dayworks can be used as the agreed method of payment. The contractor inserts a percentage against each of the daywork schedules for labour, plant and materials in the tender to cover for profit and overheads.

In order to present a level playing field, the nationally agreed rates produced by the BCIS are used for the base rate on which the contractors add a percentage to cover all their on-costs, added to the tender document.

Pricing dayworks: labour rates

Labour 'all-in' rates for craft workers, skilled, unskilled and gangs need to be established so the estimator can price the tender works and accurately calculate all the costs of directly employing labour.

Several factors to be taken into account when calculating the all-in cost of labour are:

- the basic rate of pay per hour that has been agreed between the employer and the worker
- annual holiday entitlement and public holidays
- employer's National Insurance contribution
- the weather (possible loss of production)
- staff sickness
- CITB (Construction Industry Training Board) levy
- travelling time
- bonuses.

▶ What are daywork rates?

Application of labour costs in unit rates

To apply the labour cost per hour, one further piece of information is required: the output rates for the labour, such as the speed a bricklayer can lay bricks. Once this is established, then the unit cost of labour per unit can be calculated and applied to the tender prices. Output rates can be established in two ways:

▶ timing operatives using work study (watching the operative work on a known quantity and seeing how long it takes them to finish that work)

▶ using output tables from historical works or price books, which will provide information on unit output rates (how long things took to construct).

Worked example

Calculate the annual cost of labour for a general operative using the following data.

Data:

- Basic salary = £15,000.00

  Holiday pay: 4 weeks at £288 per week = £1152.00 Employer's National Insurance contribution at 12.5% = £1875.00

- Lost production time, estimated, 2 weeks at £288 = £576.00
- Public holidays: 8 days at £58 = £464.00
- Sick pay, estimated, 2 weeks at £288 = £576.00
- CITB levy at 2.5% = £375.00
- Travelling, estimated, 2 weeks at £288 = £576.00
- Bonuses, estimated, 48 weeks at £30 = £1440.00

Total annual cost = £22,034.00

Unit rate calculation:

To calculate a unit rate of labour, we need to establish the total number of productive hours on site. Total hours in one year:

- 4 days per week × 8 hours + 7 hours for Friday × 52 weeks = 2028 hours
- Less holidays: 4 weeks × 39 hours = 156 hours
- Less public holidays: 7 × 8 + 1 × 7 = 63 hours
- Less sickness: 2 weeks × 39 hours = 78 hours
- Total = 1731 hours

The hourly rate of labour is a simple calculation, as follows:

£22,034 ÷ 1731 hours = £12.73 per hour

This is the labour rate that an estimator would use to price the works contained within a tender.

Pricing dayworks: Materials costs

When calculating the cost of materials, each item should be broken down into common units, so they can be added together to establish a unit rate for the material delivered to site.

Worked example

The site production of brickwork mortar involves three materials: cement, sand and an additive. Calculate the unit rate for this brickwork mortar using the following data handed to you by the estimator.

Material data:

Mortar is mixed in the proportion specified by the architect or engineer, which is usually a ratio of 1:4 for load-bearing brickwork.

- Cement is delivered in a 25 kg bag, which costs £4.25.
- Sand is delivered in 20-tonne wagons costing £80 per load.
- Additive for workability is £9.00 for 2.5 l; 300 ml covers 100 kg of cement.

Unit rate calculation:

Each 1 m³ of mortar will weigh approximately 2 tonnes. Therefore, we require:

2000 kg = 400 g per unit

5 sets of units per 1 m³ mortar

Worked example

Cement = 1 unit × 400 kg = 400 kg

$\frac{400}{25} \times £4.25 = £68.00$

Sand = 4 units × 400 kg = 1600 kg

$\frac{1600}{20,000} \times £80 = £6.40$

$\frac{4 \times 300 \, ml}{2500 \, ml} \times £9.00 = £4.32$

Therefore, the total cost of 1 m³ brickwork mortar is

£68 + £6.40 + £4.32 = £78.72.

Pricing dayworks: Plant rates: Calculation of fixed and operating costs

Fixed plant costs are those that are not time-related and include:

▸ delivery

▸ erection

▸ removal.

Information for calculating delivery and removal charges could be obtained from the plant supplier. For example, if delivery was £35 for a dumper, this would be multiplied by two and added onto the total duration time-related costs on site, and then divided by the number of hours to give an hourly rate:

▸ Delivery collection charges = 2 × £35 = £70

▸ Time on site = 150 hours × £9 = £1350

▸ Total cost = £1420 / 150 = £9.47 per hour

Variable or operating costs are time-related to the length of hire on site and would include the coverage of:

▸ The cost of fuel – to calculate this, the consumption rate of the plant when working will need to be established.

▸ Puncture repairs – these will occur randomly and a certain percentage can be added in to recover the costs.

▶ Operator costs – the number of hours that an operator is required specifically to use the plant will need to be established.

Pricing dayworks: Plant rates: Calculation of hourly rates

Construction plant can be very expensive to own and operate. There are generally two methods of providing plant for construction sites:

▶ purchase, service, operate and maintain the plant
▶ hire in the plant either externally or internally.

The unit rate calculation for an item of construction plant will depend on several factors, including:

▶ ground conditions
▶ trained operators
▶ productivity
▶ height and reach
▶ breakdowns and reliability.

Remember that there is often a hidden cost associated with large pieces of construction plant – the cost of delivery to site using alternative transport such as a low loader.

Worked example

Calculate an hourly rate for a 1.5 tonne dumper using the data provided.

Data:
Hire rate = £150 per week
Driver costs = £300 per week (50% driving, 50% working on tasks)
Fuel requirements = 150 litres per week at £1.50 a litre
Working week = 39 hours

Unit rate calculation in hours:
Hire = £150 / 39 hours = £3.85 per hour
Driver = £300 × 50% = £150 / 39 hours = £3.85 per hour
Fuel = 150 × £1.50 = £225 / 39 hours = £5.77 per hour
Total cost per hour = £13.47

Application of plant costs in unit rates

The application of plant costs within unit rates for bill of quantity or measured quantity items must be carefully considered. Some plant items are used for several different measured items, activities and operations on site and so cannot easily be included within unit rates. They are therefore often placed within the appropriate section of the preliminaries section of the bill of quantities or specification section (see section A1 of this unit). Examples of types of plant that would be included within the preliminaries section include tower cranes, vans, skips, scaffolding and temporary lighting.

Calculation of unit rates for various classes of construction work

To undertake the calculation of unit rates for use in estimating, you need to know the output rates for two constituent factors – labour and plant. Output rates, as we have seen, can be obtained by actual measurement or from historical work study tables. It is worth looking at the factors that affect the output of labour on a construction site. Output can be affected by several factors, including poor motivation on site, excessive breaks or lack of supervision, dangerous or complicated work, poor working conditions or weather conditions.

The output of plant and equipment on site is affected by:

▶ the type of ground conditions
▶ ground or surface water levels
▶ the height to be reached or distance to move
▶ the amount of maintenance and servicing required
▶ the skill of the operator
▶ the age of the plant and whether it is hired or owned and its capacity for the work.

To calculate a unit rate for construction work, you need to break down the rate into the three main constituent parts of labour, plant and materials and calculate each element before totalling the unit rate. If we use a typical bill of quantities description, it will show the process of compiling the rate that would be applied to each item. Finally, profit can be applied to the rate against each bill of quantities item.

Worked example

Excavation work

Calculate the unit rate for the excavation (see below) using the data provided.

Data:
Excavate to reduced level, maximum depth not exceeding 1 m (unit: m^3)
Excavator with operator = £35 per hour
Output = 6 m^3 per hour, which includes loading the dumper, waiting for it to tip and then starting the cycle again. Four tonne dumper with driver = £55 per hour. One dumper is required and tips soil 10 m away.

Unit rate calculation:
Cost of excavation = £35.00 / 6 m^3 = £5.83 per m^3
Cost of dumper = £55.00 / 6 m^3 = £9.17 per m^3
Total cost to excavate and tip = £15.00 per m^3

Worked example

> **Brickwork masonry**
>
> Calculate the following unit rate for the brickwork masonry using the data provided.
>
> *Data:*
>
> Facing brickwork one side half brick thick in cement mortar 1:4 built in stretcher bond (unit: m^2)
> Brickwork gang (two bricklayers/one labourer)
> Brickwork rate per hour = £12.50
> Labourer rate per hour = £9.50
> Output per bricklayer = 45 bricks per hour
> 1 m^2 of brickwork contains 60 bricks.
> Brick costs = £335 per 1000
> Wastage = 5%
> Mortar delivered to site = £52.50 per 1 m^3 tub
> Mortar use = 0.04 m^3 per m^2 of brickwork
>
> *Unit rate calculation:*
> Labour (2 + 1 gang cost)
> Bricklayers = 2 × £12.50 = £25.00
> Labourer = £9.50
> Total hourly rate = £34.50
> Output for the gang = 60 bricks per m^2 / 2 × 45 bricks per hour = 1.5 m^2 per hour
> Unit rate = £34.50 / 1.5 = £23.00 per m^2
>
> Materials
> Bricks = £335 × 60 / 1000 = £20.10
> Add wastage @ 5% = £1.01
> Mortar = 0.04 × £52.50 = £2.10
> Total cost = £23.21
> Unit rate = £23.00 + £23.21 = £46.21 per m^2

 PAUSE POINT   Do you understand how to carry out calculations to create estimates?

    Look back through the examples provided and try to follow the workings.

    Put together a possible job in your local area – how would you begin to cost it?

Pricing preliminary items

Preliminaries (or prelims) is a description of a project that allows the contractor to assess particular costs which, while they do not form a part of any of the package of works required by the contract, are required by the method and circumstances of the works.

The Code of Estimating Practice published by the Chartered Institute of Building describes preliminaries as, '... the cost of administering a project and providing general plant, site staff, facilities, and site based services and other items not included in the rates'.

The National Building Specification suggests 'the purpose of preliminaries is to describe the works as a whole, and to specify general conditions and requirements for their execution, including such things as subcontracting, approvals, testing and completion'. Preliminaries and work sections together describe what is required to complete the works required by the contract. Standard preliminary items that will need to be considered are:

- employer's requirements and any accommodation required on site
- any costs and requirements of the management and staff
- site establishment and its security, safety and protection
- providing temporary services to the site, such as water
- providing safety and environmental protection, including possible cleaning costs
- providing fixed plant and any scaffolding or temporary works
- insurances, bonds, guarantees and warranties
- allowance for fixed or fluctuating price.

Completion of the priced bills of quantities

When completing the final bills of quantities, there are several considerations that need to be included to finalise the bill of quantities:

- Collections are the items for each trade or operation involved in the work. Summaries are the totals for each stage. For example, erecting an internal stud wall might involve carpenters for the stud work, electricians to put in the cabling for 13A sockets and switches, plasterers to finish the wall and decorators to paint the wall. Quantities will be needed for each trade (collections) and then totalled (summary).
- Check the procedures used to formulate the bills of quantities. In other words, how were these quantities arrived at? Were the methods used accurate and effective and were they based on the current versions of the design and specification?
- Ensure that the final bills of quantities are produced in accordance with the standard methods of measurement such as NRM2 (Royal Institute of Chartered Surveyors New Rules of Measurement 2) and CESMM4 (Civil Engineering Standard Method of Measurement 4).

Note: The Royal Institute of Chartered Surveyors and the Institution of Civil Engineers are professional bodies to which surveyors or civil engineers can belong. They set standards and provide training, news and information for the professions.

With these last actions completed, the various prices need to be added together and presented as the estimate for the works.

B4/B5 Analysis of the estimate

It is essential to get the estimate right. If the estimate is too high then the contractor might not get the project. If the estimate is too low, then the contractor will not make a profit. You need to show your client how you have arrived at your cost, which helps them to see the price is fair and professionally provided.

Breaking down the estimate into key cost centres helps you to analyse the considerations involved in making a professional project estimate as well as to show how to set it out clearly so the potential client can understand the breakdown; this includes different costs as below:

- Preliminaries – have the costs of preparation works such as temporary supplies, site accommodation and fencing been considered?
- Labour cost – are the hours allowed for the works sufficient and have the costs incurred been calculated accurately to cover these hours? Do these labour costs include overheads such as National Insurance?
- Materials cost – are all materials allowed for? Have discounts been applied and reflected in the quote?
- Plant cost – would there be any gain from buying plant rather than hiring? Are insurance, fuel and any training (for use) costs included in the estimate?
- Overhead allowance – have hidden items such as insurance, administration and departmental costs been included in the quote?
- Total of **prime cost (PC) sums** – has it been made clear that these sums are, in fact, provisional and not finite costs?
- Total of provisional sums and contingencies – have extras, changes to the design, breakages, stoppages from bad weather or other unseen problems been considered and costed into the estimate? Has it been made clear that extras will need to be negotiated with the client and are outside the original estimate? Have transport and fuel costs and other hidden expenses been included in the quote?
- Suggested alternative tenders or variant bids – where appropriate, have the quotes and method statements from alternative suppliers been considered and included, and have variant bids been included, for example quotes for alternative materials, construction or installation methods?

Key term

Prime cost (PC) sums – the likely price of items difficult or impossible to price accurately at the time of estimation.

Assessment practice 3.2

You have been asked to complete an estimate for the police station project. You have been given the following information on the local police stations:

Bungalow construction: traditional brick construction with breeze block load-bearing internal walls; stud and plasterboard non-load-bearing internal partitions.

Roof: traditional pitched roof; wooden frame; tiles. Windows and doors: uPVC double glazed; custom made with added security features.

Wall dimensions:
- length 25,000 mm
- width 17,000 mm
- height to eaves 4000 mm
- height to roof peak 6000 mm.

Parking area:
- width 23,000 mm, including entrance and exit roadway that runs along the side of the building to car park at back
- width of car park minus roadway 17,000 mm
- length of car park from rear of building to perimeter 18,000 mm.

Other features:
- CCTV cameras
- intruder alarms
- car park at the rear of the building
- gas, electricity and water supplies needed
- interior fittings to be specified separately.

Produce an estimate that:

1 Builds up unit rates:
- description – demolition of existing buildings (unit of measurement: m^2 area cleared)
- facing brickwork in type A facings; external cavity wall; 1/2 brick thick in coloured mortar type 1 (unit of measurement: m^2)
- retaining excavated topsoil on site in temporary spoil heaps (unit of measurement: m^3); imported hardcore fill in level beds exceeding 50 mm, not exceeding 500 mm, overall thickness 375 mm (unit of measurement: m^3)
- damp-proof membrane, 2000 g Visqueen, over 500 mm wide horizontal (unit of measurement: m^2)
- 75 mm cavity wall insulation built in as the work proceeds (unit of measurement: m^2)
- 100 mm blockwork internal facing in sand and lime mortar (1:1:6) (unit of measurement: m^2). **(7 marks)**

2 Complete a draft bill of quantities with measurements. **(4 marks)**

3 Complete preliminary calculations and pricing to suggest an estimated cost. **(5 marks)**

Plan
- What are the tasks I am being asked to do?
- How can I condense the information into an effective presentation?

Do
- I know where to research the information needed.
- I can sift through the information to identify the key points needed.
- I can present the information and the report in a clear and logical manner.

Review
- I can explain the task and how I approached it.
- I can see how I would approach this differently next time.

C Commercial decisions

There are many factors that influence the final tender decision. However much effort a firm puts into preparing the design stage or the cost of elements, risk will always exist when working with project information at the planning stage.

Winch (2010) believes the lower the level of information available at the earlier stages, the higher the possibility of risks and uncertainties; as project information increases, risk is expected to decrease.

C1 Application of risk analysis to make commercial decisions

Risk is an uncertain event that can have either a positive or negative effect on the project objectives such as cost, time, quality and scope. Risk is a major factor that a contractor will need to review when deciding to make a tender for a project.

The principal impacts to consider when examining risk are impacts on:

▶ Time – will the risk make the project longer to complete than expected, or cause it to miss a deadline required by the client?

▶ Cost – will the risk make the project more expensive to complete? Will it become more expensive than the client can afford, or will the extra outlay start to endanger the contractor's profit on the project? Budget overrun on a construction project can be affected by:
- complexity of design and construction and its scale and scope
- method of construction
- tender period
- market conditions and client's financial position
- availability of materials and labour
- buildability and site constraints
- attitudes and risks of the local community (such as crime).

A business will need to carefully review the risks that could affect time and budget before making a decision about whether or not to make a tender for the work.

Types of risk

There are two main types of risk: quantifiable and unquantifiable.

▶ Quantifiable risks are those where you can make a measurement of the likely level of risk. The appropriate costing for the project can then be applied to the cost estimate to cover this. For example, if a very tight timescale has been applied to a project, a small delay in the work programme could put this timescale at risk. Contingency for delays will need to be included in the estimate.

▶ Unquantifiable risks are those where it is impossible to make a clear estimate of the level of likely risk. In these cases the amount is based on the management's perspective and perception of the possible level of risk. For example, work held up by the discovery of asbestos (requiring specialist removal), in a building undergoing renovation.

The addition for risk on the project is made in one or more of the following ways:
1 A percentage in the profit margin
2 A separate percentage on all the costs
3 A lump sum in the preliminary bill
4 A percentage in one bill if the risk is in that bill alone.

Link

We looked at use of commercial risk analysis earlier in this unit in learning aim A. This section looked at the many areas where risk could arise, including the workload of the contractor, market conditions and economic factors, programme factors connected to the client's aims and the contract and financial issues.

Case study

A large computer centre received an Improvement Notice from the local authority, stating that comfort cooling was required in the main office areas.

These were major works. The suspended ceiling would have to be removed, scaffolding erected and runs of large pipework and electrical supplies installed in the ceiling space. Diamond drilling would also be required as the walls were reinforced concrete.

The main problem was that the office required staffing 24 hours a day, seven days a week. Arrangements would have to be made to move operations from each floor for a complete weekend, starting at 5:30 p.m. each Friday and then restoring service by 8:00 a.m. on Monday. Round-the-clock working would be required to meet the deadline. If any problems arose that held up the work the penalties would be high.

Check your knowledge

Discuss the possible risks to successful completion of this project and how these could be overcome.

C2 Use of commercial intelligence

Many tenders are likely to be competitive, with multiple companies bidding to take on the work for the client. The more knowledge that a business has about its market, and the expected rivals it has in that market, the more prepared it will be for dealing with competition and making a unique offering to the client. It will also be in a better position to present a clear case for why it is a better option for a client to select than its rivals.

The main aim of understanding competition in the industry and on a tender is to improve the effectiveness of the company's investment. Competition can encourage companies towards better quality, time management and efficiencies, as companies compete to offer the most cost-effective, efficient service. The more competition there is, the harder it will be for over-priced and badly organised companies to survive.

When preparing tenders, bidders seek every cost-effective factor to provide a low estimation of the cost for undertaking the work, giving a good deal to the client while still allowing the company to make a profit.

> **Link**
>
> We looked at use of commercial intelligence earlier in this unit in learning aim A.

Ⅱ PAUSE POINT Do you understand how risk can affect a decision on whether to tender for a project or not?

Hint When you are deciding to do something risky or not what stages do you go through before you make a decision?

Extend Research construction projects in your area. What risks were there about these projects before they decided? How might the contractors have planned to deal with these risks?

C3 Tender adjudication and settlement meetings

The tender price consists of both direct and indirect costs elements.
▶ Direct costs are traceable to a specific work item, such as labour, plant, materials and subcontractor costs.
▶ Indirect costs are not traceable to a specific work item and cannot be allocated to a cost centre, such as site overheads, general overheads, profits and allowances for risks.

When indirect costs exclude site overheads they are often termed 'mark-up'.

Despite the difficulties in estimating, direct cost estimating is usually similar between firms because they all have access to the same labour supply, type of tools and equipment and suppliers. They essentially have the same opportunities for supervision and mobilisation of labour and materials.

The cost variations in competitors' bids are therefore due to their selected mark-ups. Competitive tendering processes result in high tendering cost and high general overhead costs so any tools that can increase the efficiency and productivity of the tendering process will be beneficial to every individual firm.

Adjustments to the final sum

Overheads and profit margin

General overheads are often determined by expressing the budgeted annual overheads as a percentage of budgeted turnovers and applied as a proportion of the cost to individual contracts. Examples of overheads include insurance, administration, vehicle tax and fuel costs. General overheads can also be a fixed percentage of the tender sum; this can be determined by senior management. The general overheads can be estimated as an annual budget for the company based on the previous year's expenses adjusted for inflation.

All contractors have a budgeted turnover target; this includes current work in hand, future work and anticipated tenders to be won. Based on this and other factors, the senior management decides on the amount of money or percentage of the contract to be added as profit. The major factors which affect the profit include type of work, complexity of work, experience, market situation, duration of contract and its completion.

Potential for buyers to generate further profit margins

When buying materials and services, there are opportunities for cost saving, and increased profit margins. For example, prices can be negotiated and brought down from the standard charges levied by the supplier. Buying in bulk can also result in discounted prices.

Current market conditions and economic climate

The final sum may need to be adjusted to take into account economic conditions. In a time of prosperity, with markets riding high, clients might be more willing to spend what is required to get the work completed as completion may be more important to them than cost. During times of adverse economic conditions, cost becomes more of an issue and the profit margins are adjusted to be much tighter.

The need to maintain workload

The major aims of contracting firms are survival, growth and profitability within an increasingly competitive environment. These aims can only be achieved by ensuring that an adequate and appropriate workload is obtained. This means contractors must continue to obtain construction work from preparing bids or tenders in accordance with the Code of Estimating Practice.

Maintaining the workload of the company will:
▶ contribute to covering the cost of overheads for the company's day-to-day operations
▶ avoid the need to make staff redundant
▶ continue to bring a steady cash flow into the business.

Retaining skills and experience

A steady supply of work makes it easier for a company to retain staff, allowing them to train and develop their employees and keep that expertise within the business, both in craft and construction specialists and in the management of the business. This helps the business to excel.

Previous experience of similar projects

Experience of the type of project the contractor is tendering for may be reflected in their final quotation. For example, if particular problems were encountered in the past, the possibility they may do again will be taken into account as a contingency.

The time taken to prepare the tender may also be shorter, as the methodologies for completing all or part of the work have already been established. A selling point for the tender may be that the timescales for completing the project may be shorter than those of competitors with little or no experience of this type of work.

Planned expansion and acquiring new expertise

If a company can build on and develop its level of workload, it will be in a stronger position to expand and develop. This will help make the company larger and able to take on more work. A larger company will be in a better position to increase its profits.

A larger company will also be able to recruit new, more highly skilled employees, increasing its level of expertise within the business, making it a stronger company.

Commercial risk and competition

Risks and the level of competition will be a major consideration when deciding on the level of adjustment to a quote that a company will make. Companies must take into account the number of other likely tenders, and whether competitors are likely to tender.

Discounts for the main contractor or subcontractor

There may be opportunities for discounts to be applied, as a negotiating point. The main source of a discount is likely to be from the suppliers of the materials needed for the work. This can then be passed on to the customer, but it can also be a source for increased profit margin for the contractor or subcontractor.

Conversion of the estimate into a bid

CIOB (2009) Code of Estimating Practice highlighted two key review meetings that occur during the tender process to help prepare the tender for a final bid for the work.

1 **The mid-tender review** – the estimators explain their approaches and others in the company can share their ideas and expectations. The review should cover several different areas including:
▶ contractual (subcontractor arrangements, terms and conditions and amendments)
▶ administration (responsibilities, staff allocation, site visit, presentation documents, subcontract arrangements)
▶ financial (insurance, cash flow, bonds)
▶ technical (temporary works, design appraisal, plant, risks, safety, alternatives)
▶ programme (type, alternative methods and sequence, input from specialists)
▶ queries for the client (decision based on raising clients' queries, details of the amount of information, consideration of qualification to tender, consideration of alternative tender, extension of tender).

2 **The final tender review** – this is the final review meeting and an opportunity for the contractor to make a final review of any risks and reach a final decision about whether to tender for the work or not. It should cover:
▶ site and location factors (location and ground conditions, hazard and security, project description alternatives, report of site visits and all the photographs)

- the different parties involved (client, consultant, subcontractor, contractors)
- the contract (format of contract, insurances, payments and detention, bonds)
- the programme of work (method statement, specific requirements such as duration, commencement date, specific contract conditions that could affect the intended method of working, restrictions, access to the site, duration of the project, sequencing of the project, costs):
  - the estimates (labour, plant, materials, subcontractors, provisional sums and prime cost)
  - any overheads
- a final summary (price and review firm adjustment, value related item) mark-up (cash flow, scope, risk, qualifications, overheads, profits, discount for main contractor and VAT).

Director's adjustment

After the final review the company will convert an estimate into a tender with the estimate reviewed through a detailed examination of its components. These include rate and quotations methodology. The firm's directors will then consider commercial matters and risk, prior to a decision on an appropriate profit mark-up. This project mark-up will then be added into the final tender, before submission to the client.

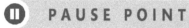

PAUSE POINT   Do you understand the various factors that feed into the tender adjudication process?

Hint   What are these factors and how do they affect the final decision?

Extend   Look at three of the factors and apply them to the computer centre case study above. How would they affect decisions made in this case study?

C4 Communication skills

The object of tendering for work is to win a contract. The continued prosperity of a contracting company depends on a steady flow of work so the more tenders they win, the more secure their position. Clear and effective communication is key to winning the contract. In all cases, the contractor will be putting forward why they should be selected for the tender.

Today's business and workplace communication uses a variety of different tools and skills. Traditional forms of communication such as written communication (letters and reports) and oral communication (face-to-face and presentations) have been used for many years. There are also many new forms of communication using technology such as email, instant messaging, text messaging and types of social media such as Twitter and LinkedIn.

Presenting tenders

If the bid package is to be submitted in writing, it must be direct, clear, well presented and compelling. All costs should be clearly presented and where necessary clarified. The style should be economical, affirmative and free of superlatives.

Some bids may also be presented to the client as part of a verbal presentation. In this case the person presenting the bid must be confident, enthusiastic about the project as well as knowledgeable and able to engage the audience. PowerPoint slides may be used and can be helpful, but should not be a crutch for the presenter. They also work best when kept uncrowded, clean, clear and economical. Handouts should be provided for the client. Any notes the presenter uses should be prompts, ideally short notes on a set of cards. The presentation must not be read from either a script or the screen – it should be a conversation between the presenter and the audience.

Imagine you are putting forward a tender for some home repairs in a large private house owned by someone working in banking. They have little experience in construction and only a short time to listen to a series of tenders. How would you prepare a presentation and what tone and level of information would you want to include in it?

Whatever the form of communication you are using, you must make sure of several key factors:

▶ Is the message, and the way you have chosen to deliver it, appropriate for the audience?
▶ Have you used the appropriate level of technical language for your target audience? Remember some clients may not be familiar with the detailed terminology used in the construction industry.
▶ Is the message you are delivering clear and concise? Do not use too many words where a few will do. If you take too long to make your point, the key issues you wanted your audience to concentrate on may get lost.
▶ Have you used the right tone for your audience? If you are presenting to a potential client you will speak very differently from how you would if talking to a friend. You will want to present an impression of professionalism and expertise, so the client feels confident in entrusting their work to you.
▶ Is the style you have used appropriate? Remember an appropriate style will hold interest. A technical style for a non-technical audience may well lose attention or may not be understood.

Assessment practice 3.3

Now you have completed your tender application for the police station project, you need to justify the tender sum you have recommended.

Write a short report for the company, explaining the decisions you have taken and justifying them. (5 marks)

Your report should include a proposed allowance for overheads and profit and a director adjustment to allow for a final total.

Plan
- What are the tasks I am being asked to do?
- How can I condense the information into an effective presentation?

Do
- I know where to research the information needed.
- I can sift through the information to identify the key points needed.
- I can present the information and the report in a clear and logical manner.

Review
- I can explain the task and how I approached it.
- I can see how I would approach this differently next time.

Further reading and resources

Chartered Institute of Building (2009) *Code of Estimating Practice*, seventh revised edition, New Jersey: John Wiley and Sons.

Cong, J., Mbachu, J. and Domingo, N. (2014) 'Factors influencing the accuracy of pre-contract stage estimation of final contract price in New Zealand', *International Journal of Construction Supply Chain Management* Vol. 4, No. 2, pp. 51-64.

Winch, G. (2010) *Managing Construction Projects*, second revised edition, Chichester: Wiley-Blackwell.

Getting ready for assessment

This section has been written to help you to do your best when you take the assessment test. Read through it carefully and ask your tutor if there is anything you are still not sure about.

About the test

This is an externally set unit and you will be provided with a date for the assessment window. You have a fixed amount of time to undertake the assessment, which will be under external exam conditions. In this section, we will examine the report sections of the assessment that cover the descriptive assessments and not the take-offs and unit calculations.

Pearson's website provides some sample assessment materials and previous external papers which can be used to revise from. The two sections we will examine can be accessed as a PDF you can download and revise from at **https://qualifications. pearson.com/content/dam/pdf/BTEC-Nationals/Construction-and-the-Built-Environment/2017/specification/Unit-3-SAM-Tendering-and-Estimating.pdf**.

The sample assessment looks at a project for 20 new units on an existing industrial estate.

Make sure you plan your time effectively. Some questions require more time than others to complete: this is especially true for any mathematical calculations where analysis is required to produce solutions.

Look at each of the questions and read them to gain an understanding of what is required. Each question has an indicated time against it. This helps you understand how long should be spent on each. Take note of these timings as they will help your overall progress. Read through the whole paper so you know the full extent of the content to cover. Note the data at the rear of the paper for unit rate calculations.

Activity 1

Produce a commercial risk report for your commercial director that details the potential commercial risks the project may expose your company to, with justification of your recommended tendering strategy. You are advised to spend 1 hour and 45 minutes on this activity.

A1, at the start of this unit, provides the information that you need to analyse to understand the commercial risk for tendering on this project. Commercial risk is identifying the factors that will affect the level of profit this project may produce.

In this case the commercial risks are as follows:
- The high water table level, indicated on the borehole log, will require pumping for the construction of the excavations for drainage and services at a depth of 1.2 metres.
- The client has only been trading for the past 15 months and has not filed any accounts and has limited share capital. Will we get paid?
- Savings are indicated already if the tender is over budget.

- There is evidence of vandalism and fly tipping which may cost your company money if it occurs when work starts.
- The location of a primary school will affect the delivery times for resources on site twice a day as when parents drop off and collect pupils congestion will occur at the school. Speed limits are also enforced.
- 360 metres of temporary fencing are required which is extensive.
- The three trees will possibly need to be protected and carefully maintained during the contract.

These are just a few of the points that you will need to draw out from reading the external assessment. It involves reading carefully each line of text and asking yourself:

- Would this have a commercial effect on our tender?
- What are the risks associated with it in financial terms?
- Do we need to make provisions within the tender?

Activity 4

Produce a report for the tender adjudication committee that considers all information in the set task brief to recommend and justify a tender sum. The report should be accompanied by a completed final section of the bill of quantities, with a proposed allowance for overheads and profit, and director adjustment, to give the final total.

Tender adjudication is the process of deciding what profit or margin to apply to a tender so it can be finally submitted. When you read the external assessment some points are raised as follows:

- Your company is based in the town of Toleigh and is currently operating at 65 per cent capacity.

The significance of this is that you have the capacity to undertake the work should you wish it along with the other tender won.

- The company wants to go through a period of rapid expansion as the present directors wish to retire in a few years' time and sell their majority shareholdings for the best possible price.

This will have a bearing upon a lower margin rather than a higher one applied to the tender.

- When the company was unsuccessful, the winning tender was between 1 per cent and 3.5 per cent lower than their bid.

This gives us an indication of the current market level of tenders and how much to try and shave off our tender to win the contract.

- Contractor A has a reputation for 'buying' work, i.e. submitting very low bids when they near the end of current contracts.

Contractor A will submit a low bid as they buy work but they appear to be busy at the moment.

- Contractor B appears to have been winning tenders with very competitive pricing but have a reputation for submitting contractual claims whenever they identify a legal loophole.

> Contractor B will price competitively.

- Contractor C, however, has been successful on a number of tenders recently and is thought to be working close to full capacity.

> Contractor C will not have the capacity to do the work and will price higher.

You can see from this analysis how to tease out the implications from the text and how the competition will react and the margin they may place on the net tender value. You have to read between the lines, using commercial understanding to assemble your adjudication report. Start with an introduction section that details the project. The main section needs to detail the factors identified above. The final section may make a recommendation as to the margin that you would propose for this tender. Make sure that your work is typed or clearly handwritten in a neat formal style.

Answer: Activity 1

The company is operating at 65 per cent of its capacity and therefore has 35 per cent available with which to take on work. The company is a regional contractor and so has been used to tendering across their region only and not nationally. We do not know what type of work they specialise in but they carry bricklayers, carpenters, ground-workers and general construction operatives. The specialist trades used are subcontractors, which would support this type of work.

The client may wish to use a serial tendering method to see how the company performs on the first new unit then offer subsequent units to the company. The commercial risk that can occur with this tender if successful would be:

- Current workload – we need more work as we are below capacity which may mean a lower tender margin is added to net costs. Any risks on site would not leave much capacity to absorb any issues that may affect costs.

- Need to maintain turnover of work – this is vital to the profitability and continuance of the company which has steadily grown from small beginnings into a larger net worth of £11m.

- Market conditions and economic climate – we do not know the current economic climate conditions that we are tendering on. Construction is starting to increase in strength after recent recessions. This may mean that tender margins will increase as companies get busier.

- Project considerations – we do not know the complexity and scope of each new industrial unit and can only assume that they will all be a standard design and similar. This means after the first one is constructed we will have developed a learning curve and the next will run much easier in terms of already knowing the issues that might occur.

My recommendation for the tendering strategy would be to use serial tendering as this provides scope for the client to maximise a schedule of rates that can then be applied to the other new units. The schedule of rates will need to cover the full specification for the unit.

The lowest schedule of rates will need to be checked to confirm that they are viable and that preliminaries are valued on a similar basis.

You have made a great attempt at this task. You have provided some depth of analysis against the state of the company and its ability to tender for the units. You have picked out relevant points that would need to be considered in the decisions that the company's management would need to make. You have identified some factors that would strongly influence the commercial risks associated with this tender.

Answer: Activity 4

Having examined the documentation with the tender and taken into account the commercial risks I have recommended a tender mark-up of 2 per cent. The reasons for this are as follows:

- We are at 65 per cent capacity and need more work in order to keep all employees actively engaged and working.
- We do not know what the competition is on this tender and have no commercial knowledge.
- The police stations are on a serial tendering basis and we will, with a good relationship with the client, be able to win all six projects.
- The serial tender should contain a section where we can price for the contract preliminaries which we do not know about at the minute.
- We may be able to make savings with contract preliminaries sections of the tender as they are regional police stations; some places will be cheaper to operate within than others
- If we successfully complete one police station we may be able to negotiate the subsequent police stations to ensure a successful outcome across all six stations.

You have provided justification of the factors that have influenced your tender mark-up of 2 per cent with the limited knowledge that the tender documents contain. Without any commercial information on the competition you have made a wise decision on the low mark-up recommended.

Construction Technology 4

Getting to know your unit

Knowing how a building has been assembled, installed or constructed enables you to understand how you can adapt, extend or decommission it when changes are needed or required by clients.

This unit examines the technologies associated with the construction of low-rise domestic and commercial buildings. Many of these modern methods of construction involve a high degree of prefabrication. You will explore the foundations required to support their superstructures, and the subsoil investigations undertaken to provide data for foundation design. Superstructure design will be explored in terms of the details required to construct floors, walls and roofs in forming the external envelope of a building.

Finally, you will learn about the external elements of a building's infrastructure, the drainage systems that serve them, the roads and pathways for access to and egress from the building, and the provision of utilities.

This unit can help you progress into a career in construction management and supervision in the construction sector as a contracts manager or site management or to Higher Nationals in Construction and degrees in construction specialisms.

How you will be assessed

Throughout this unit, you will find assessment activities that will help you work towards your assessment. Completing these activities will not mean that you have achieved a particular grade, but will mean you have carried out useful research or preparation that will be relevant when it comes to your final assignment.

To achieve the tasks in your assignment, you should check that you have met all of the Pass grading criteria. You can do this as you work your way through the assignment.

If you are hoping to gain a Merit or Distinction, you should also make sure that you present the information in your assignment in the style that is required by the relevant assessment criterion. For example, Merit criteria require you to analyse and discuss, and Distinction criteria require you to assess and evaluate.

The assignment set by your tutor will consist of several tasks designed to meet the criteria shown below. This is likely to consist of a written assignment but may also include activities such as:

▶ a report on the forms of low-rise construction for a client to consider
▶ designing a foundation within given parameters
▶ designing external works to service a building.

Assessment criteria

This table shows you what you must do in order to achieve a **Pass, Merit** or **Distinction** grade.

Pass	Merit	Distinction
Learning aim A Understand common forms of low-rise construction		
A.P1 Explain the different structural forms used in the construction of low-rise buildings.	**A.M1** Discuss the use of different structural forms for use with a given low-rise buildings project scenario.	**A.D1** Evaluate the effectiveness of different structural forms for use with a given low-rise buildings project scenario.
Learning aim B Examine foundation design and construction		
B.P2 Explain the different types of investigation used to provide information required for the design of foundations for low-rise buildings.	**B.M2** Discuss the principles of foundation design and how they impact on the choice of foundation type for low-rise buildings.	**BC.D2** Evaluate the construction of new low-rise buildings.
B.P3 Explain the different types of foundation used for low-rise buildings.		
B.P4 Describe the principles of foundation design and how they impact on the choice of foundation type for low-rise buildings.		
Learning aim C Examine superstructure design and construction		
C.P5 Explain the construction details used in the construction of walls, floors and roofs on new construction projects.	**C.M3** Analyse the different details and finishes used in the construction of new construction projects.	
C.P6 Summarise the use of internal finishes for floors, walls and ceilings on new construction projects.		
Learning aim D Examine external works associated with construction projects		
D.P7 Summarise the design and construction of external works on new construction projects.	**D.M4** Discuss the design and construction of external works for new construction projects including the incorporation of a Sustainable Urban Drainage System.	**D.D3** Analyse the design and construction of external works for new construction projects including the incorporation of a Sustainable Urban Drainage.
D.P8 Explain the use of Sustainable Urban Drainage Systems within new construction projects.		

A Understand common forms of low-rise construction

A1 Forms of low-rise construction

Several types of construction methods can be employed for the construction of low-rise dwellings and other structures. These can be classified as:

▸ framed structures

▸ traditional construction

▸ modular construction.

Each method has different characteristics based upon the building's final use, the transfer of loads, sustainability and energy conservation. Each of the three methods will be examined in detail, evaluating the benefits of each and their limitations in terms of their application to low-rise construction.

The term **modern methods of construction (MMC)** is often applied to those methods involving modular construction. This unit also considers the traditional methods that may still be employed and which you are likely to find in any conversion or adaptation.

Framed structures

Steel skeleton frame

Steel is the most common material adopted in modern buildings, and includes light gauge steel with prefabricated components. A steel structure starts with pad **foundations**. Steel columns are fabricated off site and assembled and attached to the pad foundations using bolted connections. The bottom of the columns (the base plate) is then grouted into place so no movement is possible once the building has been levelled and plumbed. Each column is secured into position by horizontal beams, all with bolted connections. Diagonal wind braces, to prevent and control wind loading, are added to the external elevations of the steel **skeleton frame** as it is erected. These can be seen at the rear of the structure in the picture below. Load transfer with a steel frame is from the floor loads onto the beams which are then connected to the columns or stanchions. These transfer the loads down to apad foundation which transmits the loads to the ground.

Key terms

Modern methods of construction (MMC) – a range of techniques involving off-site manufacture or assembly, or more efficient on-site methods, to minimise construction time on-site.

Foundation – the lowest, load-bearing part of a structure, usually below ground level.

Skeleton frame – a rectangular framed structure made of several elements joined together.

▶ Steel skeleton frame consisting of a network of columns and beams

Advantages

There are several advantages to using steel skeleton frames:

▶ fast erection phase
▶ off-site quality-controlled fabrication
▶ standardised sections enable economies of scale during production
▶ recycling and sustainability benefits
▶ site-bolted connections do not require any welding
▶ creates a strong structure.

Disadvantages

However, there are also some disadvantages, including that:

▶ steel loses 50 per cent of its strength if exposed to high temperatures such as fires
▶ it needs a secondary fire-protection finish
▶ it needs to be treated to prevent corrosion from rust
▶ the maximum length of columns is limited by transport restrictions
▶ repainting is required in external exposed locations.

In-situ concrete skeleton frame

A system called formwork and falsework is set up for low-rise buildings with more than one floor. The formwork is the sheet materials that the concrete is directly poured against or onto. Falsework is the support system that holds the formwork into position. Mould oil is brushed onto the surface of the formwork to prevent the concrete from binding to it, making it easier to remove. Wet concrete is then poured into the formwork moulds. Concrete must be left to 'cure' for 28 days so that it can steadily achieve strength by allowing the chemical processes in the cement to bond to the aggregate mix. After seven days, most formwork systems are **struck**, cleaned and re-fixed into the next position.

If a beam were to be placed under concrete's dead and imposed load, it would bend downwards, causing the top of the beam to go into compressive stress and the bottom into tension. In-situ concrete does not perform well in tension and tends to crack. Therefore, steel reinforcing bars are placed in the zone of tension at the bottom of beams and held in place by wet concrete compacted around the steelwork. Spacers are placed between the surface of the formwork and the steel reinforcing bars to ensure that an adequate cover of concrete is maintained around them. The picture below shows an example.

Research

Investigate the design of the bolted connections within a steel skeleton frame. Evaluate their advantages in terms of strength and efficient erection.

Key term

Struck – the process of removing formwork and falsework by taking it away from the set concrete.

Loading transfer is the same as for steel, except for the type of foundation that may be supported. A concrete frame is heavier and the foundation may need to be bigger to transmit the loads over a wider area. One solution is to use piled foundations with a pile cap on which the columns sit.

▶ Steel reinforced in-situ concrete

Advantages

There are some advantages to an in-situ concrete frame:

▶ Fire protection is built into the material.
▶ Post-tensioning systems can be added to increase loading capacities.
▶ Concrete can be moulded into any shape.
▶ Thin shell construction can be achieved.

Disadvantages

However, there are also some disadvantages:

▶ It requires longer construction periods due to the time needed for curing before formwork can be removed.
▶ Heavier foundations are required.
▶ A crane is required on site for the duration of the works.
▶ A concrete pump is required.

 PAUSE POINT What method of construction would you recommend for an office with three floors?

Hint What factors would you need to consider?

Extend What alternative method could you recommend?

Prefabricated concrete skeleton frame

This system uses a factory-produced unit that forms a prefabricated (or precast) element for the superstructure. The elements are unique systems that are assembled on site by craning them into position. The joints are secured with concrete or other filler systems.

Virtually any component can be prefabricated, for example, flooring units, wall panels, beams, columns and lintels. The benefits of producing such elements within a factory far outweigh those of using in-situ methods.

Each manufacturer has their own system for jointing the units on site: some involve a bolted system, while others may be site-concreted into their final position. Load transfer with precast concrete is the same as for steel and in-situ concrete but the foundations are prepared with a pocket to accept the precast column which connects to the foundation.

The illustration below shows an example of customised slabs that have been precast and are then installed on site as required.

▶ Precast concrete wide floor units installed on site

Advantages

There are several advantages to using prefabricated concrete skeleton frames:

▶ Fire protection is built into the material.
▶ Curing periods can be accelerated in the factory using steam.
▶ Elements can be erected quickly, as no need to wait for curing.
▶ A range of surface finishes can be accommodated.
▶ They are manufactured in a factory, which is a controlled-quality environment.
▶ The system is unaffected by weather.
▶ Multiples of the same type of unit can be produced efficiently.
▶ The amount of concrete used to cover the reinforcing steel is factory-controlled so there is less wastage.

Disadvantages

However, there are also some disadvantages:

▶ Jointing is required on site.
▶ Heavier foundations are required.
▶ A crane is required on site for the duration of the works.
▶ Maximum sizes are limited by transport restrictions.
▶ There are issues with site assembly in high winds.

Portal frame

A portal frame is a unique shape formed using sloping beams and columns. A portal design, and the components that form it, can clearly be seen in Figure 4.1.

> **Research**
>
> Investigate the different methods of site connections employed by various manufacturers for prefabricated concrete components.

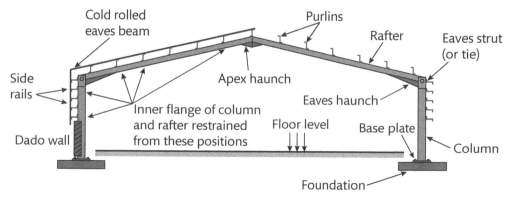

Cold rolled eaves beam — Purlins — Rafter — Eaves strut (or tie) — Side rails — Apex haunch — Eaves haunch — Inner flange of column and rafter restrained from these positions — Floor level — Base plate — Dado wall — Foundation — Column

▶ **Figure 4.1:** An example of a portal-framed building

Research

Investigate the different types of portal frames made from each material and compare their advantages and disadvantages.

The eaves connection is often referred as the 'knee' of the portal frame. The beauty of this system is that a large clear span can be formed within the building without any obstructive columns. This can aid, for example, the operation of fork trucks. A portal frame can be manufactured from a variety of materials including steel, laminated timber and prefabricated concrete.

The rafters transfer roof loads to the stanchions, which are then bolted to the pad foundations. As there are no internal columns, the corners and the ridge are reinforced with additional web plates.

Timber frames: Prefabricated platform frames

This method gets its name from the storey-height panels headed by a horizontal binder (header plate) to form a platform that the intermediate floor sits upon. This header plate is attached to the top rails of each platform panel. As with all timber frames, the accuracy of the construction is initially determined by the starting levels that the frame sits upon. The sole (bottom) plate that is fixed to the substructure needs to be perfectly level so that the bottom rails of the platform panels are also level and vertical. This avoids any distortion within the frames and ensures that the installation will be problem-free. Figure 4.2 shows typical timber framed wall panels and their components.

Theory into practice

A client wishes to relocate to a new industrial inner-city development and has approached your design and construction company. It requires a building that does not contain any internal columns because an automated picking system will operate on the storage racks. You have been asked to produce and present a bid for this at a tender meeting.

Produce a series of slides as part of the main presentation that detail the type of building that could accommodate the client's needs. Ensure you include:
- foundation details
- type of structural frame
- access and egress into the building
- a height of 5 metres at the eaves
- some natural lighting.

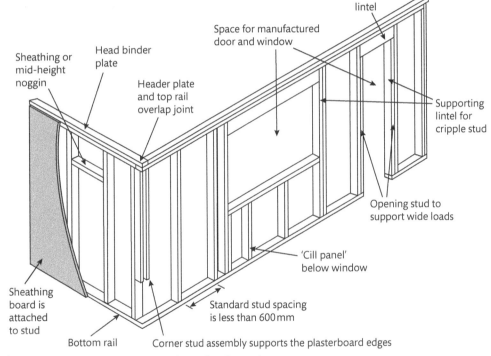

▶ **Figure 4.2:** The components of a timber framed system

Advantages

There are several advantages to using prefabricated timber platform frames:
▶ The method of erection is faster than traditional methods.
▶ Structures can be occupied sooner as drying time is not required.
▶ They are not affected by the weather as the completed construction is waterproof.
▶ Not as much skilled labour is required on site.
▶ The structure provides spaces for the installation of services conduits.
▶ Engineered timber is a higher quality product.
▶ Its load-bearing system is instantly assembled.
▶ The lighter construction means savings on foundations.
▶ It has life cycle recycling potentials.

Disadvantages

However, there are also some disadvantages:

▸ A crane is required on site to lift panels, floor sections and roofing into place.
▸ Installation in high winds can be dangerous due to relatively light weight and large surface area ratio.
▸ The damp-proof membrane may become penetrated prior to plasterboarding by trades on site.
▸ Lead-in time is required for design of panel layout.
▸ Fire rating detailing is required to ensure conformity.
▸ Timber decay with moisture ingress can occur.

These advantages and disadvantages apply to all the timber framed systems that follow.

Timber frames: Open-panel systems

In an open-panel system, the panels are not closed by covering or insulation to the inside. This enables electrical and plumbing work to be undertaken and for the finishes to be installed along with a vapour control layer. Panels are usually delivered with the plywood and breather membrane attached and installed using manual and crane techniques. In the picture below, you can clearly see that the back of the panels remains open. Insulation is cut and placed within the panels after the services have been installed. The panels are then covered with a layer of damp-proof membrane material, and plasterboarded, and then either dry-lined or skim finished.

▸ An open-panel timber frame system

As well as the advantages above, an open system means that the services can be easily installed, hidden from users' view and tested before the final finish.

Timber frames: Closed-panel systems

This type of panel system is closed with plywood, and a membrane layer is built in. The panels in the picture (right) have been battened to accept a dry-lined finish. This provides an air gap behind the plasterboard for services to be run through.

▸ A closed-panel timber frame system

Additional advantages

There are some additional advantages to a closed-panel timber frame system:

▶ The insulation is factory-installed, which reduces on-site wastage and disposal costs of off-cuts.
▶ Panels can be erected by unskilled labour, reducing on-site labour costs.
▶ Conduits can be factory fitted for service installations on site.
▶ Sealed factory-produced panels meet u-value requirements.

Additional disadvantages

There are a couple of additional disadvantages:

▶ The services need to be designed and conduits installed so cables and pipework can be threaded through the panels if required.
▶ Openings will need to be sealed to protect the air integrity test for Building Regulations.

Structural insulated panels (SIPs)

These panels are constructed using OSB board and a foam insulation to bond the two plywood sheets together, which forms a very strong composite panel. A variety of different systems are available to bond the insulation panels to the plywood. A recess is formed at the edge of the panel by leaving the insulation short. This enables a softwood timber key to be inserted between the two panels, fixing them together in a solid joint. The picture below shows an example.

▶ An example of a building constructed from SIPs

The joints that are commonly used are shown in Figure 4.3. Each must be strong enough to transfer any loads, prevent moisture ingress while construction is completed, and not bend or weaken over time. The thermal continuity must also be maintained across the joint so no cold bridges are formed. The thermal core is normally a choice of polyurethane (PUR), polyisocyanurate (PIR) or polystyrene (PS) insulation. Each system must meet air tightness and fire rating standards in order to achieve Building Regulations approval.

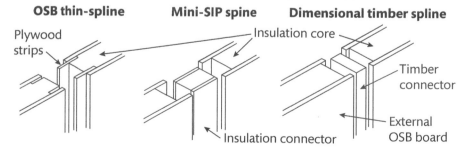

▶ **Figure 4.3:** Common joints used in SIPs

Traditional construction

Link

For cavity and solid masonry walls, see Unit 10: Building Surveying in Construction.

Roofing

A traditional roof is also known as a 'cut rafter' or 'cut' roof. This means it has been hand cut from solid timber rafters and assembled on site to form a traditional roof. Often it incorporates **purlins**, which transfer the loads from the roof to the gable external walls and down to the foundations. A purlin provides some intermediate support because it is centred in the middle of the rafters. This enables the rafters to have a smaller section size, as they do not have to span from the ridge to the wall plate, which would require a greater depth to resist the deflection. Figure 4.4 shows an example of a cut roof and its components.

A traditional roof is more expensive to construct than a modern trussed rafter roof, but does provide a stronger hand-built installation.

Key term

Purlin – a horizontal beam that runs the length of a dual pitch roof with gable ends.

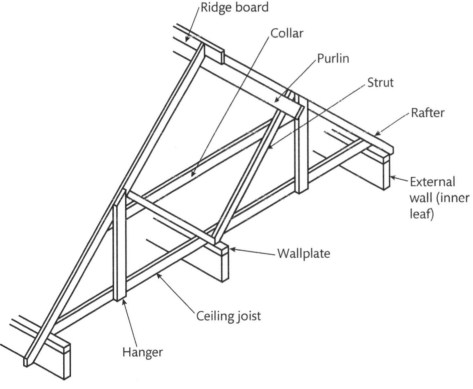

▶ **Figure 4.4:** Components of a cut roof

The cut roof is connected to the load-bearing external walls and internal walls using the following features:

▶ Purlins transfer the loads from the rafters to the external gable walls.
▶ Wall plates transfer the loads from ceiling and rafters down to the foundations.
▶ Collars bind the purlins together and prevent roof spreading.
▶ Struts stabilise the structure.
▶ Ceiling joists laterally tie the roof together at each rafter.

Research

Find out the following:
• What is a gable wall?
• What size should a wall plate be?
• What size is a ridge board?
• How are the purlins installed into the gable walls?

Advantages	Disadvantages
Faster method of construction	Box-type constriction with no curves or features
Greater levels of insulation can be incorporated	Possible issues with resale value
Services can be installed in the factory	Cannot install in high winds
Faster delivery and installation	
Less space required on site as modules are lifted and installed	
Better quality	
Environmental benefits	

Four-sided modules

Four sides form a volume, which is volumetric construction in terms of modern methods of construction (MMC). These modules are enclosed by four walls which may contain a window or an entrance door. They can be mass produced in a factory, transported to site, lifted into position and installed. A hotel bathroom can be completely modulised, including all the sanitary ware, tiling, flooring and wall finishes. The final connections are all that is needed to be done on site.

Open-sided modules

Open-sided modules can be partial or corner-supported, as you can see in the picture below. Each application depends on the arrangement required by the final design. For example, a corner module will require three completed sides and one open side to accept the next module.

It is worth remembering that modules need to be standardised. It is not efficient if several modules are different sizes and configurations as this increases costs and slows down programme times.

Partially open-sided modules

This picture shows a partially open-sided module. As you can see, most of the finishes have been fitted. When as much as possible can be installed in a factory, this speeds up production considerably.

▶ An example of a partially open-sided module

A client has approached your conservation company to undertake some work on their farm buildings. The outbuildings are more than 100 years old and need some renovation. The roof construction is traditional with clay rosemary roofing tiles.

You have been asked to explore options for the client, and compare the costs of installing an alternative modern roof truss replacement against the costs of repairing the existing traditional roofs of the outbuildings.

Produce a report that compares the two roofing methods and include an indicative idea of costs.

Flooring

Link

For suspended timber floors and in-situ solid floor construction, see Unit 10: Building Surveying in Construction.

Modular construction

Modular construction is also known as volumetric construction. In this method, three-dimensional units are prefabricated in a factory and then assembled on the construction site. This method has been successfully developed so that factory-assembled units can be used to construct a multi-storey housing project in less than a week. China is the leading country for this type of development.

Research

Find and watch a time-lapse video of the Chinese modular construction technique, which uses floor modules and external wall units to form a multi-storey building. Make a note of the method's advantages and disadvantages and compare with a partner.

Modules are constructed from many different materials, including:

▶ timber framing
▶ steel channels
▶ composite materials
▶ precast concrete.

Table 4.1 describes the advantages and disadvantages of modular construction in general.

▶ **Table 4.1:** Advantages and disadvantages of modular construction

Advantages	Disadvantages
Wastage is kept in the factory, not on site	One unit looks the same as another – there are no variations in appearance which restricts flexibility

Corner-supported modules

A corner-supported module has a structural post in each corner that connects from the floor unit to the roof unit. These can be used for larger areas where uninterrupted floor spaces are required. The photo shows an example.

▶ An example of a corner-supported module

Stair modules

The development of quality precast concrete composite construction means that a stair module could be prefabricated off site and simply winched into position, such as that shown in the photo. A set of stairs can be installed very quickly as part of the floor panel installation.

This demonstrates that half-space landings can also be designed into the module and provision for a temporary handrail can be designed into a cast in-situ fixing. This enables the stairs to be used as the rest of the construction installation continues, as a safe means of access to each floor.

Table 4.2 evaluates the advantages and disadvantages of different elements of modular construction.

▶ A precast stair module

▶ **Table 4.2:** Evaluation of modular construction elements

Element	Advantages	Disadvantages
Cavity masonry walls	• Aesthetic finish • Blends in with the street scene so helps with planning approval • Can be retrofitted with insulation	• Difficulty in achieving u-values • Cannot be prefabricated
Solid walls	• Faster construction	• Cold construction • No insulation
Cut roof	• Aesthetic when open • Stronger construction • Deeper rafters provide insulation capacities	• Takes longer to construct • Heavier components • Skilled labour required
Suspended timber floors	• Void created for services installations • Easier access for service repairs	• Void requires suspended insulation • Rot if ends of joists touch external wall below damp-proof level • Intermediate supports required
Solid in-situ floors	• Insulation can be incorporated to form a warm floor • Underfloor heating can be easily accommodated • Harder wearing than timber floors	• Provision for services needs to be designed into concrete layer

What factors influence the decision to use traditional or modern methods of construction?

Hint — Consider the impact of stakeholders and existing materials.

Extend — How can a traditional building be modernised by retrofitting?

Lift modules

The advent of reasonably small efficient lift systems led to the need to quickly install a lift shaft into a building by installing a precast concrete set of lift shaft modules. These contain the opening for the doors and fixing positions integrated into the concrete, as shown in the photo.

▶ A precast lift shaft module

Non-load-bearing modules

These are mainly purely aesthetic and used where the application is self-contained so suited to modulisation, for example a bathroom, washroom, reception or toilet module. Kitchen modules are another common example – the client's choice of kitchen is then installed within a module with all services plumbed in so that all that needs to be completed on site are the final connections.

Assessment practice 4.1

A.P1   A.M1   A.D1

A client who wishes to develop a new out-of-town housing estate has approached your design company for advice and guidance. They have not decided what type of construction to employ on this prestigious new build. The land is a greenfield site, has reasonable access and egress roads and is free from any planning constraints but sustainability must be a core theme.

Produce the following in support of the procurement of the design for your company:

- an explanation of a range of different structural forms that could be used for two-storey family homes
- the potential benefits of each type of structural form
- an evaluation of the effectiveness of different structural forms for these types of houses.

Plan
- Do I know how to carry out research on the types of structural form that can be used for two-storey construction?

Do
- Have I covered a range of structural forms that contrast with each other?
- Have I used examples from current industry practices as evidence?

Review
- Does the evidence contain an evaluation of each technique?
- Have I made a recommendation and justified it?

Examine foundation design and construction

B1 Subsoil investigation

Investigation methods

Before any construction work can start, it is often necessary to undertake a thorough site investigation. This is where the majority of the budget should be spent, in order to remove many of the uncertainties around constructing a foundation on unknown ground. This can be defined in terms of buildability, where the maximum amount of information can be obtained about a site to remove any unforeseen works and any related costly contract variations after the award of the contract.

A thorough site investigation is therefore an essential requirement of any proposed construction project. This can generally be achieved using a combination of two methods:

1 A desktop survey, involving a paper-based and/or internet search, including looking at online maps, aerial photos, local authority archives and other records

2 A site survey, involving the use of trial pits and borehole records, and a walkover.

Online maps

The availability of online satellite maps has made it easy to access an aerial photograph of a potential site. This will often reveal:

▶ topographical details – heights and levels can often be established by shadows within the photograph

▶ existing features, e.g. boundary fence lines, rights of way, existing buildings

▶ watercourses – strict planning regulations are enforced on any development within reach of flood plains

▶ trees with preservation orders, which can be identified by talking to the local conservation officer and/or referring to site maps and photographs

▶ archaeological features – any historical assets, such as the remains of notable old buildings, are protected under planning law. If a desk study identifies such assets, archaeological digs are required, which will delay the project and increase costs as the developer often has to pay for the work to be undertaken.

Aerial photographs

A survey using a light aircraft can produce a scaled photograph of a potential site from which a drawn plan can be extracted. Specialist contractors provide this service.

Historical maps

Many local authority libraries have copies of or even original local maps, which may also be available online. These can reveal the history of a potential site and the industrial applications it has been subjected to. This can provide valuable information on possible:

▶ types of contamination by identifying the industrial process that used to take place on the site

▶ underground obstructions by identifying the size of earlier buildings

▶ boundary locations, footpaths and roadways.

Ordnance Survey maps 1:1250 and 1:500

The Ordnance Survey is the UK's national mapping agency. Its maps are available in a range of scales right down to a 1:500 scale series. These give details of any buildings, structures, boundaries and surrounding roads, footpaths and properties. Again, these are also available online.

Local authority archives

These archives are a great source of local historical information, including photographs, old maps, street index records and other information that can be used to research a particular property or site. You may need to visit the local archive centre to see these records.

Building and planning applications

This is a valuable desktop resource. Many local authority websites have an online planning application database where you can search for planning applications relating to a particular address or site, including any historical planning applications. This can provide valuable information about the site with regard to:

▶ what type of application has been previously passed for the site

▶ existing drawings for the current use

▶ any planning conditions or restrictions associated with the potential site

▶ the level of neighbourhood support for the potential site (for example the type and number of objections).

Online address search

If you enter the address of the site into a search engine, you can locate public documents connected with that address, such as:

▶ county court judgements

▶ historical values of the property and average values for the area

▶ previous use of a site or building

▶ previous site investigation reports.

Walkover survey

This is a physical walk over the proposed site, enabling you to pick up visual information that will contribute to the desktop survey such as:

▶ existing structures
▶ existing services
▶ trees, hedges and boundaries
▶ indication of topography and levels
▶ evidence of groundwater levels and flooding
▶ overhead obstructions, for example pylons and telephone lines.

The walkover survey can be recorded on a sketch or detailed plan as Figure 4.5 illustrates.

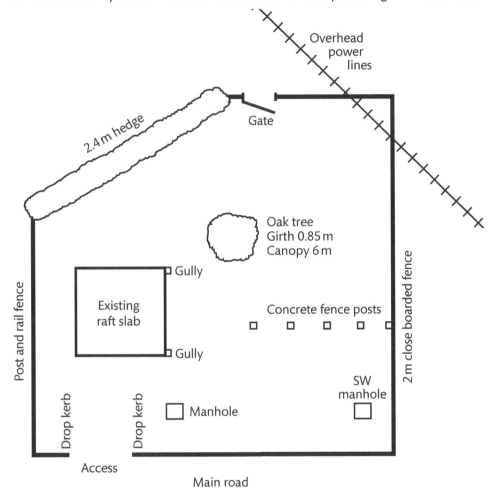

▶ **Figure 4.5:** An example of a sketch generated from a walkover survey

Ⅱ PAUSE POINT   How can a walkover survey contribute to your understanding of the likelihood of a site to flood?

Hint   Walkover surveys can give you signs of flooding, watermarks, flood deposits, levies, closeness to water courses and evidence of the water table on the land.

Extend   How can the walkover survey report be used in conjunction with desk research to make recommendations about the impact of a flood risk?

Trial pits

Trial pits and slit trenches are used to examine the structure of the soils that the proposed development will rest on. A trial pit is a square pit normally 2m x 2m x 2m deep whereby an engineer can visually inspect the layers of the strata, establish the groundwater level, and take samples from the wall of the pit for further testing. Trial pits are normally excavated using a backhoe or back actor excavator.

Slit trenches cover a longer area than trial pits and provide a soil engineer with a greater range of opportunity to visually inspect the subsoils.

The location of any trial pit or slit trench must be recorded accurately on a scale plan or map and numbered so any records obtained through soil testing are identified as having been taken from that trial pit or trench. Figure 4.6 shows an example.

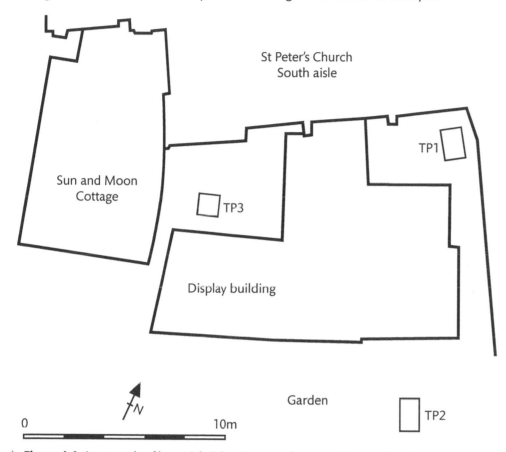

▶ **Figure 4.6:** An example of how trial pit locations can be recorded on a plan diagram

A detailed record should be made of the trial pit in a log including such information as:
▶ the depth and type of the different strata
▶ the level of the groundwater
▶ any testing undertaken and its depth.

Figure 4.7 shows an example.

Care must be taken with the health and safety associated with entering the confined space of the trench or trial pit. Additional earthwork support should be provided in unstable ground.

The requirements of the Confined Spaces Regulations and the Health and Safety at Work Act (see Unit 5: Health and Safety in Construction) must be adhered to with regard to any excavation work.

URS

TRIAL PIT LOG

Project Name and Site Location		Client		TRIAL PIT No
Orchard Retail Park		**Defran Investments Ltd**		**TP1**

Job No	Date		Ground Level (m)	Co-Ordinates ()	
49327965	Start Date 18-02-09 End Date 18-02-09				

Contractor	Method / Plant Used	Sheet
Whippet Plant	JCB 3CX	1 of 1

SAMPLES & TESTS				STRATA			
Depth	Type & Ref No	Test Result	Water	Reduced Level	Legend	Depth (Thickness)	DESCRIPTION
						(0.50)	Brown clayey SAND. Frequent rootlets. (TOPSOIL)
0.5	B 1					0.50	Light brown slightly gravelly medium to coarse SAND. Gravel is subrounded and coarse quartzite. (GLACIAL SAND AND GRAVEL)
1.0	B 2					(1.30)	
1.5						1.80	
2.0	B 3					(1.00)	Brown and light grey sandy subrounded and rounded fine to coarse GRAVEL. Frequent subrounded and rounded cobbles. (GLACIAL SAND AND GRAVEL)
2.5						2.80	
3.0	B 4					3.00	Firm to stiff grey mottled orange brown silty CLAY. Occasional black woody fragments. (GLACIAL SAND AND GRAVEL)
						(0.80)	Firm red brown slightly sandy CLAY. (MERCIA MUDSTONE)
3.5						3.80	
							End of trial pit

URS GEOTECH TRIAL PIT ORCHARD RETAIL PARK TRAIL PITS.GPJ AGS3 ALL.GDT 27/03/09

TRIAL PIT INFORMATION				GENERAL REMARKS
Dimensions (m)	Stability	Groundwater Observations	Remarks	
x	Unstable below 1.30 m Support: None	Seepage at 0.70 m into north end of pit		

All dimensions in metres Scale 1:31.25		Logged By LJ	Approved By AH

▶ **Figure 4.7:** Example of a trial pit log

Percussion drilling, sampling and auger holes

A borehole is a method of investigating the subsoil strata by using a drilling rig such as the one shown in Figure 4.8. This is used to record information from greater depths than could be established using a trial pit and is normally towed behind a four-wheel drive vehicle and erected on site as a tripod.

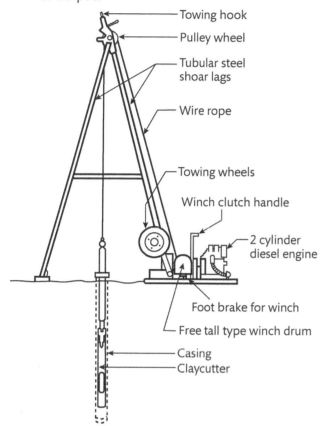

Towing hook
Pulley wheel
Tubular steel shoar lags
Wire rope
Towing wheels
Winch clutch handle
2 cylinder diesel engine
Foot brake for winch
Free tall type winch drum
Casing
Claycutter

▶ **Figure 4.8:** A borehole drilling rig

The rig operates by using a winch and cable attached to a motor and a winch clutch. When the clutch is released the cable drops the heavy cutting head, which gradually works its way into the ground. This method is known as percussion drilling using a shell and auger. Various samples can be taken at specified depths and loose samples can be taken from the cutting head as it is emptied.

This method enables a borehole log of the subsoil strata to be quickly established. Several boreholes will be required across a potential site to check for the consistency of the strata and to provide information for a final foundation design.

Boreholes can be made at shallow depths using a hand auger. This is turned by hand and drills a hole from which disturbed and in-situ soil samples can be taken.

Plate bearing test

This is a **field test** that involves the following sequence of operations.

1   The area of ground is levelled using dry sand to ensure full contact with the plate, which is circular and of a known diameter (usually 600 mm).

2   The plate is moved across the ground surface to ensure full contact with the test area.

3   A second bearing plate is placed on top of the first plate at its centre.

4   A load cell is then placed on top of the second bearing plate.

5   A couple of spacers may be used to support the jack, which pushes against a physical restraint, such as kentledge or the rear axle of a vehicle.

6   A horizontal support is placed across the plate to hold the dial gauges, which will measure the settlement of the plate as the load is applied. Three dial gauges obtain an average settlement of the 600 mm plate.

7   Dials are calibrated to zero.

8   A hand hydraulic pump applies a load in incremental stages and the settlement is recorded.

9   The timings associated with the test are recorded, along with its location and conditions.

When the test results have been analysed, the bearing capacity of the soil can be established and compared against the design value that needs to be met in terms of the allowable settlement when a structure is built on the soil strata.

> **Key term**
>
> **Field test** – a test carried out on an actual site, under real conditions.

> **Research**
>
> Watch a video showing a typical plate bearing test. When you understand its mechanics and how settlement is measured, explain the process in your own words.

Information used for foundation design

A structural engineer is normally the person who will advise on the most suitable foundation design for the type of bearing strata on-site. They will use the data obtained from the visual inspection, such as from trial pits, and also from any ground investigation reports obtained from a specialist testing contractor. These reports show the different ground layers and enable informed decisions to be made upon the choice of foundation. Information that will be used to inform the design of a foundation includes some or all of the following:

▶   historical soil investigation reports
▶   trial pit visual analysis
▶   borehole in-situ testing
▶   Building Regulations Approved Document A
▶   previous site use.

Bearing capacity

The capacity of a soil to bear the weight of the foundation and superstructure it supports needs to be established so that the structural engineer can design the foundation to safely spread the loads of the building across a large enough area to prevent the shear failure of the soil. Shear failure is when soil settlement exceeds the bearing load of the soil, which can lead to movement, cracking and damage to the substructure and superstructure of a building. Table 4.3 demonstrates typical bearing capacities of a range of soils.

▶ **Table 4.3:** Typical soil bearing capacities (Source: **www.uwe.ac.uk**)

Category	Types of rocks and soils	Presumed bearing value
Non-cohesive soils	Dense gravel or dense sand and gravel	> 600 kN/m²
	Medium dense gravel, or medium dense sand and gravel	< 200 to 600 kN/m²
	Loose gravel, or loose sand and gravel	< 200 kN/m²
	Compact sand	> 300 kN/m²
	Medium dense sand	100 to 300 kN/m²
	Loose sand	< 100 kN/m² depends on degree of looseness
Cohesive soils	Very stiff boulder clays and hard clays	300 to 600 kN/m²
	Stiff clays	150 to 300 kN/m²
	Firm clay	75 to 150 kN/m²
	Soft clays and silts	< 75 kN/m²
	Very soft clay	Not applicable

As you can see, the denser gravels packed with sand can provide the highest bearing capacity while soft clays are at the opposite end of the scale, with very little bearing capacity.

Subsoil classification

The British Soil Classification System is based upon BS 5930:2015. The primary classification is shown in Figure 4.9. This illustrates the titles, descriptions and characteristics used to classify a soil under this standard. This then enables a common understanding across ground investigation contractors, structural engineers, architects and other designers of substructures, for example a piling contractor.

FIELD DESCRIPTION OF SOILS in accordance with BS5930:2015

D N David Norbury — Engineering Geologist

SOIL GROUP	Very Coarse soils			Coarse soils						Fine soils			
PRINCIPAL SOIL TYPE	BOULDERS		COBBLES	GRAVEL			SAND			SILT			CLAY
	Large boulder	Boulder	Cobble	Coarse	Medium	Fine	Coarse	Medium	Fine	Coarse	Medium	Fine	
Particle size (mm)	>630	630 - 200	200 - 63	63 - 20	20 - 6.3	6.3 - 2.0	2.0 - 0.63	0.63 - 0.2	0.2 - 0.063	0.063 - 0.02	0.02 - 0.0063	0.0063 - 0.002	<0.002

Visual identification
- Very Coarse soils: Only seen complete in pits or exposures. Difficult to recover whole from boreholes.
- GRAVEL: Easily visible to naked eye; particle shape can be described; grading can be described.
- SAND: Visible to naked eye; no cohesion when dry; grading can be described.
- SILT: Only coarse silt visible with hand lens; exhibits little plasticity and marked dilatancy; slightly granular or silky to the touch; disintegrates in water; lumps dry quickly; possesses cohesion but can be powdered easily between fingers.
- CLAY: Dry lumps can be broken but not powdered between the fingers; dry lumps disintegrate under water but more slowly than silt; smooth to the touch; exhibits plasticity but no dilatancy; sticks to the fingers and dries slowly; shrinks appreciably on drying usually showing cracks.

Density/Consistency
- Very Coarse soils: No terms defined. Qualitative description of packing by inspection and ease of excavation.
- Coarse soils: Classification of relative density on the basis of N value (Table 10), or field assessment using hand tests may be made (Table 11).

Fine soils consistency:

Term	Very soft	Soft	Firm	Stiff	Very stiff
Field test	Finger easily pushed in up to 25 mm. Exudes between fingers	Finger pushed in up to 10 mm. Moulded by light finger pressure	Thumb makes impression easily. Cannot be moulded by fingers. Rolls to thread	Can be indented slightly by thumb. Crumbles in rolling thread. Remoulds	Can be indented by thumb nail. Cannot be moulded, crumbles

Discontinuities

Describe spacing of features such as fissures, shears, partings, isolated beds or laminae, desiccation cracks, rootlets etc.
Fissured: Breaks into blocks along unpolished discontinuities.
Sheared: Breaks into blocks along polished discontinuities.

Term	Scale of spacing of discontinuities		
Mean spacing (mm)			
very widely >2000	widely 2000 - 600	medium 600 - 200	closely 200 - 60	very closely 60 - 20	extremely closely <20

Bedding

Describe thickness of beds in accordance with geological definition. Alternating layers of materials are inter-bedded or inter-laminated and should be described by a thickness term if in equal proportions, or by a thickness of and spacing between subordinate layers where unequal.

Term	Scale of bedding thickness		
Mean thickness (mm)			
very thickly bedded >2000	thickly bedded 2000 - 600	medium bedded 600 - 200	thinly bedded 200 - 60	very thinly bedded 60 - 20	thickly laminated 20 - 6	thinly laminated <6

Colour

HUE can be preceded by LIGHTNESS and/or CHROMA 33.4.4.2
Red / Pink / Orange / Yellow / Cream / Brown / Green / Blue / White / Grey / Black
Light / - / Dark
Reddish / Pinkish / Orangish / Yellowish / Brownish / Greenish / Bluish / Greyish
Colours may be mottled
More than 3 colours is multi-coloured

Secondary constituents

For mixtures involving very coarse soils see 33.4.4.2

Term in coarse soils	slightly (sandy) Note 2	(sandy) Note 2	very (sandy) Note 2	Term in fine soils	slightly (sandy) Note 4	(sandy) Note 5	very (sandy) Note 5
Proportion secondary Note 1	<5%	5-20% Note 3	>20% Note 3	Proportion secondary Note 1	<35%	35 - 65% Note 6	>65% Note 6

For example: SAND AND GRAVEL About 50% Note 3 / Silty CLAY / Clayey SILT
For example: slightly (glauconitic) / (glauconitic) / very (glauconitic)

Mineralogy

Terms can include: glauconitic / micaceous / shelly / organic / calcareous
Carbonate content: carbonate free = no reaction to HCl / slightly calcareous = weak or sporadic effervescence / calcareous = clear but not sustained effervescence / highly calcareous = strong, sustained effervescence.
Organic soils contain secondary finely divided or discrete particles of organic matter, often with distinctive smell, may oxidise rapidly.
For example: slightly organic – grey / organic – dark grey / very organic – black

Particle shape

Very angular / Angular / Sub-angular / Sub-rounded / Rounded / Well rounded
A dominant shape can be described, for example: Cubic / Flat / Elongate

PRINCIPAL SOIL TYPE	LARGE BOULDERS	BOULDERS	COBBLES	GRAVEL			SAND			SILT			CLAY

Tertiary constituents

Example terms include: shell fragments / pockets of peat / gypsum crystals / pyrite nodules / calcareous concretions / flint gravel / brick fragments / rootlets / plastic bags
Qualitative proportions can be given: with rare / with occasional / with numerous / frequent / abundant.
Proportions are defined on a site or material specific basis, or subjectively

Geological Unit

Name in accordance with published geological maps, memoirs or sheet explanations.

Notes:
1) Percentage coarse or fine soil constituents excludes cobbles and boulders.
2) gravelly or sandy and/or silty or clayey.
3) Or described as fine soil depending on mass behaviour.
4) gravelly and/or sandy.
5) gravelly or sandy.
6) Or described as coarse soil depending on mass behaviour.

▶ **Figure 4.9:** Field description of soils in accordance with BS 5930:2015

Groundwater levels

When a hole is excavated for a foundation, any groundwater will be quickly visible as the excavation will fill up with water. Ideally, the **water table** should be below the level of the foundation. Where water tables are high then surface water is visible on the ground.

Groundwater can be managed by installing subsoil drainage, which removes water from a foundation and discharges it into a surface water sewer. High water tables have a marked effect on a soil's bearing capacity as the moisture lubricates the soil, making it less cohesive.

Chemical analysis of soil samples and presence of sulphates

Many construction sites are in **brownfield** locations where soil may be contaminated if chemical processes have leached into the ground and settled. If a concrete foundation is constructed in such soils, **sulphates** in the ground, which can also occur naturally, can attack the cement bond. The cement expands, weakening the foundation and potentially breaking it down.

Presence of obstructions

Obstructions present on a site may include:

▸ boulders that might have been deposited during the Ice Age, especially in clay soils
▸ existing substructures from demolished buildings on brownfield sites
▸ services that may obstruct new foundation proposals, such as drainage pipes, electrical cables and water services
▸ natural water features such as springs
▸ high water tables
▸ rock formations.

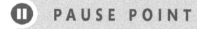

PAUSE POINT   What type of foundation best overcomes the issues of groundwater and chemicals in the soil?

Hint   Consider vertical columns.

Extend   Can soil contamination be dealt with in situ?

B2 Subsoil improvement

Many techniques can be used to improve subsoil so as to increase its bearing capacity to sustain the loads imposed upon it. This enables the use of development land to be maximised in areas where there are shortages of available plots.

Vibroflotation

This process uses a vibrating probe as illustrated in Figure 4.10.

Vibro-replacement is similar to vibroflotation but does not use water. Instead, the replacement aggregate is poured into a hopper at the surface and travels down the hollow probe to the base, where it is compacted into a vertical column under vibration.

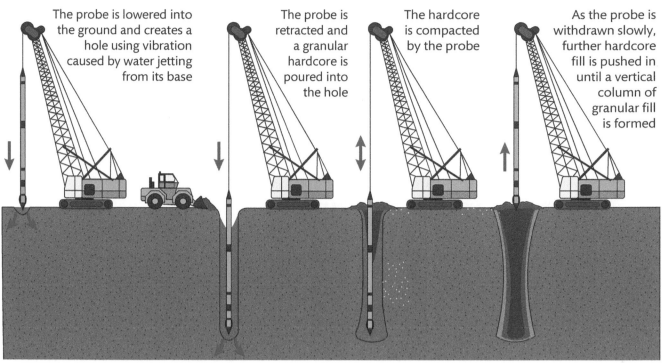

The probe is lowered into the ground and creates a hole using vibration caused by water jetting from its base

The probe is retracted and a granular hardcore is poured into the hole

The hardcore is compacted by the probe

As the probe is withdrawn slowly, further hardcore fill is pushed in until a vertical column of granular fill is formed

▶ **Figure 4.10:** The four stages of vibroflotation

Grouting

This process injects a cement-based product into the subsoil strata under pressure. This expands and fills any voids, eventually setting solid to improve the permeability and increase the strength of the soil.

Land drainage

The basis of this method is to remove a high water table and lower it so it does not have an effect upon the substructure and allows the ground to dry out and improve its bearing capacity. This may involve the water entering flexible, perforated drainage pipes beneath the ground and directed away to a stream or soakaway. The process may be used in conjunction with a layer of shingle to help the water drain into the pipes from waterlogged soil.

B3 Design principles

The design of a foundation depends upon several factors, including:
▶ the design load that has to be carried, including any safety factors
▶ the capacity of the soil to bear this load with acceptable settlement.

The balance is important in ensuring that the soil can support this load without any movement of the building or structure that would cause more serious issues other than aesthetical cracking which can be decorated over.

Factors used during design to minimise settlement

Table 4.4 shows some of the factors that designers should take into account when considering how to minimise settlement.

▶ **Table 4.4:** Factors to consider when minimising settlement

Factor	Reason
Building load	The design of the building that the ground has to support can be varied by using a lighter superstructure, for example timber framing. The lighter structure can be supported by lower bearing ground conditions.

Factor	Reason
Soil bearing capacity and type	Selecting a location which has better bearing capacity soils.
Foundation depth	The deeper a foundation, the better the capacity of the ground to support a heavier load.
Groundwater	A low water table reduces **permeability** of soils and enables better bearing capacities.

Key term

Permeability – the ability of a material to transmit water. For example, a soil with many voids can allow more water to percolate through it and will therefore provide better drainage.

Design to minimise other movement

A number of other variables can have an effect upon a foundation. Table 4.5 illustrates four factors that a foundation may have to resist.

▶ **Table 4.5:** Factors to consider when minimising movement

Factor	Reason
Soil shrinkage	When wet cohesive soils dry out in the summer, they compact under the action of gravity as the soil shrinks in volume.
Ground heave	Heave can be caused by expansion of water freezing into ice, excessive loading, and expansion from chemical reactions in the soil.
Differential settlement	This settlement may differ across the whole area that supports the foundation, which can result in non-horizontal settlement of the substructure and superstructure.
Effects of tree growth and tree removal	The roots of a growing tree will absorb the moisture within a soil, which can dry out the soil and cause settlement. If a tree is cut down then the moisture removal ceases and the soil expands with the increased moisture content. The roots themselves can also grow under foundations and destabilise them.

Building Regulations Part A

Approved Document Part A covers the design of substructures and includes a number of different guidelines:

▶ The recommended minimum widths of strip foundations are specified in Figure 4.11. A typical width is 600 mm.
▶ The thickness of strip foundations must be equal to the projection of the foundation from the wall perimeter. This is normally 150 mm for a 600 mm-wide foundation.
▶ Where foundations are stepped due to being constructed on a sloping site, the overlap should be equal to twice the depth of the step.

Figure 4.11 illustrates these principles of simple strip foundation design.

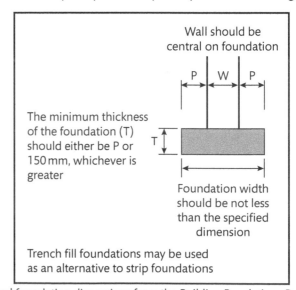

Wall should be central on foundation

P W P

The minimum thickness of the foundation (T) should either be P or 150 mm, whichever is greater

T

Foundation width should be not less than the specified dimension

Trench fill foundations may be used as an alternative to strip foundations

▶ **Figure 4.11:** Ideal foundation dimensions from the Building Regulations Part A

B4 Types of foundation

A number of different foundations can be used to support the loads from buildings. The choice of foundation is based upon the site investigation and subsoil report. The bearing capacity of the soil determines the type of foundation that needs to be constructed.

We will now examine a range of different foundations that can be used to overcome difficult soil conditions, outlining the advantages and disadvantages of each type.

Link

See Unit 10: Building Surveying in Construction for more about different types of foundations.

Strip

This traditional foundation consists of a strip of concrete that an external wall sits upon. It is normally excavated to a metre depth so it is unaffected by frost and will not be subjected to moisture changes within the soil. Modern strip foundations use trench blocks which are wall thickness concrete blocks that raise the wall off the foundation up to ground level.

Trench fill

This is a variation upon a strip foundation. The trench is normally excavated to 450 mm wide and filled with concrete to just below the ground surface. This can be done in one day so is a fast method of constructing the foundation.

Raft

A raft foundation in essence floats onto the soil strata and consists of a concrete slab with edge thickenings. The slab provides a platform to ensure that the building is supported over the whole of the area of ground beneath it. It can therefore cope with differential settlement so is best suited to unstable ground.

Pad

A pad foundation is square, rectangular or sometimes circular and supports a localised load such as a column. It spreads the load to a level where the soil can support the loads placed upon it. Pad foundations can be reinforced to resist the punching shear forces exerted by the steel columns with a holding down bolt box that casts the holding down bolts into the foundation. The steel base plate attached to the column is then bolted onto shims, which allow for a grout to be used to set the whole arrangement into a solid configuration. Grouting is usually done when all the steelwork has been levelled and plumbed.

Pile

A pile is a vertical column usually constructed of concrete or steel. It is driven into the ground and relies on either the friction of the ground against the driven pile or the end bearing of the pile. Piles can be bored into the ground using a pile drilling rig. Piling can therefore be classified as follows:

▶ Replacement piles – the ground is removed by boring and replaced by pouring concrete into the void and lowering a reinforcing cage into the concrete to form a reinforced concrete pile.
▶ Displacement piles – a falling weight drives a pile to push aside the ground.
▶ Pile caps – structures that attach to the pile foundations to form a cap. This prevents any movement of the pile and provides a suitable platform on which to attach columns.
▶ Ground beams – often pile caps are linked together using ground beams which are reinforced with steel cages on which the external wall sits.

ⓘ PAUSE POINT   What factors affect the choice of foundation?

    Hint   Consider the function of foundations and what might affect them below and above the ground.

    Extend   How are foundations influenced by modern methods of construction (MMC)?

Link

See Unit 10: Building Surveying in Construction for more about raft, strip, pad and pile foundations.

Different foundations are suited to different conditions and building types. Table 4.6 compares the types of foundation.

▶ **Table 4.6:** Advantages and disadvantages of each foundation type

Foundation type	Advantages	Disadvantages
Strip	• Economical for the depth with limited excavation. • Simple to construct. • Supports a two-storey domestic structure.	• Wider strips required to support greater loads on poor ground. • Trench side may require supporting. • Requires backfilling. • Limited loading capacity.
Trench fill	• Can cope with groundwater, which can be displaced by poured concrete. • Fast method of construction. • No trench support required. • No backfilling required.	• High costs associated with bigger concrete volumes. • Low sustainability in terms of high cement content.
Raft	• Economical solution with minimal excavation depth. • Allows for differential settlement. • Can span over obstructions. • Floor and foundation are cast as one, saving time.	• Requires formwork and steel reinforcement design. • Edge of slab has to be protected from the elements.
Pad	• Economical foundation for steel columns. • Can be reinforced to withstand greater axial loads. • Can be linked with a continuous beam that provides external wall foundation.	• Large pad may be required to resist uplifting forces from wind.
Pile	• Economical solution to reach greater bearing capacity soils. • End and friction bearing capacities. • Can be prefabricated off site.	• Noise during installation.

Assessment practice 4.2

B.P2 B.P3 B.P4 B.M2 B.D2

The site investigation report for the new housing estate site has been completed. Your design company has been asked to produce a foundation design that can be used to support the two-storey and three-storey town houses. Produce a report about the site, detailing what types of investigation have been undertaken and the types of foundation that could be used, and recommend a foundation type to use. Your report should include:

• a description of the types of site investigation that can be used for obtaining foundation design data
• details of the types of foundations that could be proposed for the site
• the principles of foundation design and how these would influence the design and build of the houses
• an evaluation of each type of foundation with a final justified recommendation.

Plan
• Have I sufficiently researched the types of foundations and their advantages and disadvantages?
• Have I identified the site investigation techniques used to inform assessment decisions?

Do
• Have I covered a range of substructure foundations?
• Have I evaluated each type?
• Have I defined the principles of foundation design and the factors that affect the choice of foundation?

Review
• Does the evidence contain an evaluation of each technique?
• Have I made a recommendation and justified it?

Examine superstructure design and construction

Several of the following superstructure elements have been covered in other units so links to this information are provided.

C1 Walls

Walls transmit loads from the floors at intermediate level and from the roof down to the foundations. They form the elevations of a building and can be constructed using a variety of traditional and modern methods.

Internal walls and partitions

Blockwork partitions

Solid block partition walls provide better acoustic and fire insulation than timber **stud walls**, as well as being a stronger base for hanging shelves and cabinets (although they can also be built up within a timber frame). However, they are heavier so usually require extra support at floor level, which may mean reinforcing the floor or foundations. Partition walls are not integrated into the walls at each end, so require metal wall profiles to connect them and keep the wall vertical. Long blockwork or brick walls may also need a control joint filled with sealant to ensure they can flex without cracking.

Timber stud partitions

These internal walls are made of a timber framework, which may not be able to bear loads as effectively as solid walls but can be strengthened with double uprights. Timber framed houses are likely to use this form of partition for all their internal walls, while brick-built houses may just use them upstairs, especially if a trussed roof transfers its load to the external walls. The frame is usually filled with plasterboard and skimmed with plaster.

Metal stud partitions

This system is used mainly for commercial applications in locations such as hospitals. It consists of cold-formed metal channels which are fixed as stud uprights and horizontal sole and header plates. They are clad with plasterboard to provide a ridged non-load-bearing wall. Services can be run through the internal void, which can also provide **patresses** to support electrical equipment and be filled with sound insulation if required.

Demountable partitions

A demountable partition can be taken down and relocated to expand and contract different spaces in an office environment so it is well suited to the ever-changing business environment. Specialist subcontractors normally manufacture their own systems, which can include aluminium or timber framing with large aspects of glazing. Demountable partitions need to be secured as they tend to be lighter than a traditional partition. An offset wall should often be included that forms a T-shape as this will provide lateral stability. Filing cabinets and desks can also be fixed to the wall to support it as they increase its width and stability. To reduce noise, dense materials need to be provided within the partition and the void in the ceiling above the partition closed so noise cannot travel via the ceiling.

Table 4.7 compares different types of partition.

▷ **Table 4.7:** Advantages and disadvantages of different types of partition

Element	Advantages	Disadvantages
Blockwork partitions	• Strong and sturdy. • Good acoustic insulation. • Good fire protection. • Can be load-bearing if required.	• Slower construction. • More difficult to demolish or move.

> ### Link
> See Unit 10: Building Surveying in Construction for more about external cavity walls: traditional brickwork and blockwork with waterproof external skin, and solid wall with rainscreen cladding.

> ### Key terms
> **Stud wall** – an internal wall that divides rooms, constructed as a frame with elements resting on studs or fixings.
>
> **Patress** – container for the space behind electrical fittings, such as light switches and plug sockets.

▶ **Table 4.7:** *Continued …*

Element	Advantages	Disadvantages
Timber stud partitions	• Faster construction. • Lightweight. • Easier to demolish or move.	• Require additional acoustic and fire protection. • Need to be protected from damp.
Metal stud partitions	• Faster construction. • Services can be hidden within the void. • Curved walls possible. • Joints on studs can be crimped.	• Additional supports are required to hang items on the wall. • Access panels required. • Specialist dry wall screws required for plasterboards.
Demountable partitions	• Can be repositioned easily. • Flexible. • Lightweight. • Services can be integrated. • Range of finishes available.	• Not as strong as a traditional stud wall. • Leaves holes in structure when moved that need to be maintained.

Theory into practice

A client is undertaking a refurbishment programme across several regional offices across the UK. They are considering using either metal stud partitions or demountable partitions and have asked you to provide a comparative report on each.

Produce a report for the client that:

- contrasts each system
- outlines the advantages of each system
- outlines the disadvantages of each system
- references examples to manufacturers' details.

Prefabricated timber frame construction

External and internal wall details

Earlier in this unit (section A1) we examined the different types of timber framed construction for the fabrication of a load-bearing element.

The details that form an external wall include the elements that are adjacent to a timber frame. Figure 4.12 illustrates the elements that form the completed external wall. In this case a brick outer skin is used as the final finish.

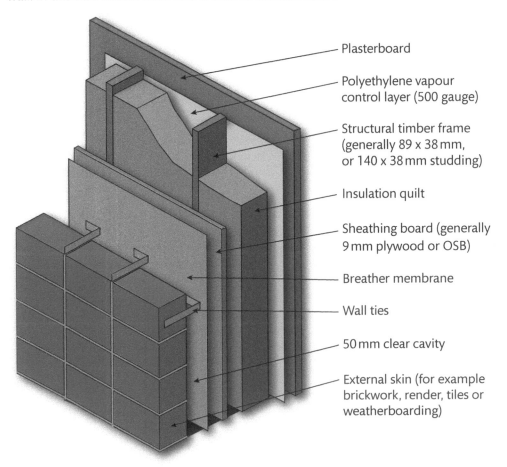

Plasterboard

Polyethylene vapour control layer (500 gauge)

Structural timber frame (generally 89 x 38 mm, or 140 x 38 mm studding)

Insulation quilt

Sheathing board (generally 9 mm plywood or OSB)

Breather membrane

Wall ties

50 mm clear cavity

External skin (for example brickwork, render, tiles or weatherboarding)

▶ **Figure 4.12:** External timber frame wall

The timber frame construction from the outside to the inside consists of:

1 an external skin that protects the wall from the weather
2 a cavity that is void of insulation to allow drainage
3 a breathable vapour barrier
4 plywood boards covering vertical timber studs
5 timber studs
6 insulation layers between studs
7 a vapour control layer
8 plasterboard finish
9 internal decorations.

Cladding options

Many cladding options are available for this type of timber framed construction, such as:

▶ cedar boarding
▶ facing brickwork
▶ feature boarding
▶ tiling
▶ render.

Openings in walls

Head detailing and lintels

Openings in walls are formed for windows and doors. The type of support above the opening depends upon the load and the span that it has to carry. Steel or concrete lintels can be used to support the weight above openings. In steel lintels, and in head detailing above doors and windows, the central void is filled with insulation to avoid the damp associated with **cold bridging**.

Figure 4.13 shows some typical lintel details.

Key term

Cold bridging – where a cold spot occurs between the warm internal structure and the colder external structure of a wall.

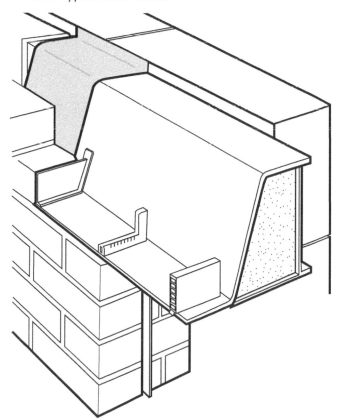

▶ **Figure 4.13:** An example of lintel detailing in a cavity wall

Figure 4.14 shows some typical timber framing details.

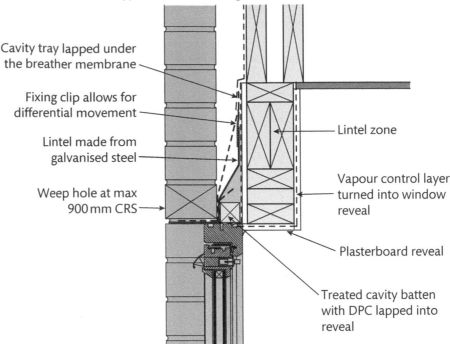

Cavity tray lapped under the breather membrane

Fixing clip allows for differential movement

Lintel made from galvanised steel

Weep hole at max 900 mm CRS

Lintel zone

Vapour control layer turned into window reveal

Plasterboard reveal

Treated cavity batten with DPC lapped into reveal

Key term

End bearing – the area that the lintel rests upon on the supporting structure.

▶ **Figure 4.14:** Typical timber framed lintel supports for brickwork and the internal studwork

The **end bearing** of a lintel is a minimum of 150 mm long to meet with the requirements of Part A of the Approved Documents for Building Regulations.

■ P A U S E   P O I N T   Are metal stud partitions better than timber studwork? Provide reasons to back up your answer.

Hint   Look at a manufacturer's website and technical details.

Extend   How are doors fitted into partitions?

Research

Investigate a thermal cavity closer and describe how it is fixed in position.

Jamb detailing

The sides of any opening need to be thermally efficient to prevent cold bridging. This can be achieved by using a thermal cavity barrier which sits in the cavity and closes it.

Cill and threshold detailing

Window cills need to discharge the water that runs down a window away from the building face. This can be achieved in several different ways, such as:

▶ facing brick headers with a damp-proof course below
▶ quarry tiles
▶ uPVC window cills
▶ timber window cills.

Link

For information about windows and doors, see Unit 10: Building Surveying in Construction.

The cavity can be closed with a thermal barrier or a damp-proof membrane (DPM) placed over the insulation to prevent any moisture been driven into the cavity. In timber framed construction a treated batten is placed to close the cavity and the DPM is wrapped around this.

C2 Floors

There are two types of floors: ground and intermediate. The methods used to construct them differ according to the materials used and the provision for services that have to be routed through them. Floors can be constructed using:

> timber joists with floor coverings – these can be **engineered timber** or structurally tested joists

> solid concrete, both precast and in situ.

Ground floors

Link

For information about solid and suspended timber floors, see Unit 10: Building Surveying in Construction.

Key term

Engineered timber – timber that has been engineered and reconstructed into a component.

Beam and block

This modern method of construction uses precast concrete beams at set centres that are shaped as an inverted 'T'. The space between the beams is filled with a concrete block and the whole floor is grouted to lock the blocks into place. A structural topping of screed or concrete is poured over the top to finish the floor construction. A beam and block floor can be used at ground and intermediate floor levels. End bearing needs to be maintained and on ground floors a telescopic air vent has to be installed to ventilate the subfloor void. Figure 4.15 shows an example.

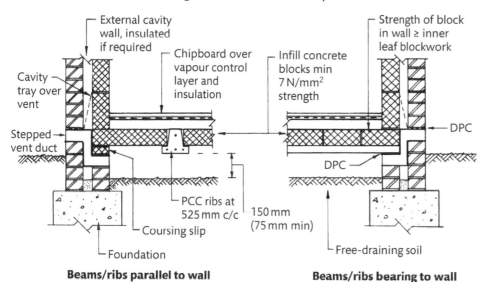

Beams/ribs parallel to wall   **Beams/ribs bearing to wall**

> **Figure 4.15:** A beam and block floor

Prestressed concrete

Prestressing is a method of tensioning reinforcing wires into wet concrete to spread the compression load into the surrounding concrete to increase its load-bearing capacity. The prestressing causes the precast concrete beams to curve, which is compensated by the loads imposed upon the units. They take the form of a plank that spans between supporting walls or steelwork. The plank can be configured so that voids are formed within its length to reduce the volume and hence the weight of the concrete. This type of unit can be used for both ground and intermediate floors.

Research

Investigate a prestressed concrete plank manufacturer and examine the structure of flooring units. Draw a diagram to show how the load is carried.

Intermediate floors

Timber

An intermediate floor constructed of timber differs from a ground suspended timber floor in the way it is attached to building structure. The intermediate floor joists are secured at their ends with a joist hanger. This galvanised steel hanger is built into the internal skin of a wall and accepts a floor joist, which is then nailed into place through the hanger. Engineered timber can be used for the intermediate floor and can be built into the internal skin of the cavity walls so that joist hangers are not required.

Research

Investigate the use of engineered timber floor joists in modern house construction.

Platform floors

With the advent of prefabrication techniques in timber framed housing, a floor can be prefabricated as a unit then lifted and fixed into place. These floor units are fabricated in the factory and include the primary floor finish. They instantly form a floor that is secure to walk upon.

Table 4.8 shows the advantages and disadvantages of different flooring systems.

▶ **Table 4.8:** Advantages and disadvantages of each flooring system

System	Advantages	Disadvantages
Beam and block floors	• Fast method of installation. • Lighter than planks. • Service openings easily formed. • Beams' loading capabilities are pre-tested before installation. • Factory production reduces waste.	• Requires a secondary finish. • Crane required for beam installation.
Softwood timber	• Sustainable. • Recyclable. • Can be engineered. • Lightweight. • Reduced environmental impact.	• Rots and degrades under moisture. • Requires treatment to resist insect attack. • Long spans cannot be achieved with solid timbers due to deflection.
Prestressed planks	• Large areas can be covered for a small number of units. • Greater spans achieved. • Floor can be immediately used. • Acoustically insulating.	• Crane is required for installation. • Heavy units. • Forming service holes is difficult. • Ceiling finishes required.
Platform floors	• Prefabricated unit is a faster method of installation. • Some primary services can be factory fitted.	• Ceiling finishes cannot be installed. • Top surface of floor can become wet when installed in bad weather.

Openings and stairs

Forming openings and stairs

Stairs are used to gain access and egress to a building that has a basement or more than one floor. A set of stairs has to comply with Approved Document Part K of the Building Regulations. The **rise** and **going** of each stair should be within minimum and maximum dimensions, so that when you climb a stair they all feel the same. The width of a stair is dependent upon the use of the building in terms of fire evacuation.

A set of stairs needs an opening within the floor to enable access from the intermediate floor and onto the next level. When a timber floor is used then trimmer joists are installed around the opening to carry the ends of any joists that are affected (see Figure 4.16). This trimmer may be a double joist to reinforce this structural element.

When the floor is constructed of precast concrete or beam and block then a structural detail is required. If the beams cannot be run the length of the stairs, a supporting steel edge beam may be required to support the ends of the beams. Floor planks again will run the length of the stair if possible and again may require supporting steel beams if this cannot be accommodated in the design. With both systems, the position of the stairs must be clearly indicated on the architectural drawings that are sent to the supplier for their floor design.

> **Key terms**
>
> **Rise** – the vertical height of a stair.
>
> **Going** – the horizontal distance of a stair

Trimming to stairwell

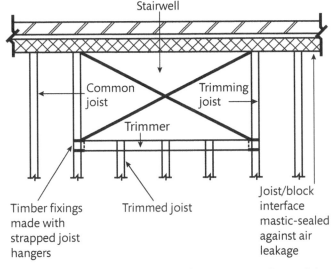

▶ **Figure 4.16:** Forming an opening for stairs using trimmer joists in timber

Stairs can be manufactured from:

▶ timber – a traditional type of construction that uses strings for each side of the stair with housings to hold the steps and risers in position by gluing wedges into place

▶ precast concrete – formed in a mould and cast as a monolithic unit which can be lifted and set into position. These specialist units have to be designed to fit the dimensions of the main structure

▶ in-situ concrete – cast on site, which enables a more customised design and changes to be made if necessary

▶ galvanised mild and stainless steel – these can be lightweight and manufactured off site. Designs can be flexible – for example spiral or curved – and treads may be timber for an aesthetically pleasing finish.

C3 Roofs

The roof is the top element of a building's construction. It is essential in keeping the weather away from people inside the building, for example by directing water off the structure and into the drainage system. There are various different types of roofs; each is best suited for a different purpose.

> **Link**
>
> Unit 10: Building Surveying in Construction, learning aim A, describes types of roof in detail, including pitched roofs, trussed rafter construction, traditional timber roofing and the effects of poor detailing.

> **Research**
>
> Go to the National House Building Council website (**www.nhbc.co.uk**) and find the most recent NHBC Standards for roofs (Part 7). Use it to research:
>
> - possible materials
> - possible components
> - methods of support
> - detailing
> - finishes
> - performance requirements
>
> for both flat and pitched roofs. Summarise your findings in a presentation.

Steel lattice frame

Steel lattice roofs can span wide areas without compromising strength, by their design of interlocking triangles or hexagons. These shapes are often incorporated into a visible design but their industrial appearance makes them more common in commercial and public buildings than domestic homes.

Flat

A flat roof is a quick, economical and easy method to close the envelope of a building. A flat roof can be used for small extensions to existing buildings, for commercial applications and for sustainable roof designs. Flat roofs have two types of construction:

▶ Warm deck – the insulation is on the roof finishes side and retains the warm air within the roof void to prevent condensation issues.

▶ Cold deck – the insulation is within the roof void just

above the internal finishes ceiling and any warm air passing through may condense on the underside of the roof.

Figure 4.17 compares the two types.

Traditional roof

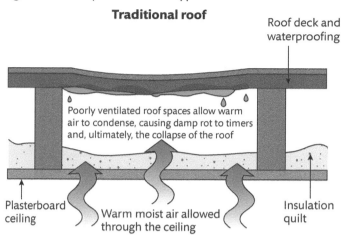

Warm roof

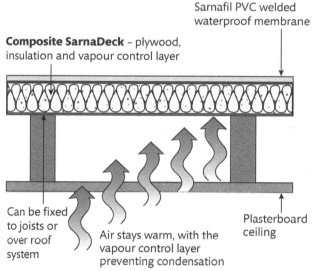

▶ **Figure 4.17:** The top diagram illustrates a cold roof deck where any warm moist air will pass into the cold region and condense on the underside of the roof finishes. A warm deck prevents such condensation

Roof details

Both a pitched and flat roof have eaves details to protect the structure from the elements. This includes the following features:

▶ a soffit – a flat horizontal board that projects out from the face of the wall

▶ a fascia board to support the guttering and rainwater goods

▶ soffit vents to provide ventilation to the roof space if required

▶ insulation continuity – roof insulation meets the vertical insulation in the external wall.

Figure 4.18 shows typical eaves detail for a pitched roof, including the insulation that continues down into the cavity wall.

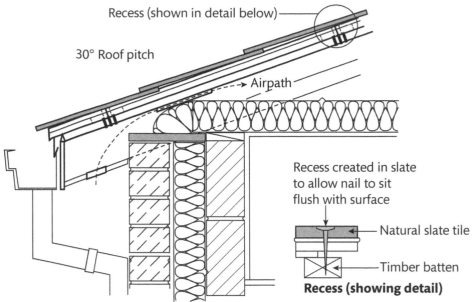

▶ **Figure 4.18:** Detail of typical eaves for a pitched roof

PAUSE POINT   What are the potential issues with a cold deck system for a flat roof?

Hint   Consider the effect of high moisture content on untreated timber.

Extend   How could you reduce the problems?

Method of achieving required falls

A flat roof requires some fall so water can be directed off its surface and into adjacent guttering and then into a surface water drain. To create a fall the following can be installed onto the existing structural elements of the flat roof:

▶ **Firrings** – sloping fillets of timber are fixed to the tops of joists to create a fall. They are usually covered with plywood decking.

▶ **Laser-cut tapered insulation** – the roof joists are covered with a plywood deck and then a warm deck roof is installed. The insulation layer is cut to falls and is laid over the plywood to form a sloping surface.

▶ **Screed** – a sand and vermiculite lightweight insulating screed can be laid to a fall across a solid roof structure such as precast planks.

Table 4.9 compares pitched and flat roofs.

▶ **Table 4.9:** Advantages and disadvantages of pitched and flat roofs

Type of roof	Advantages	Disadvantages
Pitched	• An attractive shape. • Water and snow runs off more efficiently. • Roof void can provide extra room(s). • Storage space formed.	• Heavy when tiled. • May require cross battens and extra insulation in front of joists to achieve u-values.
Flat	• Single layer membranes can be used as a covering. • Fibreglass coverings can be installed. • Less visual impact. • Ideal economical solution for extensions.	• Prone to roof leaks when life cycle ends. • Maintenance of surfaces required.

Theory into practice

You have been asked to design a roof for a single-storey garden room extension. The room is 6 m wide by 3.5 m deep. The width runs parallel to an existing two-storey cavity wall.

Produce two designs that could be used for the roof, outlining the advantages and disadvantages of each. Ensure both designs incorporate some natural lighting through the roof.

C4 Internal finishes

Wall finishes

The internal surfaces of walls have to be finished to protect the surface and to provide decorations to suit users' needs and tastes. Table 4.10 provides an outline of common finishes, their application and their advantages and disadvantages.

▸ **Table 4.10:** Comparison of different wall finishes

Type of finish	Application	Properties	Advantages	Disadvantages
Traditional two-coat plasterwork	• A wet finish applied to solid backgrounds and trowelled smooth. • Plaster has to set using a hydration process.	• Hard-wearing. • Provides thermal and acoustic insulation. • Impact-resistant. • Lime plaster required for older properties.	• Solid finish which is hard-wearing and long-lasting. • Provides a smooth surface for further finishes such as paint or tiles.	• Specialist skills required for a good finish. • Time required to dry out. • Can crack when drying out. • Higher cost than paint.
Dry lining	• Attaching plasterboard to walls, frames or insulation for a consistently smooth surface.	• Provides thermal and acoustic insulation. • Easily fixed. • Can be skimmed. • Lightweight. • Available with different properties.	• Fast to install. • Reduced drying times. • Lighter construction. • Easy to paint.	• Not as impact-resistant as wet plaster. • Sanding required.
Ceramic tiling	• A tile is bonded to a plaster background using adhesive, spaced using tile spacers then finished with a grout to the joints.	• Hard-wearing. • Impervious to water.	• Waterproof installation ideal for bathrooms.	• Labour-intensive. • Cannot be recoloured. • Repairs difficult if cracked. • Grout needs maintenance.
Wood panelling	• Cross battens are fixed to walls and wood panelling is fixed vertically across the battens.	• Water repellent if finish is applied. • Stainable. • Sustainable.	• Panels and frames can be feature of a wall. • Can be painted.	• Labour-intensive. • Timber has to be finished.
Paint/wallpaper	• Paint is liquid finish applied with a brush or roller as an emulsion or a gloss. Available for internal and external applications. • Wallpaper consists of rolls of thick paper applied with paste.	• Flexible. • Range of colours, patterns, textures and finishes.	• Many decorative possibilities. • Easily changed if new colours or patterns are required.	• Has to be maintained. • Preparation of receiving surface required. • Wallpaper repairs are difficult.

Ceiling finishes

Plasterboard and skim

Plasterboard is fixed horizontally to stud partitions using dry wall screws to the underside of the ceiling or intermediate floor joists. The ceiling is then skim coated with 3 mm of plaster ready for decoration.

Suspended ceilings

Suspended ceilings tend to be a commercial installation and consist of a 600 x 600 mm grid in aluminium with an edge angle detail that fixes to the perimeter walls. The main grid is suspended by wires attached to the supporting structure above and fibre ceiling tiles are dropped into it. The tiles are available in a variety of patterns to create different visual effects.

> **Key term**
>
> **Suspended ceiling** – a ceiling finish that appears to hang from the ceiling structure. Wires hold the ceiling tracks in place and are fixed to the supporting floors or roof structure.

Investigate a website of a suspended ceiling manufacturer, such as Armstrong. Recommend suspended ceiling types and designs for:

- a primary school
- a doctor's surgery
- the office of a small start-up company
- a fashionable clothes shop
- a swimming pool.

Justify your choices.

uPVC ceiling cladding

This is a hygienic application because it can be washed down and maintains a sterile environment. The ceiling cladding can be boards or large uPVC sheets. The joint detail also needs to be cleanable. Kitchens are an ideal location for this type of ceiling finish.

Timber boarded ceilings

This is a feature ceiling that uses softwood timber boards with a moulded edge to form a feature joint. The nails are not visible. The timber can be treated to leave a natural finish or painted.

Table 4.11 compares the different types of ceiling finish.

▶ **Table 4.11:** Advantages and disadvantages of different ceiling finishes

Ceiling finishes	Advantages	Disadvantages
Plasterboard and skim	• Lightweight. • Fireproof properties. • Void formed for services.	• Curved work is difficult to construct.
Suspended ceilings	• Covers up poor surfaces or ugly roof voids. • A range of different tiles and styles available. • Increases sound reduction between floors. • The grid can house lighting.	• Not suitable for domestic applications. • Tiles cannot be cleaned.
uPVC ceiling cladding	• Provides a food-safe finish that can be cleaned. • Lightweight. • Large panels available.	• Poor fire rating. • Joints require detailing. • Panels sag under heat.
Timber-boarded ceilings	• Sustainable. • Attractive appearance. • Joints can be decorative; detailed with mouldings.	• Requires sealing to prevent discoloration. • Issues with flame spread and fire rating.

Floor finishes

A range of common floor finishes is used in domestic and commercial applications. Table 4.12 details some options, along with the advantages and disadvantages of each type.

▶ **Table 4.12:** Comparison of different floor finishes

Type	Installation methods	Advantages	Disadvantages
Natural timber	Timber is machined into floorboards which are nailed into place perpendicular to the direction of the joists.	• Natural sustainable product. • Can be recycled. • Can be laid directly onto floor joists.	• Has to be sealed and treated to prevent damp and insect damage. • May be damaged if need to be lifted for services.

▶ **Table 4.12:** *Continued ...*

Type	Installation methods	Advantages	Disadvantages
Laminates	A system that snaps together as it is installed onto a foam cushion.	• Clean finish available in a range of different colours and timber features. • Easy to install.	• Flat floor surface required. • Expansion gap required at perimeter. • Many not be waterproof. • May not look as attractive as timber.
Carpets	A natural or artificial fibre is woven onto a backing. Underlay is initially laid then the carpet is fixed to grippers at its perimeter.	• Many types available. • Easy to install. • Insulating and warm to walk on.	• Difficult to clean and make hygienic. • Grippers difficult to fix to solid floors.
Ceramic tiling	Tiles can be directly bedded onto sand and cement or using flooring adhesive onto concrete or plywood-lined timber floors.	• Attractive appearance. • Hard-wearing. • Effective in damp conditions such as bathrooms.	• Labour-intensive to install. • Repairs are difficult. • Cold finish. • Joints discolour.
Sheet materials	Flexible sheets of vinyl, rubber or linoleum are bonded to the floor surface.	• Easy to clean. • Slip resistant. • Easily removed. • Can be hard-wearing. • Range of colours and textures.	• Easily discoloured. • May be damaged if need to be lifted for services. • May require specialist installer.

Assessment practice 4.3

C.P5   C.P6   C.M3   C.D2

As the architectural technician on the new town house development you have been asked to:

1 in the form of a presentation, explain the construction details used in the construction of walls, floors and roofs on new construction projects such as this one

2 produce a report that summarises the following aspects of wall finishes and includes an analysis of such techniques with recommendations:
   - internal first floor walls
   - intermediate floors
   - solid ground floors
   - ceilings
   - main pitched roof.

Plan
- What finishes will I include to cover a range of appearances for each application?
- Have I researched at least three manufacturers' websites?

Do
- Have I used photographs and diagrams to enhance my evidence?
- Have I referred to manufacturers' websites?
- Have I analysed the relative benefits of the different finishes?

Review
- Will my recommendations for finishes be suitable for the occupiers?

Examine external works associated with construction projects

D1 Foul and surface water drainage

The drainage connected to buildings is essential in removing foul and surface water from toilets, bathrooms and kitchen activities, as well as rainwater. The UK drainage system is built around two methods:

▶ Combined drainage – both foul and surface water goes into one sewer.
▶ Separate drainage – foul and surface water is separated.

The advantage of a separate drainage system is that the need for water treatment is reduced as the surface water does not require treatment (see Figure 4.19). This reduces costs associated with water treatment and the impact upon the environment.

The disadvantages are that two sets of resources are required as each drain run is doubled. The other issue is that, if care is not taken, foul drainage may be connected to a surface water drain.

A new drainage system is constructed in accordance with Building Regulations Approved Document H. The requirements of this document include:

▶ the falls required for drainage so it flows correctly
▶ means of access such as inspection chambers and manholes so any blockages can be dealt with
▶ rainwater drainage
▶ building over drains
▶ foul water drainage.

Local water authorities maintain drainage systems. The main sewer eventually ends up at a treatment works where biodigesters break down the sewage into non-harmful products that can be disposed of elsewhere.

Research

Investigate the construction of an inspection chamber and manhole for a foul sewer. Draw an annotated diagram of the components and main features.

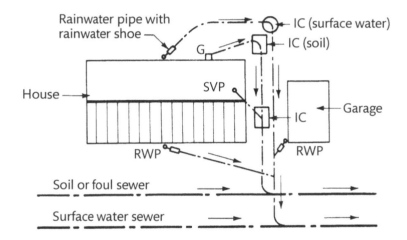

▶ **Figure 4.19:** Typical layout of separate drainage systems

D2 Utility services

Utilities are distributed nationally using power lines, gas terminals and local water services. They are incorporated within the pavements adjacent to the main roads at depths where they will not be subject to any damage from traffic.

Service outlets and cables are colour coded as follows:

▶ water – blue
▶ gas – yellow
▶ electricity – black with markers
▶ telecommunications – green.

Service entry into domestic buildings is normally via an external meter cupboard box which is built into the outside skin of a cavity wall. Figure 4.20 illustrates such arrangements for a gas service. The electrical service is similar and is often located next to the gas service entry box.

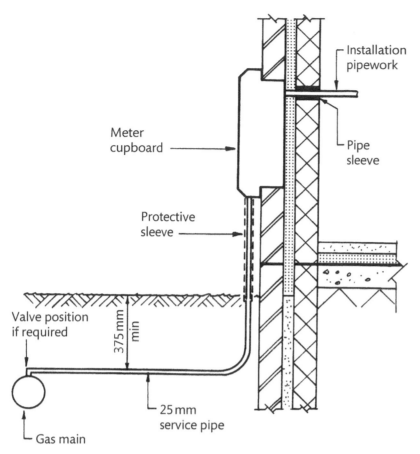

▶ **Figure 4.20:** Service entry for a gas main

A water service entry has to be protected from frost so that it does not freeze in the winter months. The depth of a water pipe is a minimum 750 mm as it enters a building. Figure 4.21 illustrates the component parts of the water service entry, including a stop valve (which commonly also houses the water meter), ductwork into the building and the stopcock isolation valve.

Typical cold water service layout

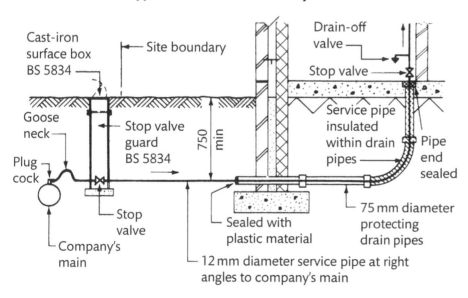

▶ **Figure 4.21:** Service entry for a water supply pipe

The electrical service entry is a minimum 450 mm deep. It uses armoured cable surrounded by a metal strand to protect the insulated conductors of live and neutral. A duct is provided to direct the cable into the external meter cupboard. Figure 4.22 shows an electrical service entry.

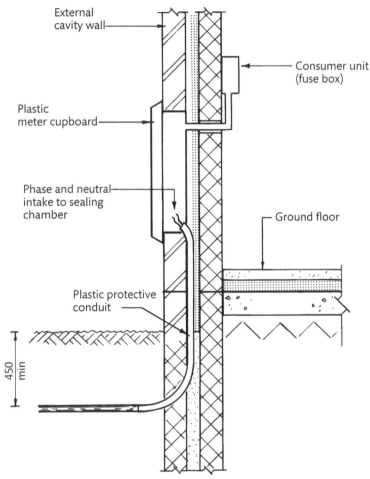

▶ **Figure 4.22:** Service entry for electric cables

Telecommunications are data and phone lines. Some of these may be on overhead telegraph poles with a link to each property. The modern installation uses a cable duct from a relay unit. The duct is normally coloured green and set at a depth of 450 mm below the pavement level.

D3 Roads and footpaths

The infrastructure that pedestrians and road users rely on can be constructed as a solid or flexible pavement. The latter relies upon bitumen-based layers to allow some flexibility and movement under load.

Tarmacadam to footpaths

Footpaths are safe zones where pedestrians can walk and avoid the traffic on the road. They can be constructed using flexible tarmac or paving slabs. Figure 4.23 illustrates both methods.

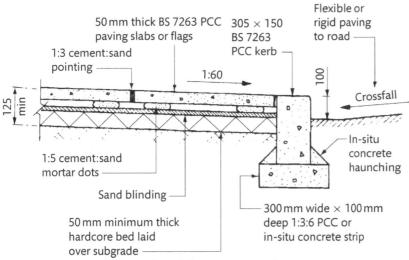

Paving slabs on mortar dots

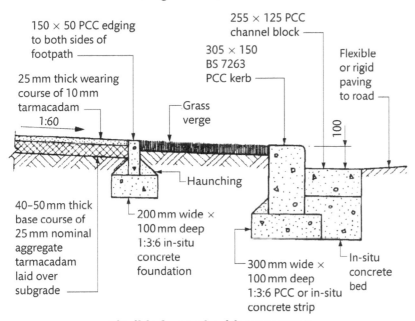

Flexible footpath with grass verge

▶ **Figure 4.23:** Typical footpath construction details

Footpaths have edging kerbs to retain the tarmac finishes and to delineate a grass verge. A footpath requires a subgrade of hardcore to distribute the loads from traffic. A base layer is finished with a wearing course to complete a tarmac footpath.

Tarmacadam to vehicular areas and road

This road surface requires a more detailed layered system, which may consist of the following, sequenced from the bottom through to the top:

3   Subgrade – existing layer after excavation has taken place.

4   Sub-base – layer of varying depth of compacted hardcore.

5   Base course – layer of bitumen and aggregate roadbase.

6   Wearing course – the layer that incorporates an anti-skidding surface of rolled stones.

The depth of each layer is varied according to the classification of the road, its required durability and the traffic loadings.

Block paving

This attractive finish is often used to enhance the driveways around houses and can incorporate different design patterns and colours of blocks. Blocks are normally manufactured from a dense clay or concrete process to form a durable block. Standard herringbone patterns are normally used which interlock the blocks in place. The typical sequence of layers is as follows:

1 Subgrade.
2 Geotextile membrane to prevent punching settlement of blocks.
3 Type 1 granular hardcore rolled and compacted.
4 Paving sand layer.
5 Blocks vibrated into place using kiln dry sand.

In-situ concrete

This is a solid method of construction for roads. It requires road forms to be installed to allow the roadway to be poured and cast into place.

A solid road has a sub-base layer that is compacted and blinded then covered with a slip membrane to allow for thermal movement of the roadway. A wearing layer can be applied over the concrete surface if required. Movement joints have to be accommodated within the length of the roadway to allow for expansion and contraction. Figure 4.24 shows a typical cross section through a solid road.

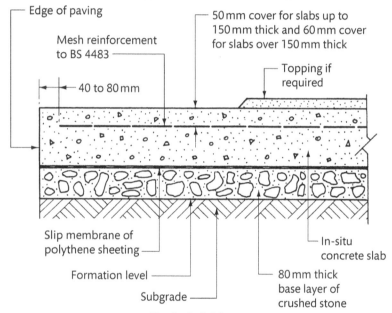

Typical rigid pavement

▶ **Figure 4.24:** Typical cross section through a solid road

Precast concrete paving

This is used for footpath construction and normally paving slabs are 900 x 600 mm and 50 mm thick, although other sizes are available. The paving slabs or flags are laid onto a bed of sand, which is supported by a hardcore layer. A road kerb retains the edge of the paving slabs and acts as a kerb line for water drainage to road gullies.

Table 4.13 compares the different types of road and footpath paving types.

▶ **Table 4.13:** Advantages and disadvantages of different paving types

Paving type	Advantages	Disadvantages
Tarmacadam footpaths and roads	• Ease of repair. • Falls created within structure. • Can work around obstructions such as trees.	• Repair appearance. • Melts in hot summers. • Anti-slip finish has to be applied.

▶ **Table 4.13:** *Continued ....*

Paving type	Advantages	Disadvantages
Block paving	• Sustainable drainage system built in. • Range of sizes and colours available.	• Regular maintenance to prevent weeds growing. • Foundation required to support sand bedding and road loads. • Chewing gum adheres to surfaces.
In-situ concrete	• Hard-wearing. • High loading capacity. • Can span soft spots.	• Noisy surface. • Requires curing to achieve strength.
Precast concrete paving	• Economical application. • High strength.	• Limited range of colour options.

PAUSE POINT   What paving type would you recommend for a market square in a small town which is used for stalls and stallholders' vans on market day and is pedestrianised the rest of the time?

> **Hint**   Think about the type and volume of traffic the area will experience.

> **Extend**   What factors, other than durability, might influence the choice of road or footpath surface in a public place?

D4 Sustainable urban drainage systems (SUDS)

SUDS are sustainable methods of managing surface water by simulating natural flow patterns such as permeation. The combination of systems allows the surface water to be stored until it can naturally flow away from the hard urban surfaces. SUDS enable the speed and strength of the water to be controlled to prevent flash flooding and to remove pollutants.

Methods of temporary storage of excess surface water

A number of different methods can be used for the temporary storage of excess surface water:

▶ **Swales** – these are run-off areas that collect water and then allow it to drain slowly into the substrata.

▶ **Infiltration basins** are larger areas that are covered in vegetation that fill up with storm water and then slowly dissipate it into the ground.

▶ **Extended detention basins** collect water and allow it to build up to a given level and then run out from a discharge outlet or spillway.

▶ **Wet ponds** – these are permanent ponds where the level fluctuates during the year as storm water enters.

▶ **Infiltration systems** - these can be as simple as a trench that has been filled with gravel so that, when surface water falls into the trench, it fills up the voids within the gravel, allowing for a given capacity of water which is then absorbed into the substrata.

Methods allowing natural percolation to groundwater

These allow water to seep into the ground via gaps, voids or the porosity of the manufactured product and include:

▶ filter strips – mini trenches that are filled with a granular material such as gravel that allows surface water to run off and filter through into the ground below. The gravel allows for some run-off capacity to be retained

▶ porous surfaces:
   • porous block paving – the joints between the block pavers are filled with a dry sand that is vibrated into the joint. The joint allows water to filter through into the subsoil
   • permeable tarmac – this modern new technology allows water to drain through the surface of the tarmac and into the subsoil

> **Link**
>
> For more about SUDS, including their features and benefits, see Unit 2: Construction Design.

- porous concrete – this is a 'no fines' concrete where only larger aggregates are used. Water therefore filters through the voids and into the soil below
- gravel – the voids between the aggregates allow water to percolate through and into the subsoil.

Advantages of SUDS

There are some advantages to SUDS:

▸ It uses a mixture of simple technologies to encourage water flow and evaporation.

▸ The systems can be a feature of an urban environment, for example by providing attractive ponds.

▸ The natural systems can enable biodiversity by being attractive to local wildlife.

▸ It reduces the environmental impact of urban water management problems, such as flooding and pollution.

Disadvantages of SUDS

However, there are also a few disadvantages to be aware of:

▸ Its capacity is limited to the rate at which water seeps into the ground, which might vary according to local conditions.

▸ The system requires some maintenance to retain its optimal effectiveness.

▸ It uses land which might have a high commercial value.

Research

Examine each of the methods used for temporary storage of water and contrast the differences between them.

Theory into practice

Design a sustainable urban drainage system for a small development of six houses where surface water currently discharges from driveways and roof guttering. The design should be a feature that will attract potential buyers to the development.

Assessment practice 4.4   D.P7   D.P8   D.M4   D.D3

You are the designer of the external works package on the new housing estate.

1 Summarise options for designing and constructing suitable external works.

2 The client has asked that sustainable drainage techniques are used where possible. Produce a SUDS design brief that:
- forms an attractive feature
- encourages biodiversity
- does not overflow into drains
- is self-contained
- will clean surface water.

Plan
- What format will I use for my design brief?
- Have I carried out enough research on different SUDS?

Do
- Have I justified my recommendation for the design within the brief?
- Have I covered all the criteria for the brief?

Review
- Does my design reduce the environmental impact of groundwater?

Websites

Wiki on building design: **www.designingbuildings.co.uk/wiki/Home**

National House Building Council: **www.nhbc.co.uk**

THINK ▶FUTURE

Ali

Ali is a manager for a large construction company that specialises in the prefabrication of domestic housing. This requires detailed knowledge and understanding of construction technology so that he can detail, design and assemble modules for on-site installation. The company fits all services, fixtures and fittings into the factory-produced module.

Ali is responsible for the production of the units to the design specification to ensure that they arrive on site in time and meet the needs of the client. A variety of options are available, from flat green roofs to pitched roofs. The foundations are designed and installed by a specialist groundworks company.

Ali's career started after he completed his degree in Construction Management at university. A large design and construction company employed him because of his specialist skills in the design of prefabrication of modules for the construction of living spaces.

Ali's skills include detailing the connections of the modules to ensure a perfect fit on site and enabling each module's services to be connected by unskilled site labour. This requires knowledge of a range of different materials and services and an understanding of how they are assembled.

Focusing your skills

A design and construction manager must demonstrate a range of skills when they are designing and co-ordinating different aspects of construction technology. These include:

- attention to detail, for example by considering ways of keeping the weather out from the inside of a building

- planning and co-ordinating different types of contractors
- understanding materials technology
- research skills in components technology
- information technology skills in the production of digital drawings and schedules.

Getting ready for assessment

Linda has started a BTEC Extended Diploma in Construction and the Built Environment and works one day a week as an apprentice site manager with a local construction company. For her Unit 4 assessment Linda has to complete a report to a client covering the use of different structural forms for a proposed project, considering the effectiveness of each.

The report needs to include a range of structural forms that could be used to support the loads from the proposed project. This will involve undertaking some research to establish suitable methods that could be used for the assignment scenario.

How I got started

As I work on a construction site I showed my line managers on site the tasks for the assignment. Using their experience my managers were able to point me to jobs the company had undertaken in the past. These projects all used different structural forms to support the functions of each building. The site manager suggested asking the contracts manager if they could provide details on each of the historical projects our company had undertaken.

I looked at the Unit 4 content from the downloaded specification we had all been given. This provides some guidance as to the coverage of content. There is a further section at the back of the unit, which tells you how to achieve a pass, merit or distinction standard. This is really useful in building up the level of evidence and the extension across the grading from the pass through to distinction.

I decided to split my structural forms into domestic and commercial applications. I could then contrast traditional cavity walls and timber framed construction with modern methods of construction, including modular applications. Many manufacturers' websites are available and provide information that I could use for my assignment.

How I brought it all together

I wrote an introduction then asked my site supervisor to take a look and read it over to see if I was on the right track. When this was confirmed he gave me some pointers as to how it could be improved. The introduction should set the scene, not be over complicated and should tell the reader what is coming up in the main body of work.

I split the main body into domestic and commercial forms and structural frames. This made it easier to check the unit contents guidance against each type of application. For example, a portal frame is mentioned in section A1 but it doesn't have a domestic application at all. It is similar with timber framed construction, where commercial opportunities are applied less than domestic ones.

I used a bibliography to reference all the sources I had used. This meant that I declared all the work as my own and I had put speech marks around all referenced quotations and used footnotes to indicate the sources used.

What I learned from the experience

Working in construction provided me with hands-on opportunities to view different types of structural forms used to construct buildings for domestic and commercial applications. I was able to talk to construction professionals, which helped considerably.

It is difficult to try and be concise with the evidence for the main assessment. You need to use and appraise information so it is expressed in your own words. Another time I would spend longer planning my work to make sure that I was delivering the correct level of information for the assessment in a clear and concise way.

Think about it

▶ How will you find out about the different structural forms?

▶ What sources should you reference?

▶ Should you use diagrams and photographs?

▶ Can you use evidence obtained from working with a company?

Health and Safety in Construction 5

Getting to know your unit

Health and safety should be an integral part of all construction operations, from initial planning and design to the build itself and then to the use, maintenance and eventual demolition of the structure. Construction workers, and those who live in, work in or visit the building, should not be put at unnecessary risk. Everyone has a part to play in ensuring that health and safety is a priority.

This unit describes the relevant legislation and regulations relating to construction health and safety, and considers the procedures for designing a safe system of work. You will consider how such systems should be monitored, reviewed and amended if necessary. By the end of the unit, you should be able to evaluate how safety systems are maintained and improved to reduce accidents and ill-health.

This unit can provide a basis for moving into health and safety management in construction, or for progressing to specialist health and safety qualifications.

How you will be assessed

This unit will be assessed by a series of internally assessed written tasks set by your tutors. Two assignments are normally set to split the assessment criteria into manageable portions. Throughout each section of this unit you will be given opportunities to test your knowledge and understanding of the assessment criteria for each learning outcome. Assessment activities have been included to help you to practise.

To achieve the tasks in your assignment, it is important to check that you have met all the Pass grading criteria. In this unit there are seven Pass criteria spread over learning aims A, B and C. There are three Merit criteria and three Distinction criteria which extend and stretch your learning by requiring you to justify, evaluate, discuss further and optimise the work that you did for the Pass criteria.

Evidence that you produce for assessment may take the form of presentations, leaflets, formal reports, case studies, statistics, safety surveys and risk assessments.

Assessment criteria

This table shows what you must do in order to achieve a **Pass**, **Merit** or **Distinction** grade, and where you can find activities to help you.

Pass	Merit	Distinction
Learning aim **A** Understand how health and safety legislation is applied to construction operations		
A.P1 Explain the legislative duties of employers and employees in the current legislation.	**A.M1** Discuss the impact of health and safety-related legislation, education and training in controlling health and safety in construction.	**A.D1** Evaluate the effectiveness of health and safety-related legislation, education and training in controlling health and safety in construction.
A.P2 Explain how the application of health and safety-related legislation controls health and safety in construction.		
A.P3 Explain how education and training improves standards of health and safety		
Learning aim **B** Carry out the development of a safe system of work for construction operations		
B.P4 Explain methods used to identify hazards and assess risks.	**B.M2** Optimise the safe system of work for a construction operation.	**B.D2** Justify the optimised safe system of work for a construction operation.
B.P5 Produce a safe system of work for a given construction operation and risk assessment to include a method statement with effective control measures.		
Learning aim **C** Understand the need for the review of safety systems for construction operations		
C.P6 Explain how safe systems of work are reviewed.	**C.M3** Discuss how safety systems are improved following the reporting of accidents and review of procedures.	**C.D3** Evaluate how safety systems are improved following the reporting of accidents and review of procedures.
C.P7 Explain the procedures that follow an accident to improve future safety.		

A Understand how health and safety legislation is applied to construction operations

The UK government has produced a range of health and safety laws and regulations used to control an employer's activities, reduce **risks** and **hazards**, and ensure that all construction workers are kept safe and secure when undertaking construction of our built environment.

A1 The Health and Safety at Work etc. Act 1974

The Health and Safety at Work etc. Act (1974; often called HASWA or HSWA 1974) is one of the main pieces of health and safety legislation in the UK. It defines the responsibilities of everyone in any workplace for ensuring that the workplace is kept safe 'as far as is **reasonably practicable**' and that everyone ensures they are working safely.

Duties of the employer

The employer has specific responsibilities under this primary legislation, which covers all the vital elements to provide safe environments for employees, including to:

▸ ensure the health and safety of all employees
▸ provide **safe systems of work**, including handling, storage and transport
▸ provide information, instructions and training
▸ provide a health and safety policy if there are five or more employees
▸ consult and co-operate with employees on safety measures
▸ observe the regulations on safety committees
▸ not charge for anything provided for safety
▸ ensure that persons not in their employment are not exposed to health and safety risks.

Duties of employees

The employee has specific responsibilities under this primary legislation, including to:

▸ act with due care for themselves and others, e.g. to walk rather than run down a corridor
▸ co-operate with the employer, e.g. taking part in **toolbox talks**
▸ correctly use anything provided for health and safety in accordance with any instruction or training
▸ not misuse or damage equipment provided for health and safety purposes.

Duties of self-employed workers

Those who are self-employed are contracted by employers to undertake specialist work on a construction site. Under the Health and Safety at Work etc. Act, they have to:

▶ conduct their work in a safe manner so it does not affect anyone exposed to their work

▶ provide any information relating to their work which supports the health and safety of anyone exposed to their working.

Duties of designers and manufacturers

There are duties for people who design and manufacture anything that will be used at work. These are to:

▶ ensure the design and construction is safe to use and operate without risk to health and safety

▶ provide information and instructions to ensure health and safety during operation

▶ ensure the installation, erection and operation of the designed item is done safely

▶ test and research the operation and use of designed items to ensure that they are safe.

Discussion

The designers and manufacturers of items clearly have responsibilities under HASWA 1974. Does this cover the architect who designs a building?

Research

Find out how the different professionals can work together on a construction site to ensure they are fulfilling their duties under HASWA 1974. You can do this either by speaking to people who work on a local construction site or by using the internet to research a recent construction project.
• How successfully did everyone fulfil their duties?
• What do you think they could do differently?

Health and Safety Executive (HSE)

The HASWA 1974 legislation ordered the formation of the Health and Safety Commission and the Health and Safety Executive, which were merged into a single body, the Health and Safety Executive (HSE), in 2008.

The main contractor has a duty before any major construction work commences on site to inform the HSE through a document called an **F10**. An F10 has to contain information on the following:

▶ address of the site

▶ details of local authority

▶ a description of the project

▶ details of the start date, duration, number of operatives working, number of contractors on site

▶ details of principal designer, principal contractor, client, designer and contractors

▶ signed and dated declaration.

Key term

F10 – a formal document submitted to the Health and Safety Executive that advises what, where and when a construction project will start and who is undertaking the work, with key contact details.

Powers of the HSE

The HSE has many powers at its disposal to ensure that construction sites are safe. These are shown in Table 5.1.

Action	How the employer or employee has to answer the action
Inspection	Co-operate with the HSE inspector and provide access as required.
Improvement notice	Improve the unsafe item or system within the time frame specified on the improvement notice. Work can continue.
Prohibition notice	Immediately stop work as employees are in serious and imminent danger of risks from hazard. The work cannot start until the unsafe actions have been completed and signed off by the HSE.
Documentation	Provide all requested documentation for the inspector to be kept under statutory provisions.
Investigation	Comply with all aspects of an investigation, providing evidence and witnesses, copy any documentation, secure any accident site undisturbed.
Entry	Comply with inspectors who have the power to enter any premises at any reasonable time.
Seizure	Comply with HSE seizure of any article deemed dangerous and make it safe.
Prosecution	Submit to court action by HSE for enforcement of legislation, with penalties involving fines and prison sentences for serious breaches by employers.
Advisory	Listen to HSE advice on any aspect of health and safety legislation and how to apply it on site.

Penalties for non-compliance

The HSE, when investigating a breach in health and safety legislation, will decide what enforcement action is necessary. This can be one of the following routes or penalties for non-compliance of any regulation or Act of Parliament:

▶ prohibition notice is served giving a stop notice until the corrective work is completed
▶ reference to the magistrate's court or to Crown Court
▶ unlimited fines
▶ corporate manslaughter
▶ imprisonment for up to two years.

An employer can be further penalised when a serious accident occurs on a construction site through:

▶ **Sanctions imposed by clients** – for example they may be removed from a tender list and prevented from obtaining any future work with that client.
▶ **Loss of reputation** – a reputation in the local construction sector is only as good as the last project.
▶ **Adverse press** – critical press reports on accidents can have an effect on obtaining work if clients read them.
▶ **Loss of work** – when fines and other HSE penalties are published, client searches on a business will highlight these enforcements.

 **PAUSE POINT**   Could you explain the legal duties of your employer under the Health and Safety at Work etc. Act 1974?

 Hint   Look at the HSE website for information and guidance on employers' general duties.

Extend   Are the self-employed covered in the Act?

A2 Construction (Design and Management) Regulations 2015

These regulations, known as the CDM Regulations or CDM 2015, are used to ensure that a design for a building makes it safe to construct, use and maintain. European Union research established that 35 per cent of construction accidents could have been prevented at the design stage while the building was still on the drawing board.

The CDM Regulations are divided into five parts.

▶ **Table 5.2:** Parts of the CDM Regulations

Part 1	The application of the regulations and definitions of terms.
Part 2	The duties of clients, both commercial and domestic, for all construction projects.
Part 3	The duties of specific people under the regulations.
Part 4	The general requirements for all construction sites.
Part 5	The transitional arrangements as the regulations came into force.

Phases to be followed

Pre-construction information

This important phase gives the client duties to provide all information about the proposed site and development to the other duty holders. Before the project starts, the client must ensure that:

▶ all pre-construction information has been handed over to the design team and appointed contractor. This may be drawings, specifications, previous health and safety files, asbestos logs, site surveys and desktop research. Pre-construction information is defined under the regulations as having relevance, an appropriate level of detail and being proportionate to the risks involved

▶ a construction phase plan is drawn up by the contractor

▶ the principal designer prepares a health and safety file for the project.

Construction phase safety plan

The construction phase health and safety plan covers the arrangements for this phase, the site rules and any specific measures concerning work that involves particular risks.

The principal contractor has the responsibility of setting up the plan and file. The principal designer must assist in providing all information that they hold. The construction phase health and safety file contents cover aspects such as:

▶ safety arrangements on site

▶ site rules and **site induction** arrangements

▶ welfare facilities

▶ fire, emergency and evacuation procedures

▶ arrangements for co-operation and co-ordination between teams.

> **Key term**
>
> **Site induction** – the presentation that makes new visitors or workers on a site aware of the rules of the site and the hazards that it contains.

> **Link**
>
> Section B2 of this unit details the construction phase health and safety plan.

The health and safety file

This file is a historical record of the project. It is retained so that any future work such as maintenance, cleaning, refurbishment, adaptation, extension or demolition can be examined with regard to any risks. A health and safety file might contain any of the following:

1 A brief description of the work carried out.

2 Any hazards that have not been eliminated through the design and construction processes, and how they have been addressed (e.g. surveys or other information concerning asbestos or contaminated land).

3 Key structural principles (e.g. bracing, sources of substantial stored energy – including pre- or post-tensioned members) and safe working loads for floors and roofs.

4 Hazardous materials used (e.g. lead paints and special coatings).

5 Information regarding the removal or dismantling of installed plant and equipment (e.g. any special arrangements for lifting such equipment).

6 Health and safety information about equipment provided for cleaning or maintaining the structure.

7 The nature, location and markings of significant services, including underground cables; gas supply equipment; fire-fighting services etc.

8 Information and as-built drawings of the building, its plant and equipment (e.g. the means of safe access to and from service voids and fire doors).

Source: *Managing Health and Safety in Construction* (HSE)

It is only required for projects where there is more than one contractor.

Duties of parties under CDM

Research

Why are a principal contractor and a contractor required within the regulations? What are the differences between the two? Research the differences and report back to the rest of the group.

Client

A client must ensure that they make suitable arrangements for the management of their project. This is normally achieved by hiring an expert from the industry who knows how to apply the CDM Regulations to their project. The client must ensure these arrangements are maintained throughout the project. The client must provide all pre-construction information before the project commences and ensure a construction phase health and safety plan is in place. The client is responsible for appointing the principal designer and for notifying the relevant enforcing authority (HSE) about the project.

Domestic clients, for example those commissioning work on their own home rather than for a business, have the duty to appoint a principal designer, and also a principal contractor if there is more than one contractor. However, in practice, these duties are automatically transferred to the contractor or designer they appoint.

Principal designer and designers

Regulation 9 covers the duties of designers for a client's project. A designer is the organisation or individual that prepares or modifies a design for any part of a construction project, or who instructs someone else to do it.

One organisation or individual should take on the role as principal designer to plan, manage, co-ordinate and monitor all the design duties during the pre-construction phase.

The principal designer has to:
▶ assist the client in identifying, obtaining and collating the pre-construction information
▶ provide pre-construction information to other designers, the principal contractor and contractors
▶ ensure that any designers they appoint have the relevant skills, knowledge and experience
▶ ensure that designers comply with their duties and co-operate with each other
▶ liaise with the principal contractor to help in the planning, managing, monitoring and co-ordination of the construction phase
▶ prepare the health and safety file.

Other designers have to:
▶ ensure that clients are aware of their duties
▶ reduce and, where possible, eliminate any risks associated with carrying out the construction, maintenance, cleaning or use of the final project
▶ control any risks through the design process and communicate these risks to everyone

▶ ensure information is included in the health and safety file that adequately assists clients, other designers and contractors

▶ co-operate and co-ordinate with others working on the project.

Principal contractor

The principal contractor is the contractor in overall control of the construction phase on projects with more than one contractor. They are appointed by the client and there should only be one principal contractor for a project at any one time. Their duties start to take effect in the construction phase of a project and cover:

▶ the planning, management, monitoring and co-ordination of the project to ensure its safe construction

▶ organising co-operation between contractors

▶ providing a suitable site induction

▶ preventing any unauthorised access onto the construction site

▶ providing welfare facilities in accordance with Schedule 2 of the CDM regulations

▶ liaising with the principal designer on all aspects of safety

▶ consulting and engaging with workers.

Contractors and workers are also duty holders, and have a responsibility to co-operate and co-ordinate with others to ensure that the site and structure remain safe.

General requirements

Welfare facilities

Schedule 2 in the regulations gives details of the minimum requirements for welfare facilities that have to be provided on a construction site.

> **Key term**
>
> **Suitable and sufficient**
> – a phrase often used in legislation to mean that enough of something (e.g. toilets) should be provided depending on the number of people on site.

▶ **Table 5.3:** Welfare requirements for all sites under CDM

Item	Requirements
Sanitation	**Suitable and sufficient** WCs should be provided; the facilities should be well ventilated and lit. Facilities must be kept clean and orderly. Separate rooms do not need to be provided for the exclusive use of men or women if each facility is lockable from the inside and can be used by one person at a time.
Washing	Suitable and sufficient washing facilities including showers must be provided adjacent to sanitation facilities or changing rooms. Washing facilities must include hot and cold water, soap and towels or drying equipment. Rooms must be ventilated and lit, and kept clean and orderly. Men and women can use the same facilities if the washrooms are lockable from the inside and can be used by one person at a time.
Water	A supply of drinking water must be provided with cups or a jet.
Changing rooms	Suitable and sufficient lockable changing rooms must be provided. They should be provided with seating, contain drying facilities and have storage lockers for hanging clothing and equipment and locking away personal effects.
Rest facilities	Rest rooms must be equipped with a number of tables and seating, have arrangements for preparing and eating meals, boiling water and be maintained at an appropriate temperature.

General principles of prevention to be employed on site

Part 4 of the CDM Regulations covers many of the aspects of safety that must be implemented on a construction site. These are:

▶ construction sites must be safe places of work

▶ good order and site security must be established

▶ structures whether permanent or temporary must be stable

▶ any demolition or dismantling must be planned

▶ excavations must be supported, inspected, barriered off, and maintained

▶ measures to prevent drowning must be in place when working over water

▶ traffic routes must be organised

▶ movement of vehicles must be controlled

▶ emergency procedures and arrangements must be in place

- emergency routes and exits must be designated and maintained
- fire detection and fighting arrangements must be in place
- fresh air supplies must be provided
- temperature and weather protection measures must be in place for workers
- adequate lighting must be provided.

> **Discussion**
>
> Fire detection and fire-fighting arrangements must be in place on a construction site. How could this be achieved? As a group, discuss this problem and make some suggestions for how fire-fighting tools could be placed.

PAUSE POINT   What pre-construction information needs to be made available before the start of the construction work?

Hint   Review Part 2 of the CDM Regulations for more on this process.

Extend   Is the principal designer part of this process?

A3 Management of Health and Safety at Work Regulations 1999

These regulations are applicable to all areas of industry and not just construction. Many aspects apply to construction, such as those in Table 5.4.

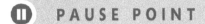 **Table 5.4:** Details of the Management of Health and Safety at Work Regulations relevant to construction

Regulation	Duties
Risk assessments	Regulation 3 covers this vital practice. Every employer has a legal duty to make a suitable and sufficient assessment of the risks to the health and safety of their employees and persons not in their employment such as visitors, the self-employed and subcontractors.
Health and safety arrangements	Every employer has to make arrangements for health and safety and where they employ five or more people these must be formally recorded in a document for all to access.
Health surveillance	All employers shall undertake health surveillance if the nature of their activities poses risks to workers' health.
Health and safety assistance	Every employer shall appoint a competent person to assist with health and safety matters balanced against the size of their undertaking. They shall ensure that this person(s) is informed, trained and experienced. They must co-operate and assist this person in all duties.
Co-operation and co-ordination	Employers must co-operate with other employers on site and co-ordinate all measures and activities. All employers must be made aware of any risk to health and safety so employees can be informed.
Competency	This covers the capabilities and training of employees. Every employer must provide health and safety training that is suitable and sufficient to the activities that they will be undertaking. Training must be updated with new equipment, technology or operatives and repeated where appropriate.
Specific duties	All employees will use the equipment provided and any safety device in accordance with the training and instructions given by the employer. Any employee must inform their employer if the training and instructions would lead to serious and immediate danger.

A4 Work at Height Regulations 2005

A lot of construction activities are undertaken above ground level. Any activity above ground level, even if it is not high up, places risks on workers using temporary platforms to gain access to, and work on, the exteriors of the buildings as well as those erecting steelwork and installing floors. The same applies if you are working at ground level next to an excavation. These regulations are used to prevent fatalities from working at height.

Ideally, the need to work at height should be eliminated at the design stage of the building but often it is hard to avoid. In any case, it should be taken into account when organising and planning construction work, with risk assessments and method statements in place before work starts.

The regulations impose duties upon employers and employees as follows.

Duties of employers

For all employees, visitors and subcontractors, employers must:

▶ ensure that all work at height is planned, supervised and carried out in a safe manner where reasonably practicable in suitable weather conditions
▶ ensure the competence of all workers at height
▶ not carry out work at height where alternative measures could be utilised at ground level
▶ provide all measures to prevent any person falling a distance liable to cause personal injury
▶ provide training and instruction in working at height
▶ ensure no worker passes over or near fragile surfaces
▶ prevent the fall of any materials or equipment from height
▶ clearly indicate any areas of danger
▶ inspect all equipment provided for working at height and maintain, repair replace this as required
▶ inspect all places where working at height is taking place.

Duties of employees

Employees have duties as persons at work to:

▶ report any defect on a measure provided for working at height to the supervisor
▶ use all provided work equipment for working at height
▶ follow the training provided
▶ use all instructions given for working at height.

> **Research**
>
> 'The Work at Height Regulations require employers to take all reasonable measures to undertake work at ground level.' Is this statement true?

> **Discussion**
>
> How can you safely install an element of a building, such as a door or window, that needs to be lifted onto a multi-storey building?
>
> What is prefabrication? How is this safer?

▶ What personal protection equipment (PPE) would you require if you were working on this site?

Work equipment requirements for operatives

The Work at Height Regulations 2005 set out a hierarchy of fall protection measures. If such work cannot be avoided, the appropriate type of work equipment and other measures should be in place to prevent falls from height. Where the risk of a fall cannot be eliminated, work equipment or other measures should be used to minimise the distance and consequences of any fall:

▶ If it is not possible to avoid work at height, use collective means of prevention (one that does not need the person working at height to act to be effective), such as guard rails.

- If this is not possible, use personal means of prevention (one that applies to only the individual), such as work restraint.
- If this is not possible, use collective means of protection (one that does not prevent a fall but reduces the effect of a fall), such as air bags.
- If this is not possible, use a personal fall protection system (one that requires the individual to act to be effective), such as a fall arrest system or harness.

Equipment must be suitable, stable and strong enough for the job, maintained and checked regularly.

Employers should also provide training and instruction and take other measures to prevent workers being injured by a fall.

Requirements for working platforms

A working platform is any platform that is going to be used to access or carry out work at height. This includes any scaffold, cradle, mobile platform, trestle, gangway and stairway which is used as a place of work. Whatever its form, a working platform must be of sufficient strength and of suitable composition for the task it is going to be used for and may contain the following elements:

- guard rails to prevent falls at a height with no gap of more than 470 mm between guard rails
- toeboards to prevent materials falling
- barriers to prevent any materials passing through guard rails
- supported by a suitable and stable structure.

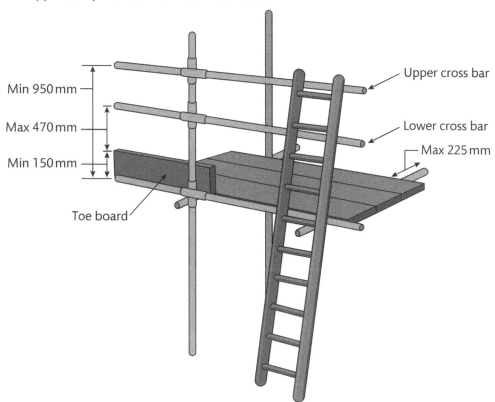

▶ **Figure 5.1:** A finished working platform

Requirements for personal fall protection

Schedule 5 of the regulations details the requirements for personal equipment that prevents employees falling while working at height. Before any work commences when using a fall arresting method the following points must be covered:

▶ A risk assessment must be undertaken to ensure that no other means of fall prevention could be used instead of personal protection equipment (PPE).

▶ It must be checked that it is of suitable and sufficient strength for the intended use, fitted correctly and designed to minimise injury to the user.

▶ Any anchor points should be of suitable and sufficient strength.

▶ All users must have received appropriate training in the use of the personal fall protection equipment.

The use of ladders

Ladders are in common use on construction sites – used by everyone from electricians wiring lights in ceilings to construction workers climbing scaffolding. Ladders must conform to the following aspects covered within Schedule 6 of the regulations:

▶ used for short durations because no other suitable equipment could be justified

▶ only used after a risk assessment shows it is not reasonably practicable to use any other means of work, and that the risks are low

▶ be placed upon surfaces that are firm, stable and of sufficient strength

▶ be stable during use

▶ secured at the top if applicable

▶ protrude past the place of landing

▶ have available secure hand and footholds.

Ⅱ PAUSE POINT   The requirements for protection of workers have to be equally applied to a mobile tower scaffold. What provisions does the working platform have to contain to ensure that operatives are safe when working upon it?

Hint   Look at Schedules 2 and 3 of the regulations for more information.

Extend   What is the minimum distance (gap) between guard rails?

Working at height

You have been asked to advise a client on the requirements for the safety of workers who are cladding new offices, working on mobile tower platforms. This involves looking at any risk assessments, method statements, and health and safety plans that are available. You will need to compare this information with what you see on site. The client is concerned that the platform that they are working on is not suitable or sufficient and an accident may occur. The principal contractor has erected a mobile tower scaffold, which is pushed along the site with operatives at the top of it, running along boards placed on the ground. The health and safety officer at the company has not seen this yet. Large sheets of cladding are passed up the side of the tower and then moved by hand over and onto the wall, where they are fixed using a nail gun.

Check your knowledge

1 What advice would you give the client?

2 What legislation would cover this work?

3 What is an acceptable means of access onto the working platform?

4 Can the workers carry materials up this type of access?

5 What are the requirements for workers operating on the platform?

6 Who is responsible for inspecting the platform before and during its use?

A5 Control of Substances Hazardous to Health (COSHH) Regulations 2002

The COSHH Regulations are designed to control any substance that is used or found on a construction site. Hazards may arise from:

▶ fumes from chemicals and solvents such as paints, thinners and glue

▶ dust created by construction work

▶ direct contact with materials that can burn, poison or react with the skin, such as cement

▶ biological hazards such as animal waste already on site.

The risk is not created by the substance itself but by how it is used, so the emphasis should be on the activity rather than the substance itself.

COSHH places duties upon an employer to ensure that:

▶ a risk assessment has been completed on all substances used in the workplace, highlighting precautionary methods to be employed before and during their use

▶ control measures on the use of substances have been put into place and are maintained

▶ substances have been examined and tested to reduce the risk to an acceptable level

▶ employees using substances at work are monitored

▶ all information, instruction and training of employees in the use of such substances has been given and completed before their use.

Theory into practice

The joinery team on site is fixing the second fix joinery items, such as skirtings and architraves, using a product called No-nails. It is a bonding compound that fixes by adhesion any material to another material. It has a high bonding strength when dry. You have looked at the writing on the tubes of No-nails and there are several hazard warning signs.

What must you do in accordance with the COSHH Regulations?

- First you must contact the manufacturer of the product and request their safety data sheet, or find it on their website.
- Read the data sheet and specifically the control measures required for using the product.
- Produce a COSHH risk assessment for the product.

Discussion

Many different substances used on a construction site have to be controlled. Can you name them? How would you find out if they were hazardous?

A6 Training and education

Employers have a legal duty to train, educate and provide instructions for operatives to operate safe systems of work in the workplace. This may involve training courses for specific pieces of equipment and plant. It also covers the use of construction materials.

HASWA 1974 requires that employers must train and educate all operatives to ensure that they are **competent** in their work activities on a construction site. **CSCS cards** provide employers with a means to certify that individuals working on construction sites have the required qualifications and training (including health and safety awareness) for the job they do on site.

CDM 2015 puts less emphasis on competence than previous versions of CDM, stating in Regulation 8 that anyone appointing a contractor must ensure that they have received the necessary information, instruction and training, and has appropriate supervision. In practice, this may mean that the person or organisation appointing these workers performs a training needs assessment.

Training can be conducted on site or off site depending on whether it is a practical operation or requires theoretical knowledge. A simple toolbox talk may suffice. Records of all toolbox talks and who attended should be kept in the event of any accident investigation. Off-site training is often provided by specialist contractors who are certified to deliver this specific training. It may cover topics such as abrasive wheel, working at height, COSHH, noise, confined spaces and the use of particular plant or equipment. If those working on the site know about such topics in detail, they are more likely to understand how they link to control measures and therefore follow the rules and processes in place that reduce the risk of accidents.

Key terms

Competency – the combination of training, skills, experience and knowledge that a person has and their ability to apply them to perform a task safely.

CSCS card – Construction Site Safety Scheme card that provides proof of qualifications and training. Most sites require operatives to hold a card before allowing them to work on site.

Research

- Research how the CSCS card system operates.
- Investigate the Health and Safety Passport Training scheme and see what is available nationally (Tip: search for the term 'CCNSG').
- Compare the advantages and disadvantages of both schemes and report them back to your group.

Construction Skills Certification Scheme (CSCS) card

▶ What colour CSCS card would an apprentice have?

Key term

CITB – the Construction Industry Training Board, which provides support and training to the construction industry, and takes a levy from larger companies to redistribute as training grants.

What do you need to do in order to obtain a CSCS card? This skills certification scheme operates nationally in the UK. Most construction sites will not allow you on site unless you have a valid CSCS card. Which card you can apply for depends on your occupation and the qualifications you hold.

In the majority of cases, applicants must pass the relevant construction-related qualification for their occupation, and pass the appropriate **CITB** Health, Safety and Environment Test (note that some occupations require additional training before a CSCS card can be issued). Table 5.5 shows the different types of CSCS cards.

▶ **Table 5.5:** Types of CSCS cards

Colour of card	Type of card	Status of holder
Green	Labourer	Completed any one of the approved labourer qualifications such as QCF Level 1 or SCQF Level 4 Award in Health and Safety in a Construction Environment.
Red	Provisional	Someone on probation in their job or on work experience (16+).
Red	Apprentice	Registered on a recognised apprenticeship.
Red	Trainee	Registered for a qualification (usually an NVQ) that leads to a skilled CSCS card.
Red	Experienced worker/experienced technical, supervisor or manager	A temporary card (for those with experience but no qualification) while a construction-related qualification is being studied.
Blue	Skilled worker	Achieved a construction-related NVQ or SVQ level 2 or completed an employer-sponsored apprenticeship which included the achievement of a City and Guilds of London Institute Craft Certificate.
Gold	Advanced craft	Achieved a construction-related NVQ or SVQ level 3 or completed an approved indentured apprenticeship. Completed an employer-sponsored apprenticeship that included the achievement of a City and Guilds of London Institute Advanced Craft Certificate.
Gold	Supervisory	Achieved a construction-related supervisory NVQ or SVQ level 3 or 4.
Black	Manager	Achieved a construction-related NVQ/SVQ levels 5, 6 or 7 in the relevant construction management level qualification or holds an NVQ level 4 in construction management.
White	Academically qualified person	Completed a specific construction-related degree, HND, HNC, CIOB Certificate or NEBOSH diploma.
White	Professionally qualified person	A competence-assessed member of a CSCS-approved professional body, such as an architect, engineer, consultant or surveyor.

Client Contractor National Safety Group (CCNSG) Safety Passport

The CCNSG Safety Passport Scheme is for the engineering construction industry. You may need this instead of or as well as a CSCS card if you specialise in, for example, the construction of power stations or a particular engineering skill like welding or pipefitting. It passes a basic knowledge of health and safety to all site personnel so that they can work more safely on site. Three types of course are available under this scheme:

▶ the National Safety Course
▶ a renewal course
▶ a supervisor's course.

The courses are delivered by training providers across the UK. The passport covers a wide range of different modules and is valid for three years. A photo card again provides evidence of your competency in basic health and safety.

Fire safety

Fire safety training is an important aspect of fire control measures on a construction site. The following points, for example, must be considered as part of the construction phase health and safety plan:

▶ good housekeeping to remove potential fuel for fires
▶ a no-smoking policy on site
▶ operation of **hot work permits** if required, or alternative methods to hot work where possible
▶ training in the use of fire extinguishers
▶ temporary alarm systems
▶ appointment of fire wardens.

Fire safety should be covered in inductions and toolbox talks.

Provision and Use of Work Equipment Regulations (PUWER) 1998

These regulations control the risks associated with the use and operation of plant and equipment by operatives.

Employer duties are:

▶ to assess the risks of using particular equipment
▶ to ensure that provided work equipment is suitable for the intended use
▶ to inspect and test all equipment and maintain it in a serviceable condition, including carrying out portable appliance testing (PAT) of electrical devices
▶ to provide all information, instructions and training for employees on equipment
▶ to ensure that all guards and safety measures operate and are maintained around any dangerous parts
▶ to provide stop controls where appropriate.

Employee duties are:

▶ to operate the equipment in accordance with the training and instructions provided by the employer
▶ to follow all control measures identified within a risk assessment
▶ to use all specified PPE for the operation of the work equipment.

Ⅱ PAUSE POINT   The requirements of the PUWER regulations involve testing of equipment. What form of test could you perform?

Hint   What is a PAT test?

Extend   What other items would you check on a piece of equipment?

Competence

Saira has started work on site as a trainee site manager after completing her BTEC National Extended Diploma in Construction. The site manager has told her to investigate which CSCS card she will require as a trainee site manager, as she must hold the card to work on site.

Saira has looked at the CSCS card website and does not know which one applies to the trainee position. Not wanting to appear foolish in front of the site manager she is unsure of how to proceed.

Check your knowledge

1   What should Saira do?

2   Who should she ask?

3   Is there a trainee site managers' card?

4   Would a Safety Passport be enough?

5   Do Saira's existing qualifications count towards a CSCS card?

6   Does the Health and Safety Unit she took on her Diploma exempt her from needing a card?

Assessment practice 5.1

`A.P1` `A.P2` `A.M1` `A.1` `A.D1`

You are working with a construction company's safety department as a graduate site manager. You have been asked to prepare a leaflet for operative site inductions that will take place on the various construction sites they operate on.

Prepare the contents for the leaflet detailing the following:

*   The duties of employers and employees across the health and safety regulations and laws covered in this unit.
*   How such legislation controls the safety of workers, citing a range of examples.
*   The impact of such legislation on the company's site operations.
*   How statistics can be used to indicate how effective the legislation is in reducing the accident rates on site.
*   How training and education in specialist skills, such as using, operating and changing blades on an angle grinder, should improve standards of health and safety on site.

Plan
*   What is the task?
*   Am I confident about producing a leaflet?
*   Do I know enough about the relevant legislation to create the required leaflet? If not, what else do I need to find out?

Do
*   I know how to design my own leaflet, and gather information about the legislation.
*   I can identify where my leaflet may have gone off topic and adjust my thinking to get myself back on course.
*   I can explain how training can reduce risks to myself and others.

Review
*   I can explain what the task was and how I approached the development of my leaflet.
*   My leaflet makes sense and has been checked for correct grammar and spelling.
*   The evidence in my leaflet fulfils the assessment criteria.
*   I can explain how I would approach the more difficult elements differently next time.

Carry out the development of a safe system of work for construction operations

Safe systems of work are essential for the operation of construction sites. Many different contractors will work on site with many employees. Planning safe systems prevents people, tools, plant and materials causing hazards for others.

B1 Health and safety preparation

Notifications to HSE

The CDM Regulations make some projects notifiable to the HSE. An online form called an F10 has to be completed.

Health and safety construction phase plan

This plan begins at the start of the project and information is added to it as the contract progresses. It must outline the measures and site arrangements for managing health and safety. The initial plan needs to detail the following:

▶ a description of the project so the uninformed person knows what is being constructed on site
▶ details of each of the project team: the client, principal designer, designer, principal contractor, contractors, other consultants if employed
▶ any information available from the health and safety files from previous work on site
▶ the structure and organisation of the site management
▶ how safe systems of work (SSW) will be monitored and reviewed
▶ the construction site rules
▶ how co-operation arrangements between all parties on site will be conducted and how the workforce will be consulted
▶ how the design for the project will be communicated and design changes will be managed
▶ the security of the site such as fences, gates and access arrangements
▶ the site induction process and content
▶ welfare facilities on site
▶ fire and emergency procedures
▶ site waste management arrangements
▶ how working near live services will be handled
▶ excavation precautions
▶ procedures for work associated with asbestos
▶ the health and safety file contents
▶ design and construction hazards.

Site induction

Under CDM 2015, the principal contractor is responsible for organising a site **induction** for all operatives on site as well as visitors, the self-employed and contractors to ensure that they are aware of hazards, how these are controlled and any evacuation or emergency procedures. The induction may include:

▶ access arrangements to the site and parking for vehicles
▶ PPE rules
▶ fire alarms and evacuation procedures
▶ welfare facilities
▶ use of mobile phones and details on supervision and breaks
▶ principal hazards and controls
▶ permits to work systems

Link

Section A1 of this unit outlines the information that needs to be filled in on an F10.

Research

The *notifiable* element of a project must meet certain conditions under CDM. What are the conditions for a project to become notifiable?

Key term

Induction – an introduction to the site and the initial information that you need to enter.

▶ safe systems of working
▶ risks specific to the site.

Health and safety inductions can be delivered using different methods, such as a video with a presenter explaining the site, its hazards, layout, key personnel, first aid facilities and expected behaviour. A manager delivering the induction in person using PowerPoint can answer any questions that come up. It is also best practice to give inductees a short test at the end of the induction to ensure that they have learned and retained information about the site and its hazards.

Site Waste Management Plan (SWMP)

The Site Waste Management Plan Regulations 2008 placed duties on employers to manage construction waste. The aim was to reduce the volumes that are sent to landfill and to reduce the impact on the environment from construction development. The regulations were repealed in December 2013; however, it is still a good idea to produce an SWMP before work commences. It could contain the following aspects:

▶ identification of the client, principal contractor, person who wrote the plan, site location and cost of the project
▶ description of waste type that will be produced
▶ estimate of the quantity of each
▶ identification of the waste management action against each type of waste.

Aspects of reuse, recycling and disposal must be considered; landfill should be the last resort for construction waste after all other options have been explored. For example, British Gypsum can take containers of plasterboard offcuts and recycle them.

Site access

The entrance to a construction site needs to display gate and entrance signage. This should cover the safety signage for PPE when entering the site. The F10 also has to be displayed within the site entrance. The health and safety law poster ('What you need to know') must be displayed within the site cabins.

The site access is often a temporary arrangement until the permanent one is constructed. It needs to be clearly visible for vehicles on the road and adequately lit and visible to all road users. This may mean placing temporary signs on the main road warning other users of the temporary site access. Traffic lights systems may have to be installed on the main road while the new entrance is constructed. Consideration should be given to lighting of the entrance during the winter months.

▶ What key information should be provided at the entrance to a construction site?

The photo shows that the gates have safety information clearly displayed upon them with the mandatory site rules.

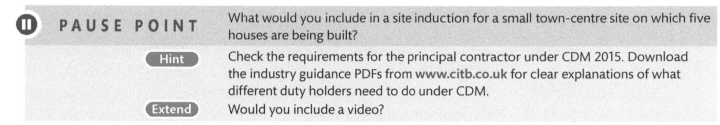

⏸ PAUSE POINT   What would you include in a site induction for a small town-centre site on which five houses are being built?

Hint   Check the requirements for the principal contractor under CDM 2015. Download the industry guidance PDFs from **www.citb.co.uk** for clear explanations of what different duty holders need to do under CDM.

Extend   Would you include a video?

B2 Construction phase health and safety

This forms part of the Construction (Design and Management) Regulations 2015 and is a legal requirement on notifiable projects. It is formally recorded in the construction phase health and safety plan and must be communicated to all operatives and contractors on site to ensure that they are kept safe while working. The construction phase safety plan may contain the elements shown in Table 5.6.

▶ **Table 5.6:** Contents of a construction phase health and safety plan

Content	Description
Site inductions	Records of the site inductions undertaken with operatives' names and the date of their induction. This provides evidence if there is an accident that they were made aware of the site rules and risks
Hazards	Methods of hazard identification, direct observation, using checklists, safety audits and inspections, toolbox talks, safety committees
Risk assessments	Responsibility for writing risk assessments, standard format, primary risks identified, risk rating adopted, acceptable levels, control measures and review
Method statements	Responsibility for writing, content, sequencing, resources to be used on the activity
Toolbox talks	Frequency of talks, topics, timing, method, who should be present
PPE	Register of issue and date, care and storage of PPE, frequency of issue, expiration dates
First aid	Facilities against workforce size, storage, qualified first aider, training, refresher courses
Traffic management	The management of traffic both pedestrian and vehicular on site, signals, one way, no entry routes
Fire	Temporary fire and evacuation procedures
Communications	Managing subcontractors' safety information, site meetings
Waste	Sorting of waste, following a SWMP, waste segregation, recycling, good housekeeping

Items can be added to the plan as and when they develop during the construction phase.

Hazard identification methods

These can be summarised as shown in Table 5.7.

Table 5.7: Hazard identification

Method	Description
Direct observation	This requires practice in spotting hazards that are not obvious. Use photographs to check the image later to identify or confirm a hazard in the workplace.
Use of accident data	Statistical analysis of accident data is another tool to help spot potential hazards. A pyramid of injuries, with the minor ones at the base and the major ones at the top, can help identify the causes of the injuries and hence a control method.
Analysing risk assessments	Looking through several risk assessments can locate a common hazard that can be collectively dealt with by global control measures. In effect, you are using this data as a set of fresh eyes to assess the situation and to point out something you may have missed.
Checklists and method statements	Checklists are identified by a safety audit or inspection, which involves walking around the site and ticking off the hazards from the pre-set list. New hazards are added to the existing checklist to build up a more comprehensive checklist. Method statements are produced as part of the CDM Regulations. They state the methods to be used to construct a particular item. Method statements enable you to analyse the correct and safest way to undertake a task.
Safety committees	Safety committees should contain a mixture of operatives, senior managers and health and safety representatives. The meetings provide the opportunity to discuss health and safety matters with employees, and allow them to bring up any hazards that concern them.

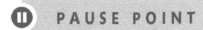

PAUSE POINT

Can you explain methods used to *identify* hazards, rather than just list the hazards themselves?

Hint How would you go about it? What equipment could you use? Is there a formal process to follow?

Extend How could you use software to help identify hazards?

Risk assessment processes

The risk assessment process follows five steps:

1 Identify the hazards.
2 Decide who might be harmed.
3 Evaluate the risks and decide on control measures.
4 Record findings.
5 Review and monitor.

Risk rating of hazards

Risks can be rated using numbers or words for the likelihood that the hazard will cause harm and the severity of that harm if it occurs.

Risk = Likelihood x severity

Now you can use numbers to judge the likelihood and the severity of the risk and produce a risk matrix which looks like the one shown in Figure 5.2.

The key to the ratings may take the following descriptions against the rating 1 to 4:

▸ 1 = slight
▸ 2 = possible
▸ 3 = very likely
▸ 4 = certainty

And severity as:

▸ 1 = no injury
▸ 2 = minor injury
▸ 3 = major injury
▸ 4 = fatal

Rate the severity and likelihood of a given risk from 1 to 4 and then multiply them together to produce a risk rating. The table colouring means:

▸ green – existing controls are sufficient
▸ yellow – needs further measures
▸ red – the work being risk assessed is too dangerous to continue without further controls and measures.

The more complex the work, for example on a power station reservoir dam wall, the bigger the matrix can become. High ratings must have further control measures. These are reassessed to check that the rating has been reduced to an acceptable level.

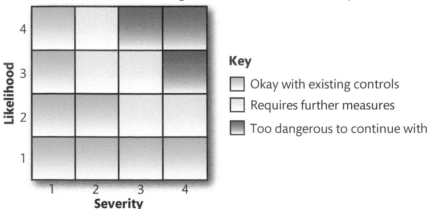

▸ **Figure 5.2:** A risk matrix

Ⅱ **PAUSE POINT**   If a rating for a particular situation is 24 on a 4 x 4 matrix what action would you take?

    Hint   What would be the implications of continuing work?

    Extend   What recommendations would you make to a supervisor?

Assessment practice 5.2   `B.P4`

The construction phase is well established on site with a large number of operatives and contractors working on it. This week's activities are the construction of the first floor using precast concrete planks onto steel columns. You have been asked to review the hazards on site.

- Explain two methods that could be used to identify hazards on the site.
- How would you assess the risks from these hazards in terms of severity?
- How would you ascertain the opinions of those involved in hazardous activities?

Plan
- How will I structure my evidence?
- What research do I need to produce?
- Have I got all the information?

Do
- Have I outlined the appropriate methods of hazard identification?
- If not, what additional information do I need?

Review
- Have I checked that I have used the correct spelling and grammar?
- Does the evidence I have written meet the assessment criteria that is targeted?
- Have I covered enough for the Merit and Distinction?

B3 Health and safety file

This file is a legal document under CDM 2015 that has to be produced and maintained throughout the project's life if more than one contractor has been involved. It informs any person, individual or contractor who is working upon a client's building about the methods of construction, materials used and any inherent hazards associated with the maintenance of the building.

It should include enough detail to allow the likely risks to be identified and addressed by those carrying out the work and be proportionate to those risks. Information must be in a convenient form that is clear, concise and easily understandable, for example printed or in an electronic format.

The health and safety file should contain as a minimum:
- a brief description of the work carried out
- any residual hazards which remain and how they have been dealt with, e.g. information concerning asbestos, contaminated land, buried services
- key structural information, e.g. bracing, sources of substantial stored energy – including pre- or post-tensioned members
- safe working loads for floors and roofs, particularly where these may prohibit placing scaffolding or heavy machinery
- hazardous materials used, to include manufacturer's data sheets, e.g. pesticides, special coatings which should not be burnt off
- information regarding the removal or dismantling of installed plant and equipment, e.g. any special arrangements for lifting, special instructions for dismantling
- health and safety information about equipment provided for cleaning or maintaining the structure
- the nature, location and markings of significant services, including underground cables; gas supply equipment; fire-fighting services
- information and as-built drawings of the structure, its plant and equipment, e.g. the means of safe access to and from service voids, fire doors and compartmentalisation.

The file should not include things that will not help when planning future construction work, such as pre-construction information, the construction phase plan, construction phase risk assessments or contractual documents.

Standard indexes can be used to ensure that all items have been covered. The file should be continually reviewed as the construction phase progresses to ensure it is kept up to date and reflects the current best practice on the site. This ensures that the file remains a relevant and authoritative resource for the site.

The completed file should be distributed to the client who must keep it, or ensure the owner of the building keeps it, so that it is accessible to anyone who wants to see it, and that it continues to be updated if necessary.

Assessment practice 5.3   B.P5   B.M2   B.D2

An excavation to the deep strip foundations has been undertaken and supporting shores have been placed as the ground is weak and the trench sides will not support themselves. It has rained during the night, filling the trench partly with water. The concreting operations are due to start today. Produce the following safety documentation.
- Develop a safe system of work for the pouring of the foundation concrete in the form of a method statement.
- Produce a risk assessment for the work with all control measures in place.
- Optimise your safe system of work.
- Justify the optimised safe system of work in terms of your selection of control measures.

Plan
- What does a method statement need to contain?
- Is there a standard format?

Do
- Have I used a risk assessment template?
- Have I justified the control measures selected?

Review
- Have I checked and corrected my spelling and grammar?
- Does the evidence I have written meet the assessment criteria of optimising the safe system of work for the Merit and Distinction?

C   Understand the need for the review of safety systems for construction operations

All safety systems used on site need to be constantly reviewed to ensure that they:
▶ are current
▶ are effective and working correctly
▶ meet technological advances
▶ meet current legislative requirements.

The reviews are often undertaken as part of the risk assessment monitoring process. Supervisors may identify revisions to safety systems of working through discussions at safety committees and toolbox talks. These must not be ignored as they are essential improvements raised by the operative who is using the systems provided.

C1 Accident reporting procedures

There are requirements regarding the classification of unplanned events in terms of reporting procedures.

Definitions

The Reporting of Injuries, Diseases and Dangerous Occurrences Regulations 2013 (RIDDOR) is the law that requires employers to report and keep records of work-related accidents which cause death or certain serious injuries (reportable injuries), diagnosed cases of certain industrial disease and certain '**dangerous occurrences**'

> **Key term**
>
> **Dangerous occurrence** – a specific, reportable adverse event, as defined in the Reporting of Injuries, Diseases and Dangerous Occurrences Regulations 2013 (RIDDOR).

(incidents with the potential to cause harm). Previous versions of RIDDOR made the distinction between major and minor injuries but now the classification of 'major' injuries has been replaced with a shorter list of 'specified' injuries, including:

▶ a fracture, other than to fingers, thumbs and toes

▶ amputation of an arm, hand, finger, thumb, leg, foot or toe

▶ permanent loss of sight or reduction of sight

▶ crush injuries leading to internal organ damage

▶ serious burns

▶ scalpings (separation of skin from the head) which require hospital treatment

▶ unconsciousness caused by head injury or asphyxia

▶ any other injury arising from working in an enclosed space, which leads to hypothermia, heat-induced illness or requires resuscitation or admittance to hospital for more than 24 hours.

You can read more about RIDDOR below.

In any case, all **accidents** that result in injury or death, and **near misses** that could have done so, should be recorded and reported if appropriate.

Key terms

Accident – an event that results in injury or ill health.

Near miss – an event not causing harm, but having the potential to cause injury or ill health.

Accident procedures

The following procedure indicates how you must proceed following an accident.

▶ Shout for help and ask the responder to ring for emergency services.

▶ A trained first aider will administer first aid to the casualty.

▶ Rescue teams should be on site to assist if it is too dangerous to approach the casualty.

▶ The accident scene must be left intact if possible.

▶ The casualty's immediate supervisor must be informed.

▶ The company health and safety department must be informed.

▶ The HSE must be informed by phone if it is a fatal accident or if it is reportable under RIDDOR.

▶ An accident investigation must be carried out in full.

The accident must be reported to a supervisor as they have a legal obligation to record all the details of the accident, for example in the accident book, if it requires reporting under RIDDOR or results in a worker being away from work or

incapacitated for more than three consecutive days.

Internal accident reporting procedures will also need to be followed. This may mean referring the accident to the health and safety officer of the company.

An internal report may review the accident and its causes and make recommendations on what systems or procedures could be put in place to ensure that the accident cannot happen again, or that the risk of it reoccurring is reduced and controlled.

Reporting of Injuries, Diseases and Dangerous Occurrences Regulations (RIDDOR) 2013

Schedule 1 of the Regulations states the reporting procedures that must be followed by an employer:

▶ Where a person is incapacitated for more than seven consecutive days because of an injury caused by an accident at work, the responsible person must send a report within 15 days of the accident.

▶ Notify the relevant enforcing authority of the reportable incident by the quickest practicable means (for example online) without delay.

▶ Send a report of an incident causing injuries, fatalities or dangerous occurrences in an approved manner to the relevant enforcing authority within ten days of the incident.

Reporting can be done online or by the telephone.

Reporting dangerous occurrences

Schedule 2 of the Regulations lists the dangerous occurrences. These are construction-specific examples:

▶ collapse, overturning or failure of load-bearing parts of lifts and lifting equipment

▶ plant or equipment coming into contact with overhead power lines

▶ electrical short circuit or overload causing fire or explosion

▶ collapse or partial collapse of a scaffold over 5 metres high, or near water where there could be a risk of drowning after a fall.

HSE and internal investigations

When a fatality is reported under RIDDOR the HSE will want to investigate it. This is the process of analysing what went wrong following a specified injury or a fatality. This process must be thorough and methodical and may be conducted by the HSE. This is necessary:

▶ so that the accident may not reoccur, injuring another person

▶ because it is a legal requirement under RIDDOR

▶ to provide evidence in a civil claim from the injured party.

Accident at work

A construction worker is fixing a window replacement unit into the first floor of a domestic two-storey house using a working platform on a mobile tower scaffold. The window frame has been hauled up the outside of the tower and the glass passed to the worker from the inside of the building. While installing the last piece of glass, the worker steps back and falls through the open door of the access platform, falling to the ground and breaking a bone in their foot.

Check your knowledge

1 How would you proceed with this investigation?

2 Would you have to report it using RIDDOR 2013?

3 What control measures would you revise to prevent this occurring again in the future?

Employers need to provide the employees' training records, risk assessments and method statements for the activity, construction health and safety plans, and any other relevant documentation. Then the HSE accident investigation will take the following procedures in sequence.

1 Decide if an investigation is required.

2 Inform the employer.

3 Visit the scene.

4 Communicate with employees.

5 Interview the injured worker.

6 Gather facts about the accident.

7 Produce a report.

8 Decide on enforcement.

It is also good practice to carry out an internal investigation of any accident or near miss, even if there was no serious injury, so that methods of work can be reviewed to prevent it happening again.

 **PAUSE POINT**

A construction operative has just tripped up and fallen at ground level in front of you. What do you do?

 Hint

What company procedures and training might you need to follow?

Extend

Do you understand who you might need to contact about the accident or what steps you might need to take in the future to help to stop it happening again?

C2 Reviewing safety systems

It is important to constantly review safe systems of work (SSW) to ensure that the control measures that have been put in place are still operating effectively to reduce risks to an acceptable level. When substantial changes are made to the SSW, then the interval between reviews may be shorter and reviews may be repeated several times to ensure all systems are safe.

Analysis of accident information

One method of supporting a review of safety systems is to check and screen the data on accident rates and near misses on site. Any trends can be identified and further control measures applied to prevent their reoccurrence. Statistics can also be used to justify any suggestions and recommendations made for safety improvements, as they give measurable evidence and can provide a wider context for any SSW choices, beyond the immediate workplace. For example, comparing the accident and near-miss rate on a particular site with the national average would help to indicate the success of the SSW on that site.

Accident trends

The construction industry employs some 2.2 million people each year and still accounts for over 35 fatalities every year. The biggest cause is through falling, but there are other reasons why employees are killed while working in construction.

Figure 5.3 shows the major causes of accidents plotted from 2009–14.

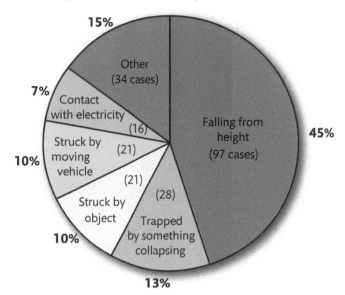

▶ **Figure 5.3:** Causes of death of construction workers

Figure 5.3 shows the four principal causes are:

▶ falls from height
▶ trapped by something
▶ being struck by an object
▶ being struck by a moving vehicle.

Figure 5.4 illustrates the current trend in fatalities over more than 40 years. Fatalities have fallen but remember: one death is one too many – the effect on friends, employer and families is devastating.

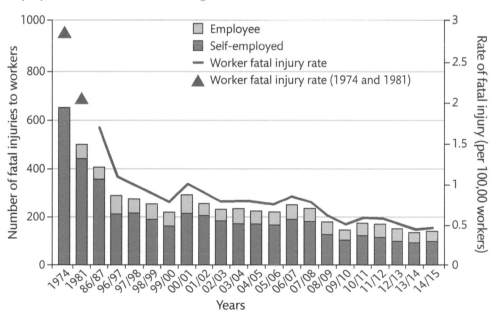

▶ **Figure 5.4:** There has been a fairly steady downward trend in the number of fatal injuries to workers since 1974

Discussion

It is essential that employers record all accidents and incidents in construction-related activities. What information would you record to analyse trends in the data?

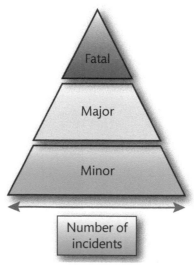

Figure 5.5: The number of incidents that occur is represented by the width of the triangle at the base

Observable trends

Keeping records of accidents and near misses helps to prevent accidents. For example, if the number of reported accidents increases in a certain area or location, an investigation may be required. The accidents could be due to a change in a process or procedure or a site that does not have sufficient SWW. Accident trends at a minor level require acting upon so they do not increase into a major or fatal accident. Many trends in this pyramid diagram (Figure 5.5) may cause a larger number of major accidents. The wider the base, the more danger that a fatal accident may occur.

 **PAUSE POINT**   What would you do if you noticed a trend of foot injuries on construction projects under your supervision?

 Hint   Consider safe methods of work and PPE.

Extend   What action would you take as a result of an increasing accident rate?

Working with workforce to work safely

Employers must communicate on health and safety matters with their employees, contractors, visitors and any other personnel on a construction site. This can be a major factor in reducing the risk of accidents. Here are some methods for communicating health and safety information to a workforce:

▶ Site safety meetings, in which workers are gathered to discuss their ideas for safety, can encourage workforce contributions to safe systems of work.

▶ Safety committees in which workers, managers and other relevant levels of staff are all represented. Any suggestions for improvements should be debated in terms of how practicable and effective they would be.

▶ Confidential interviews with members of staff who have a safety concern.

▶ Trialling safety systems to provide evidence of whether the improvements are working.

▶ Toolbox talks where revised procedures are discussed.

▶ Safety notices in a prominent place, such as on-site notice boards adjacent to the rest room.

▶ Attaching memos to wage slips.

Benefits of safety reviews

Using a company's safety data can establish where unplanned events occur and how these can be reduced to provide a company with several benefits as shown in Table 5.8.

▶ **Table 5.8:** Benefits of safety reviews

Benefit	Description
Costs	Safety reviews can reduce both direct and indirect costs. Every incident has unforeseen invisible costs, for example, the cost of finding a temporary worker while an employee is off work with an injury, or higher insurance premiums as a result of a poor accident rate.
Reputation	A company uses its reputation to attract clients to gain more work. A bad reputation would have an adverse effect on the amount of work a company receives from the market. It may improve a company's reputation and attract clients if it can state on its marketing materials that it undertakes regular safety reviews.
Morale	Worker morale would be higher if there are low incident and accident rates as the workforce will be more confident that their employer is doing everything to ensure their safety at work.
Performance	Production is increased on site as safe systems of work operate effectively, and work does not need to be stopped following an accident or near miss.
Contracts	A good health and safety record is likely to attract clients, increasing pre-contract enquiries and obtaining work by getting onto clients' tender lists.

Ⅱ PAUSE POINT   Your employer has changed the site mixer for brickwork mortar mixing. How would you review this change?

Hint   You could review the manufacturer's instruction leaflets for more information.

Extend   In terms of the safety of operatives using the new mixer, how would you review its use on site?

C3 Changes to systems and procedures

After reviews of your SSW procedures, it is likely that you may need to make changes or improvements, or to adapt systems and procedures to meet new requirements or situations in the workplace. Any changes to systems of working have to be completed by closing the **safety cycle**.

The HSE promotes Plan, Do, Check and Act. As you can see from Figure 5.6 the cycle is closed as you run through each section.

Key term

Safety cycle – a continuous process of developing and maintaining safety systems, for example by using a plan, do, check and act system.

▶ **Figure 5.6:** The HSE promotes the plan, do, check, act approach

Plan

▶ Revisit the original plan, which could be a method statement, a health and safety plan, a written safe system of work or original training.

▶ Review the plan to ensure that it will operate effectively for the existing system of working.

▶ The plan must consider the implementation of the safe system of working.

Do

▶ Identify hazards and risks then evaluate and control them using suitable and appropriate measures to ensure that they are at the lowest achievable level.

▶ Risk assessments will need reviewing and revising if required, and signed and dated as a revision or rewritten as completely new risk assessments if the SSW has major changes to it.

Check

▶ Discuss the implementation of the new SSW with all employees, supervisors and managers to ensure that the control measures are working as intended.

Act

▶ Review the discussions and react to any changes that have not worked.

▶ Decide on revised control measures and implement them into the cycle of review.

▶ Run the cycle again to ensure that the impact is effective in controlling hazards and risks.

The plan, do, check, act process is a safety cycle that must be closed and signed off when you have evaluated all safe systems of work. They should be reviewed annually or at appropriate intervals because legislation and regulations may change. You may need to attend training to establish if any regulations involve a change to your systems.

Reviewing control measures

The effectiveness of control measures must be monitored. It is no good providing a control measure if it does not effectively control a hazard. Similarly all employees working with health-related hazards should undergo health surveillance. Standard hearing and respiratory tests are recommended for noisy and dusty environments.

All risk assessments should have a review box so a responsible person reviews the hazards and ratings and the current control measures. New processes or materials

may affect an existing measure that now needs revision. The review should be signed and dated and the next review date inserted. All reasonable practicable measures should be provided for employees, contractors and site visitors.

> **Discussion**
>
> If the severity of an accident is fatal but the likelihood is zero is this okay? Discuss this as a group – what other implications could there be?

Revisions to risk assessments and risk ratings

A review may require a risk assessment to be re-considered and the rating evaluated in light of the new changes. This should be undertaken to ensure that:

▶ the risk rating is still at an acceptable level

▶ the severity has not increased

▶ control measures can be reviewed and revised

▶ there is no other method that could be utilised.

Risk ratings

If the risk rating (likelihood x severity) has increased, the new process needs to be reviewed completely and new control measures established to reduce risk ratings to acceptable levels. This may mean:

▶ a change to PPE

▶ isolation of the activity (e.g. preventing workers from accessing the area)

▶ amendments to training

▶ toolbox talks about the risk

▶ revised site induction.

Revisions

All revisions must be communicated to the workforce that will be dealing with the risk assessment hazards and control measures to ensure they know about the review and its revised measures.

When reviews have been undertaken and the safety cycle is closed, findings have to be communicated to all operatives involved in the reviewed activities. This can be done through:

▶ in-house training

▶ safety instructions such as leaflets, toolbox talks or verbal directions

▶ external training

▶ recertification when training has expired (e.g. ensuring workers renew their CSCS cards)

▶ issuing new risk assessments.

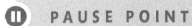

PAUSE POINT   When would you revise a risk assessment?

Hint   Think about what might change on a construction site.

Extend   How would you indicate and record that a risk assessment has been revised?

C4 Skills, knowledge and behaviours

When you start work in a construction sector you will need to develop a set of essential skills. Many of these will be related to communication, whether verbal or written, or using computers. The required skill set also involves interpersonal skills such as building a professional relationship with clients, managers and other workers on site. This takes some time to learn and to practise until you become confident.

Behaviour

When conducting yourself on a construction site you need to behave in a professional manner at all times. This demonstrates your attitude to the work itself and health and safety on the site and in construction offices. You should promote health and safety at all times by wearing PPE on site. Health and safety is of vital importance and should be top-down driven from the senior management team right down to general operative level.

Etiquette is essential in respecting other people's views on site. For example, you should take time to listen to a senior manager and respect their experience, responsibilities, wishes and views. Any clashes in hierarchy in management again should be dealt with in a professional manner.

Individual responsibility and accountability

This is very important when conducting health and safety provisions on site as errors can lead to serious accidents. As a manager, you will be given responsibilities regarding health and safety. This could be the completion of risk assessments, method statements, giving toolbox talks or running the safety inductions. If you are not sure about something it is always better to ask. 'No questions' is not sensible when it comes to safety. Always challenge practices that you consider to be unsafe and report these to a supervisor.

Safety evaluations

As a manager you will have to evaluate risks and hazards on a site to reduce them to what you judge to be an acceptable level. This includes evaluating what is reasonably practicable in terms of the provision to control the severity of the hazard. You will have to use the soft skills learned on communication in order to achieve this, writing out recommendations which are justified if you are challenged.

Media and communication skills

You must be able to communicate clearly, whether this is written or verbal, so that your message does not get confused or misunderstood. It is vital to ensure that any information you communicate about health and safety is easy to understood and act upon.

The format, or media, that you use to convey information can take many forms such as posters, videos, safety signs, documents or verbal discussions. The media used must convey the meaning that it intends and be simple, concise and informative.

The two main forms of communication you are likely to use are written and verbal communication:

▸ **Written communication:** You must demonstrate clear and accurate handwriting or typing when writing or evaluating risk assessments. The language that you use must be understood by the person who will read the details of the assessment of the risk from hazards. Proofread and check all information to ensure that it makes sense, can be understood and is valid.

▸ **Verbal communication:** This skill comes with the confidence of the health and safety knowledge that you know and understand. A well-prepared manager will therefore have to have the confidence to know what they are saying is right and applied correctly. Verbal communication skills can be difficult at times on a construction site with all the noise and activities that occur. You should learn to verbally communicate at the right time, in the correct manner and at the right level so people listening understand you.

Tone and language

The tone and language that you use should be in terms of equality and access for all employees regardless of ethnic origin or disability. It should respect the ability level of employees in being able to understand written English so they can write responses on documentation for you. The tone you take should be measured so it does not offend, respects all and is polite in its approach. Tone must make a positive impact on the intended audience.

Assessment practice 5.4   C.P6   C.M3   C.D3

An operative has been trained to drive the 2 tonne dumper on site and is using it to transport concrete into the newly excavated foundations. As they approach the trench they over run and the front of the dumper drops into the trench. The operative is thrown from the dumper and lands heavily, fracturing a collar bone and receiving cuts to the face, head and hands.

- The casualty has been treated by the site first aider. As the safety manager, explain the procedures that must be followed from the point of reaching the casualty.
- Explain how the system of working could be improved to prevent a reoccurrence of this accident.
- Produce a revised safe system of work in the form of a method statement.
- Evaluate the success of your SSW in producing an acceptable risk rating.

Plan
- What do I need to explain for the accident procedures?
- What does the revised method statement need to contain?

Do
- I can write a method statement.
- I am able to evaluate this in terms of risk rating.

Review
- I will read what I have produced and correct it for spelling and grammar.
- Does the evidence I have written meet the assessment criteria of accident reporting and review procedures?

Further reading and resources

HSE (2013) *Managing for Health and Safety* (HSG65) downloadable from **www.hse.gov.uk/pubns/books/hsg65.htm.**

HSE (2013) *Managing Health and Safety in Construction* (HSG150).

CITB and HSE (2015) *Construction (Design and Management) Regulations 2015: Industry Guidance.*

CITB (2017) *Construction site safety* (GE 700).

CITB (2015) *Risk Assessment and Method Statement Manager* (RACD 0015).

THINK ▶▶FUTURE

Callum

Callum is a site manager for a large construction company that specialises in the refurbishment and construction of education projects for local authorities in the north west of England. The company provides a maintenance service help desk where any centre can ring in and report a maintenance issue. His role is to organise and develop a new school building programme, which designs and constructs new schools on the same site as the existing school then demolishes and recycles the original building.

The role of the site manager covers every aspect of health and safety on the sites that they manage. The hazards on the site must have suitable control measures in place to protect all employees and contractors working on the site. Callum has to take special care because the site is on a school that is still operating. This means that deliveries and access and egress need to be timed with pupils arriving and departing from the site.

Callum conducts regular safety talks with all employees and contractors to which the school's estates manager is also invited. All safety on the new site has to be co-ordinated so minimum disturbance occurs to the school, which has to keep its pupils and staff safe and to conduct exams and other assessments. Noise is therefore an important consideration.

Callum has to operate all activities in accordance with the legislation and regulations. All employees are checked for competence, are given training and are inducted before they commence any activity on the site.

Focusing your skills

There is a wide range of skills that a site manager has to use when they are managing a large construction project.

- Organisational skills – to co-ordinate all activities on site in a safe manner ensuring that all are safe on the site.

- Communication skills – ability to communicate clearly, giving instructions to all, and to be able to read and understand drawings, regulations and specifications. To be able to keep accurate and detailed records of all safety activities on site.

- Listening – to listen to what people have to say and react positively to this especially in toolbox talk discussions and safety briefings.

- Accurate – a site manager has to ensure that all work is set out to the drawings and specification agreed with the client.

- Respect – to be able to interact with all levels of employees, ensuring that they are safe, secure and have high morale.

Getting ready for assessment

This section has been written to help you to do your best when you take the assessment test. Read through it carefully and ask your tutor if there is anything you are still not sure about.

Jamile is in the first year of his Extended Diploma in Construction and the Built Environment. For this unit he has been given an assignment to prepare a formal report, with case studies, that explains the impact of legislation and regulations in upholding and improving health and safety on construction sites. His report needs to include statistical evidence to support the effectiveness of legislation.

How I got started

I started this assignment by looking at the sources of information that the tutor had provided on the last page of the assignment. These sources are very useful and are a great start to help you research and gather information on helpful features and items.

I looked back through my notes and put together a list of all the legislation and regulations that I needed to cover in my report. I wanted to make sure that my report included all the key command words ('justify', 'evaluate', for example) within the assessment, so I downloaded a copy of the unit from the specification on Pearson's website which gave me lots of guidance on what I needed to make sure I was doing.

I carried out a lot of research online and in books to find out as much as I could about the legislation and regulations. I also wanted to find some interesting case studies, but this was difficult as it's such a huge topic!

How I brought it all together

To start with I wrote a very short introduction that explained what the purpose of my report was. I remembered my tutor telling us that it was important our introductions outlined what we were planning to say in detail in the report, so made sure my introduction did this.

After this I decided to cover each piece of legislation in turn in the report. I wanted to write a brief section explaining each one before I could.

I had found a lot of European and UK health and safety statistics online and these were really helpful. I tried to link these as closely as possible with different pieces of legislation, and to explain what the numbers meant.

Because I used lots of books and websites to support the points I was making in my work, I put together a bibliography and references so that all my points could be checked. I asked my parents to read over my report to make sure that what I was writing was correct and to check

for writing errors. I also proofread it against the assessment criteria to make sure that it was spot on with the command word use from the grading criteria in the units table.

What I have learned from the experience

I'm glad I had done some work beforehand planning my research and identifying where to look for information about legislation and useful statistics. I had found supporting case studies hard to find and I think next time I would try to approach local businesses as well as looking at larger construction companies to try and get some more information.

I think I spent a little too much time focused on explaining the legislation and statistics and not enough time linking it together with the case studies from the construction industry. Next time I would look to spend more time planning my work out, using my notes from the unit, to put together a structure for the report before I started writing it. This would help me to make sure I was covering everything I needed to cover.

Think about it

▶ Type the assessment criteria into an internet search and see what you find. This is a good starting point. From this you can then develop some useful sources of information that summarise legislation and health and safety on construction sites.

▶ Where will you find health and safety legislation?

▶ Does the unit content help?

Surveying in 6 Construction

Getting to know your unit

The surveying of land is concerned with the measurement of existing features of the natural and built environment and the presentation of data in a format suitable for architects and engineers to use when designing construction projects. It plays an important role in the early stages of the design process and links with the setting-out phase of construction projects

In this unit, you will learn about the methods and technologies that underpin surveys, including the techniques and instruments used to record survey data, potential sources of systematic errors and how to minimise them to produce accurate data. You will undertake fieldwork surveys to collect data for drawings, for example carrying out practical surveying tasks and completing linear survey and level booking sheets to demonstrate accurate recording of surveying measurements. You will also develop drawings from completed fieldwork surveys using manual or computer-aided design (CAD) drawing techniques.

How you will be assessed

This unit is internally assessed by your tutor. The varied activities will help you to understand land and construction surveying. These include both the practical use of surveying equipment as well as linked calculations and the production of drawings or models. You will need access to a range of basic surveying equipment you should seek guidance and support from your tutor in obtaining the appropriate kit and equipment.

To achieve the tasks in your assignment, check that you have met all the minimum Pass assessment criteria as you work your way through the assignment.

If you want to gain a Merit or Distinction, make sure that you present the information in your assignment to the required depth and style. For example, Merit criteria require you to analyse and discuss, and Distinction criteria require you to assess and evaluate.

The assignments set by your tutor will consist of several practical and research tasks designed to meet the criteria in this unit. This is likely to consist of a written report covering the methods and technologies used to conduct linear, levelling and angular measurement surveys, and also showing an understanding of how errors can affect your results. Another major assignment will include undertaking practical surveying fieldwork and producing a range of outputs such as drawings, contours maps and closed traverse details.

Assessment criteria

This table shows what you must do in order to achieve a **Pass**, **Merit** or **Distinction** grade, and where you can find activities to help you.

Pass	Merit	Distinction
Learning aim **A** Understand the methods and technologies that underpin surveys		
A.P1 Explain the methods and technologies underpinning linear, levelling and angular measurement surveys.	**A.M1** Discuss the methods and technologies underpinning linear, levelling and angular measurement and surveys.	**A.D1** Evaluate the methods and technologies underpinning linear, levelling and angular measurement and surveys.
A.P2 Explain systematic errors in surveying measurements.		
Learning aim **B** Undertake fieldwork surveys to collect data for drawings		
B.P3 Perform systematic checks and adjustments to equipment and instruments appropriate for the fieldwork surveying activity.	**B.M2** Justify the selection of equipment, methods used, the application of systematic checking, instrument adjustment and accuracy of calculations to provide accurate fieldwork survey information.	**B.D2** Evaluate the methods used to produce accurate fieldwork survey information for the development of accurate drawings.
B.P4 Perform linear, levelling and angular measurement surveys using appropriate equipment and booking methods.		
B.P5 Perform accurate calculations to support fieldwork activities.		
Learning aim **C** Develop drawings from completed fieldwork surveys		
C.P6 Produce plans of land and section detail drawings from completed fieldwork surveys.	**C.M3** Produce plans of land and section detail drawings from completed fieldwork surveys to a high level of technical skill and accuracy.	**C.D3** Evaluate the production of drawings from completed fieldwork surveys.

Getting started

Using a tape measure, record the overall size of the room you are in, including the position of its windows and doors. Produce a sketch plan view of the room using your measurements. What assumptions did you make while carrying out this survey? How accurate do you think the final sketch is in representing the room layout? Why?

 A # Understand the methods and technologies that underpin surveys

A1 Linear, levelling and angular measurements

Survey frameworks

The first stage of any measured survey is to carry out a full reconnaissance of the area. The basic process involves developing a **survey framework** that covers the whole area, which can be broken down into smaller parts to record all the necessary details and features. Surveyors need to adopt a methodical way of recording their measurements and carrying out suitable checks to make sure they have not made any mistakes. Any errors must be examined to check if they are within acceptable limits.

A survey framework is a series of interlinking lines where the lengths of the lines and the angles formed uniquely fix the shape of the framework, enabling the position of all the points to be found and plotted. For linear surveys, the framework is usually based on a series of interlinked **well-conditioned triangles** as these are the most stable shape, and for traverse surveys angular measurement is also incorporated to create **polygonal frameworks**.

Figure 6.1 is an example of a simple survey framework using linear measurements and four **survey stations**. The framework A, B, C and D is formed of two well-conditioned triangles with a **baseline** AB and **check line** CD. All the points can be uniquely fixed when the lengths of the lines are known. Then from this framework all the physical features of the site such as trees, buildings and roads can be observed, measured, recorded and finally plotted. This process is commonly known as 'working from the whole to the part', because it requires surveyors to take an overview of the whole survey area and think carefully how it could be split up into manageable parts.

> **Key terms**
>
> **Survey framework** – a series of interlinking lines, the lengths and the angles of which uniquely fix the shape of the framework.
>
> **Well-conditioned triangles** – points on a survey that make roughly equilateral triangles with small internal angles ranging from 30° to 120°.
>
> **Polygonal framework** – a many-sided closed shape with measured angles and plan distances.
>
> **Survey station** – a point at the end of a survey line.
>
> **Baseline** – the longest line between two points on a survey framework.
>
> **Check line** – a line provided to check the accuracy of surveying measurements.

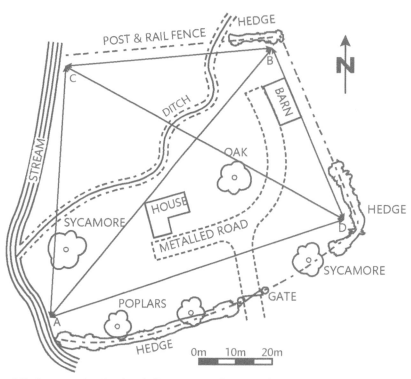

▶ **Figure 6.1:** An example of a simple linear survey framework

Chainage

Distances measured along a line are referred to as **chainage** or **running measurements**. These horizontal distances are measured from one fixed point at the start of the line and run to the end of the line. The measurements are printed on the line between two parallel lines which represent the **survey line**. This is shown in Figure 6.2 in the typical booking (recording) sheet for a survey line G–C, where Station G has a chainage of 0 m and Station C has a chainage of 75.42 m.

Note that in the field, distances are sometimes written with the decimal point shown as a '/' to avoid it being confused with dirt on your surveying notes.

All the physical features, like the edge of the car park that cuts the line G–C in two places, have their own running measurements measured from G, such as 29.5 m and 46.34 m. The advantage of running measurements like this is that it is quicker and you are less likely to make errors.

You will note that the corner of the factory building does not touch the line G–C but has a dimension of 3.75 m next to the chainage of 28.0 m. This fixes the corner of the factory building by measuring a running dimension of 28 m along then measuring a distance of 3.75 m as a **perpendicular offset**. This is a distance measured at right angles (90°) to the survey line G–C. Similarly at a running measurement along G–C of 15.12 m the concrete access road cuts into the perpendicular offset at 5.52 m. This method of measuring at right angles to the survey line is effective and accurate for distances of less than 10 m where the right angle can be judged by eye. Alternatively, you can swing the tape in an arc and take the shortest measurement. The right angle is located where it just touches the survey line.

You can also fix the position of physical features from a survey line using a pair of **tie lines**. This is shown in Figure 6.1 where the oak tree is fixed in relation to the line G–C. Two tie lines of 14.38 m and 15.97 m are measured from two points on line G–C whose chainage is known (46.34 m and 55.0 m respectively). These three measurements form a well-conditioned triangle and enable the tree's position to be correctly located.

> **Key terms**
>
> **Chainage** or **running measurement** – a horizontal distance measured along a line.
>
> **Survey line** – any direct plan distance measured between survey stations.
>
> **Perpendicular offset** – a distance measured at right angles (90°) to the survey line.
>
> **Tie line** – survey line connecting a point to other lines, often to check accuracy or to locate a feature. Also known as a check line.

Generally, tie lines should be used in situations where perpendicular offsets exceed 10 m, to ensure good accuracy.

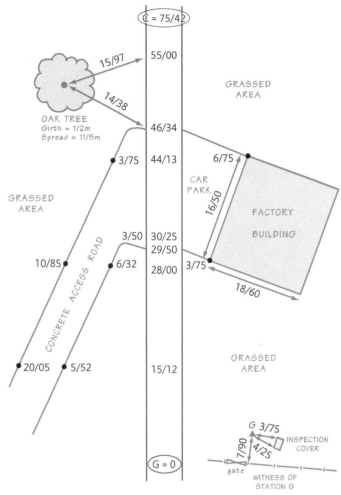

▶ **Figure 6.2:** Booking sheet

Measuring horizontal and slope distances

All of these methods of measuring distances require that the distance is measured horizontally, i.e. 'flat' and not along the ground. As you can see from Figure 6.3, a slope distance is longer than the plan distance and so would cause problems when we try to plot the true plan view position of a point on the ground. In general if the slope rises less than 600 mm over an approximate distance of 30 m there is minimal difference in the plan distance and slope distance but anything greater than this will need to be reviewed.

Angular measurements

You need to know about the basic units for measuring angles in order to make surveying calculations. All angles used in land and setting-out surveying use the sexagesimal system of measurement:

▶ There are 360° in a circle.
▶ Each degree comprises 60 minutes.
▶ Each minute is split into 60 seconds.

For example, an angle of 47 degrees, 35 minutes and 13 seconds is written as 47° 35′ 13″. This system is common where a high degree of accuracy is needed. It is important to know how to add and subtract these angles, both manually and with a scientific calculator.

There are two basic methods of fixing points by measuring angles: by using either

Link

Section A3 of this unit looks at errors in taping distance.

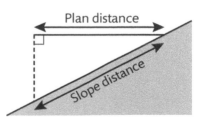

▶ **Figure 6.3:** Slope distances measured along the ground are longer than the corresponding plan distance

triangulation or polar measurements. Both involve identifying the direction of a point from a fixed reference and are measured as seen vertically from above, as flat **horizontal angles**. Figure 6.4 shows how we can fix the position of a brick wall XY in relation to an aerial mast at Z by using angles.

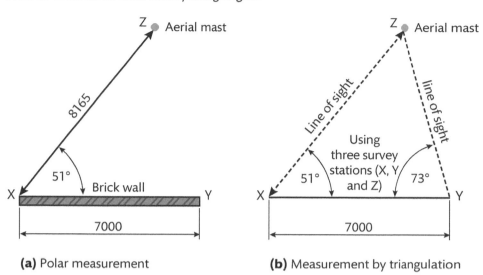

(a) Polar measurement　　　　**(b)** Measurement by triangulation

▶ **Figure 6.4:** Angular methods of measurement

Polar measurement

Polar measurement requires an angle to be measured in relation to a reference direction. In Figure 6.4(a) above this is the angle of 51° relative to the wall. To fix the position of the aerial mast at Z, we need to measure the distance XZ which, in this example, is 8165 mm.

Measurement by triangulation

This method is ideal if the point to be fixed cannot be physically reached. In Figure 6.4(b) above, the aerial mast and wall may be separated by a deep excavation. The only requirement is that there are clear lines of sight between the ends of the wall at points X and Y and the aerial mast at point Z. The only linear measurement that needs to be recorded is the length of the wall. Hence point Z is found where the two lines of sight intersect at an angle of 51° and 73° from points X and Y respectively, separated by a distance of 7000 mm. Again, the principle of well-conditioned triangles applies here.

Key terms

Horizontal angle – an angle measured between three fixed points within a horizontal plane.

Datum – a reference point or set of reference points from which measurements are made.

Ordnance Survey bench mark (OSBM) – a point of known or arbitrary height used as a reference for other height measurements.

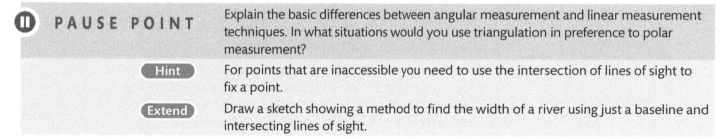

Ⅱ **PAUSE POINT**　Explain the basic differences between angular measurement and linear measurement techniques. In what situations would you use triangulation in preference to polar measurement?

Hint　For points that are inaccessible you need to use the intersection of lines of sight to fix a point.

Extend　Draw a sketch showing a method to find the width of a river using just a baseline and intersecting lines of sight.

Levelling measurements

Levelling is a process that compares heights of points on the Earth's surface. When carrying out a level survey, surveyors should refer to a fixed point, or **datum**, of known height. A reference point with an assigned height is a **bench mark**, and is established by levelling from the datum.

Datum terminology

Levels can be bench marked in different ways, depending on the scale of the project.

▶ On large civil engineering projects such as roads, railways and bridges, levels are linked to the Ordnance Survey Bench Marks (OSBM) system, which is referenced to the height of mean sea level at Newlyn in Cornwall. These heights appear on terrestrial models, maps and small-scale plans published by Ordnance Survey.

▶ On smaller local construction projects, it is usually sufficient to relate heights to an arbitrary fixed point established on site called a **temporary bench mark (TBM)**.

▶ These bench marks enable **reduced levels (RL)** to be calculated.

Purpose of levelling

Levelling and height control have the following purposes:

▶ to measure vertical heights of points or stations located on the ground

▶ to produce a grid of spot heights to indicate changes in the ground surface; this information is often shown on maps and plans as a series of contour lines that join points of equal height

▶ to set out profiles on site to give the heights of new construction works, such as depths of foundations relative to the ground slab level or damp-proof course height

▶ to produce a cross-sectional drawing (a long section), which helps plan the construction of underground services, such as setting out sewer gradients.

The principle of levelling

The basic method of undertaking levelling measurement involves measuring vertically up or down from a level plane called the plane or line of **collimation** (see Figure 6.5). This level line is identified using an optical, digital or laser level, and the vertical measurements are measured with a graduated measuring rod called a levelling staff. The height of the plane of collimation is relative to a fixed point of known height (a datum).

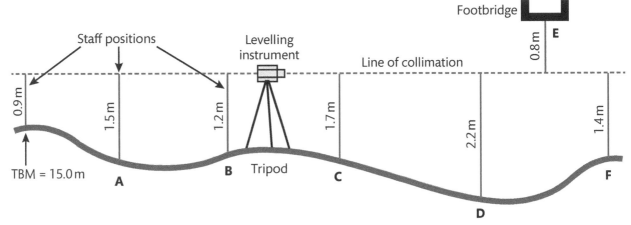

▶ **Figure 6.5:** The principles of levelling

If a level is set up correctly and sighted on to series of points on the ground and a TBM of known height, we can work out the reduced level of any of these points so that their heights can be compared. For example, as can be seen in Figure 6.5, we can find the reduced level of points A to F:

▶ point A = 15.0 m + 0.9 m − 1.5 m = 14.4 m

▶ point B = 15.0 m + 0.9 m − 1.2 m = 14.7 m

▶ point C = 15.0 m + 0.9 m − 1.7 m = 14.2 m

▶ point D = 15.0 m + 0.9 m − 2.2 m = 13.7 m

▶ point E = 15.0 m + 0.9 m + 0.8 m = 16.7 m

▶ point F = 15.0 m + 0.9 m − 1.4 m = 14.5 m.

Note that point E has the level staff turned upside down, to take an 'inverted' staff reading; the reduced level of the underside of the footbridge is obtained by adding the staff reading to the height of the instrument.

There are two methods for booking (recording) level readings to enable the reduced levels of points to be found:

▶ the height of the plane of collimation (HPC) method
▶ the rise and fall method.

The HPC method

This uses the horizontal line of sight through the level instrument as a reference for all the individual staff readings.

$$\text{Height of plane of collimation (HPC)} = \text{Given bench mark height (BM)} + \text{Backsight reading (BS)}$$

Once the HPC is found, the reduced levels (RL) of the ground can then be found:

$$\text{Reduced level of ground (RL)} = \text{Height of plane of collimation (HPC)} - \text{Any staff reading (SR)}$$

This is as shown in Figure 6.6(a).

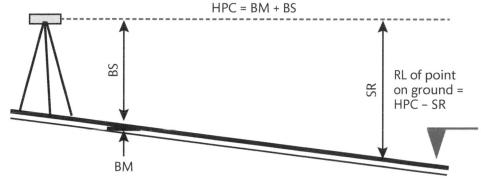

HPC = BM + BS

RL of point on ground = HPC – SR

▶ **Figure 6.6(a):** The HPC method

The rise and fall method

This method does not use the height of collimation but instead compares the difference in height between adjacent staff readings. In Figure 6.6(b), the staff reading at the TBM is 1.56 m and the next staff reading at point A on the ground is 1.75 m.

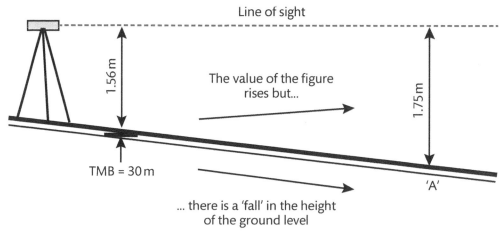

Line of sight

The value of the figure rises but...

1.56 m

1.75 m

TMB = 30 m

... there is a 'fall' in the height of the ground level

'A'

▶ **Figure 6.6(b):** The rise and fall method

The difference between these two readings is found by subtracting the second reading from the first reading, i.e. 1.56 m – 1.75 m = -0.19 m. Because it is a negative value, it is a 'fall'.

If the TBM is at a known height of, say, 30 m, then the reduced level of point A is 30 m − 0.19 m = 29.81 m. This can be seen from looking at the staff readings compared with the level line of sight. The figures recorded on the staff rise but the ground levels actually fall. The process is repeated for each new staff reading, where each adjacent staff value is compared to see if it is a rise or a fall. The reduced level of these points is then calculated by adding the rises or subtracting the falls from the previous point's reduced level.

Levelling applications

Fly levelling (flying levels)

This is the transferring of levels from one point to another using only **backsight** and **foresight** readings; **intermediate sight** readings are not used. It is a quick and convenient way to establish a new TBM from an existing bench mark. However, the flying levels must be returned to close back at the original bench mark, that is, the levels will need to fly back to close the survey (finish the survey at the first datum point) and to carry out all the necessary accuracy checks. The distance between readings should not exceed 50 m for the accurate establishment of a TBM using flying levels.

> **Key terms**
>
> **Backsight (BS)** – First staff reading of a levelling operation or the first reading after the instrument has been moved.
>
> **Foresight (FS)** – Last staff reading of a levelling operation before the level is moved.
>
> **Intermediate sight** – Any staff reading taken between the backsight and foresight readings.

PAUSE POINT Why do you need to start and finish all level surveys from a fixed point of known height?

Hint Close the book and sketch the relative position of the height of collimation, the ground level and a fixed datum.

Extend Explain the main differences between the rise and fall method and the HPC method of recording level readings.

Whole circle bearings

Whole circle bearings are like polar measurements, as the angles are measured in a clockwise direction from north.

If the direction of north is not known, an arbitrary reference direction is chosen to a distant object and consistently applied throughout the whole survey. This is the reference object (RO).

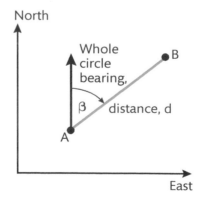

▶ **Figure 6.7:** Whole circle bearings

Worked example

From Figure 6.8, calculate the angle α made between two WCBs measured from point O to points W and X.

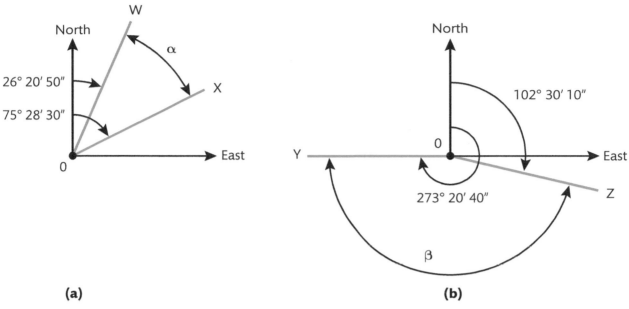

Figure 6.8: Examples of simple whole circle bearings

The angle α is measured between the WCB of W (26° 20′ 50″) and the WCB of X (75° 28′ 30″) as shown in Figure 6.8(a). To find α we have to subtract 26° 20′ 50″ from 75° 28′ 30″, using a calculator or manually. The manual subtraction is shown here:

$$
\begin{array}{r r r}
 & & +60'' \\
 & 27' & 90'' \\
75° & \cancel{28'} & \cancel{30''} \\
26° & 20' & 50'' \quad - \\
\hline
49° & 7' & 40'' \\
\end{array}
$$

α = 49° 7′ 40″

Note that, starting with the seconds, we have to subtract 50″ from 30″, which cannot be done. Therefore we need to 'borrow' 1′ from the 28′ in the next column and adding that 60″ to 30″ to make 90″.

Similarly, to find the angle β between the two WCBs of Y (273° 20′ 40″) and Z (102° 30″ 10″) as shown in Figure 6.8(b), we have to borrow 1° from the 273° and add 60′ to give 80′ from which 30′ can be subtracted.

$$
\begin{array}{r r r}
 & +60' & \\
272° & 80' & \\
\cancel{273°} & \cancel{20'} & 40'' \\
102° & 30' & 10'' \quad - \\
\hline
170° & 50' & 30'' \\
\end{array}
$$

β = 170° 50′ 30″

PAUSE POINT

Can you state how many angular seconds are contained within a whole circle bearing of 5° 10′ 30″?

Hint   An angle of 1 minute is equivalent to 60 seconds.

Extend   Can you identify which direction is north from your current location? Using this as a reference, what compass point (N / NE / E / SE / S / SW / W / NW) is a whole circle bearing of 135°?

Key terms

Traverse – a series of connected lines with known lengths and directions.

Open traverse – a traverse with different start and end points.

Closed traverse – a traverse that encloses a defined area, with the same start and finish point.

Traverse types

Earlier in this unit, we looked at trilateration or tie line measurement, a framework of triangles with known side lengths, based on the same baseline. This enables the positions of the survey stations (points) at the ends of the lines to be recorded then plotted because we know the lengths of the three lines.

However, it is a more efficient method to combine distance measurement with horizontal angles as an **open or closed traverse**. This involves measuring the horizontal distance between survey stations, and the angle that each line makes with its adjacent line. Thus we can calculate the rectangular co-ordinates of each of the survey stations and plot them very accurately – these calculations and techniques are covered in section C3 of this unit. Figure 6.9 gives examples of both types of traverse.

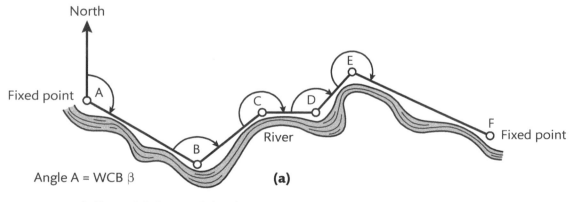

▶ **Figure 6.9:** Open and closed traverses

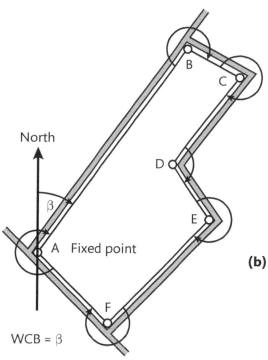

North

β

A   Fixed point

(b)

D

B

C

E

F

WCB = β

▶ **Figure 6.9:** *Continued ...*

An open traverse is used where a linear feature such as a river or canal needs to be surveyed. The surveyor goes from a fixed point to a known finish point.

By contrast, a closed traverse is ideal for mapping an enclosed area, as the survey starts and finishes at the same point.

All traverses must start and finish at a known fixed point to enable the positions to be checked and any errors corrected. This method is very useful as it enables us to determine the rectangular co-ordinates of the line positions and survey stations.

Link

Section B5 of this unit has more on rectangular co-ordinates.

A2 Equipment used to perform fieldwork surveys

Some of the equipment required to carry out a traditional linear survey is described in Table 6.1.

▶ **Table 6.1:** Equipment required for a linear survey

Equipment	Description
Synthetic tape	• Made from fibreglass or nylon and coated with PVC. Cheaper than steel tape, but only graduated down to the nearest 5 mm and can become permanently stretched if pulled too hard. • Has a metal prong on its end to hook into position – avoid bending this flat. • Wipe before rewinding, and avoid it becoming twisted and jammed into the case.
Steel tape	• Durable and more stable than synthetic tape. • Can be enamelled or plastic coated for added protection and available in 30- or 50-metre lengths. • Graduated at every millimetre and calibrated to read true lengths at a standard temperature and pull; both are printed on the tape at the start, and are typically 20°C and 50 Newtons force respectively. • Has a metal prong on the end to hook into position. Tape should be dried and wiped with an oily rag before being rewound.
Steel band	• Similar to steel tape but is carried on a four-arm, open-frame winder and has oval handles connected directly to the tape. Standard length is 100 m. • It is more durable than steel tape but also more expensive and is less commonly used than other methods.
Arrows	• 400 mm-long steel pins with a point at one end and a ring at the other to aid carrying. • Used for marking out points on the ground or marking survey station positions before fixing with a stout wooden peg.

Equipment	Description
Arrows	• A strip of red cloth or plastic tape is attached to the ring so the arrow can be easily observed. • A variation is the dropping arrow with a weighted point (plumb bob) used for step measurements of slopes.
Ranging rods or poles	• Circular timber poles, usually 2.5 m long, coated with distinctive red and white bands 500 mm in length. • Only used to mark survey stations and for ranging straight lines. • Points are encased in steel shoes which allow them to be pushed into the soil. • Ranging rods can also be used with lightweight stands for hard surfaces such as tarmac or concrete.
25 mm folding wooden rules	• Two main types: 1 metre long boxwood folding rule, with a swivel hinge, and longer two-metre multi-lath rule. • Both graduated to the nearest 1 mm and used mainly for internal building surveys, but can also be used for short offsets. • Care must be taken when opening or closing the hinges to avoid bending into the wrong position.

> **Safety tip**
>
> The current minimum statutory requirements for site workers' personal protective equipment (PPE) are to wear a hard hat, a high visibility jacket and safety boots or shoes; this applies to site engineers and surveyors as well as operatives. Even when undertaking land surveys prior to construction work, you should always wear PPE.

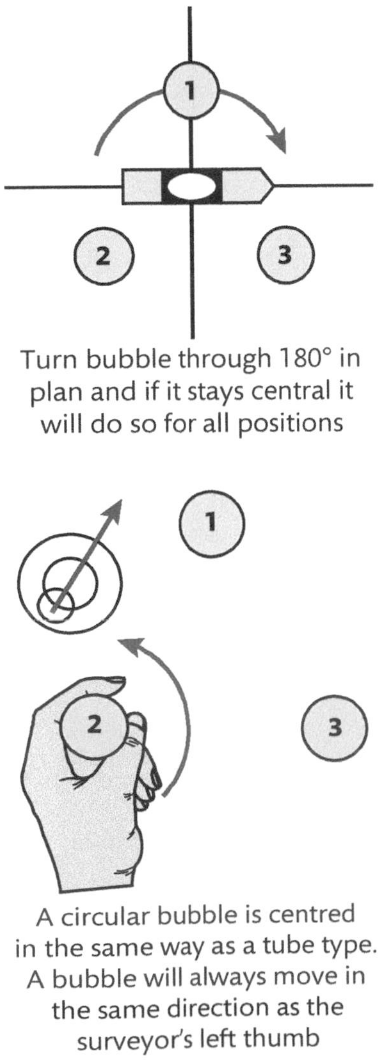

Turn bubble through 180° in plan and if it stays central it will do so for all positions

A circular bubble is centred in the same way as a tube type. A bubble will always move in the same direction as the surveyor's left thumb

▶ **Figure 6.10:** Automatic levelling

Basic levelling equipment

Levelling instruments are very versatile and can be used both for setting out points and for recording heights of existing features. The basic set-up of these main elements and linked terminology is shown in Figure 6.10. They are accurate and, if properly maintained, should provide a long and useful service. All levels comprise two basic parts:

▶ a high-resolution telescope with magnification of ×20–×24 which rotates around a vertical axis

▶ a spirit bubble tube set parallel and horizontal to the telescope.

If the level is adjusted correctly, the line of sight through the telescope – the line of collimation – will be parallel to the horizontal level line given by the spirit bubble tube, with both these lines at 90° to the vertical axis of the level. A simple two-peg test checks that both line of collimation and level line are parallel (see A3 Sources of systematic errors, later in this unit).

The automatic level

The automatic level has a mechanism which allows the instrument to be levelled automatically once it is roughly levelled using a small circular spirit bubble on its body. The method by which this bubble is levelled varies from instrument to instrument. Some favour a central knuckle joint while others have a pair of sliding circular wedges; some also use three foot-screws such as used in a dumpy level.

The tilting level

Tilting levels are most often used for high accuracy work but can be used in all types of level surveys. They can also assist in setting out slopes and sewer gradients with the aid of the tilting screw. The rough levelling of the tilting level is sometimes done by a circular spirit bubble. Every time the instrument is pointed in a new direction it is accurately levelled by centring a plate bubble with a fine vertical adjusting screw. This does not affect the line of collimation.

Dumpy or builder's level

The dumpy level is a traditional optical level levelled by three foot-screws and a single plate, or tube, spirit bubble. It is a common device used extensively in theodolites and total stations (see digital theodolites, later in this unit). The dumpy level can give an accurate reading up to a tolerance of 3 mm over a distance of 150 m.

Laser levels

Laser levels are increasingly used for setting out levels in construction work such as foundation depths for buildings and the vertical alignment of road and rail works. They create a horizontal plane from which measurements up or down can be recorded or checked depending on whether you are surveying or setting out points respectively. They are very accurate and have a wide range – some as long as 300 m. They do not require any visual sightings to be taken because they rely on an audible signal to indicate if the level is too high or too low. They are easy to set up and quick to use. Most instruments also provide an automatic warning signal if the laser gets knocked or is unintentionally moved.

There are many types of laser level, including pipe laying lasers and rotating lasers. The rotating laser is fixed to a tripod which, when switched on, automatically finds its level and generates a red horizontal line that can be picked up by a suitable sensing device to give an audible or visual reading of the height. A typical application is in the control of excavation depths and grading level surfaces where the sensor is fitted to the excavating bucket or blade as shown in Figure 6.11.

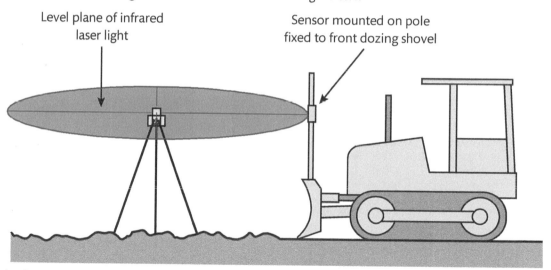

Level plane of infrared laser light

Sensor mounted on pole fixed to front dozing shovel

▶ **Figure 6.11:** Use of the laser level for earthworks

The digital level and bar-coded staff

Digital levels are automatic levels that read the bar-coded staffs electronically. The basic characteristics are shown in Figure 6.12. This is a relatively new development and although more expensive than normal optical automatic levels, they are quicker to read and, when used properly, are less prone to human error. They can also be set for tracking continuous measurement, for use in setting out height.

After the digital level is aimed and focused on the bar-coded staff using optical sights, the reading is taken electronically and then displayed. They also incorporate facilities to store and then download levelling measurements to a variety of computer software programs for working out cross-sectional areas and volumes.

(a) A bar-coded staff in use

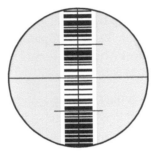

(b) View through telescope: if three horizontal cross hairs are indicated, the central one is used for height measurement

(c) Electronic display of difference in height and reduced level

▶ **Figure 6.12:** Surveyor using bar-coded staff

Surveying 'E' staff

These are lightweight, aluminium, telescopic staffs available in 4-metre and 5-metre lengths and graduated in distinctive E-shaped black, red and white blocks where each horizontal bar is equal to 10 mm, making the height of the 'E' 50 mm. Staffs are generally used to measure the distance up to the level line of sight. However, they can just as easily be turned upside down to record the heights of bridge soffits or gable walls by measuring the distance down to the level line of sight. These readings are known as inverted staff readings and are booked as negative values in a level booking form.

To help the person holding the staff to keep it vertical, there is usually a circular spirit bubble on the back face of the staff. However, these can get dirty or damaged so may not always be reliable. An alternative is to gently sway the staff backwards and forwards in line with the instrument, and the surveyor observing the staff reads the minimum distance observed on the staff. With good eyesight, you should be able to estimate the staff reading to the nearest 1 mm, one-tenth of one of the 10 mm coloured blocks.

The way in which the E staff is read is shown in Figure 6.13. The following readings were sighted through the telescope of an optical level A, B, C and D: A is 1.033 m, B is 0.630 m, C is 0.970 m and D is 0.215 m.

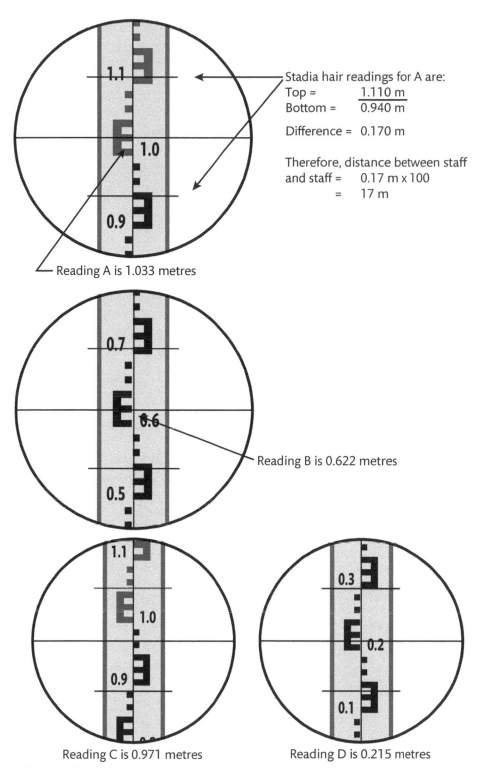

Stadia hair readings for A are:
Top =   <u>1.110 m</u>
Bottom =   0.940 m

Difference =   0.170 m

Therefore, distance between staff
and staff =   0.17 m x 100
           =   17 m

Reading A is 1.033 metres

Reading B is 0.622 metres

Reading C is 0.971 metres   Reading D is 0.215 metres

▶ **Figure 6.13:** Reading a levelling staff: the top and bottom cross hairs have been used to calculate the horizontal distance

Digital theodolites

The theodolite measures horizontal and vertical angles. Its basic components are a telescope and two protractors which measure angles in the vertical and horizontal plane. Digital theodolites are fitted with an electronic system for reading both the digital horizontal and vertical angles. Figure 6.14 shows the essential parts of a digital theodolite together with its main controls and moving parts.

A theodolite has two horizontal digital circular plates that rotate independently, called the upper and lower plates The lower plate is calibrated in degrees, minutes and seconds in a clockwise direction and is either free to rotate or can be clamped to the base (tribrach). The upper plate has an indicator or pointer, enabling angles to be read on a direct reading digital display depending on the type of instrument. Control knobs called clamps include horizontal fine motion screws (tangent screws) for fine adjustment to both plates. By adjusting these clamps, the theodolite's cross-hairs within the telescope can be set to zero or any given angle and pointed at the required survey target. The same principles apply to the digital vertical circle.

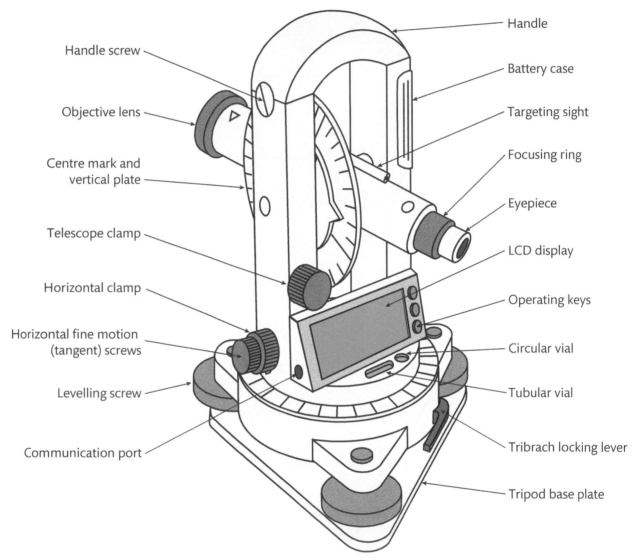

Handle screw

Objective lens

Centre mark and vertical plate

Telescope clamp

Horizontal clamp

Horizontal fine motion (tangent) screws

Levelling screw

Communication port

Handle

Battery case

Targeting sight

Focusing ring

Eyepiece

LCD display

Operating keys

Circular vial

Tubular vial

Tribrach locking lever

Tripod base plate

▶ **Figure 6.14:** Basic digital theodolite

The theodolite is set up correctly when it is:

▶ levelled in all directions, which implies that it is truly vertical
▶ centred over the desired survey station point.

This is achieved in two main stages – roughly and finely.

▶ **Roughly**
1 Set up the tripod of the theodolite over the survey station. This can be a survey nail head or fine cross made with a site marker pen. Make sure the tripod is pulled up to at least shoulder height and that the top of the tripod is roughly level and centred above the nail.

2 Screw the theodolite firmly to the tripod, first checking all the foot screws are at the mid-thread position.

3 Check the instrument is centred. Most theodolites have an optical or digital plummet where you can see the vertical axis of the theodolite. There you can see if the plummet target which represents the vertical axis is centred on the nail. If not, adjust the foot screws slightly to get the axis at its base to coincide with the nail, or, if there is a large distance, two legs of the tripod can be moved while still looking at the plummet target.

▸ **Finely**

1 Once the theodolite has been roughly levelled and centred, finely level it using the three foot-screws. This is the same method as levelling a dumpy level.

2 Check the plummet to see if the instrument's vertical axis has moved off the nail. If so, bring it back on centre by loosening the tripod screw and gently sliding the whole instrument across the top of the tripod while looking at the plummet target.

3 Now recheck the fine levelling by levelling the three foot-screws.

The theodolite is now centred over the survey nail, levelled and ready to use.

Electronic distance measuring (EDM) devices

Electronic distance measurement (EDM) determines the distance between two points using electromagnetic wave energy. Light energy travels from the device to the target prism at the end of a survey line and is reflected back. The measurement works on the principle of measuring both the number of whole wavelengths and any fractional part of the wavelength (the phase difference) that is reflected back from the target. Depending on the type of instrument, various visible light or infrared wavelengths are compared and, once their speed and the phase difference of the reflected light is known, the direct distance is automatically calculated and displayed on the LCD screen of the device.

There are many types; some used by building surveyors or estate agents are hand-held and quick and accurate to use. Typically, their range approaches 50 m with an accuracy of + or – 2 mm. Some, like the one shown, have a continuous reading which allows maximum distances for measuring room diagonals, and minimum distances for finding right-angle or offset distances.

Total stations

When EDM's capabilities are combined with a digital theodolite, the instrument becomes a 'total station' which can measure both angles and distances to provide a variety of useful 3D positional information such as co-ordinates, eastings and northings (see section B5, later in this unit) or elevation values. It can also be used to identify the 3D co-ordinates of any targeted station within the range of the instrument so is ideal for setting out. Total stations are usually used with reflecting prisms placed on tripods over adjacent survey stations or on reflecting targets stuck onto a building, although some instruments are also designed to be reflectorless.

Global positioning systems

The global positioning system (GPS) is one of the global navigation satellite systems (GNSS) which are being developed. GPS was originally developed by the American military but is now a worldwide radio-navigation system formed from 24 orbiting satellites and their ground stations. GPS uses these satellites and ground stations as reference points to calculate positions and is accurate to a matter of metres.

GPS receivers are becoming smaller and more economical so are ideal for survey mapping and for other construction and civil engineering applications. In its basic form, GPS measures the travel time of radio signals to three satellites at any given time and because the speed of radio waves is known (approximately 186,000 miles per second) the distance to the satellite can be calculated.

▸ Electro-distance meter

The simultaneous distance measurements to the three reference satellites allows the position of the GPS receiver to be calculated. However, a fourth satellite is needed to resolve problems caused by small time delays in signal transmission from the other satellites.

Detailed relative position on the ground can be obtained more accurately by using two receivers, one stationary and the other carried by the surveyor who is making position measurements. There are two techniques:

▸ Differential GPS (DGPS) can be used on relatively low-cost receivers to give submetre precision.
▸ The more advanced real time kinematic (RTK) system utilises additional measurements and can achieve an accuracy at the centimetre level.

Ⅱ PAUSE POINT   Can you describe the types of equipment used for accurately measuring distances or changes in ground level height or angles?

Hint   For a project that involves extensive excavation to build a car park, identify the equipment that ensures that the correct level of ground is maintained across the whole area.

Extend   Discuss the issues involved in the care and maintenance of surveying equipment to ensure accurate measurements.

A3 Sources of systematic errors

Measurement of distances, angles or heights will always be prone to inaccuracies. As a surveyor, you must be aware of where errors can occur and more importantly how they can be minimised. There are three basic classifications of errors:

▸ **Gross or human errors** are caused by carelessness, tiredness or inexperience. For example, mistakes made in reading or booking (recording) measurements are often due to miscounting, wrongly reading the tape or E staff, or writing down the wrong measurement. These errors can be large and can occur at any stage. If not picked up early, the survey may need to be abandoned and redone. The best way to avoid making mistakes is to develop a clear method of reading and booking results. For example, double-check all measurements after recording measurements to ensure that they are the same as what you have just written.

▸ **Systematic errors** occur in the **calibration** of equipment and will have a gradual cumulative effect on the survey. For example, plastic tape can become permanently stretched but the only way to check this type of error is to compare the tape against standard steel tapes kept for that specific purpose. If the error is found after fieldwork has been completed, the distance can be corrected by applying a suitable correction factor, for example adjusting measurements according to the amount the tape has stretched.

▸ **Random errors** are generally small errors that compensate each other and are due to the limitations of the surveyor or the equipment. For example, a person with a slight visual impairment may misread the tape's graduations. Although these errors tend to be small, they can be avoided by taking suitable check measurements and repeating readings.

Key term

Calibration – the process of ensuring that measuring equipment is correctly adjusted to give true readings by comparing the equipment that is going to be used with a standard version.

Taping errors

In normal practice, linear surveys should try to achieve an accuracy of 1 in 5000, or 6 mm over a distance of 30 m. To do this, you must consider the following errors that might affect finding the true horizontal plan distances:

▸ *Slope* – the measured distance was recorded along the sloping ground. If there is a difference of height between the ends of a 30 m tape of less than 600 mm, then an accuracy in plan distance of 6 mm over 30 m is achievable. Where the drop is greater than 600 mm, use other methods to find the true plan distance such as step measurements.

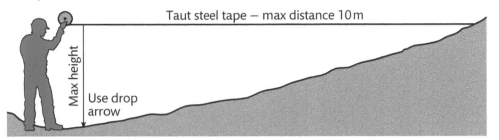

▸ **Figure 6.15:** Step measurements

▸ *Sag* – the measured distance was in an arc due to the weight of the tape. To obtain an accuracy of 1 in 5000, the centre of a 30 m steel tape should not sag by more than 300 mm from the horizontal.

▸ *Temperature variation* – the temperature of the steel tape caused it to expand where it was warmer than the standard temperature of 20°C, or contract where cooler. A temperature of 2° will cause a 6 mm contraction and temperature of 38° will cause a 6 mm expansion based on the standard length at 20°. If the temperature is within these 18°C parameters, then an accuracy of 1 in 5000 for a 30-metre tape is possible.

▸ *Tension* – a steel tape is pulled to reduce sag but as a result stretches. These errors are generally very small and can be ignored; however plastic tapes can be permanently elongated so should be compared with standard steel tapes for calibration.

Levelling errors

The main source of error in optical levelling is when the line of collimation of the levelling instrument is not parallel to the horizontal level line. This can be detected and adjusted using the two-peg test, which should be carried out before any major level survey. The method for carrying out this test is as follows:

▸ Select two points, X and Y, approximately 60 m apart and set the level up midway between them. Take readings for the staff at point X and point Y. The readings when subtracted will give the true difference in height between the points as the collimation error will cancel itself out.

▸ Move the levelling instrument beyond point X to a position at its minimum focusing distance and take readings on points X and Y again. If the level has no collimation error, the difference between the readings of the two points should be identical to that previously measured.

▸ If the result shows a difference of more than 5 mm, the levelling instrument should have a calibration service. BS 5606: *Guide to Accuracy in Building* states the error should be less than 5 mm over 60 m.

Errors using theodolites or total stations

The theodolite or total station can measure very small angles; in fact, some theodolites can read down to 1 second of a degree, which is equivalent to 1 in 216,000th of a whole circle. An error of just 60″ of an angle roughly equates to an error of about 15 mm over 50 m.

To understand how to reduce errors, you need to understand the main axes of the theodolite and how they relate to each other. These are illustrated in Figure 6.16 and are as follows:

▸ The vertical axis – at right angles to the horizontal plane and transit axis, which is considered to be truly vertical when the theodolite has been levelled.

▸ The horizontal circle – at right angles to the vertical axis.

Theory into practice

Keep the distance of backsights and foresights approximately equal as this will reduce collimation errors. Why is it easier to estimate the staff reading to the nearest 1 mm if you keep staff readings to less than 30 m distance?

- The transit axis – at right angles to the line of sight and the vertical axis and forms the pivot for rotation of the telescope.
- The vertical circle – at right angles to the transit axis and used for measuring vertical angles.

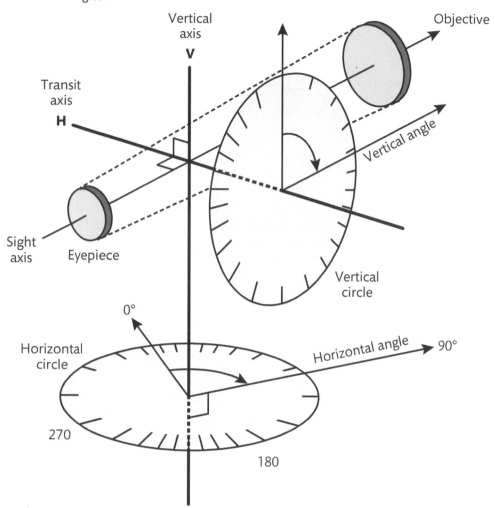

▶ **Figure 6.16:** Main axes of a theodolite

The axes of a theodolite or total station must be in good adjustment with each other to ensure the instrument is level and rotates perpendicularly about its various axes. It is advisable to have the theodolite serviced by a certified calibration and testing company at the start of major contracts and at frequent intervals as they receive rough handling, particularly under site conditions. Therefore it is important to undertake basic checks for errors before any fieldwork, including:

- **Vertical axis or circle check** (V) – after levelling the instrument on the three foot-screws, simply rotate the instrument 180° and the plate bubble should stay central. If not, there is an error in the circle reading where the centre of the measuring circle is not the same as the centre of the physical instrument (a circle error).
- **Transit axis check** (H) – the transit axis should be perpendicular to the line of collimation; if you book readings on both **face right** and **face left** (see B3 Reading and recording horizontal angles of a closed traverse, later in this unit), this error, if any, will be cancelled out when the mean is calculated (a horizontal collimation error).
- **Spire check** – the transit axis (H) should be truly horizontal. It involves sighting a target high up, such as a church spire, which is then brought down to ground level by dipping the telescope and marking a point on the ground. The instrument is then changed onto a different face and the spire is sighted again. If this reading,

when brought down, coincides with the first mark, there is no error. Booking readings on both face right and face left can eliminate this error but the instrument should be checked and calibrated.

▶ **Plummet test** – the simplest way to check whether the plummet is true is to use a plumb bob to set the instrument over the survey point. Remove the plumb bob then look through the optical plummet or switch on the digital plummet laser. If the cross-hairs or laser point are not directly over the survey point, it must be off centre and will need re-calibration to correct the off-centre error.

▶ **Vertical angle collimation** - theodolites measure vertical angles, usually down from the zenith direction (straight up). However if the true zenith direction does not coincide with the 0° reference on the vertical circle there is a constant error. To check whether this error is present:

• On face left a raised target is sighted and the vertical angle recorded (α) T.

• Rotate the instrument 180° about the vertical axis, and flip the telescope 180° about the transit axis.

• Site the same target again and record the result (β). (Note that in this case β is always the largest angle recorded.)

• When both of these angles are added together they should equal exactly 360° if there is no error. Otherwise, the average of the two recorded angles α and β will give the correct angle as the error will be cancelled out.

True vertical angle $= \dfrac{\alpha + (360° - \beta)}{2}$

To avoid transit or vertical axis errors, it is always advisable to measure any horizontal angle twice, once on each face (see B3 Reading and recording horizontal angles of a closed traverse, later in this unit) as any error is cancelled out by calculating the mean value. Vertical circle errors cannot be compensated for in this way since the circle moves with the telescope and more sophisticated techniques are required to recalibrate the plate bubble.

Human errors include sighting the wrong target, adjusting the wrong clamp, or reading/booking the angle incorrectly. These errors can be significant and the only way to avoid them is to fully familiarise yourself with the instrument's controls, practise setting up and use a standard methodical way to read and book your results.

Errors in using EDMs

There are two common basic types of systematic errors in electronic distance measurements.

▶ Scale error – generally caused by atmospheric conditions at the time of the readings, where the variability of the refractive index of the air can bend light, changing the distance travelled between the device and the target. The refractive index can be affected by air temperature, pressure, humidity or gaseous composition.

▶ Index error – usually due to the optical/electrical centres of the instrument and target reflector combination not being in the same place as their physical centres. Instrument reflectors all have reflector constants which is the distance between the centre of the target and the centre of the reflector but this still does not compensate for the index error that may be present.

EDMs have to be calibrated to ensure that they read the correct standard length. RICS recommends that EDMs should be checked against a local baseline which has been measured with an EDM calibrated on an international or national standard baseline. Local baselines should also be re-observed and **verified** annually.

Key term

Verification – the determination of whether or not an instrument conforms to a published (normally the manufacturer's) specification.

Hint | How might you minimise systematic errors in your surveying work?

Extend | What are the most problematic types of error that can occur in undertaking measuring surveys – systematic, gross or random errors?

Assessment practice 6.1

A.P1   A.P2   A.M1   A.D1

You are working with a surveyor as part of your work experience. Your manager has asked you to put together a report on the techniques and instruments used to record survey data.

To prepare this report, choose a small area of land close to your place of study that you could survey which includes a number of key features. Select the methods and techniques you would need to perform a linear and levelling survey of this site. Undertake a reconnaissance of the site making appropriate notes and sketches.

Your report should include a:

- brief explanation of what 'working from the whole to the part' means when carrying out surveying work
- description of the possible systematic errors that may arise when carrying out your linear and level survey measurements for the above
- discussion of the methods that you would use to reduce the systematic, gross or random errors identified in terms of the temporary and permanent instrument adjustments and undertaking the surveying processes
- justification of why a traverse angular survey using a total station would be more accurate than a taped linear and level survey. Clearly show your reasoning in your response to the equipment, processes and error detection.

Plan

- When planning your surveying work, think about what activities to carry out and the accuracy of the measurements that you are going to take.
- Write a list of equipment you need and match that with the job you need to carry out.

Do

- When producing your report on errors do you appreciate the relative size of the errors that can affect your survey results?

Review

- When reviewing and comparing the different surveying methods can you demonstrate the need for a methodical approach to the surveying process?
- What are the most likely errors to occur when undertaking your survey work? Are they systematic, random or gross errors – can you explain the differences?

B | Undertake fieldwork surveys to collect data for drawings

The first stage of any survey is to carry out a full reconnaissance of the area to be surveyed. The basic process involves creating a survey framework covering the whole area, which can then be broken down into smaller parts to record all necessary details and features of the area.

Once the survey framework has been decided, the methods by which linear and angular measurements can be planned and decided will depend on the equipment and time available to carry out the work, and the accuracy required of the survey. For example, a cadastral survey to identify field boundaries for land management may require an accuracy of $+/-250$ mm, while a large steel frame building construction may need a far greater tolerance, usually around $+/-5$ mm to allow for the higher tolerance of the components to be erected on site.

The surveyor should walk over the site to select the most suitable positions for the survey stations (the key reference points). These stations should be clearly marked and referenced so their positions can be found at any time. This process is called 'witnessing'; points should be sketched with clear dimensions referenced to at least three existing fixed features such as the corner of an inspection chamber or a steel fence post.

When deciding the positions of the stations and survey lines that form the surveying framework the surveyor needs to confirm what equipment and techniques they will use. If they are using traditional linear measurement techniques like taping they need to consider the following points:

▸ Survey lines should be kept to a minimum, but sufficient to form a well-conditioned triangular framework over the site.

▸ The location of the baseline from which to form all the other survey triangles.

▸ The triangles should not have small internal angles.

▸ Check lines should be used.

▸ Survey lines should pass close to the boundary and any features of the site and be positioned over level ground.

▸ Stations in any triangle should be intervisible (in sight of each other).

▸ Obstacles to ranging and chaining should be avoided, e.g. trees and ponds.

B1 Linear surveys

Any measured data should be recorded in a special field book, traditionally known as the chain book, which has two thin lines drawn down the centre of the page. These represent the tape line laid out between the survey stations. On either side of these double lines, draw the features to be measured with offsets or tie lines (see Figure 6.2 for an example of this).

You, as the surveyor, should walk down each survey line and sketch the physical features on either side of the line. Do not start measuring anything until after you have drawn all the features to be measured. In this way, you are concentrating first on what and how you are going to measure and then you can devote your energies to measuring and correctly recording your measurements. This clear and methodical system will help you to avoid missing details or errors when you measure.

ⅠⅠ **PAUSE POINT**	Describe the main principles for surveying a known area of land that has irregular boundaries and contains a building, using linear methods.
Hint	Sketch on a piece of paper the area of land chosen, then lay out a survey framework showing examples of how you would record perpendicular offset measurements and tie lines.
Extend	Discuss the source of likely linear measurement errors and how you would reduce them when carrying out your measurements.

Measuring and booking the details

After you have drawn all the features you plan to measure on your booking sheet or chain book, record the running measurements or chainage along the tape from one station to the point on the line where the perpendicular offset distances occur to selected features; for example, the location of boundary of the site, paths and trees. The offset distances can be measured with a synthetic tape by estimating the point where the 90 degrees occurs, as explained earlier in this unit.

The main points to be considered when recording information in the booking sheets are as follows:

▸ Offset measurements should be as short as possible; they should not normally exceed 10 m. You can estimate a right angle by eye if the distance is less than about 3 m or by swinging the tape in an arc and taking the shortest measurement. The offset dimension lines should not be drawn in the book.

▸ Where offset distances are greater than 10 m, it is best to use a pair of tie lines as this is more accurate. Both tie line measurements should be shown in the booking sheet.

- Information need not be drawn to scale in the booking sheet, but should be clearly set out so that it can be understood when the survey is plotted in the office. It is best to write the offset distance close to the feature drawn in the booking sheet.
- Several pages of booking may be needed for long or complicated tape lines, and explanatory notes should be added where necessary. It may be very difficult and time-consuming to return to the site to re-do a measurement or check up on information, especially if the site is a long way from the office.
- Straight lines may be continued beyond a survey station to the site boundary, if necessary, and the information recorded so that detail in the corners of the site is included.
- In the booking sheet, circles should be drawn around the survey stations so they can be seen clearly, and the overall distance of the survey line included within the circle.
- The circumference or girth of a tree should be measured 1m above the ground and recorded in brackets by the tree. The spread of the tree's branches should also be recorded.

B2/B3/B4 Levelling surveys, read and record horizontal angles of closed traverse and basic arithmetic operations

There are two methods for booking level readings to enable the reduced levels of points to be found:
- the height of the plane of collimation (HPC) method
- the rise and fall method.

Each method requires the use of standard booking sheets, which can be purchased in pre-printed, bound notebooks, or can be drawn up on squared paper or on a spreadsheet.

Often the levelling instrument needs to be moved during the levelling circuit. A staff is held at a **change point** (CP), a known point on the ground, and a foresight is read. The staff remains at the change point and the level is moved and set up again. A backsight is then read to the change point and the survey continues. In this way, the reduced levels relative to the original bench mark can be transferred around the circuit.

The staff readings are recorded by putting the backsight reading in the left-hand column, intermediate sight in the second column and foresight in the third column. It is important to remember that the staff reading for any new point on the ground needs a new line in the booking table, and the only time two staff readings will appear on the same line is at a change point. This is where the foresight from the previous level position and the backsight from the new level position coincide.

Once the fieldwork has been completed the first check is to ensure the adding and subtracting is correct. For both the HPC and rise and fall methods the checks are similar:

The difference between the sum of the backsights and the sum of the foresights $=$ The difference between the first and last reduced level $=$ The difference between the sum of the rises and the sum of the falls

When all these checks balance then the calculation is correct. However, this is not an indication of how accurate or good the survey work was. The levelling operation is undertaken as a circuit of levels between points of known height, usually a bench mark such as a TBM or OSBM. A circuit will start and return to the same point – a closed level survey – or it will go between two different bench marks of known height. Either way, the series of levels can be checked for accuracy so that the closing errors between the first and the final transferred levels can be determined. This survey error is called the misclosure and can be compared against an allowable value given in BS 5606: *Guide to Accuracy in Building* which is often quoted as:

▸ $\pm 10\,mm\ \sqrt{(\text{kilometres travelled around circuit})}$ or
▸   for small surveys, less than a kilometre
    $\pm\ 10\,mm\ \sqrt{(\text{number of instrument positions})}$.

If the misclosure is outside the allowable value, the survey should be repeated. If it is within the allowable value, the error should be distributed equally between the readings.

Research

The Channel Tunnel had to achieve an accuracy of 100 mm in levels over a distance of 18 km. Carry out some research to find out the accuracy of other notable recent projects.

Safety tip

Make sure you do not position the level instrument so that it blocks access on any footway or is vulnerable to being knocked. When reading the instrument, do not lean on the tripod or the level, and also be careful when moving around it. Finally, do not leave the instrument unattended – opportunist thieves may steal your equipment, and your prize level may end up at the local car boot sale!

Case study

Level surveying

A sewer is to be constructed for a new industrial warehouse on the Premiere Industrial Estate. Part of the initial design work requires finding the profile of the existing ground level. A series of levels is taken along the length of the proposed sewer at regular intervals.

Details of the fieldwork that has been undertaken are shown in Figure 6.17. There were three instrument positions to pick up the ground levels at 15 m chainage intervals, and the circuit of levels started from an OSBM on an existing building.

This case study will guide you through the recording or booking of the levels for the Premiere Industrial Estate using first the HPC method then the rise and fall method.

The HPC method

The staff level recordings are placed in the HPC booking table shown in Figure 6.18. In the table the readings have been labelled A, B and C to help you understand how each value was calculated based on the instrument position. The survey is accurate to 4 mm.

A is the calculated reduced levels of the ground.

B is the first level instrument position staff readings and the HPC height is found.

HPC = BM + BS from section A1 of this unit (The HPC method) therefore,

HPC = 65.5 + 1.351 = 66.851 m

and the RL at 15 m chainage = 66.851 − 1.455 = 65.396 m etc.

C is the second level instrument position staff readings and thus the new HPC height is calculated
e.g. New HPC = 64.339 + 0.435 = 64.774 m
and RL at 60 m chainage = 64.774 − 1.52 = 63.254 m etc.

D   is the third level instrument position staff readings and HPC height
e.g. New HPC = 63.312 + 3.261 = 66.573 m
and finally the RL to close back onto the starting OSBM = 66.573 − 1.077 = 65.496 m as shown.

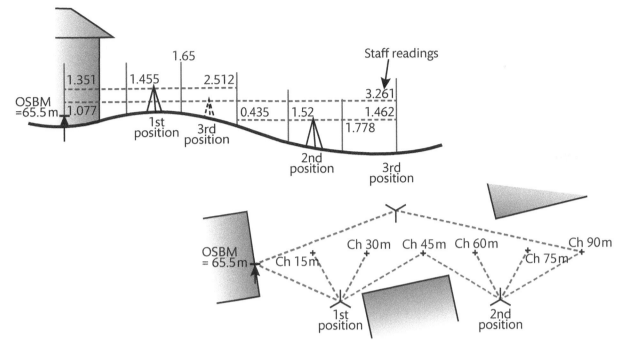

▶ **Figure 6.17:** Premiere Industrial Estate plan

Job description: Ground levels along sewer line					**Surveyor:** MJH	
Name of site: Premiere Industrial Estate					**Job Reference:** 1/19TT	
Address: Eastwood Road, Southend-on-Sea, Essex					**Date:** 27th Feb 2017	

Backsight BS	Intermediate sight IS	Foresight FS	Height of plane of collimation HPC	Reduced level RL	Distance	Remarks
1.351(B)			66.851(B)	65.500(A)	-	OSBM 65.5 m
	1.455(B)		66.851(B)	65.396(A)	15 m	
	1.650(B)		66.851(B)	65.201(A)	30 m	
0.435(C)		2.512(D)	64.774(C)	64.339(A)	45 m	Change Point (CP)
	1.520(C)		64.774(C)	63.254(A)	60 m	
	1.778(C)		64.774(C)	62.996(A)	75 m	
3.261(D)		1.462(C)	66.573(D)	63.312(A)	90 m	Change Point (CP)
		1.077(D)	66.573(D)	65.496(A)	-	OSBM close
5.047		**5.051**		**65.500**		
		5.047		**65.496**		
		0.004		**0.004**		

Booking checks

Difference in BS – FS = Difference between 1st & last RL

Therefore levels have been corrected correctly, however there is a misclosure of 4 mm

▶ **Figure 6.18:** The HPC booking table

Now we have to check the maths is correct for the booked levels. Totalling all the backsights and foresights and finding their difference gives an answer of 4 mm.

Finding the difference between the first and last reduced levels also gives 4 mm. Therefore, the backsights and foresights have been correctly booked.

The rise and fall method

This method compares the relative height of two consecutive staff readings (see Comparing the HPC and rise and fall methods, earlier in this unit). In Figure 6.19, note that coloured arrows have been added to show which staff readings are being compared to give the rise or fall and the calculated reduced level of the ground.

The first readings to be compared are the backsight to the OSBM and the staff reading at the 15 m chainage:

$1.351 \text{ m} - 1.455 \text{ m} = -0.104 \text{ m}$ 'fall'

Giving a reduced level of the ground at the 15 m chainage as:

$65.5 \text{ m} - 0.104 \text{ m} = 65.396 \text{ m}$

The next two figures to be compared are the 15-metre and 30-metre chainages:

$1.455 \text{ m} - 1.65 \text{ m} = -0.195 \text{ m}$ 'fall'

Which when added to the reduced level of the 15-metre chainage gives the reduced level of the 30-metre chainage

$65.396 - 0.196 \text{ m} = 65.201 \text{ m}$ etc.

At the change point you cannot compare figures written on the same line as these are the same point on the ground. You must compare staff readings written on different consecutive lines, so that at the 45-metre chainage point you have to compare the 0.435 m backsight with the 1.52 m intermediate sight at the 60-metre chainage:

$0.435 \text{ m} - 1.52 \text{ m} = -1.085 \text{ m}$ fall

This then gives the reduced level of the ground at the 60-metre chainage as:

$64.339 - 1.085 \text{ m} = 63.254 \text{ m}$

The process of comparing consecutive staff readings continues. If it is a fall in numerical value in the booking it is a 'rise' in the ground level. When the survey is finally closed back to the starting bench mark we can then apply our mathematical checks and calculate the misclosure or accuracy of the survey.

As we can see from Figure 6.19 the maths checks are correct with all the differences equalling 4 mm. However this means that the misclosure is also 4 mm as the OSBM started at 65.500 m but finished at 65.496 m which means we have an error of 4 mm.

As with the HPC booking survey we can check to see if the misclosure is acceptable by applying the formula:

$\pm 10 \text{ mm} \sqrt{\text{(number of instrument positions)}}$.

Job Description: Ground levels along sewer line **Name of site:** Premiere Industrial Estate **Address:** Eastwood Road, Southend-on-Sea, Essex						**Surveyor:** MJH **Job Reference:** 1/19TT **Date:** 3rd Nov 2016	
Backsight BS	**Intermediate sight IS**	**Foresight FS**	**Rise**	**Fall**	**Reduced level RL**	**Distance**	**Remarks**
1.351(B)					65.500(A)	-	OSBM 65.5 m
	1.455(B)			0.104	65.396(A)	15 m	
	1.650(B)			0.195	65.201(A)	30 m	
0.435(C)		2.512(D)		0.862	64.339(A)	45 m	Change Point (CP)
	1.520(C)			1.085	63.254(A)	60 m	
	1.778(C)			0.258	62.996(A)	75 m	
3.261(D)		1.462(C)	0.316		63.312(A)	90 m	Change Point (CP)
		1.077(C)	2.184		65.496(A)	-	OSBM close
5.047		5.051 –	2.500	2.504 –	65.500 –		
		5.047		2.500	65.496		
		0.004		0.004	0.004		

Booking checks

Difference in BS – FS = Difference between 1st & last RL

Therefore levels have been corrected correctly, however there is a misclosure of 4 mm

▶ **Figure 6.19:** The rise and fall booking table

Comparing the HPC and rise and fall methods

- The HPC method requires less calculation but it does not have a built-in check on the intermediate sights, unlike the rise and fall method.
- The rise and fall method is best suited to when a level is being transferred over a distance via a number of change points where no intermediate sights are used (flying levelling). The HPC method is best used when a large number of readings are taken from the same instrument position, such as plotting or setting out a small levelling grid.

PAUSE POINT   Can you produce a list of precautions to take when carrying out a levelling survey to ensure accurate results and a low value of misclosure?

> Hint   Identify the difference between a backsight, a foresight and an intermediate sight when carrying out a levelling survey.

> Extend   If the booking checks with your survey balance, does this mean that the survey is accurate and all the ground levels are correct? Why is it important to close the levelling survey?

Using digital theodolites

Digital theodolites have direct reading electronic displays which clearly state the values of angles measured in degrees, minutes and seconds.

When using a digital theodolite, you need to know what 'face' the instrument is on. This can be found by seeing on what side the vertical circle lies in relation to the telescope:

- face left is when the vertical circle is to the left of the telescope
- face right is when the vertical circle is to the right of the telescope.

Moving from face right to face left or vice versa is a simple matter of flipping the telescope over to point in the opposite direction, then turning the instrument 180° to get the telescope to point back to its original sighting. This is called transiting the telescope.

It is common practice in surveying to take at least two readings for a single angle, one on face left and one on face right. This uses the whole 360° of the circle and eliminates any centring errors.

Booking procedures for a single horizontal angle

Remember that a horizontal angle is an angle measured between three fixed points within a horizontal plane. In the survey book, the number or letter used to denote the survey station over which the instrument is placed is written in the 'At' column, and the numbers or letters of the two stations which are being sighted are placed in the 'To' column. It should also be noted whether the instrument is on face left or face right. This is shown in Figure 6.20.

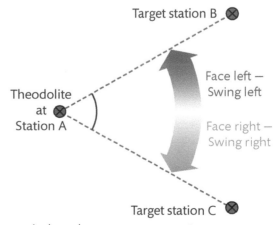

▶ **Figure 6.20:** Plan on a single angle survey measurement

Worked example

The standard booking table has five columns to show where the instrument is located 'At' and at what sights you are targeting 'To'. The next two columns are 'Face right' and 'Face left' and the final column records the 'Mean angle' of the two angles.

Table 6.2 shows the readings from Figure 6.20.

▶ **Table 6.2:** Example of a booking table

At	To	Face right	Face left	Mean angle
A	B	0° 0′ 0″	180° 00′ 15″	82° 20′ 05″
	C	82° 20′ 15″	262° 20′ 10″	
		82° 20′ 15″	82° 19′ 55″	

Here is an explanation of how the fieldwork was done and the readings booked correctly.

- Set up the instrument levelled and centred over the required station A.
- Using the lower plate clamps and tangent screws, set the index to a low value of angle, preferably 0° 00′ 00″, and put the instrument in the face right position.
- Sight onto the first station B making sure you are always striking out in a clockwise direction. Using the lower plate clamp and its fine adjacent tangent screw, set the 0° 00′ 00″ onto station B.
- Release the *upper plate clamp* – not the lower plate clamp – and swing right to the second station at C. Sight accurately onto C using the tangent screws with the upper plate clamps on and, after adjusting the optical micrometer, book the reading, which in this case was 82° 20′ 15″. The instrument is on face right and pointing at station C, so this is where the reading is booked in on the booking table.
- Calculate the angle made at A between C and B by subtracting the top from the bottom to give a value of 82° 20′ 15″.
- Transit the telescope by flipping it over and unclamping the upper plate – *but not the lower plate!* – and turn the telescope to point back to station C.
- Sight onto C using the upper plate clamps and tangent screw, and accurately book the reading which in this case is 262° 20′ 10″. The instrument is on face left and pointing towards C, so that is where this reading is booked in the booking table. Also note that, ideally, if there were no errors, this value should be 180° greater than the previous angle.
- Release the upper plate clamp and swing left back onto the original station B and book the reading of 180° 00′ 15″ in the booking table. This reading is on face left and pointing at B so this is where it goes in the booking table.
- Calculate the angle made at A between B and C by subtracting the 'top from the bottom' to give a value of 82° 19′ 55″.
- Where the difference between the angles measured is less than 30″, the mean of the pair can be calculated and taken to be the true angle subtended at A measured between points B and C. The mean of the face left and face right angles is 82° 20′ 05″.

Discussion

Can you now apply what you have learned to find the mean horizontal angle XYZ given below? Table 6.3 shows the angular measurements that were recorded on face right and face left.

▶ **Table 6.3:** Complete the booking table

At	To	Face right			Face left			Mean angle
Y	X	0°	0′	0″	179°	55′	45″	
	Z	79°	50′	30″	259°	40′	15″	

Booking procedures for a single vertical angle

A similar procedure is used for measuring vertical angles, except that the zero is generally fixed in one position, either pointing straight up – known as the zenith – or set at the level horizon. Angles that dip below the horizon are often called angles of depression while those that rise above the horizon are called angles of elevation.

It is good practice to record single vertical angles on both faces, but be aware that once the values go beyond 360°, they revert back to 0°, just as a clock hand moves from 12 o'clock to 1 o'clock.

Establishing survey stations and recording measurements

As we saw earlier, a traverse survey is an irregular shaped open or closed series of interlinked lines. The position of the survey stations needs to be decided, then both the angles between adjacent stations and the plan distance between them recorded. When deciding the positions for the survey stations and survey lines that form the traverse, the surveyor needs to consider the following points:

▶ Survey lines should be as few as possible but sufficient to form a closed or open traverse framework over the site.
▶ The traverse should always start and finish at a known fixed point so the survey can be properly checked for errors.
▶ The adjacent survey stations for the traverse should be intervisible (in sight of each other), and where possible positioned out of the way of proposed construction works or other activities that could block the sight lines between stations.
▶ Where a traverse is to be used for either measuring the positions of existing features or for future setting out, these points should be clearly visible from the traverse survey station.
▶ Survey stations should be positioned away from hazards such as busy roads or deep excavations.
▶ The position of the traverse stations is fixed with a survey nail to form a reference point during the works.

The equipment required to carry out the traverse measurements is usually an electronic total station, which combines a digital theodolite and a direct reading electro-distance measurement (EDM) device. Also for measuring the angle between three points, two additional tripods

with reflecting prisms should be positioned over the survey stations.

The total station is set up centred and levelled over the survey station. The two adjacent traverse stations have reflecting prisms centred and levelled over the survey nail or marker. The horizontal angles are then measured at least three times each on face left and face right to enable the mean angle to be calculated. Finally the telescope is sighted onto the target prism and the horizontal plan distance recorded. This measurement should also be repeated at least three times and the mean value recorded.

B5 Application of applied mathematical techniques

Calculation of rectangular co-ordinates: Eastings and northings

If a huge sheet of graph paper was unrolled on the ground it would be possible to find the position of any point in that area by working out the x and y co-ordinates. In surveying, the x and y co-ordinates are known as eastings and northings. Figure 6.21 shows this, with A the instrument station and B the target station.

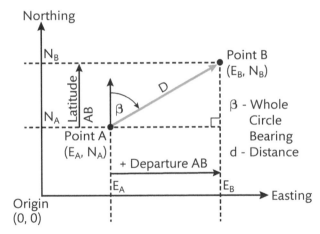

▶ **Figure 6.21:** Rectangular co-ordinates: eastings and northings

The survey line AB can be defined by its whole circle bearing β and its distance d. Point A has easting co-ordinates of E_A and northing co-ordinates of N_A, while point B has co-ordinates E_B and N_B.

⏸ PAUSE POINT

Can you set up a theodolite including all temporary adjustments and record a horizontal angle?

Hint Borrow a theodolite and practise levelling and centring it over a survey point. Choose two distant points and then measure and book the horizontal angle between them using the booking correct format and calculations.

Extend Discuss why it is important to always record on face right and face left when booking angular measurements.

Using right-angle trigonometry we can calculate the relative easting and northing distances, known respectively as the departure and latitude, shown in Figure 6.21:

▸ Departure AB = d x Sine β
▸ Latitude AB = d x Cosine β

By converting angles to whole circle bearings (WCB) and knowing the measured plan distance we can work out the departures and latitudes and find the exact position of any target station in terms of its eastings and northings because:

▸ EB = EA + Departure AB
▸ NB = NA + Latitude AB

Note that easterly departures are always positive and westerly departures are always negative; and using similar logic, northerly latitudes are always positive and southerly latitudes are always negative.

Worked example

A survey station F has easting and northing co-ordinates (72.000,50.000) and the plan distance between it and the adjacent station G is 67.350 m with a whole circle bearing of 65° 20′ 50‴.

Calculate the easting and northings for station G.

Solution:

- Departure FG = $67.35 \times$ Sine 65° 20′ 50″ = 61.211 m
- Latitude FG = $67.35 \times$ Cosine 65° 20′ 50″ = 28.093 m

Therefore, the eastings and nothings of station G can be found:

- $E_G = E_F +$ Departure FG = 72 m + 61.211 m = 133.211 m and
- $N_G = N_F +$ Latitude FG = 50 m + 28.093 m = 78.093 m

Therefore, the easting and northing co-ordinates of station G are (133.211,78.093)

Closed traverse surveys: Calculation and adjustment

The closed traverse forms the basis of most survey frameworks because it combines the high precision of modern total stations and the ability to measure and determine the co-ordinates of any point on the ground. As we saw earlier, a closed traverse starts and finishes on a point of known departure and latitude with its easting and northing co-ordinates defined at the start of the survey. It can also have a number of sides or survey stations forming a polygon to which all internal or external angles can be measured along with the plan distances between the stations.

The traverse needs to be orientated to the true or grid north direction. This can be done with a magnetic compass or to another agreed fixed reference direction.

The overall accuracy of a traverse survey depends on the purpose of the survey. If the work involves civil engineering groundwork, such as embankment construction, a fairly low accuracy, typically 1/500, would be satisfactory. For more setting out, such as a steel-framed building, a higher accuracy in the region of 1/5000 would be needed. For detailed monitoring or precise surveying, such as crack or asset measurement, accuracy of 1/50,000 would be required. Generally, surveys which achieve 1/5000 are sufficient to undertake construction work.

Angular misclosure adjustments

In any polygon the **sum of the internal angles** can be found from the formula:

$(2n - 4) \times 90°$ where n = the number of survey stations

Or alternatively where the **sum of external angles** was used a similar relationship exists:

$(2n + 4) \times 90°$ where n = the number of survey stations

Theory into practice

Work out the easting and northing co-ordinates for points K, L, M, N and P measured from a single station set up at J at (10.000, 20.000). The measured angle and distances are:

- Point K − WCB 136° 40′ 25″ Distance 34.329 m
- Point L − WCB 27° 55′ 32″ Distance 123.420 m
- Point M − WCB 292° 37′ 02″ Distance 146.224 m
- Point N − WCB 331° 49′ 17″ Distance 12.549 m
- Point P − WCB 124° 19′ 57″ Distance 76.511 m

When you have worked out these eastings and northings for these points, plot point J and point K to P using a CAD software program, where x and y co-ordinates are the eastings and northings respectively. Then measure the WCB and distances using the dimension tools of the software package to see if they check with the values given above. Were you right?

For example in a three-sided closed traverse all the internal angles will add up to:

$$(2 \times 3 - 4) \times 90°$$
$$= (6 - 4) \times 90°$$
$$= 2 \times 90° = 180°$$

It is very unlikely the angles will add up to this value no matter how carefully we try to reduce the errors that can occur while undertaking the measurements. However, the sum should be very close to this amount and should be within the following limits:

$+/-$ **40** $\sqrt{n}$ *in seconds where n = the number of survey stations*

If our survey falls within this misclosure we can distribute the error equally among all angles measured to ensure that the sum of the angles come to the correct value.

However, errors in the distance measurements have to be corrected separately (see Errors in using EDMs, earlier in this unit).

Back bearing adjustments

The back bearing is in the opposite direction to the forward bearing, a 180° difference.

Back bearing = previous forward bearing $+/-$ 180°

So if the forward bearing is 120° from A to B then the direction to return from B to A will be 300°. This is shown diagrammatically in Figure 6.22 and is set out as follows:

▸ Forward bearing A to B = 120°
▸ Back bearing B to A = forward WCB + 180°
$$= 120° + 180° = 300°$$

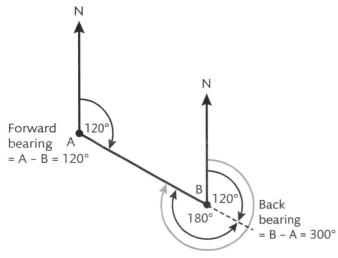

▸ **Figure 6.22:** The relationship between back and forward bearings

Remember that, like a clock face, you can never have an angle greater than 360° or less than 0°. If this occurs when working out a back bearing you need to either add or subtract 360°. So a WCB of 410° is effectively a WCB of 50° measured clockwise from north, as 410°−360° = 50°.

Step-by-step: Traverse co-ordinates and adjustment

Figure 6.23 shows a simple closed traverse, where the horizontal angles and plan distances have been measured using a total station. The five-sided closed traverse survey started at a point A with a known bearing of line AB of 56°. All the internal angles were measured, along with the plan distances of each individual survey leg.

We will use this to show how the eastings and northings of each of the five survey station points are calculated, how traverse is checked for accuracy and how these corrections are applied to give the correct easting and northing co-ordinates for the survey stations, enabling them to be plotted and fixed.

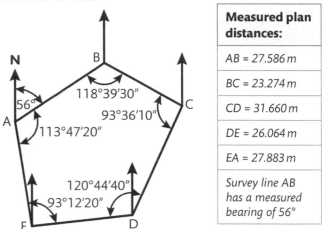

Measured plan distances:
AB = 27.586 m
BC = 23.274 m
CD = 31.660 m
DE = 26.064 m
EA = 27.883 m
Survey line AB has a measured bearing of 56°

▸ **Figure 6.23:** Closed traverse angles with north grid directions

Step 1 – Angular corrections

First, we will calculate the sum of the internal angles and compare them to the sum of the measured values recorded on site.

Sum of internal angles
$$= (2n - 4) \times 90°$$
$$= (2 \times 5 - 4) \times 90°$$
$$= (10 - 4) \times 90°$$
$$= 6 \times 90°$$
$$= 540°$$

Sum of recorded angles = (113° 47′ 10″) + (93° 12′ 10″) + (120° 44′ 30″) + (93° 36′ 00″) + (118° 39′ 20″) = 539° 59′ 10″

Therefore the angular error = 540° − 539° 59′ 10″ = 50″

To be acceptable this must be less than or equal to $+/-$ 40″ $\sqrt{n}$ where n = the number of survey stations

▸ Acceptable angular misclosure = $+/-$ 40″ x $\sqrt{5}$ = 89″ > 50″

Therefore this is an acceptably small error and can be corrected. If this was larger than 89″ then the survey would need to be repeated with more care taken to reduce the errors, as outlined in section B of this unit.

This 50″ error is then distributed evenly to all the measured angles as shown in Table 6.4.

▶ **Table 6.4:** Distributing the error among the angles

Station	Angle measured	Correction	Corrected angle
A	113° 47′ 10″	+10″	113° 47′ 20″
B	118° 39′ 20″	+10″	118° 39′ 30″
C	93° 36′ 00″	+10″	93° 36′ 10″
D	120° 44′ 30″	+10″	120° 44′ 40″
E	93° 12′ 10″	+10″	93° 12′ 20″
Totals	539° 59′ 10″	+50″	540° 00′ 00″

Step 2 – Finding the whole circle bearings

The next stage is to find the whole circle bearings (WCB) of each leg of the traverse going clockwise around the stations. This is shown graphically in Figure 6.24.

Starting from the leg AB the measured forward bearing AB is 56°.

▶ Therefore the corresponding back bearing
B−A = 56° + 180° = 236°

▶ Then to find the forward bearing
B−C = 236° − 118° 39′ 30″ = 117° 20′ 30″

▶ With a corresponding back bearing
C−B = 117° 20′ 30″ + 180° = 297° 20′ 30″

▶ Then to find the forward bearing
C−D = 297° 20′ 30″ − 93° 36′ 10″ = 203° 44′ 20″

▶ With a corresponding back bearing
D−C = 203° 44′ 20″ + 180° = 383° 44′ 20″

▶ BUT as all WCB must be greater than 0 or less than 360° we have to SUBTRACT 360° making the correct back bearing D−C = 383° 44′ 20″ − 360° = 23° 44′ 20″

▶ Then to find the forward bearing
D−E = 23° 44′ 20″ − 120° 44′ 40″ = − 97° 0′ 20″

▶ Again all WCB must be greater than 0 or less than 360° so we have to ADD 360° making the correct forward bearing D−E = − 97° 0′ 20″ + 360° = 262° 59′ 40″

This sequence of calculations continues, finding the back bearing E−D, then the forward bearing E−A, then the back bearing A−E, then finally as a check the forward bearing A−B should be exactly 56°, which is where we started from.

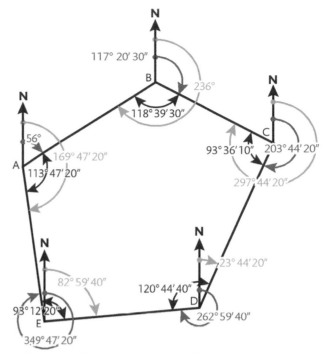

- All angles in black are the corrected internal measured angles
- All angles in green are the calculated back bearings
- All angles in red are calculated forward bearings

▶ **Figure 6.24:** Graphical representation of the process of finding whole circle bearing

> **Theory into practice**
>
> Can you continue working out these WCBs E−A, A−E and A−B to see if you can close the traverse at 56°?

Step 3 – Finding the partial co-ordinates

The next stage involves calculating the partial co-ordinates of each leg of the survey using the WCB and the measured distance (d) between each station. This is shown in Figure 6.25, where we calculate the departures and latitudes for each leg of the survey. The calculations for each leg are often called partial co-ordinates as these distances are relative to the previous station.

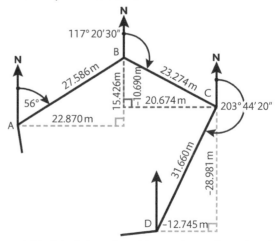

▶ **Figure 6.25:** Calculation of partial co-ordinates for a closed traverse

We already have:

▸ Easting departure = d × Sine WCB
▸ Northing latitude = d × Cosine WCB

Therefore for the first leg of the survey A–B:

▸ Easting departure = 27.586 m × Sine 56° = + 22.870 m
▸ Northing latitude = 27.586 m × Cosine 56° = +15.426 m

And for the second leg of the survey B–C:

▸ Easting departure = 23.274 m × Sine 117° 20′ 30″ = + 20.674 m
▸ Northing latitude = 23.274 m × Cosine 117° 20′ 30″ = −10.690 m (note that the − sign indicates a southerly direction)

And for the third leg of the survey C–D:

▸ Easting departure = 31.660 m × Sine 203° 44′20″ = − 12.745 m (i.e. going west)
▸ Northing latitude = 31.660 m × Cosine 203° 44′20″ = − 28.981 m (i.e. going south)

This continues until we complete all the partial co-ordinates for the entire survey, finishing back to the starting point at A where the survey closes.

Step 4 – Checking and adjusting the traverse using the application of Bowditch's method

The next stage is to check and adjust or correct the survey. This is done using Bowditch's method which distributes the easting and northing misclosure corrections in proportion to the relative leg length compared to the total length of all the survey lines traversed. This is best done using a traverse table or spreadsheet as shown in Table 6.5. For ease of identification each column has been given a reference number 1 to 10.

▸ **Table 6.5:** Completed closed traverse calculations

1			2	3	4	5	6	7	8	9	10	
WCB			**Plan distance**	**Partial co-ordinates**		**Corrections**		**Corrected co-ordinates**		**Total co-ordinates**		
deg	min	sec	(m)	Easting	Northings	Departure (E)	Latitude (N)	Eastings	Northings	Eastings	Northings	Station
Line										0.000	0.000	A
AB 56	0	0	27.586	22.870	15.426	0.003	−0.004	22.873	15.422	22.873	15.422	B
BC 117	20	30	23.274	20.674	−10.690	0.002	−0.003	20.676	−10.693	43.549	4.730	C
CD 203	44	20	31.66	−12.745	−28.981	0.003	−0.004	−12.742	−28.985	30.807	−24.256	D
DE 262	59	40	26.064	−25.869	−3.179	0.003	−0.003	−25.867	−3.182	4.940	−27.438	E
EA 349	47	20	27.883	−4.943	27.441	0.003	−0.004	−4.940	27.438	0.000	0.000	A
		Totals	136.467	−0.014	0.017							

The partial co-ordinates in columns 3 and 4 are tabulated in relation to the line references. If the survey was without any errors, if we totalled up all the eastings and northings (both positive and negative) the sum would be zero because we started and finished at A (0,0). However, because we may have made errors in measuring the distance and the angular adjustments in Step 1, the survey will not close back on (0,0) and in fact there is a misclosure in the eastings of −14 mm and in the northings of 17 mm.

We can work out the combined misclosure error because the 17 mm and 14 mm are two sides in a right–angled triangle where the misclosure is the hypotenuse. So using Pythagoras's theorem:

Total misclosure = $\sqrt{(17^2 + 14^2)}$ = 22 mm

The total distance we have traversed summing up the leg lengths of the survey in column 2 is 136.467 m. So the accuracy of the survey can be stated by comparing the total misclosure to the total length surveyed, which in this case is 22 mm in a total 136.467 m traversed. This can be expressed as a fraction:

$$\text{Traverse accuracy} = \frac{\text{Total miscolosure}}{\text{Total distance}} = \frac{22}{136467} = \frac{1}{6203}$$

Therefore the accuracy of the survey is 1/6203. As this is less than 1/5000, this is sufficiently accurate for most standard construction works so we can spread this misclosure over each of the survey station easting and northing co-ordinates.

▸ The error of −14 mm in the eastings needs to be corrected by distributing +14 mm among the partial easting co-ordinates. With the previous angular corrections in Step 1 we distributed the error evenly; however with Bowditch's method we distribute the error by proportion to the length of the survey leg compared to the overall distance traversed. That is for example for line A−B:

- easting correction $= \frac{\text{Leg length}}{\text{Total length}} \times \text{total correction} = \frac{27.586}{136.467} \times (+0.014\,\text{m}) = +0.003\,\text{m}$

Similarly:

- northing correction $= \frac{27.586}{136.467} \times (-0.017\,\text{m}) = -0.004\,\text{m}$

Note that the correction is always the opposite sign to the error to adjust the survey!

In this way the corrections in columns 5 and 6 are tabulated. Then in columns 7 and 8 these corrections are added to the original partial co-ordinates in columns 3 and 4 to give the corrected partial co-ordinates. Finally, the total co-ordinates relative to the starting point at station A are found by cumulatively adding all the corrected partial co-ordinates to obtain the values in columns 9 and 10.

If we consider all the easting co-ordinates:

▸ For station B,
total easting co-ordinates = Eastings of A + Corrected partial co-ordinate of A−B
= 0 m + 22.873 m = 22.873 m

▸ For station C,
total easting co-ordinates = Eastings of B + Corrected partial co-ordinate of B−C
= 22.873 m + 20.676 m = 43.549 m

▸ For station D,
total easting co-ordinates = Eastings of C + Corrected partial co-ordinate of C−D
= 43.549 m + (−12.742 m) = 30.807 m

▸ For station E,
total easting co-ordinates = Eastings of D + Corrected partial co-ordinate of D−E
= 30.807 m + (−25.867 m) = 4.940 m

▸ For station A,
total easting co-ordinates = Eastings of E + Corrected partial co-ordinate of E−A
= 4.940 m + (−4.940 m) = 0 m

The process is repeated for all the northing total co-ordinates until the closing value of 0 m is obtained for the northing at A.

> **Theory into practice**
>
> Once the co-ordinates of the stations are known you can use simple right-angle trigonometry to calculate other useful geometrical information. For example, from the traverse we have just completed, could you calculate the length of the line that would join station A to station D, and its WCB from A to D? As a hint, look back at Figure 6.21: Rectangular co-ordinates: eastings and northings.
>
> If station A had co-ordinate values of (200,300) instead of (0,0) then what would the co-ordinates of all the other stations be? Would this change the WCB from A to D or the length of the line A−D?

Develop drawings from completed fieldwork surveys

C1 Conventions used in survey drawings

Final survey drawings must be neat and accurate. The effort and time taken to produce comprehensive field notes and measurement data will be wasted if a survey drawing is badly presented. Your client will not want to pay for a poorly presented and inaccurate drawing!

The scales to which a drawing is plotted vary according to the purpose of the survey. The ranges of available scale used are given in Table 6.6.

Table 6.6: Types of scale

Small-scale maps	Large-scale maps	Site plans	Detailed plans
1:1 000 000 to 1:20 000	1:10 000 to 1:1000	1:500 to 1:50	1:20 to 1:1

After the points are plotted, the detail is drawn in using standardised symbols. The north point is drawn and any necessary lettering, including a suitable title block, scales and annotation, is added.

C2 Production of survey drawings

Plotting linear survey lines accurately to scale

When plotting the survey manually, it is good practice to draft out the survey lines on tracing paper. By overlaying tracing paper on the paper to be used, the survey may be properly centred on that sheet. The north point should always be shown, preferably pointing towards the top of the sheet.

▶ The final survey plot should progress from the 'whole to the part', just as the survey progressed.
▶ The lengths of all the survey lines are taken from the booking sheets and the baseline is plotted first to the desired scale.
▶ By striking arcs with a large radius compass, the other survey stations are established and the network of triangles drawn.
▶ Check lines are scaled off and compared with actual distances.
▶ Once the framework has been plotted, the details of all the features of the site can be plotted. Offsets and ties are systematically plotted in the same order they were booked; work from the beginning to the end of each line. The right angles for offsets may be set out by set square.
▶ The accuracy of plotting needs to take into account the final scale to which the drawing is to be plotted. In manual plotting the thinnest line width that can be plotted by a sharp pencil or ink pen is approximately 0.25 mm. If the final plot of the survey is to be 1:200, then the smallest dimension that can be measured during the survey and accurately plotted will be 0.25 mm × 200 = 50 mm.

This process is slightly different when using a CAD software drawing package as all surveys are drawn to a scale of 1:1 before plotting on paper.

Link

Unit 7 has more details on using CAD software.

Spot levels

These are points on the ground whose reduced level has been surveyed relative to a local level datum. The location of these points is decided by identifying where the ground level changes. Then the position of these points is surveyed and plotted together with the spot height levels.

Grid levels

This is where a square grid is marked out on an area of ground. The staff readings at the grid intersections are measured and the reduced level of these points calculated. This method can be used to indicate a contour on the ground surface, where a contour line shows a height above a given datum. Plotting a series of contour lines gives an indication of the topography of the land. Figure 6.26 shows a hilly area with a definite contour pattern. Height measurements like these can be used to calculate the volume of soil to be excavated from a certain area of ground, or the amount of imported material needed to build up a road embankment.

Contours

Grid levels are where a square or rectangular grid is set out on the ground and the reduced levels determined. These grid levels can be manipulated manually to give contour levels at regular vertical intervals or used to calculate volumes of excavation. An example of an interpolated contour map is shown in Figure 6.26.

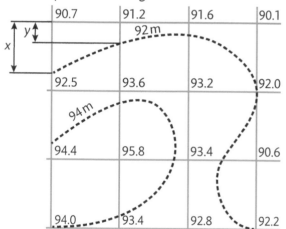

▶ **Figure 6.26:** A plan on grid levels with contours

The position of the 92 m contours can be plotted by interpolating between the recorded grid spot heights. In Figure 6.26 the plotted distance x is found from applying the principle of similar triangles:

$$\frac{x}{12\,m} = \frac{(92 - 90.07)\,m}{(92.5 - 90.7)\,m} = \frac{1.3\,m}{1.8\,m}$$

Therefore $x = \left(\frac{1.3\,m}{1.8\,m}\right) 12\,m = 8.667\,m$, and similarly

$$y = \left(\frac{(92 - 91.2)\,m}{(93.6 - 91.2)\,m}\right) 12\,m = \frac{0.8\,m}{2.4\,m} 12\,m = 4\,m$$

Alternatively, digital level data can be downloaded directly into software packages to produce virtual 3D models. These can be manipulated for contours and volumes as required.

Site cross sections

Vertical cross sections are taken along the length of the proposed construction project such as a sewer or a highway. Staff readings are taken from points in the ground, such as the centre line of the proposed road or sewer, from which the existing ground levels can be recorded and compared with the proposed construction levels. Areas which need excavation, such as in road cuttings, can be shown on the vertical cross section. Similarly, any areas where imported material is required, such as road embankments, can be shown and used to aid the calculation of soil volumes.

Long sections

Long sections are vertical cross sections taken along the length of proposed sewer or road construction; because these forms of construction are often very long compared to their depth, they are plotted with an exaggerated vertical scale to clearly show the slope and changes in height relative to length. For example, a new sewer may be 800 m long, but the levels throughout may vary by only 5 or 6 m. Plotting these to the same scale would give no impression of how the gradient may change along its length. Therefore, a typical long section will have a vertical scale ten times the horizontal scale, for example 1:500 horizontally and 1:50 vertically.

Using this method, numerical information is tabulated below the long section for the new sewer and will include the cumulative distance or chainage, as well as the reduced levels of the ground, the **invert level** of the sewer, the **cover level**, the **formation level** of sewer excavation and the sight rail heights that aid the setting out. A typical long section is shown in Figure 6.27.

Key terms

Invert level – the reduced level of the lowest part of the internal diameter of a sewer, trench, pipe or tunnel.

Cover level – the reduced level of an object indicating underground services, e.g. top of an inspection chamber cover.

Formation level – the reduced level of the bottom of an excavation, such as a sewer trench.

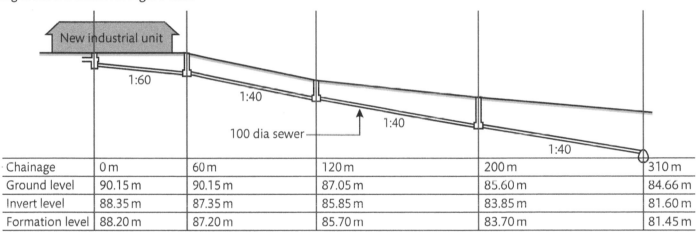

Chainage	0 m	60 m	120 m	200 m	310 m
Ground level	90.15 m	90.15 m	87.05 m	85.60 m	84.66 m
Invert level	88.35 m	87.35 m	85.85 m	83.85 m	81.60 m
Formation level	88.20 m	87.20 m	85.70 m	83.70 m	81.45 m

Long section on sewer
Scales – horizontal 1:1250 – vertical 1:200

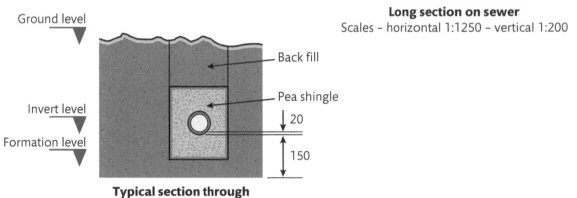

Typical section through

▶ **Figure 6.27:** Long section of a sewer

Cut and fill cross sections

In long linear forms of permanent construction such as rail or road alignments it is important to be able to predict the amount of excavation or additional soil materials required. This is done by comparing the contoured surface of the ground profile with the vertical and horizontal alignment of the proposed permanent way.

By superimposing these two profiles the net excavation and imported fill areas and related volumes required can be obtained. An example of a comparison between the two profiles is shown in the image of a specialist software earthworks program in Figure 6.28. The typical drawings for cut and fill can be shown either as one long section taken at the centre line position of the permanent way (similar to the sewer long section with exaggerated horizontal/vertical scales), or as a series of regular cross sections at right angles to the centre line.

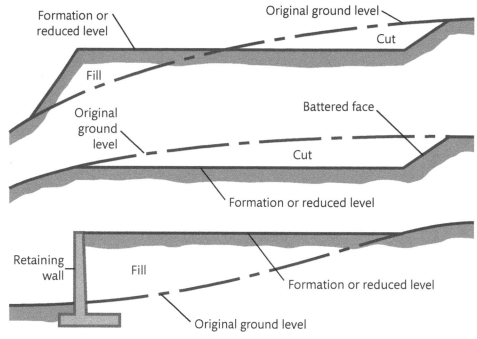

▶ **Figure 6.28:** Cut and fill areas

C3 Corrected closed traverse drawing

Traverse surveys are always plotted according to the co-ordinate easting and northing values that have been previously adjusted. This method of plotting the survey framework and fixing the points is very accurate and can be replicated by using a CAD drawing package where these values are directly inputted into the CAD drawing or model. Manual plots on graph paper can also be undertaken by plotting the easting and northing co-ordinates, however they are less accurate than CAD plots.

Where details or features have been recorded from a given survey station in co-ordinate form, this too can be easily transferred to the CAD drawing. It is important to check if these detail co-ordinates are in absolute terms relative to the grid origin or in relative terms according to the direction of the local survey leg.

PAUSE POINT   What are the main features of a good survey drawing?

> Hint   Describe the variety of views and cross sections that can be produced from linear, levelling and traverse data.

> Extend   Research some specialist CAD survey packages and comment on the advantages of this method of managing and plotting survey data.

Assessment practice 6.2

B.P3 B.P4 B.P5 C.P6 B.M2 C.M3 B.D2 C.D3

You are working with a surveyor as part of your work experience. Your manager has assigned you a series of tasks:

1 Test your levelling instrument and theodolite to check they are in good adjustment:
   a) Carry out a two-peg test on the levelling instrument.
   b) Carry out a spire test, and check for circle and horizontal collimation error checks on the theodolite.

2 Undertake the following practical surveys in groups of three using the equipment provided:
   a) Survey and record the ground levels along the line of a proposed sewer. You will need to select equipment, then peg out a straight line a minimum of 40 m long over undulating land, representing the line of the sewer. Set out pegs or arrows at regular intervals to take account of slope of the ground. Establish a suitable TBM and undertake a level survey of the ground marked by the pegs. Book field notes and staff readings in HPC and rise and fall formats.
   b) Undertake a grid levelling exercise on an area of sloping land. Then undertake suitable calculations to contour that area of land with a minimum of four equal height difference contour lines. Plot the levelling grid and contours to a suitable scale.
   c) Carry out a simple three station closed traverse and record features using linear methods:
      i. Set out three intervisible stations roughly 30 m apart. Undertake a survey to fix the positions of these three stations. You will need to undertake all necessary angular and linear checks.
      ii. Using this triangular framework undertake a simple linear survey using offset and tie lines to measure existing features in the area. Plot this survey to scale using standard symbols.
   d) Carry out a complex closed traverse survey:
      i. Undertake a closed traverse survey around your main college building, including all angular and linear measurements. Then use the traverse stations to record the position of the main walls and corners by polar measurement based on the traverse control stations.
      ii. Book the angular readings from and carry out all the necessary angular misclosure checks.
      iii. Calculate the corrected co-ordinates of your closed traverse surveyed and state the overall accuracy of the survey; and also calculate the co-ordinates of the corners of the main building walls surveyed.
      iv. Plot the corrected traverse on paper or using an appropriate CAD package. Clearly show the traverse stations, survey lines and the plotted corners of the main building walls and other permanent features surveyed.

3 Produce a detailed illustrated blog explaining the work:
   i. Include a justification of the equipment used and methods by which you checked to reduce errors.
   ii. Undertake further research to contrast and compare the different methods to improve accuracy in linear, level and traverse surveys.
   iii. Evaluate how you personally managed, planned, carried out, checked and presented the detailed accurate survey data.

Plan

- Plan carefully. A well-annotated sketch including all the calculations helps demonstrate your knowledge.
- A photographic record of your practical work can supplement any witness statements.
- Before starting, decide where you will set up your instrument to take all the necessary readings.
- Surveying is a team activity. Make sure work is shared out evenly.

Do

- It is important you demonstrate you know the right tests and checks to use for each piece of surveying equipment.
- In your write-up of the surveys remember to include labelled sketches or actual photographs of the work being carried out.
- Your tutor will want to see that you can relate a physical situation to a mathematical model of the situation.

Review

- Set out your calculations in clear easy to follow stages using standard mathematical notation.
- Remember to include all necessary arithmetic checks and comment on the accuracy of your results.
- Remember to show all appropriate units in your field notes and the final presentation drawings, e.g. metres etc.

Further reading and resources

Brinker, R. and Minnick, R. (2012) *The Surveying Handbook*, reprint of 2nd edition, New York: Springer.

Irvine, W. and Maclennan, F. (2005) *Surveying for Construction*, 5th edition, New York: McGraw-Hill Education.

Price, W.F. and Uren, J. (2010) *Surveying for Engineers*, 5th edition, Basingstoke: Palgrave Macmillan.

THINK ▶▶FUTURE

Eve

Eve works for a county council as a highways surveyor. Part of her job is to undertake built surveys of existing roads and pedestrian pavements which are used to help the highways engineers to plan traffic calming measures in built-up areas, improvements to traffic junctions to help traffic flow and developing new user-friendly cycle paths.

The last main project that Eve worked on was a survey used to design a range of speed control measures for a busy suburban road that is often used as a short cut by commuters at rush hour and commercial vehicles. The new measures included a combination of road traffic humps, automatic speed indicator signs and pelican crossings.

The skills and knowledge that Eve acquired while studying surveying in her BTEC Construction course have been invaluable in her job. She uses up-to-date surveying kit and equipment, using her knowledge of the basic principles of linear, level and angular measurement to help her to carry out her work. The knowledge of specific technical processes, their application, and how results can be checked are essential to being a good surveyor. These skills will be even more important as her career progresses into more senior roles. Team-working skills, which are vital to managing and undertaking practical surveying work, were developed while at college doing practical group surveys. The ability to communicate confidently, with both her professional colleagues as well as the public, through the use of survey drawings and clearly set out calculations, is very important and it is these skills she continues to develop.

Focusing your skills

Surveying is a varied and interesting job role that requires a wide range of skills. As well as being familiar with the up-to-date practice and tools and equipment, you will need to have a wide range of work-related skills.

- Teamwork can be crucial in surveying. Look at opportunities you have to work on projects with teams – both as part of your course and outside. What makes a team work well together? What qualities do you need to have to work well with other people?

- Communication is vital – both written and verbal. How should you present written communication so that the message is clear and easy to understand?

Whenever you prepare written work, try to remember who you are writing for and what message you are trying to get across. Try reading back to yourself what you have written – can you understand what you are trying to say?

- Presenting is very closely linked to communicating verbally. The most important factors are to prepare in detail and to present with confidence. Try to find opportunities to present to people outside your course as well as on it – what do you need to focus on to make sure that you get your message across and that your audience is engaged?

Getting ready for assessment

Roger is in the first year of his BTEC Extended Diploma in Construction and the Built Environment. For this unit he has been given an assignment to prepare a portfolio of practical surveying exercises, with linked calculations, that demonstrate his ability to apply and use linear, levelling and angular measurement surveying techniques. His portfolio needs to include field sketches and measurement and final plotted drawings, as well as calculations of linked properties, such as areas, lengths and co-ordinates.

How I got started

I started this assignment by looking at the aims of the individual survey tasks and the sources of information that the tutor had provided. The aims and source information were helpful in determining which surveying processes and equipment would best fit the tasks that I had been set.

I downloaded a copy of the unit from the specification on the Pearson website and this gave me lots of guidance on what I needed to make sure I was doing. I looked back through my notes and previous practical surveying activities to make sure my portfolio was meeting all the key verbs within the assessment.

Before starting the practical assessment tasks I made sure I fully understood the basic setting-up operations for surveying equipment and their limitations, as well as reviewing all I had learned about minimising human and systematic errors while carrying out surveying. The practical surveying work needed me to work with my peers as a team. We drew up our booking sheets in advance and planned our sequence of measurements, so as not to waste time and to make sure we did not miss anything.

How I brought it all together

We decided to take turns being the survey leader, so the tasks were shared fairly, and we all had an opportunity to use the techniques we had been studying. This system worked well as we could record the work we did by taking photos and videoing what we were doing as evidence. During the fieldwork, I made sure I booked and checked all my readings as I went along to spot errors, so that they could be corrected early in the field. This was very relevant for the levelling survey we did. Finally, I undertook the required booking checks so I could be confident my results were within acceptable tolerances.

Knowing the field measurements were correct meant I could then use the data to calculate important information like co-ordinate values and setting-out information. To show I understood the process I produced a brief written commentary explaining my thought processes. I also proofread it against the assessment criteria to make sure I had hit all the points in the grading criteria unit table.

Think about it

▶ Have you practised setting up the levelling and total station equipment?

▶ For every system of measurement you plan, have you checked whether there is an alternative measurement that could be taken to check your work?

▶ Could your own field notes be easily read and understood by another person?

Graphical Detailing in Construction

7

Getting to know your unit

Communicating and co-ordinating information is crucial to the successful design and construction of building and civil engineering projects. Graphical detailing is a key way of doing this. In this unit you will learn about the resources required to produce construction drawings and you will develop construction drawings for a given construction brief, which adheres to British Standards, using manual and CAD methods.

A drawing aims to show the size, shape and location of designed elements and their various parts. 3D virtual models are largely CAD-based, providing a wealth of useful information to the designer and contractor such as costs, layout and materials. You will also learn how to produce 2D and 3D freehand construction sketches, which are useful for formulating early design ideas.

The knowledge and skills gained in this unit are essential to prepare you for various job roles in architectural and landscape design. An understanding of graphical representation is essential in other roles too, such as site management, site engineering, planning and quantity surveying. It will also help you progress to a higher education programme in construction and related disciplines.

How you will be assessed

This unit is internally assessed by two major summative assignments. The first assignment demonstrates your knowledge and understanding of both manual and CAD graphical techniques by using your skills to create a series of construction drawings. The second assignment will focus on your ability to appreciate and undertake freehand sketching tasks, which are crucial to spatial awareness and conceptual design, as well as construction production methods.

This unit will involve the practical use of manual drawing equipment, and at least one proprietary CAD hardware and software system. Arrange with your tutor to gain access to equipment and for obtaining the appropriate access and training for the CAD system. To achieve the tasks in your assignment, check that you have met all the minimum Pass grading criteria. If you want to gain a Merit or Distinction, make sure that you present the information in your assignment to the depth and style that is required by the relevant assessment criterion. For example, Merit criteria require you to analyse and discuss, and Distinction criteria require you to assess and evaluate. For Merit, you also need to include more appropriate information in your drawings with a better attention to detail and proportionality than for a Pass, while for a Distinction you also need to demonstrate a balanced evaluation of the various methods and a superior attention to detail and quality of output.

Assessment criteria

This table shows what you must do in order to achieve a **Pass**, **Merit** or **Distinction** grade.

Pass	Merit	Distinction

Learning aim **A** Understand the resources required to produce construction drawings

Pass	Merit	Distinction
A.P1 Explain the use of media and equipment to produce manual drawings for a given building.	**A.M1** Analyse the use of manual and CAD methods to produce drawings for a given building in terms of their resource requirements, efficiency and cost.	**A.D1** Evaluate the use of manual and CAD methods to produce drawings for a given building in terms of their resource requirements, efficiency and cost.
A.P2 Describe the resources required to produce CAD drawings for a given building.		
A.P3 Compare manual and CAD methods for the production of drawings in terms of their resource requirements, efficiency and cost.		

Learning aim **B** Develop construction drawings for a given construction brief

Pass	Merit	Distinction
B.P4 Produce construction drawings for a two-storey building drawn to an appropriate scale, containing some technical information following BS 1192:2007 standards.	**B.M2** Produce good-quality construction drawings for a two-storey building drawn accurately to an appropriate scale, containing appropriate technical information following BS 1192:2007 standards.	**B.D2** Produce high-quality, fully annotated construction drawings for a two-storey building drawn accurately to an appropriate scale, containing detailed technical information following BS 1192:2007 standards.

Learning aim **C** Undertake production of two-dimensional and three-dimensional freehand construction sketches

Pass	Merit	Distinction
C.P5 Produce annotated 2D and 3D freehand sketches, using appropriate conventions, for the interior of a building.	**C.M3** Produce good-quality, annotated 2D and 3D freehand sketches for the interior and exterior of a building with convergence to vanishing points.	**C.D3** Produce high-quality, fully annotated 2D and 3D freehand sketches for the interior and exterior of a building with accurate convergence to vanishing points.
C.P6 Produce annotated 2D and 3D freehand sketches, using appropriate conventions, for the exterior of a building.		

Getting started

Take a notepad and pencil to a local building. Spend about ten minutes sketching the main features of the outside of the building and take a photo of the same view. Compare your sketch with the photo. Did you find the sketch easy? Are the relative sizes of the windows and doors correct? Did you use a 3D effect or did you draw a flat 2D sketch? List the benefits of drawing a sketch over taking a photo.

A Understand the resources required to produce construction drawings

A1 Manual methods

Equipment for manual detailing

Some of the main pieces of equipment used for manual detailing are shown in Table 7.1.

▶ **Table 7.1:** Equipment required and its use in producing construction drawings manually

Name	Purpose, function, application and use
Parallel motion drawing board 	<ul><li>The most common type of firm, smooth surface for creating technical drawings. Allows horizontal parallel lines to be drawn.</li><li>Free to tilt and usually on a free-standing frame or can be fitted with a ratchet to sit on a desk.</li><li>The drawing media is secured to the surface with drafting tape (a paper-based tape that holds drawings in place but is easy to remove), which is applied at its corners. Remove the tape every night, so the medium can expand and contract with temperature changes.</li></ul>
Adjustable set square	<ul><li>Used in conjunction with a parallel motion drawing board to draw parallel lines at any angle, including vertical lines at 90° to the parallel motion arm.</li><li>They are adjustable so are far more versatile than traditional fixed 30°/60°/90° and 45°/45°/90° set squares.</li><li>Made from clear acrylic so are transparent with an accurately divided scale for easy reading in degrees.</li><li>Available with either a 250 mm or 300 mm longest edge.</li></ul>
Scale rule	<ul><li>The main tool to help draw scaled drawings, usually a ruler with three sides each showing the most common scales, such as 1:2500, 1:1250, 1:1000, 1:500, 1:250, 1:200, 1:100, 1:50, 1:20, 1:10 and 1:5.</li><li>The 300 mm-long triangular sections are durable and easy to read.</li></ul>

▶ **Table 7.1:** *Continued ...*

Name	Purpose, function, application and use
Compass	• Used to draw circles and arcs. • Two main types: traditional spring bow and the longer horizontal beam compass, used for plotting traditional linear land surveys. • Drafting ink compasses have a screw attachment to enable an ink drawing pen to be mounted for ease of operation.
Templates, stencils and flexible curves	• Templates provide common outlines of objects such as toilet cisterns and shower units in a range of typical scales. • A flexible curve can be used to draw smooth curves such as contour lines or French curves. • Text templates ensure neat block capital printing on drawings.

Media for manual detailing

Pencils, pens, erasers and erasing shields

Pencils are the main working tool of the draftsperson and come in a range of grades or hardnesses. The following are ordered from darkest to lightest:

▶ B – for shading and texturing
▶ HB – for rough sketching
▶ F – for printing and general line work
▶ H – for dimension lines and hatching
▶ 2H – for construction lines.

Pencils can be used on cartridge paper or tracing paper. For plastic film, high polymer pencil lead is used. This is usually for use in clutch pencils (spring-loaded barrels also known as lead holders, refillable pencils or propelling pencils) with leads ranging from 0.2 mm to 0.9 mm.

Pens are used for ink drawing and allow a range of line thicknesses, which is determined by the purpose of the line drawn. Different manufacturers produce slightly different thicknesses, but typical thicknesses are:

▶ 0.25 mm – for dimension and hatch lines
▶ 0.35 mm – for general printing and linework details, and hidden details
▶ 0.5 mm – for section lines and titles
▶ 0.7 mm – for title blocks and drawing borders.

Graphic liners have a disposable plastic nib, come in a range of thicknesses, and are ideal for manual sketching on paper or film. Drafting pens have a stainless steel nib, and are durable and refillable, but are more expensive.

Different types of erasers are available, for removing both pencil and pen marks without smudging them. They are usually in the form of rubber or synthetic blocks, or as refills for eraser holders. Electric erasers (which are usually battery operated) are useful for erasing large areas of the drawing. Some erasers can damage paper so may need to be tested on scrap paper first. Erasing shields are small metal or plastic cards with standard shapes punched into them to enable small details to be erased without accidentally deleting other parts of the drawing.

Paper and other drawing media

In traditional forms of drafting there are four basic types of drawing media. The quality of paper is indicated by its weight in grams per square metre (g/m^2 or gsm)

- **Cartridge (or detail) paper** ($60-150\,g/m^2$) – ranges from very light layout paper ($60\,g/m^2$), photocopy paper ($80\,g/m^2$), thicker letter paper ($90-100\,g/m^2$) to thicker cartridge paper ($120-150\,g/m^2$).
- **Tracing paper** – semi-transparent, comes in various grades for draft work from around $80\,g/m^2$ up to $110\,g/m^2$ for master copies. Generally, pencil construction lines are drawn on tracing paper and inked in later with drawing pens.
- **Drafting film** – similar to tracing paper, but with an easily recognisable waxy, silky feel, and, as it is made from polyester, a high static electricity content. It is strong, virtually tear-proof and resistant to moisture. The copies can be reproduced like tracing paper. Drafting film needs special ink because ordinary ink will take much longer to dry and be non-permanent. Ordinary stainless steel pen nibs will wear out quickly on polyester paper and so tungsten-tipped nibs need to be used instead.

Paper sizes

All drawing media are supplied using the international paper size standard, ISO 216, which is based on the metric system. The sizes are based on the A0 sheet which has an area of 1 m^2 with a height/width ratio of 1:√2 (1:1.41). Each subsequent size (A1, A2, etc.) keeps that ratio and is half the area of the previous size. So:

- A0 = 841 × 1189 mm
- A1 = 594 × 841 mm
- A2 = 420 × 594 mm
- A3 = 297 × 420 mm
- A4 = 210 × 297 mm.

Reprographics

Photocopying is the most common way to reproduce drawings. Most office photocopiers can reproduce drawings up to A3 size but for any larger copies specialist reprographic companies offer various large-scale photocopying.

With the increased use of digital technology, electrostatic copiers are being replaced by integrated scanning and laser techniques. This is particularly useful as the quality of the image is superior, they can reproduce in colour, and the drawing can be captured electronically and transmitted by email or over the internet. Although these machines are relatively compact, they can be used to scan large drawings up to A0 size by knitting together small scan areas using advanced photographic imaging software.

PAUSE POINT What equipment will you need to source in order to create your own construction drawings? Identify a range of equipment and the associated costs of different qualities of tools and media.

Hint Use the internet to help you with your research, and record your findings in a table.

Extend What factors will you need to consider when deciding which equipment to select?

Manual drawing techniques

When deciding the best way to represent dimensional and constructional information onto paper, the draftsperson needs to consider the drawing's main purpose and who is going to be reading it. A drawing should provide:

- enough information to be useful for its purpose, without being too busy
- clear and unambiguous construction information
- individual details that are clearly referenced and labelled, such as section marks and titles
- a clear system of cross referencing between drawings and other relevant documents.

Every design office should have a consistent house style that meets all these requirements and is applied to all drawings and documents. They may also need to set up a series of standard templates for drawings, Word documents and spreadsheets which can be flexible enough to apply to both large and small jobs.

Drawing lines and shapes

All drawings should include a title block and notes, containing the information shown in Figure 7.1.

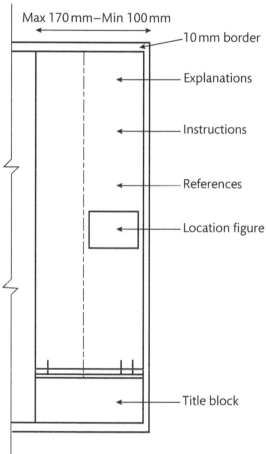

▶ **Figure 7.1:** Title block layout

Link

Details of current standards for numbering drawings according to BS1192:2007 are given later in this unit in section B1.

This information includes the following:

- Explanations – provide guidance for reading the drawing, such as abbreviations, special symbols or the dimension units chosen.
- Instructions – provide information such as the drawing should not be scaled, or on other issues.
- References – made to other relevant drawings, specifications and documents.
- Location figure – a small key plan to help identify the physical area that the drawing covers in comparison with the whole project.
- Title block – including separate boxes for the title of the drawing, the address of the project location, the client's name and the company logo panel.

Types of drawing line

The three most common types of line are continuous, dashed and chain-dotted. Each line has three relative thicknesses or line weights: thick, medium and thin. These conventions are combined to indicate specific construction elements or control information, the functions of which are described in Table 7.2.

▶ **Table 7.2:** Types of drawing line

Type of line		Function	Typical use
————	Thick	Site outline or new building	Site drawings
————	Medium	General details	
————	Thin	Reference grid, dimension lines, leader lines and hatching	
————	Thick	Primary functional elements in horizontal or vertical sections (e.g. load-bearing walls and structural slabs)	General location drawings
————	Medium	Secondary elements and components in horizontal and vertical sections (e.g. non-load-bearing partitions, windows, doors); also components etc., in elevation	
————	Thin	Reference grids, dimension lines, leader lines and hatching	
————	Thick	Primary functional elements in horizontal or vertical section (e.g. load-bearing walls, structural slabs)	Assembly drawings
————	Medium	Secondary elements and components in horizontal and vertical sections (e.g. non-load-bearing partitions, windows, doors); also components etc., in elevation	
————	Thin	Reference grids, dimensions lines, leader lines and hatching	
– – – –		Medium broken line. The purpose and position of the line should be noted in relation to the plane of section The line begins with a dash cutting the outline adjoining, and all lines should meet at changes in direction	Work not visible
Work to be removed
——/\———		Thin line with break in it or if necessary a thin continuous line with a zigzag in it	Breaks in continuity of drawings
▬ ▪ ▬ ▪ ▬		Thick chain lines	Pipe lines, services, drains
— ▪ — ▪ —		Medium chain lines	
— · — · —		Thin chain line	Centre and axial lines
————○		Indicated by a circle at the end of the line	Controlling line, grid line

Drawing to a scale

The scale of the drawing is determined by its purpose and what graphical views you want to show. The types of views are typically elevations, plan views or cross sections – see projection methods below. Typical scales for the various types of drawing are given in Table 7.3.

▶ **Table 7.3:** Typical scales in different types of drawing

Type of drawing	Purpose	Typical scales used
Site location plans	A map to show the location of the site.	1:2500, 1:1250, 1:1000
Layout of block plans	A plan showing the proposal in relation to the boundaries of the site.	1:500, 1:200, 1:100
General arrangement (location) drawings	Plans, elevations, perspectives and cross sections to show the relative position of construction elements.	1:100, 1:50
Assembly/detail drawings	Plans and cross sections of individual parts of the building showing detailed construction information.	1:20, 1:10
Component drawings	Large-scale plans and views showing fabrication information of construction components.	1:5, 1:2

In order to plan how the views will be set out, draw a rough sketch with the proposed views shown as control boxes (see Figure 7.2). The control boxes can be pencilled on the paper to give you a starting point for each individual view.

You need to know the total amount of free space available. To do this, measure the overall width and height of the drawing area to scale (e.g. 1:50). Add up the total width and height area of the control box (also to a scale of 1:50) and subtract this total from the overall width and height of the drawing area to give the total amount of free space. The free space can then be divided by two or three depending on how many equal spaces are required, as shown in Figure 7.2. It is important that plan views and cross sectional views are connected by the correct use of titles and section marks.

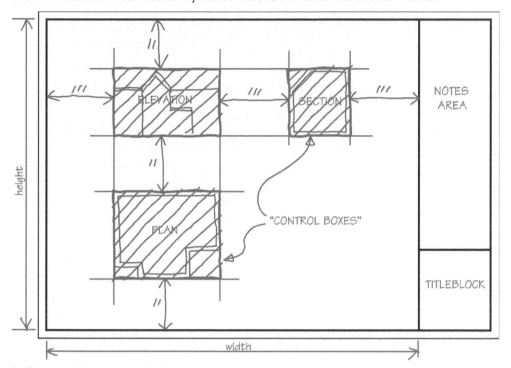

▶ **Figure 7.2:** Setting out sketch

Lettering

Refer to the house style of the organisation producing the drawings. General notes should be neatly arranged in regularly shaped panels on the drawing and have a clear printed text heading. They should be broken up into paragraphs for ease of reading, and should not be cramped or be so widely spaced as to become illegible. A simple, open print text style provides the clearest form of communication. Over-stylised or italicised fonts can be difficult to read. Text should be in block capitals with plenty of open space inside the characters. The height of the text should range from 3 mm high for general text up to 5 mm high for titles of individual sections and plans.

Dimensioning

Dimension lines can be drawn in a number of ways. Two examples are shown in Figures 7.3 and 7.4. It is important to leave a gap between the dimension lines and the detail element to avoid any confusion. Therefore:

▶ Print horizontal dimensions along the dimension line. Mark the ends of the line with arrowheads or thick oblique strike lines.

▶ Print vertical dimensions to the left-hand side of the dimension line.

▶ The dimensions should always be in millimetres. If you need to represent a different unit, for example, metres, then show the abbreviation 'm' after the dimension figure.

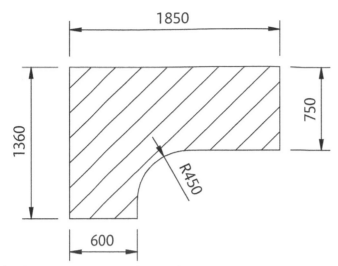

▶ **Figure 7.3:** Dimension lines with arrowheads

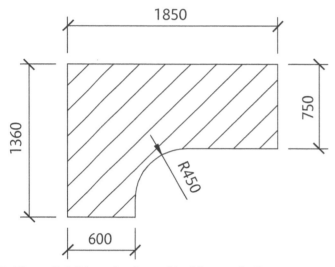

▶ **Figure 7.4:** Dimension lines with oblique strike lines

Use of graphical conventions

Drawings are a two-dimensional (2D) representation of a three-dimensional (3D) element of a building. Several graphical conventions show 3D objects on 2D flat paper media. The following are the most popular projection methods and are illustrated using the same example.

Orthographic projection

In these drawings, individual views and plans are drawn in flat profile. There are two basic types: first angle or third angle. In construction, third angle projection is used to show a building's external elevations and roof plans and is useful for showing the general arrangement of the features of a house and the relative positions of windows and doors. Figure 7.5 shows the south elevation with the plan of the roof shown from above, that is, a bird's eye view. The side elevations are the east and west elevations and are drawn next to the south elevation.

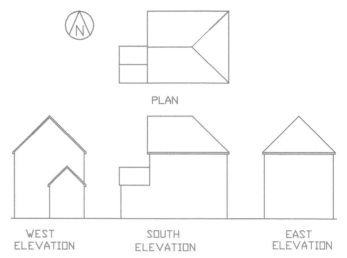

▶ **Figure 7.5:** Orthographic projection in third angle

Isometric projection

This gives a 3D image with all lines drawn at the same scale and at an angle of 30°. All the receding lines are parallel; therefore, it is not a true perspective view. An isometric projection is often used to show the overall mass of smaller construction elements such as stone window cills or special brick plinth constructions. An example is shown in Figure 7.6.

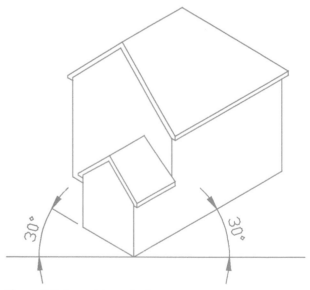

▶ **Figure 7.6:** Isometric projection

Axonometric projection

This is similar to an isometric projection, but all lines are drawn at 45°. This method is easier to use because it uses the true plan view simply rotated by 45°. The axonometric projection is most suitable for interior and office or kitchen layouts as it appears to give a higher viewpoint than isometric projection. An example of axonometric projection is shown in Figure 7.7.

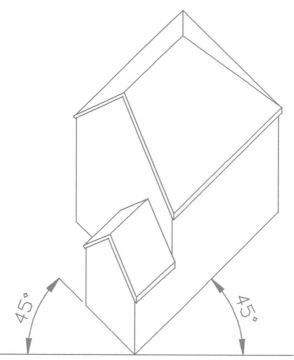

▶ **Figure 7.7:** Axonometric projection

Perspective projection

This gives the most realistic 3D view and is particularly good for elevations and external views, but can also be used for internal perspectives. For a two-point perspective, all lines converge onto two vanishing points fixed at eye level on the horizon. Perspective drawings can be drawn to scale; however, they are quite complicated to construct. The drawing shown in Figure 7.8 has been generated from the computer modelling software, SketchUp™.

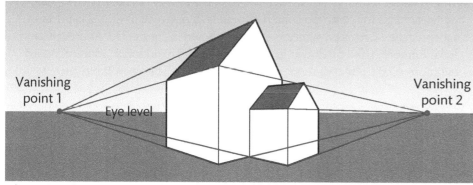

▶ **Figure 7.8:** Perspective projection

Cross-sectional elevations and plans

When deciding from where to take a suitable cross sectional view through a building, it is important to select a position that shows the most typical cross section as it applies to the majority of the proposed new work. The standard height at which the horizontal cross section is taken is at 1.2 m above the finished floor level. This means that all the main door and window positions will be picked up, together with the positions of staircases and partitions.

Standard cross-hatching patterns are used in a cross section, plan or elevation to denote the material from which construction elements are made. The most common hatching patterns are brickwork and blockwork. Other typical hatching patterns are given in Figure 7.9.

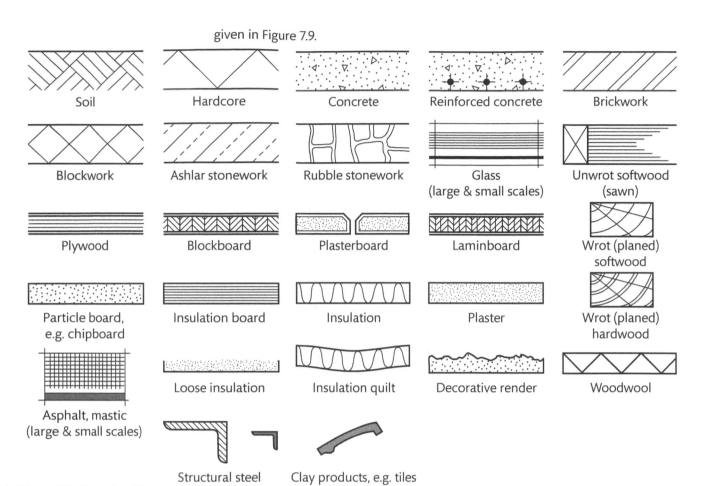

▶ **Figure 7.9:** Cross-hatching patterns

Use of standard symbols

A common convention that is used to save time as well as to aid clarity is the use of specific symbols for common constructional features, such as doors, windows and light switches. Common symbols used are shown in Figure 7.10. Abbreviations are also commonly used on drawings to aid clarity and expression. Some standard abbreviations used for text descriptions on drawings are shown in Table 7.4.

▶ **Table 7.4:** Standard abbreviations used on drawings

Item	Abbreviation	Item	Abbreviation	Item	Abbreviation	Item	Abbreviation
Airbrick	AB	Foundation	fnd	Cast iron	Ci	Rainwater pipe	RWP
Asbestos	abs	Hardboard	hdbd	Cement	ct	Reinforced concrete	RC
Bitumen	bit	Hardcore	hc	Column	col	Satin chrome	SC
Boarding	bdg	Hardwood	hwd	Concrete	conc	Satin anodised aluminium	SAA
Brickwork	bwk	Insulation	insul	Cupboard	cpd	Softwood	swd
Building	bldg	Joist	jst	Damp-proof course	DPC	Stainless steel	SS
Damp-proof membrane	DPM	Mild steel	MS	Polyvinyl acetate	PVA	Tongue and groove	T&G
Drawing	dwg	Plasterboard	pbd	Polyvinyl chloride	PVC	Wrought iron	Wi

Graphic conventions and standard symbols should be used in accordance with BS 1192:2007 (see section B1 below).

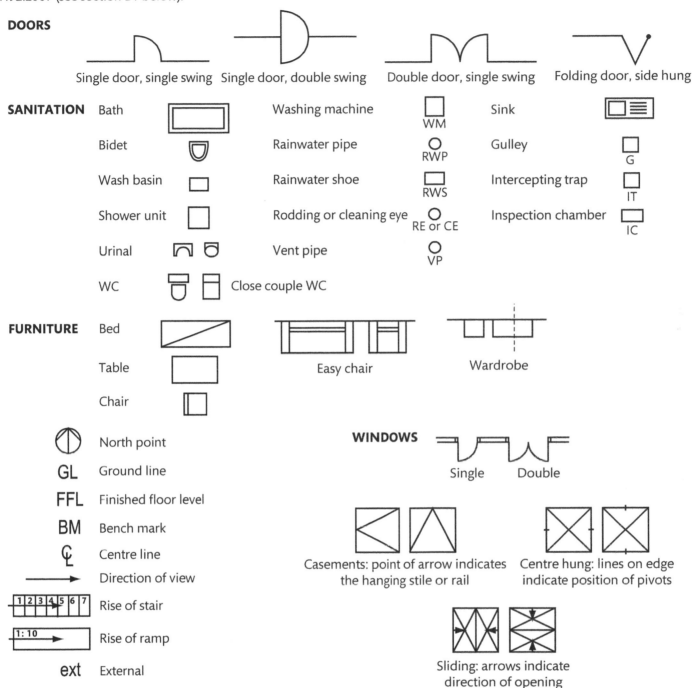

DOORS

Single door, single swing   Single door, double swing   Double door, single swing   Folding door, side hung

SANITATION

Bath	Washing machine WM	Sink
Bidet	Rainwater pipe RWP	Gulley G
Wash basin	Rainwater shoe RWS	Intercepting trap IT
Shower unit	Rodding or cleaning eye RE or CE	Inspection chamber IC
Urinal	Vent pipe VP	
WC Close couple WC		

FURNITURE

Bed

Table   Easy chair   Wardrobe

Chair

North point

GL   Ground line

FFL   Finished floor level

BM   Bench mark

℄   Centre line

→   Direction of view

Rise of stair

Rise of ramp

ext   External

WINDOWS   Single   Double

Casements: point of arrow indicates the hanging stile or rail

Centre hung: lines on edge indicate position of pivots

Sliding: arrows indicate direction of opening

▶ **Figure 7.10:** Symbols for constructional features

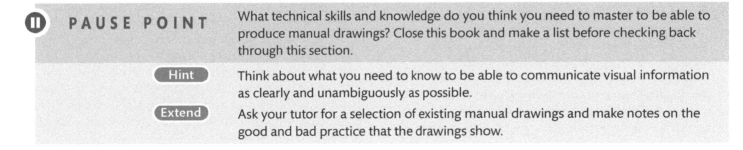

ⅠⅠ PAUSE POINT   What technical skills and knowledge do you think you need to master to be able to produce manual drawings? Close this book and make a list before checking back through this section.

Hint   Think about what you need to know to be able to communicate visual information as clearly and unambiguously as possible.

Extend   Ask your tutor for a selection of existing manual drawings and make notes on the good and bad practice that the drawings show.

A2 Computer-aided design (CAD)

An understanding of computer-aided design (CAD) is an essential requirement for today's design office. It has many advantages over traditional methods; these advantages include:

- production of high-quality graphics in a relatively short time
- facility to reuse standard house-style details
- ability to easily amend and reissue revised drawings/details
- ease of archiving and saving electronic drawings
- improved management and tracking of work.

CAD programs use powerful databases to prepare lists of materials, specifications and cost data. This integration of all types of design and construction information within a common digital environment allows projects to be modelled virtually. Prototyping designs so they can be reviewed and tested before work begins on site allows changes and issues to be resolved early on and avoids wasting time and effort. This process is known as **BIM**, which can stand for building information modelling for designers, building information management for constructors or building information manufacture for the supply chain.

Hardware requirements

Many companies rely on local servers or hard drives to house their CAD operations so when choosing hardware it is important to check the minimum specification required to run a particular CAD program and store the CAD data. It is always better to purchase hardware with a higher specification and greater storage capacity if available. Many CAD companies host the software and provide secure data storage for an agreed hosting and licence fee. With cost of software decreasing, the increased use of cloud storage for data and faster internet speeds, the need for individual companies to store the software and project data in digital format is receding.

Drawing files can be stored in a number of different ways. This could include saving versions on a computer hard drive or a USB stick or another storage device. Many businesses store files on a network or on an online area, such as the cloud. This allows multiple users in and out of the business to access the files.

Currently there are three typical hardware configurations for a hosted CAD system in a design office and they are dependent on how each configuration is networked and where the data is stored.

- **Standalone single user** – the CAD software and drawing information is held within a private domain. Information is issued by plotting off drawings and sending them by post or emailing electronic copies of the drawing files. It has limited use and is only for the small self-employed practitioner, as it relies on ad hoc data management systems that are not easily tracked.

- **Small office-based intranet set up using a local area network (LAN)** – the CAD and licensing software, together with drawing and document information, is held within a project domain which may be in a separate computer server linked to the company plotting device. The server may also contain a plot management function to prioritise the work if there are important deadlines to meet. Companies may also use project management software on this system to track drawings, record revisions and control the electronic issuing of information.

- **Large multi-location extranet set up for a wide area network (WAN)** – drawing information is uploaded or published to an internet domain which is accessible to members of the project team, even if they are in different geographical locations. This cloud storage is provided by subscription from the CAD software company and uses its project management software to monitor and flag up when and what changes are being made. It also allows the user to search for the most up-to-date drawings themselves. Local copies of the program are controlled by licence from the cloud-based management software.

Requirements to run CAD software

The following outlines the minimum essential hardware requirements for CAD work:

- graphics card – resolutions in the order of 3840 x 2160 pixels at 60 Hz with 128 bit 2 GB DDR3
- speed of processor – a central processing unit (CPU) with a speed of at least 3 GHz
- random access memory (RAM) capacity: memory – there should be an operating disk capacity of a minimum 4 GB RAM (32-bit) or 8 GB if dealing with large data sets such as survey point clouds.

The CPU should have capabilities to be networked (wired or wireless) to a local area network (LAN) and also have internet and email access via a fast broadband connection.

Theft of data and cyber crime is a problem in unsecured digital environments and precautions need to be taken. Also there should be a managed power system, both to prevent a power surge from obliterating data and also to provide standby power if required.

Input devices

An input device is attached to a computer to put information into it. A mouse and keyboard are the most common forms of input device, although some professionals prefer using digitisers and digitiser pads or light pens. Scanners and digitisers vary in size and the resolution at which they can process an image and it may be desirable to scan an image at a higher resolution if it contains complex detail which is later to be vectorised. The curser can also be controlled with a joystick or thumbwheel.

Output devices

An output device extracts information from a computer. The following output devices are required:

▶ Visual display unit (VDU) monitor with a resolution of at least 1920 x 1080. 'True color' is recommended. It needs to be free from distortion and be the largest affordable. Larger screens make it easier to see larger images and to deal with small details.

▶ A printer is the most common output device, and is used to create paper (hard) copies of drawings from the computer.

▶ A networked plotter capable of minimum A1 size drawing plots in three pen colours/thicknesses.

A useful addition for architecture and interior design offices is an additive 3D printer for developing massing models to aid the development of early design concepts. They can print 3D scale models in a variety of colours and materials such as plastic, metal and nylon. Alternatively, these output functions can easily be outsourced to less expensive third-party companies that specialise in providing various reprographic services.

CAD software requirements

When deciding what software to use, consider:

▶ Simplicity – is it easy to learn and use?

▶ Functionality – does it meet the specific needs of the organisation or discipline?

▶ Interoperability – does it work well with other software? Can it interchange document formats or convertible documents?

▶ Longevity – will it not need to be replaced until the capital outlay has been recouped, while still meeting core functionality required?

▶ Support and training requirements – is it quick, convenient and effective? Does it use both personal and electronic training programmes?

▶ Collaborative environment – does it work well with the company's hardware set-up and environment? If not, consider possible hosting services for tools and storage from a recognised software partner via a cloud-based interface.

CAD software packages and their advantages and limitations in use

There has been a great deal of development work on common platforms that can read data and information to allow useful digital collaboration between disciplines. This is shown more clearly in Table 7.5, which is based on a UK government-sponsored strategy report. As of April 2016, all new government-sponsored construction projects, such as those for schools, hospitals and prisons, are required to be designed and built using BIM Level 2 maturity. This is motivating all projects to reach this level of BIM, even within the private sector where BIM is not mandatory but where the benefits of a managed common digital environment can be felt.

Table 7.5: The levels of BIM Maturity

BIM maturity	Level 0	Level 1	Level 2	Level 3
System	CAD	2D/3D CAD	3D BIM	iBIM
Functionality	Geometric features, templates, lines, arc, and text, etc.	Models, objects, sharing information and collaboration across software platforms. At Level 2, providing agreed data drops at key milestones throughout the project through a managed data exchange (**COBie**).		Seamless linkages between 4D/5D/XD software systems
Purpose	A drawing aid to replace the traditional manual methods and systems	Lonely BIM – designer in isolation exchanging information via email/pdf file	Managed document sharing and collaboration across different organisations	Integrated real-time web or cloud service, streamlining workflow across multi-disciplines
Media	Drawing software tools and paper	File-based collaboration, allowing readability and digital importation/exportation	Rich digital file-based library management – allowing software compatibility	Integrated interoperable data managed within a web-based BIM hub

COBie – Construction Operations Building Information Exchange – an agreed protocol to share mostly non-graphical project data to support the design and construction processes, usually in the form of a simple spreadsheet.

Different types, or dimensions, of BIM have different functions:

▸ 3D BIM – Graphical and spatial information.
▸ 4D BIM – 3D BIM linked to the dimension of time. Used for construction planning and sequencing of works.
▸ 5D BIM – 3D BIM linked to material, labour and production cost. Used for tendering, estimating and cost control purposes.
▸ 6D BIM – 3D BIM linked to environmental data. Used to assess sustainability and energy design in buildings.
▸ 7D BIM – 3D BIM linked to facilities management data and maintenance regimes. Used to manage the operation of the building throughout its lifetime.
▸ XD BIM – Future applications linked to other building information data as required.

A typical Level 2 3D virtual model is illustrated in Figure 7.11. It represents a single model environment (SME) specifying the steelwork frame of the building (column layout), the setting out grid for the building (grid model) and the mechanical services (ducting model). These are brought together to form a co-ordinated composite model of the project, containing spatial and material information about all the objects and how they interlink.

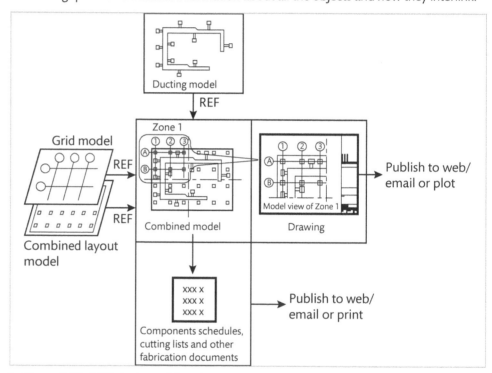

▸ **Figure 7.11:** A 3D single model environment

Currently, there are many CAD and BIM software products on the market, the most popular ones being Autodesk's AutoCAD™ with its derivative programs based on its DWG file format system, Bentley MicroStation™ suite of programs and Graphisoft's ARCHICAD™ software system. All offer capabilities for managing building data to provide secure, clear and timely management of design and construction documents for BIM systems.

PAUSE POINT   Undertake some internet research to find a selection of companies that provide CAD and BIM software.

Hint   What different services or functions do these types of software offer? Why might you choose one rather than another?

Extend   Identify some projects that used the software you researched. What influenced the decision to use that particular software for each project?

CAD techniques

Common commands and their application to produce designs

Drawing with a computer requires a different attitude as well as a different set of skills from the traditional approach to drawing. To illustrate this, we will begin with a simple 3D CAD package such as AutoCAD 2014™. The basic user interface is shown in Figure 7.12 and highlights the main command input zones.

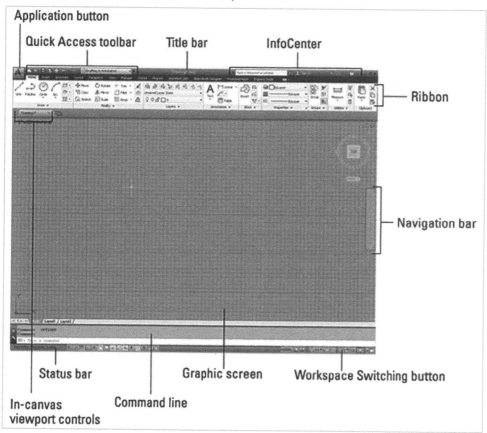

▶ **Figure 7.12:** A typical CAD user interface

Setting up the drawing and zoom

At the start of the drawing, the real size of the construction project is set up on the screen. This is known as the limits of the drawing. CAD drawings are always drawn in real size, that is, at a scale of 1:1. The limits of the drawing area or model are the x-, y- and z-co-ordinates and these are set to values which match the actual size of the project. The screen area could detail the elevation of an electrical plug socket or the complete plan view of a football stadium; it would depend on what limits were set.

How you move around the drawing is controlled by the limits of the drawing and the zoom controls. The zoom enables you both to magnify small parts of the drawing and to see an overview of the whole drawing.

Drawing and editing commands

One of the main differences between manual drafting and CAD is the way in which lines and shapes, known as entities, are created. The drawing area is gridded into x- and y-co-ordinates which define the exact positions of the drawn entities. This method is known as vector-based drawing; you need to specify an end point relative to its starting point. For example, to draw a horizontal line 3 metres long, the command would be:

```
LINE @ 3000,0
```

where the '@' indicates the chosen starting point of the line, the 3000 the x-co-ordinate and 0 the y-co-ordinate from that starting point.

This method of vector-based drawing can be a lot of work, so it is common when creating new entities to edit and modify existing ones, rather than to create from scratch the new entity. This is made easy as there are numerous different editing commands to use, such as Copy, Rotate, Stretch, Trim and Mirror. These drawing commands are assisted by customised snap features which allow you to automatically lock onto a variety of different points on the drawn lines, such as endpoints or mid points. You can also snap onto rectilinear grids to speed up the drawing process.

Small-scale projects such as a house or a building extension will often use simple 2D CAD software which mimics the processes involved in creating hand-drafted drawings such as plan, elevation, cross sections and site layout. This is shown in Figure 7.13.

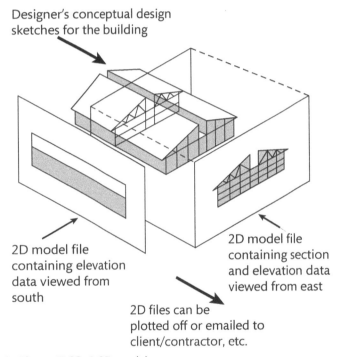

Designer's conceptual design sketches for the building

2D model file containing elevation data viewed from south

2D model file containing section and elevation data viewed from east

2D files can be plotted off or emailed to client/contractor, etc.

▶ **Figure 7.13:** A 2D model

When using this 2D CAD software to draw just lines, every change made on one view must also be changed on the other views. For example, if you change the position of the front door on the plan view then the position of the door on the elevation will also need changing to ensure consistency of information. For a small project this will not be too time-consuming and relatively easy to check. However, for larger projects, maintaining the consistency of these changes on all relevant views becomes more difficult. This can be overcome by using building information modelling techniques, as described below.

Using and applying layers in drawing production

Different constructional elements can be shown on different layers, which can be visualised as separate transparent overlay sheets containing grouped or linked objects. For example, one layer would contain all the electrical trunking and duct work and switches, which would overlay the floor plan drawn on another layer. Often hatching is shown on a separate layer so as not to slow the computer down. It also makes selection and editing of line work easier. These separate layers can have their own colour, line thicknesses and **line styles** pre-set, which can be turned off or on as well as plotted separately to aid drawing.

Selecting and applying appropriate drawing scale

Many popular CAD software packages use a model space or model view to construct the design, which is then selectively transferred to a flat paper space layout or paper view template from which it can be plotted. It is only when the drawing is plotted that the virtual drawing model is scaled down from 1:1 to fit onto the plotting paper space template. This allows different building scales to be shown on one drawing, e.g. a 1:50 floor plan on the same drawings as a 1:1250 location plan.

These views can include plans, elevations and cross sections. Cross sections can be constructed manually by transferring dimensions on the plan to vertical cross sections or by creating a 3D image of the development and selecting a cross section to be automatically generated by the software.

Using and applying line weights and their interpretation

Lines in CAD can be assigned a variety of **line weights** and styles depending on the line type such as 'chain-dotted' or the layer that the line is drawn on such as the 'setting out details' layer. These can be set at the start of the drawing or can be edited and amended at a later stage if required. Lines can also be plotted to the correct line weights or as draft lines according to the plotter settings.

> **Key terms**
>
> **Line style** – types of line in CAD; typically solid, dashed or chain-linked with same representation as lines drawn manually.
>
> **Line weight** – the thickness of a line drawn in CAD.

Plotting methods

Vector plotters are used to reproduce drawings up to A0 size, on paper quality of around 90–100 grams, from a drawing roll which is fed through a series of rollers. A variety of plotting presentations are available such as colour, black and white, or greyscale printing.

Producing a 3D virtual building model

As computing power has increased, so CAD has developed so that the project can be built up from virtual objects, such as walls, floors and windows, to form a full virtual model of the real project. Individual objects in the virtual model can be assigned attributes and properties according to their purpose and use. A typical BIM object, a steel beam, is shown in Figure 7.14, compared with a traditional 3D CAD drawing. The steel beam can be selected from a preloaded library and carries with it a set of defining data known as attributes. These could include physical information such as its material properties but could also include cost and relationships to other objects. Standard drafting software packages produce 3D CAD drawings which are just a collection of geometric lines and surfaces and so cannot be treated as 'virtual' building components. When, for example, a steel beam is erased in a plan view, it will have to be erased separately in another view. This can lead to errors. Object-based BIM does not have this problem. Figure 7.14 shows some of its data-rich advantages.

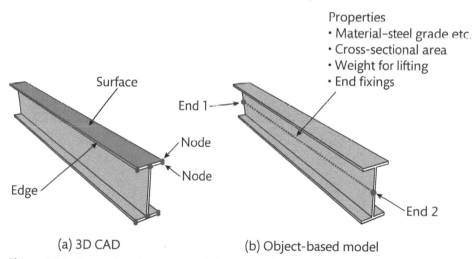

(a) 3D CAD　　　　(b) Object-based model

▶ **Figure 7.14:** Comparison between 3D CAD and an object-based element

There are a number of CAD object-based modelling systems on the market, such as Autodesk's™ Revit™ and Bentley's™ MicroStation™. They use a library of elements that can be selected from a structured database, filtered, edited and modelled. The data structure is illustrated in Figure 7.15 with an example of a column in a specific location, C ('Col C'). The category 'Column' defines the type of element. This is further defined as a generic steel universal column ('UC'); which is further defined by type ('254x254x42kg UC'); and finally allocated to the unique instance which in this case is Column position C ('Col C'). The structural engineer could specify an alternative instance such as an RC circular column with a diameter of 500 mm.

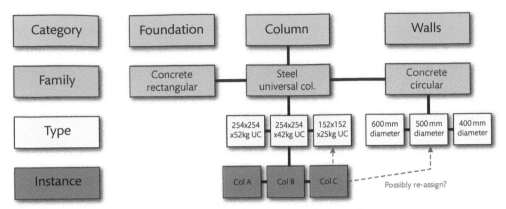

▶ **Figure 7.15:** Data structure for building information modelling software

When creating a 3D virtual building information model using objects it is important to remember that, just as with real construction, each element is supported by the previously constructed element. For example, a foundation base supports a column, which in turn supports other elements like beams and floors. This link is important because if a column is moved then the base needs to move as well. Elements such as walls incorporate doors and windows, which are said to be 'hosted' within the supporting element – so if a wall is deleted, the window is also deleted.

These key relationships of support and hosting are very important for intelligent CAD modelling. The concept is built into the current form of 3D virtual building information modelling, as shown in Figure 7.16. You'll see that the light switch is linked to the door. This is because, when you walk into a room, the light switch is sited within 500 mm of the door frame so you can see it to switch it on. If you move the door then the light switch will also need to be moved. The links between building objects enable quick and accurate sketching, modelling and modification to create complex building types and geometries.

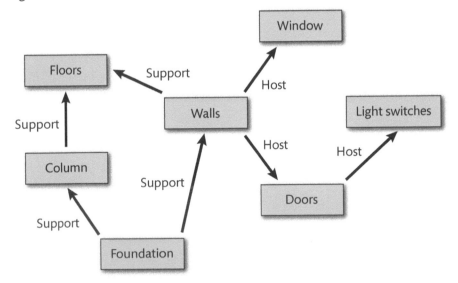

▶ **Figure 7.16:** Intelligent modelling relationships

Setting up floor and external levels

Levels are horizontal datum elements within the 3D model which relate to physical levels within the building, such as the bottom of foundations, the finished floor or roof levels. They enable wall modelling to be fixed to the appropriate datum levels, enabling the heights of the walls to be changed if required for the whole floor. Similarly, reference grids are used to create vertical reference planes that cut through the building, enabling the elements on each floor level, such as columns, to be co-ordinated correctly. Reference grids can also be used in conjunction with **temporary dimensions** which help you to roughly position the features and to refine them later.

Drawing composite elements

An example of a composite element is a basic external wall, which is a series of layered materials sandwiched together to create an assembly, i.e. a brick external skin, an insulated cavity and an internal structural skin with an internal finish material. In object-based CAD software, preloaded families of different types of elements can be modified and selected. These are then drawn at the appropriate floor level as a 3D object, just as a single line would be drawn on a 2D CAD drawing.

There are many advantages to using composite elements; for example, when walls meet, the program automatically trims and finishes them at their junction in 3D. Furthermore, information about properties of that wall, such as its thermal resistance, can be computed by the software, enabling in this case the heat loss to be determined. This is just one example of how composite elements in a drawing model represent the true properties of the actual element that will be built.

Inserting standard components

To speed up the drawing process, you can use standard components for common elements rather than creating your own each time. You can create your own standard blocks or use the pre-made entities that come with most CAD packages which can be loaded into the drawing as blocks to represent furniture, people or cars. Another option is to use the National Building Specification (NBS) series of blocks, which are at a scale of 1:1, while many product manufacturers supply CAD files of their products that can be imported and used within the virtual building model. Each component includes:

▶ information – content that defines the product
▶ geometry – representing the product's physical characteristics
▶ visualisation data – giving the object a recognisable appearance
▶ functional data – enabling the object to be hosted in the same manner as the product itself.

Producing rendered images and camera views

With the development of powerful surveying software and equipment, the 3D surfaces of both the built environment and the natural world can be accurately mapped and realistically rendered, for example by showing bright sunlight, shadows and wind-rippled foliage. In addition, some programs can map photographs of buildings and people into 3D virtual models. Figure 7.17 shows a typical rendered elevation for a major retail refurbishment project which uses this type of presentation.

There are times when textures, combined with photographic montages in 2D, are not enough for some clients, who would rather experience a more realistic physical rendering of the proposed building so that they can move around the spaces that are envisaged. Animated sequences, from the point of view of a person walking through the property (walk-through), are ideal. These are created by setting a path for the observer to walk through the model, with the view they would see shown on the screen. Further developments are exploiting gaming technologies where the viewer wears a headset so that they can explore the building by walking through it themselves.

Most software programs allow you to set up viewing points (represented as cameras) anywhere within the 3D model to capture what can be seen from that position by the observer within the model. There is often a facility to trace the path of an observer as they walk through the model. The output is normally a video from the observer's point of view as they pass through the virtual model. This can either be a walk-through as seen by a pedestrian or a fly-through as seen from an aerial observation.

> **Key term**
>
> **Temporary dimensions** – dimensions that appear on screen in a CAD program to help place elements accurately relative to a fixed reference grid or level datums.

▶ **Figure 7.17:** A 3D virtual model

Producing 2D CAD views

Object-based CAD software can be used to create a 3D virtual model and 2D CAD views can be extracted from this. It is done by drawing a sectional line through the virtual model which automatically generates both a view of all the elements that have been included and also an external view that can be seen from the position of the sectional line. These sections can include plan views, elevation views and cross sections, depending on the type and complexity of the software program employed.

A3 Comparison of manual and CAD methods of drawing

Manual drawing is a skill that takes time to learn but can produce neat, accurate and useful information to design and produce a high-quality building. The equipment costs for manual detailing are relatively slight compared to computer hardware and software requirements and the updating costs are also low. So why is there a need for investing in and developing CAD skills? The advantages of CAD when compared to manual methods are summarised below in Table 7.6.

Theory into practice

Take a look at the virtual tour through a major high-end London residential and commercial project currently being developed in East London by Ballymore. The film takes a tour around the virtual 3D model of the development and can be found at **www.londoncityisland.com**

- As you follow the walk-through, list all the different surface textures that have been created.
- Identify at least three other techniques used to create a very realistic presentation.

▶ **Table 7.6:** Advantages of using CAD/BIM systems for producing design and construction information over manual drawing

Advantage	Description
Accuracy and ease of making changes	**For the designer:** • Allows full co-ordination of various schemes and helps prevent conflict between designs. • Rapid changes can be made to graphical information by simply deleting or amending the current details, including reference grids, levels, objects, text and annotation. • If an element is repositioned in plan view, it automatically updates in all other views. In manual drawing all views must be updated manually. • Where part of the project is waiting for further information, it is ringed in a red cloud, drawing attention to its status. **For the contractor:** • Can accurately and rapidly generate an array of essential estimating information. • Allows for material take-offs to be produced automatically, which saves the quantity surveyor a great deal of time. • Permits amendments to materials quantities and costs, size and area estimates, costs and productivity projections as changes are made and/or variations are introduced.

▶ **Table 7.6:** *Continued ...*

Advantage	Description
Time and cost to produce drawings	**For the designer:** • Models and drawings can be duplicated and reproduced very quickly. New floor plans can be created quickly, based on lower levels. • Designs can be quickly developed and refined based around massing (shape, form and size) of the project and establishing elevations and floor plates. • Allows for remodelling of alternative layouts and options, and the integration of a selection of different construction techniques. It also assists with better design and space planning. • Provides the ability to identify collisions and clashes (e.g. identifying ductwork running into structural members) before construction starts on site. **For the contractor:** • Permits re-engineering exercises to be carried out within a virtual environment when looking at other construction options. • Allows a dry run of the proposed construction techniques to be reviewed to make the work cheaper and safer (such as finding the optimum position for the location of the tower crane on the construction site). • Leads to fewer on-site errors, and therefore leads to less remedial works, lowering costs. • At the end of the project it makes it easier to produce as-built drawings and operation manuals used by the facilities management team.
Training and support required	• Most large software companies offer training and support options for their product(s), including training manuals, face-to-face training, online courses and community based forums. • Training can be tailored to meet the organisation's or individual's needs and experience. • Free or inexpensive tutorials and training videos available online. • Learners can have free access to a variety of scaled down CAD/BIM software. This is a good way of learning how to use the system and most companies run these schemes.
Conversion from 3D to 2D	• It allows the automatic generation of 2D cross sections, plans and elevations based on the 3D model data that has been inputted. • It allows the automatic generation of lists of information to assist the procurement process such as window and door schedules and their ironmongery. • It provides the ability to use information in the 3D model to directly create fabrication drawings, avoiding having to manually transfer design information and drawings, which can lead to errors.
Production of rendered views	• It permits everyone involved in the design process to have a clear understanding of the concept design by visualising what is going to be built. • It helps to provide a useful and successful visualisation tool for all those involved in understanding the impact of the project, such as the local planning authority, local community, future customers/leaseholders and other stakeholders. • Providing this level of detail for each proposed scheme using manual techniques would be expensive and time-consuming.
Transfer of information	• Allows for managed co-operative digital collaboration between all parties involved in the design and construction of the building, including all the subcontractors and the supply chain. • It makes it easier to understand the dependence that various activities have upon each other. • It helps to develop more scenarios, such as looking at various sequencing options, site logistics, hoisting alternatives, etc. • Information about all aspects of the project can be stored and updated as the design and construction proceeds. • Information can be kept secure and safe, as it may be confidential and require restricted access or password protection. • The CAD management system can also create a log of all actions and the name of the persons who carried them out. The system can also detect when files are checked out and checked back. This level of security is not present in manually produced drawings and details. • Manual drawing requires a great deal of skill and checking time to ensure information from general arrangement drawings is successfully transferred to fabrication drawings. With BIM/CAD this can be done automatically.

Ⅱ PAUSE POINT   What are the advantages of using CAD to produce project information?

> **Hint**   Consider the needs of both the designer in creating the project and the contractor in building it.

> **Extend**   Using the internet, find a case study where CAD brought benefits to a development project. How were the benefits quantified?

Assessment practice 7.1

You are an architectural trainee who recently started working for a small company that specialises in residential housebuilding and domestic work. The company uses a mixture of manual sketching techniques and CAD software to produce its designs.

To explain your role to others in the company, your manager has asked you to do the following:

1 Produce a visually interesting pack describing the use of media and equipment to produce manual drawings for a proposed detached double garage.

2 Prepare a five-minute PowerPoint® presentation to describe the resources required to produce a CAD drawing for the garage.

3 Produce a table comparing the advantages and disadvantages of CAD methods with manual methods of producing drawings. Focus on the production of drawings in terms of their resource requirements, efficiency and cost.

4 Analyse and evaluate the information you have researched by writing a brief formal memo or email to your manager. What conclusion do you come to about the use of CAD and manual methods?

Plan
- Have I spent time planning my approach to the task?
- Am I aware of what type of company an architectural trainee would work in? What is the range of equipment that role would need?

Do
- Have I considered whether CAD and manual techniques are fit for purpose (i.e. suited to designing a double garage)?
- Can I make connections between what I am reading and researching and the task, and identify the important information I need?
- Am I covering CAD and manual methods with equal weight to get a balanced view?

Review
- Can I identify how this learning experience relates to future experiences in the workplace?
- Can I identify where I still have learning and knowledge gaps and do I know how to resolve them?

B Develop construction drawings for a given construction brief

B1 Construction drawings

BS 1192:2007

BS 1192 is the British Standard that sets out the methodology for managing the production, distribution and quality of construction information within a common data environment (CDE). It relates to the collaborative production of architectural, engineering and construction information and provides details of the standards and processes that should be adopted to enable efficient, structured, consistent and accurate exchange of data generated by CAD systems. This latest version of 2007 has been upgraded to a Code of Practice which recommends certain good practice.

The main element of BS 1192:2007 that you will need to know about refers to the use of the correct drawing symbols and conventions when preparing your sketches. These were covered earlier in this unit in section A1.

BS 1192:2007 also covers the management of drawings, such as the client and the design co-ordination team needing to establish standard methods and procedures for collaborative working that should be adopted by all the relevant parties involved in the project life cycle (i.e. the client, design consultants, main contractor, supply chain partners, etc.).

In most cases, digital collaboration is used to develop the design and share information. The building information management (BIM) protocol sets out, among other items, the following for the project:

▶ responsibilities and the lines of accountability
▶ level of detail and information required for the project and final built asset
▶ common digital environment/software platforms to be used.

The conventions for how construction drawings are prepared to show views, sections, and components have been covered in section A1. These conventions are published in several British Standards, as shown in Table 7.7.

▶ **Table 7.7:** Important British Standards for construction drawings

Reference	Title	Purpose
BS EN ISO 7519:1997	Technical drawings. Construction drawings. General principles of presentation for general arrangement and assembly drawings.	Establishes general principles of presentation applied to construction drawings for general arrangement and assembly, mainly for building and architectural drawings.
BS EN ISO 11091:1999	Construction drawings. Landscape drawing practice.	Establishes general rules and specifies graphical symbols and simplified representations for landscape drawing.
BS EN ISO 4157-1:1999	Construction drawings. Designation systems. Buildings and parts of buildings.	Specifies requirements for a designation system for buildings, including spaces, building elements and components.
BS EN ISO 128-20:2001	Technical drawings. General principles of presentation. Basic conventions for lines.	Establishes types of lines, their designations and configurations, and general rules for drafting of lines used in technical drawings, e.g. diagrams, plans or maps.
BS ISO 128-23:1999	Technical drawings. General principles of presentation. Lines on construction drawings.	Specifies types of lines and their application in the following drawing types: architectural, structural engineering, building service engineering, civil engineering, landscape and town planning.
BS ISO 128-30:2001	Technical drawings. General principles of presentation. Basic conventions for views.	Specifies general principles for presenting views, applicable to all technical drawings (including architectural, civil engineering, etc.).
BS EN ISO 3766:2003	Construction drawings. Simplified representation of concrete reinforcement.	Establishes a system for scheduling of reinforced bars covering dimensions and shapes.
BS ISO 129-1:2004	Technical drawings. Indication of dimensions and tolerances. General principles.	Applicable to all types of technical drawings.

⏸ PAUSE POINT

Do you know how the workflow of CAD information is managed during the process of producing detail designs?

Hint   Consider how information is shared, approved and published.

Extend   Undertake internet research into case study projects that have used the BS 1192:2207 to manage workflow information.

Types of construction drawing

Site plan requirements

A site plan drawing is a plan of the site showing the physical positions of existing site features. It is produced in the early stages of design, after the on-site survey has taken place and detailed measurements taken. Figure 7.18 shows a typical site survey drawing for a medium-sized building plot. It shows existing boundaries, trees and buildings. It also shows spot heights as crosses which indicate the slope of the land.

Link

For more details of site surveying methods and their representation please refer to Unit 10: Building Surveying in Construction.

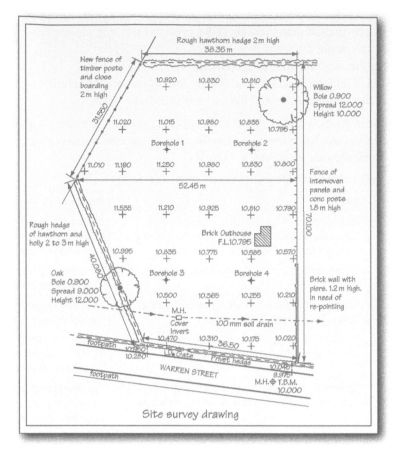

Site survey drawing

▶ **Figure 7.18:** A site survey drawing

Setting out plans

All setting out plans, whether 2D or 3D, should be created using a common project origin and orientation using a conventional Cartesian axis (x- and y-co-ordinates) and a common unit of length. The orientation should be related to a specific geospatial north or referenced to a standard named projection such as the UK Ordnance Survey grid, as shown in the example mocked up in Figure 7.19. A diagram of the project origin and orientation should be included with the project dictionary. The origin should be related to both the project grid and to the site context.

Vertical levels and height (z-co-ordinate directions) should also be indicated on the drawing to provide sufficient information to position the building.

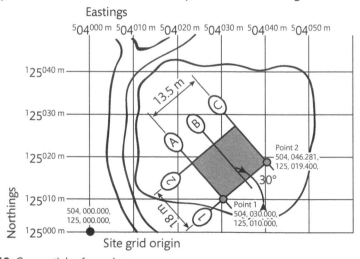

▶ **Figure 7.19:** Geospatial referencing

Two-storey building plans

Two-storey building plans are commonly used in small-scale residential or commercial buildings and are used for a variety of purposes depending on the stage of the development:

▶ sketch design to assist client in assessing their requirements
▶ town and country planning statutory approvals
▶ Building Regulations statutory approvals
▶ procurement of the contractor and supply chain companies
▶ construction and production uses to enable the main contractor and subcontractors to build
▶ as-built drawings for maintenance and archiving.

The main components of two-storey building plans include the following layouts, views or treatments:

▶ preliminary sketch drawings
▶ elevations
▶ cross sectional drawings – plan sections and vertical sections
▶ detail or component drawings.

The drawings are structured to work from the whole to the part; the general arrangement drawings (GAs) provide an overview of the project (its location, layout, overall size, and general form of construction). Elevations show the relative massing (shape, form and size) of the building and the relative positions of features and openings such as windows and doors. The GA will indicate the position of cross sections and other views to guide the contractor in finding information about the construction materials and dimensions needed. These enlarged cross sectional views through components (such as walls, floors and roofs) are shown as separate views on the same or linked drawings. They may also include written specifications or links to other documents like schedules or specifications.

For more complex structures, such as in framed buildings, separate building plans are produced which show only the structural supporting beams, columns and floors. This overall layout is often referred to as a 'general arrangement' which shows in symbolic form the key structural elements and sizes. Further structural fabrication drawings show in detail how each structural member needs to be formed in terms of materials and its dimensions and connection details. An example of an architectural general arrangement drawing is shown in Figure 7.20.

Preliminary sketches for design purposes

During the outline and scheme design stages, the building designer uses preliminary sketches to develop design ideas to help the client to understand the proposals and contribute to the design. These drawings are by their nature unrefined sketches, but provide sufficient visual and dimensional information for a client to understand the proposal. They enable both the designer and client to explore possible elevations and window/door positions, including the massing, symmetry and scale of the building.

Later in this unit, in section C2, we will be covering freehand sketches in more detail.

> **Theory into practice**
>
> Search your local planning portal for construction drawings of a new domestic building, such as a detached four-bedroom house. Imagine you are the designer and use the drawings as the basis for a short presentation to the client, a homeowner without any specialist construction knowledge. Consider how you would describe the features of the building, including materials and finishes, so that the client understands the plans.

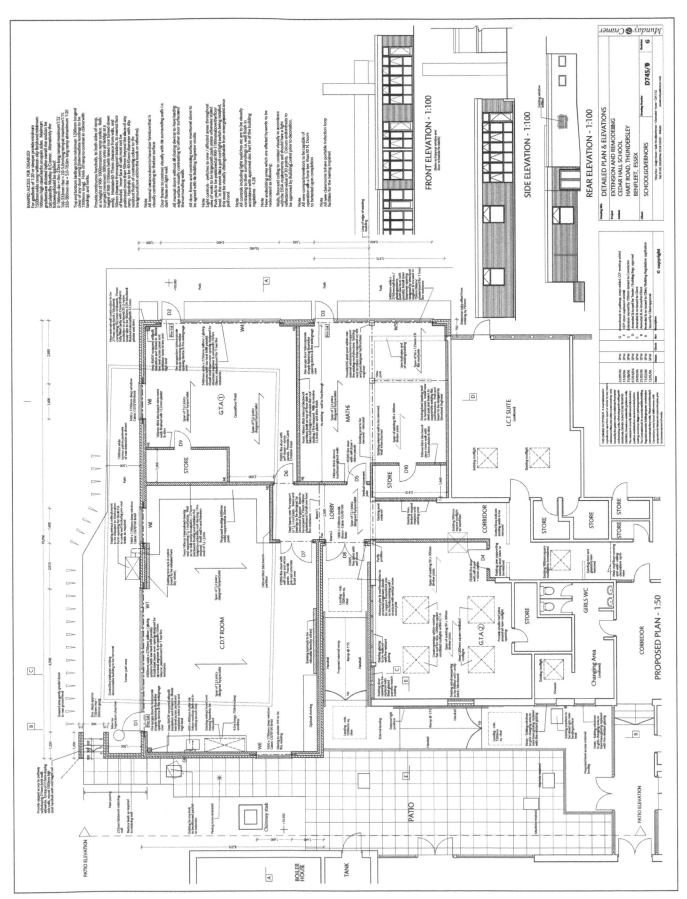

Figure 7.20: A complex GA drawing

Elevations, cross sections and component or detail drawings

Planning drawings need to be submitted for approval by the appropriate planning authorities. They need to include an Ordnance Survey tracing of the area, a site plan and proposed elevations. The information should also include the setting out dimensions of the proposed building, surface finishes, positions of external doors/windows and any permanent vehicular access that is required. Figure 7.21 is an example of proposed plans and elevations for a two-storey side extension to accommodate a dining room and toilet on the ground floor, and a master bedroom with en-suite on the first floor. It shows this by using a general arrangement plan of the proposed ground and first floor together with the proposed elevations of the two-storey extension.

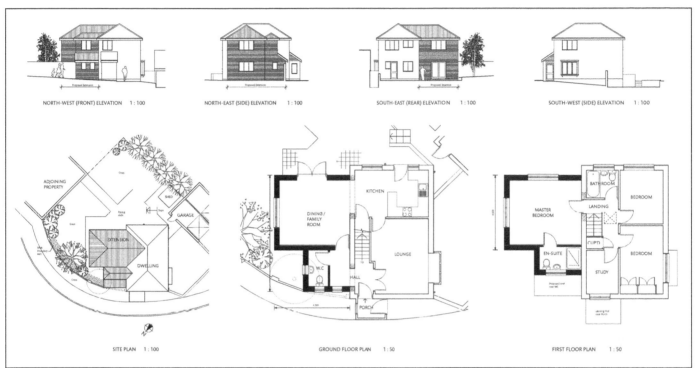

▶ **Figure 7.21:** Example of a general arrangement plan

Detail drawings for Building Regulations statutory approvals

Building Regulations involve specifying technical matters to comply with building regulations that require far more detailed information than the planners required. For example, sufficient information needs to be provided to show the provision for thermal or sound insulation, the dimensional layout required for disabled access or the minimum headroom requirements for staircases. This information is scrutinised for compliance and therefore contains a great deal of specific technical information and references to British Standards are required. It provides far more detailed technical information in the form of enlarged cross sections through some of the components.

Detail drawings for procurement and the construction processes

Once the final design information is resolved and agreed between the client, design team and relevant authorities a contractor can be sought and construction can commence. Drawings and schedules will ultimately form the working documents and contain the detailed information of how the building will be constructed in terms of the quality of materials, the type of fixings and the workmanship requirements.

Structural drawings showing general arrangements

For civil engineering projects, various GA and detail drawings identify the structural elements of the building, for example, the construction layout and the position of steel reinforcement. Figure 7.22 shows the general setting out details that the formwork

---

Theory into practice

Find a set of drawings for a building, which illustrate the use of plans, cross sections and elevations to provide dimensional and materials specifications for the construction.

For each drawing:

1 Describe the key features of information being shown that enable the buildings to be built.

2 Identify the process in which each of these drawings would be used.

contractor would use to construct the overall shape of a raised walkway platform, while Figure 7.23 identifies the exact type of steel reinforcement and where it should be positioned by the steel fixer. Note that numbers shown refer to a coded bar reference which would be used in the steel bar bending schedule.

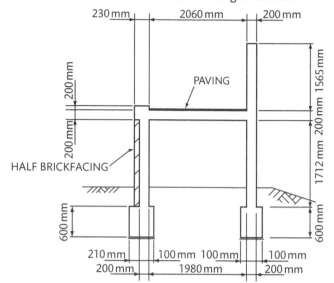

General arrangement section through raised walkway
Scale 1:50

▶ **Figure 7.22:** General arrangement drawing of a walkway

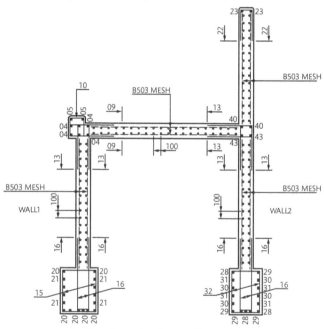

Section through raised walkway showing
position of steel reinforcing bars
Scale 1:25

▶ **Figure 7.23:** Detailed section of a walkway

Another example of a component schedule is the bar-bending schedules which specify the exact length and shape of individual steel reinforcing bars for reinforced in-situ concrete construction. In Figure 7.24 note that there are only seven different types of bar in each beam and that each has its exact diameter and steel grade noted. For example, bar mark 03 is 'H25' which means high yield steel (H) with a diameter of 25 mm. Its length is 2400 mm and it is a straight bar as '00' is its code. A 'U' shaped bar has a 21 code and an overlapping link is a shape code 51.

Member	Bar Mark	Type & Size	No. of Members	No. of Bars in Each	Total No.	Length of Each Bar	Shape Code	A	B	C	D	E/R
Ground Beams	01	H20	6	2	12	7600	00					
On Gridlines	02	H20	6	2	12	5400	00					
A to F	03	H25	6	2	12	2400	00					
	04	H25	6	3	18	5600	00					
	05	H20	6	3	18	3125	15	1450				
	06	H20	6	2	12	2875	21	1300	425			
	07	H12	6	37	222	1550	51	450	250			

Company Name: Bestend Consultants plc
Job Reference: 1298/10
Job Title: Express Foodmarket, Kings Lynn.
Bar Schedule Reference: 8 / 1298
Date Prepared: 12/03/10
Prepared By: mjh
Date revised: -
Checked By: jfs

▶ **Figure 7.24:** A reinforced concrete bar-bending schedule for a ground beam

Similarly, steel-framed buildings constructed from individual universal beam and column steel sections also need to be represented via symbols and codes.

Assessment practice 7.2   B.P4 B.M2 B.D2

In your role as an architectural trainee, your manager has asked you to produce a set of drawings for a development.

The proposed development is a small site close to the town centre's popular high street and has a road frontage of 18 m and a depth of plot of 20 m. The building line is 2 m from the rear of the front kerb. The aim of the development is to create three twin bedroom town houses with the following features:

- ground floor – integral garage, WC, hallway and access stairway to first floor
- first floor – kitchen/diner, living room with small balcony to rear
- second floor – two bedrooms, bathroom and airing cupboard
- a small rear garden.

You will need to consider the internal layout and assume approximate dimensions in order to produce the necessary drawings.

For this project you have been asked to produce the following drawings using a mixture of CAD and manual methods for each drawing to BS 1192:2007:

- site plan: to orientate site and show setting out dimensions, access to rear garden and external works/landscaping
- general arrangement plan at first or second floor for one town house: to show layout of walls, doors, windows and fittings
- elevations of town house development building: to identify the materials and massing of the building
- foundation plan: to show the structural general arrangements and setting out details
- typical cross section drawing: to show an overview of one of the house's structure and construction details
- wall details cross section: to show wall, window and lintels.

Your work should include the use of both CAD and traditional drafting techniques and contain correct detailed technical information.

Plan
- Have I researched the construction technology normally used for this type of residential building?
- Am I aware of the level of detail required to be drawn?
- Have I considered which types of drawings would be best suited to CAD or manual techniques?

Do
- Have I laid out the manual drawings clearly?
- Have I ensured I am following BS 1192:2007 standards?
- Am I utilising the most appropriate CAD software available to me?
- Have I set suitable milestones and evaluated my progress and success at these intervals?
- Am I clear about how to achieve a Merit or Distinction grade on this task?

Review
- Can I draw links between this learning and my prior learning?
- Can I explain what skills I employed and which new ones I have developed through this activity?

Undertake production of two-dimensional and three-dimensional freehand construction sketches

C1 Principles, techniques and conventions

Despite the increasing use of CAD, the ability to produce proportional freehand sketches quickly and simply remains valuable to all who design and help construct buildings, as it helps to develop spatial awareness and to understand and communicate ideas. A rapid sketch or a hand-rendered illustration can convey more information than many words, and will be quicker to produce than even the most sophisticated CAD software.

Concept of proportionality

Most sketches are made up of straight or curved lines formed into recognisable regular shapes such as rectangles, squares, circles and ellipses. When sketching, it is important to establish a framework of these main shapes and boundary features, and the position of the main axes of symmetry. Proportionality is the process of comparing the relative sizes of the basic shapes, by thinking about how big is one compared to the other: twice as big? Half as big? What about the height of the building in relation to the width of its base? You can apply these comparison judgements to gauge the rough proportionality of the various shapes to help you to reproduce them on paper as a sketch.

Theory into practice

Sketching practice

To help you develop some of the basic line work skills, take a blank sheet of A4 paper and an HB pencil and practise the set of exercises below. Practise each one until your lines are straight and parallel with the sides of the paper. Do not use a rule but try to gauge the distances and proportions manually:

1 Draw a series of parallel lines about 70 mm long, approximately 25 mm apart.

2 Repeat activity 1 but, this time, extend the length of the lines. Try to maintain the 25 mm gap between the lines.

3 Draw another set of lines and this time draw lines that cross at right angles.

4 Repeat activity 3 but this time make the lines cross at 45 degrees.

5 Draw two horizontal lines 10 mm apart to be used as guide lines. Draw capital letters H, K, L, M, N so they fit perfectly within the guidelines.

6 Repeat activity 5, but this time with letters with curves, e.g. C, O, S, etc.

Repeat these activities regularly to improve your line skills.

 PAUSE POINT   What methods are available for manual drawing techniques?

Hint   Search for videos on the internet and look at the application of manual drawing techniques.

Extend   Using manual drawing techniques copy an existing image of a building.

Oblique projection

This is another method of sketching objects in 2D to represent 3D objects. The basic method of producing an oblique sketch is to draw the front elevation of the object or building as flat orthographic view (see section A1). The object is then given a 3D depth, by drawing 'receding' lines, which are at right angles to the flat front plane, obliquely at either 30°, 45° or 60°. Each line is drawn at half its true length. Note that all the receding lines are parallel. This technique mimics what the eye sees when looking at the side of an object by foreshortening the distance.

The basic comparison between oblique and isometric projection is shown in Figure 7.25 for a cube with side length 'A', where all the receding lines are drawn to half the scale, i.e. A/2.

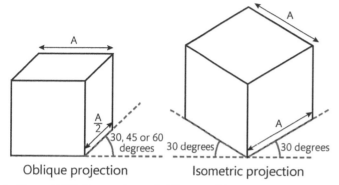

Oblique projection   Isometric projection

▶ **Figure 7.25:** Comparison of oblique and isometric projections

Oblique projections are easy to sketch freehand because the front and rear elevations are drawn flat using the simple techniques that were practised earlier. On the elevations, circles and radii keep their standard geometry, which it makes it easier to sketch.

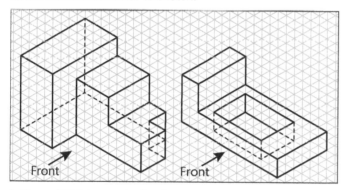

▶ **Figure 7.26:** Isometric sketches

When you are required to show a curved line or a circle on a receding face of an oblique sketch, you need to approximate it by eye or, where it is particularly complicated, produce an auxiliary elevation of the curve or circle as shown in Figure 7.27. Note that the auxiliary view is drawn to the correct size and scale then the points a to h are set out around the circumference by splitting the diameter into equal parts. The dimensions (A and B) are then measured with a ruler and transferred to the oblique view with all the A dimensions halved to give A'. All points a to h are established in this way and joined up to give the outline of the oblique circle.

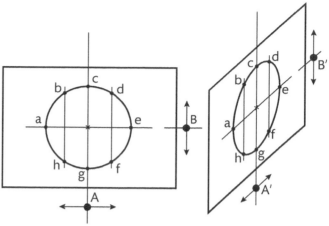

Auxiliary view of a circle   Oblique view of a circle

▶ **Figure 7.27:** Using auxiliary views to draw an oblique circle

Draw what you see – perspective drawings, horizon lines and vanishing points

Perspective is the geometry of realistically depicting 3D objects in a 2D plane. It is sometimes known as natural perspective as it most closely conveys how the human eye perceives 3D objects in real life. Their construction is generally based on the following rules:

▶ The horizon appears as a horizontal flat line.
▶ Straight lines in space appear as straight lines in the image.
▶ Lines parallel to the **picture plane (PP)** appear parallel and therefore have no **vanishing point (VP)**.
▶ Sets of parallel receding lines meet at unique vanishing points.

There are three common types of perspective projections defined by the number of vanishing points as shown in Figure 7.28, depending on the complexity of the drawing required:

▶ One-point: this is typically used for street scenes and internal room layouts.
▶ Two-point: this is typically used for external views of detached buildings.
▶ Three-point: this is typically used for external views of tall buildings where height is exaggerated.

> **Key terms**
>
> **Picture plane (PP)** – [in construction drawings] an imaginary vertical plane where the drawing is located.
>
> **Vanishing point (VP)** – [in construction drawings] a point on the horizon eye line where all parallel receding lines converge.

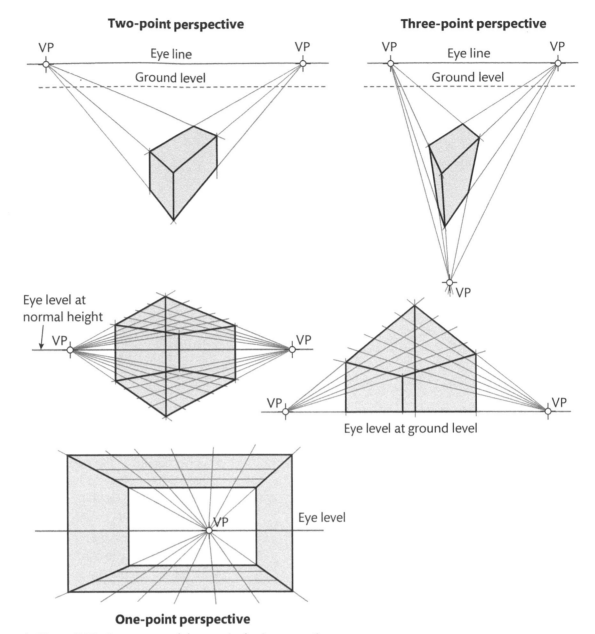

Figure 7.28: One-, two- and three-point basic perspectives

The position of the eye level relative to the ground level governs whether you look down onto the building or up at the building. The location of the vanishing points must be clearly shown as these govern the perspective view that is generated. One-point perspectives are useful for drawing interior views for rooms and enclosed spaces and the same rules apply in terms of changing the viewing position.

Step-by-step: Scaled perspective drawings

In order to understand how scaled perspectives work and how they are constructed we need to start from basic principles. Several techniques produce scaled perspective drawings but all involve establishing on a sheet of drawing paper a series of axes and planes which represent the 3D perspective constraints of that drawing.

In the following step-by-step example we will draw a scaled two-point perspective of a rectangular flat roof building. We start by drawing a series of horizontal lines on the paper to represent the **eye line (EL)**, the **ground line (GL)** and the picture plane (PP) as shown in Figure 7.29. The relative position of each of these will govern what

Key term

Eye line (EL) – [in construction drawings] the height of the observer's eye.

Ground line (GL) – [in construction drawings] a line on the horizon representing the flat ground plane.

the final 3D image will look like. The distance between the EL and GL is drawn to the scale that the drawing is to be drawn to, say, 1:50:

Step 1 – Visualising the PP

The PP is the vertical drawing sheet and can be visualised from studying Figure 7.29. This horizontal line is drawn first.

Step 2 – Drawing the first lines

The building to be drawn is a rectangular block which is first drawn in the plan to the desired scale.

Step 3 – The observer position

The **observer position (S)** is selected in front of the building with two parallel lines at 45 degrees projected to cut the PP. These two intersections represent vanishing points 1 and 2 (VP1 and VP2).

Step 4 – Cutting the EL

These VP positions are transferred to the PP by drawing them vertically to cut the EL. These will now be used to draw the scaled perspective view of the rectangular building.

Step 5 – Height line

A further line is projected at 45 degrees from the side of the building until it hits the PP, and then is projected vertically. Where it cuts the GL it becomes the position at which all scaled vertical heights are to be measured to the desired scale to give the heights of window cills etc. This is called the **height line (HL)**.

Step 6 – Drawing direct sight lines

From the observer at S, direct sight lines are drawn through the corners of the building plan to meet the PP. As before these are drawn vertically.

Step 7 – Zero height line

To locate the bottom of the left-hand side of the building in perspective, a line is drawn through VP1 on the EL and where the HL cuts the GL. This is effectively the zero height line of the building. A line is drawn through VP1 and the zero HL until it meets the vertical line drawn in step 6.

Step 8 – Locating the top of the left-hand side

To locate the top of the left-hand side of the building, the height of the rectangular building is scaled up on the HL from the GL upwards and marked. Then a line from the VP1 is drawn through this mark until it meets the upright drawn in step 6.

Step 9 – Joining the points together

Joining all four points together gives the scaled perspective view of the left-hand side of the building.

Step 10 – Tackling the right-hand side

Repeat steps 5 to 9 for the right-hand side of the building

and the roof to give the required scaled perspective drawing as shown.

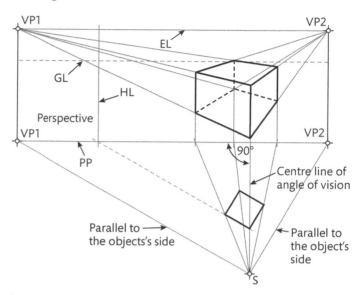

▶ **Figure 7.29:** Producing a scaled two-point perspective drawing

---

Key terms

Observer position (S) – [in construction drawings] the point at which the observer looks at the object. The height of the observer is the distance between the eye level and the ground line.

Height line (HL) – [in construction drawings] a vertical line running parallel to the picture plane that indicates the height of construction elements.

Theory into practice

You work for a construction company called Innovative Structures. One of your colleagues has produced drawings of two buildings (see Figure 7.26) in isometric view on isometric paper. You have been asked to redraw them as freehand 45-degree oblique building sketches.

Redraw the buildings, assuming the size of each isometric graph paper block is a 10 mm cube, to help gauge the proportionality of the building's features. Ensure your oblique sketch takes the front view as indicated.

You have also been asked to draw the buildings in perspective view. Draw one building as a one-point perspective and the other as a two-point perspective.

1   What would you need to consider before you start drawing?
2   What are the main differences between oblique and perspective sketching?
3   How do you maintain the correct proportions of the building to keep its parts properly scaled?

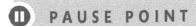

C2 Freehand sketches

The following are some techniques and examples of freehand simple perspective sketching.

Interior of a building

These sketches use perspective linework by establishing the vanishing points and the eye level horizon. Figure 7.30 is an example of a one-point perspective interior showing the foyer and circulation area of a proposed university building. These types of sketches can show the location of doors, windows, features and fittings.

▶ **Figure 7.30:** Internal one-point perspective of the Free University in Berlin. (Source: **www.fosterandpartners.com/projects/free-university**)

Exterior of a building

A simple house can be drawn as follows. Start by drawing the closest vertical corner to you. The height of that vertical line can be drawn to whatever scale you wish but you need to bear in mind the proportionality of the final sketch (see below). Using this as a guide, and your ability to judge heights and widths by relative proportion, the sides of the house can be sketched in. The centre point of the side elevations can be found by sketching some diagonal lines across the sides from top right to bottom left. These can be used to estimate the height of the gable wall and similar features as required. An example of a simple two-point perspective house is shown in Figure 7.31 and 7.32.

▶ **Figure 7.31:** Drawing a two-point perspective sketch in pencil

▶ **Figure 7.32:** Simple two-point perspective sketch by proportion

⏸ **PAUSE POINT**   Produce a perspective drawing of a detached double garage with a gable pitched roof. Assume all the necessary dimensions for the garage.

Hint   Research appropriate dimensions for the garage and use a two-point perspective view.

Extend   Add a landscape of trees, hedges and fences for a more realistic setting, using the original vanishing points as a guide.

Marking requirements to indicate vanishing points on sketches

High-rise buildings are often drawn using three-point perspectives. These give an exaggerated feel for the height of a building and are constructed in a similar fashion to a two-point perspective. However, all the parallel vertical lines will converge on the third vanishing point located centrally above (or below) the horizon eye line with its two vanishing points. Where the third vanishing point is below the horizon eye line then the tower will appear as if seen from a bird's eye view; if it is above then a high worm's eye view is obtained as shown in the example given in Figure 7.33. The scale of the building in terms of its height, breadth and width has to be judged using proportionality as before. The starting point for getting the proportionality correct is to draw the height of the front corner closest to the spectator to the desired scaled height and use this to proportion all the other main features such as floor levels and corner columns.

▶ **Figure 7.33:** Simple three-point perspective sketch of a tower by proportion

PAUSE POINT Produce a three-point proportional freehand sketch of this building, but this time draw it as seen looking down as in a bird's eye view.

Hint Set up the sketch with the third VP below ground level.

Extend Add shading and rendering to make the image more realistic.

Using and applying annotations appropriately

It is important to appropriately apply annotations to communicate details about materials, finishes, condition or any other relevant information needed for the purpose of the sketch. For example, annotations can help a client decide on which type of finishes or materials they wish to choose. There should be enough clear and unambiguous information to be useful for its purpose, but not too busy. Individual details on the drawing should be clearly referenced and labelled, such as section marks and individual titles, and there should be a clear system of cross referencing between drawings and other relevant documents.

In all cases annotation needs to be simple, clear and free from ambiguity. For an example of a complex diagram with clear annotations, see Figure 7.20 on page 328.

Research

Search on the internet to find examples of approaches to annotated sketches, and how they can be used.

PAUSE POINT Collect a range of images of architectural or landscape design sketches. Describe what each sketch is attempting to show.

Hint Use image-based websites, such as Pinterest and Instagram, to source and collect images.

Extend Compare and contrast the different styles and techniques used.

C3 Skills, knowledge and behaviours

Evaluating outcomes

It is important to evaluate your outcomes, enabling high-quality, justified decisions to be made. The best way to do this is to set measurable goals at the beginning, for both what you produce and client satisfaction. You could also make a list of what went well and what went less well, and think about why this was. What could you do differently next time?

All of those involved in communicating information for design or construction need to use appropriate written and verbal language to suit the outcomes that are required. We need to evaluate our forms of communication to comply with professional etiquette, accountability and individual responsibility. It must be clear, direct, unambiguous, suit the target audience and be fit for purpose.

Media and communication skills

When communicating any message, you need to consider several key factors, no matter what format your presentation is in:

▸ The purpose – what do you need to convey?
▸ Who your intended audience is – should you use a formal or informal tone?
▸ Its structure.

You may need to explain the technical aspects of construction to a client who does not have your knowledge and expertise. This should be done in a way that the client will understand. Using sketches, photographs and diagrams is useful.

Conveying your intended meaning

Your reports and drawings need to be understood by clients who are uninformed about construction. You should develop your ability to convey your intended meaning unambiguously in different formats, including in written form (such as design documentation, recording documentation, reports, visual aids for presentation use) and verbally (for example one-to-one and group, informal and formal situations).

When speaking to colleagues we will often use a different vocabulary than when we communicate with other people. Therefore, as well as graphical information, in the form of models, drawings and sketches, you should:

▶ understand your audience's expectations (for example, the outcome they expect, their level of understanding)

▶ use a suitable tone of language (body language, verbal language)

▶ convey your intended meaning (through choice of terms, reinforcement of message, checking and feedback).

Verbal communication

You will be required to speak with people in a range of situations. This may be one-to-one or in groups situations and in formal and informal gatherings. In each situation, consider who your audience is and what they or you need to know; make sure you speak clearly and choose words carefully so you are not misunderstood, and be prepared to answer questions. The important elements for a good presentation include:

▶ planning the content to suit the message and audience

▶ deciding on the resources to get the message across

▶ practising your delivery in terms of demeanour, tone and body language

▶ anticipating possible questions

▶ checking that the desired aims have been achieved.

Tone and language

When you are communicating verbally or in a written form, you need to ensure you are using the correct tone, and appropriate terminology.

Informal style is colloquial, it is the type of speech used between friends and what you would overhear in public; whereas formal language style uses the correct technical terminology, e.g. 'timber' versus the informal term 'wood'. In a professional context, try to avoid the use of vague or over-generalised terms such as 'sort of' or 'kind of' when formal language dictates quantifiable amounts and specific references.

Keep your communication positive and engaging to make sure you have a constructive impact on your audience and that they understand what you are saying, understand its implications and (where appropriate) are enthusiastic!

Assessment practice 7.3

Your manager wants you to develop your freehand sketching skills. He asks you to produce a set of hand-drawn sketches based on the town house development you completed earlier to explore alternatives to the external and internal finishes and layout. You will need to produce:

- a variety of 2D sketches showing:
  - at least three different external finish treatments to the front elevation of the development using a combination of different materials
  - at least two different first-floor layouts including furniture
  - a landscape garden plan
- a variety of 3D sketches to show:
  - the massing of the town house development using oblique, two-point perspective and three-point perspective sketches
  - at least two different 3D one-point perspective sketches of your chosen first-floor layout.

Your sketches should be good quality and provide details of materials, finishes, the condition and other relevant information including location of doors, windows, features, fittings and spatial layout.

Plan

- Am I clear on which views to select to best show my 2D and 3D sketching skills?
- Have I researched sufficient dimensional and constructional information to enable me to confidently lay out my sketches?
- Am I aware of the level of detail required?

Do

- Am I demonstrating my understanding by showing sufficient construction lines and vanishing point positions?
- Can I make informed creative choices based on what I have done already?

Review

- Can I explain what I have learned and why it is important?
- Can I explain how I would approach the difficult elements next time (i.e. what I would do differently)?

> ### Further reading and resources
>
> Bello, S.R. (2015) *Technical Drawing* (Amazon Kindle edition), Nigeria: Dominion Publishing Services.
>
> CADFolks (2016) *AutoCAD 2017 for Beginners*, CreateSpace Independent Publishing Platform.
>
> Circle Line Art School: **www.circlelineartschool.com**.
>
> Elys, J. (2013) *CAD Fundamentals for Architecture*, London: Laurence King.
>
> Reekie, F. and McCarthy, T. (1995) *Reekie's Architectural Drawing*, fourth revised edition, London: Routledge.
>
> The British Standards Institution (2015) BS1192:2007 *Collaborative Production of Architectural, Engineering and Construction Information*, London: BSI Standards Limited.

THINK ▶FUTURE

Rupert Elliott

Rupert works for an architectural design company in East London. He is 23 years old and has been at the company for a while. He started work straight from school as the office junior, helping with the administration and simple surveying and drawing work. Rupert has always been interested in graphics and design so, after starting his job, he went to college to study for a part-time National Certificate in Construction. After passing with a double Merit grade profile, Rupert went on to study a part-time BSc in Architectural Technology, which he is about to finish.

Rupert mostly uses specialist CAD software, although on initial design scheme phases he uses manual sketches and drawings. His skill set includes:

- freehand drawing abilities

- being able to visualise objects in three dimensions

- strong mathematical, computer, organisational and management skills

- an interest in the science and technology involved in building.

He works as part of a small team and now takes responsibility for many of the small works projects, including surveying and detailing work. He is also responsible for producing manuals and CAD details for the company's larger projects. The tasks he most commonly does are:

- preparing and presenting design proposals using computer-aided design (CAD) and manual methods

- contributing to the design process and co-ordinating detailed design information

- liaising with, and producing documentation for, statutory and local approval authorities.

Rupert enjoys his job and finds it varied. Sometimes things are quite high pressured, especially when there are client deadlines to meet, but he finds the job rewarding.

Focusing your skills

If you are looking for a similar role it is important to develop both your freehand sketching skills by compiling a portfolio of sketches, and practising your CAD skills, particularly using programs that use object-based modelling. It is useful to find and secure a work placement within a design or construction office during the holidays, as this will help you to get used to the working world. You should immerse yourself in construction technology and materials by reading and watching videos on current methods and techniques, and try to arrange a construction site visit to see construction operations in progress.

Getting ready for assessment

Jodika is in the second year of her BTEC Extended Diploma in Construction and the Built Environment. For this unit she has been given an assignment to prepare a portfolio of practical construction drawings for a two-storey building. Details must be drawn to an appropriate scale and contain sufficient technical information for the project to be specified and built. She has been asked to produce some drawings using manual methods and others using CAD and then to compare and contrast the different techniques.

How I got started

I started the assignment by examining the brief provided as well as downloading a copy of the unit specification from the Pearson website. With these I was able to identify the key things I needed to achieve the higher grades; the 'Further information for teachers and assessors' section helped me understand that for a Distinction grade I needed to provide a balanced evaluation of the methods used, and to take into account resource requirements as well as efficiency and cost.

Before starting work I made sure I reviewed my tutor's feedback from my earlier drawing exercises so I could see what improvements I needed to make. I also collected together a range of relevant British Standards and product information relating to drawing content and presentation methods, to ensure I was starting with a firm frame of reference.

How I brought it all together

I had been practising my CAD work for some time so felt fairly confident in producing floor plans using the software. It also meant that if I produced the ground floor plan I could use the same layout and edit and revise it for the first floor plan, which is just one of the advantages of using CAD that I could include in my evaluation. This meant the cross-sectional details through external walls and windows would have to be done using manual techniques. When I did this it was important to demonstrate my skills in using the correct cross-hatching patterns to show the different materials used, such as brickwork and concrete.

My summary evaluation was the main way to demonstrate my in-depth knowledge and understanding. I made sure I hit all the key criteria as set out in the assessment criteria. My uncle is a builder and I asked him to look at my

drawings once I had completed them to see if they looked acceptable. He was very impressed but made a few suggestions to improve the layout! I also proofread my written commentary to make sure I hadn't made any silly mistakes.

What I learned from the experience

I am really pleased with the final set of drawings I produced. The most difficult part was finding the right product specifications – I wish I had started researching that earlier as I had to rush the final annotation and labelling. I feel I have developed some good practical skills and techniques in the production of both CAD and manual drawings and am confident I have made the right career choice to become an architectural technologist or possibly an architect.

Think about it

▸ As a learner you can get access to a range of industrial CAD software packages for free – have you asked your tutor for information about this?

▸ Have you got yourself a set of good drawing pens and set squares?

▸ Remember that with all drawing work, the more practice you do, the better and quicker you will become!

Building Regulations and Control in Construction

8

Getting to know your unit

Today's Building Regulations have evolved over hundreds of years to improve health, safety and welfare in the buildings we use.

This unit will cover the contents of the Building Regulations and their accompanying Approved Documents. These detailed documents describe what needs to be provided in our buildings in terms of detailing and conformity.

You will learn about and examine the requirements of the Building Regulations and their related application process, including the type of information and documents needed to ensure compliance.

How you will be assessed

This unit will be assessed by a series of internally assessed tasks set by your tutor. These will cover the main theoretical knowledge of the key legislative requirements in a range of construction roles, along with their practical application. When you work through this unit you will find a series of activities that help you practise and build up your knowledge of the Building Regulations. The process is all about preparing you for the internal assessment that will be issued to you. To achieve the tasks in your assignment, it is important to check that you have met all the Pass grading criteria. You can do this as you work your way through the assignment.

If you are hoping to gain a Merit or Distinction, you should also make sure that you present the information in your assignment in the style that is required by the relevant assessment criteria. Merits and Distinctions extend and stretch the Pass criteria understanding. For example, Merit criteria require you to analyse and discuss, and Distinction criteria require you to assess and evaluate.

The assignment set by your tutor will consist of tasks designed to meet the criteria in the table. This is likely to consist of a written assignment but may also include practical activities such as the following:

▸ performing a building survey on a property

▸ undertaking a measured survey

▸ production of drawings from surveys.

Assessment criteria

This table shows what you must do to achieve a **Pass**, **Merit** or **Distinction** grade, and where you can find activities to help you.

Pass	Merit	Distinction
Learning aim **A** Understand the requirements of the Building Regulations		
A.P1 Describe the requirements of the Building Regulations.	**A.M1** Discuss the different methods of application and control for a Building Regulations application.	**A.D1** Evaluate the different methods of control for a Building Regulations application on a variety of different project types.
A.P2 Explain the different methods of control and implementation for Building Regulations applications.		
Learning aim **B** Examine the requirements of the Building Regulations		
B.P3 Describe the different methods of achieving compliance with the Building Regulations.	**B.M2** Analyse the different methods of demonstrating compliance with the Building Regulations.	**B.D2** Evaluate the different methods of demonstrating compliance with Building Regulations on a variety of different project types.
B.P4 Outline the requirements of the Approved Documents.		
Learning aim **C** Undertake a Building Regulations application		
C.P5 Describe the method and process for a Building Notice application from application to completion.	**C.M3** Produce a detailed Full Plans Building Regulations application for a new build residential project.	**C.D3** Analyse the documentation produced for a Full Plans application in relation to the requirements of the Building Regulations.
C.P6 Describe the method and process for a Full Plans application from application to completion.		
C.P7 Complete the application forms and produce outline plans for a Full Plans Building Notice application for a new build residential project.		

Getting started

There are many different areas of the construction industry that require regulation. As a group discuss these and see how any different areas you can suggest may need to be regulated. What would the aims be of regulating areas of construction?

Understand the requirements of the Building Regulations

A1 The Building Regulations

The Building Regulations developed from the Building Act 1984. This is the primary piece of legislation that formed the regulations and their Approved Documents. To understand the Building Regulations, we must first examine the Act of Parliament that created them.

The Building Act 1984 consolidated some of the previous building legislation and gave the power to make the Building Regulations. The Building Act 1984 covered some of the following provisions:

▶ power to make the Building Regulations
▶ the production of Approved Documents
▶ to allow the passing of submitted plans
▶ to give powers to inspectors
▶ penalties for breaches to the regulations.

The Building Act 1984

We will now look more closely at some of the provisions that the Act created to control construction in terms of ensuring that the completed projects are safe for the occupants. The Building Act is the primary legislation that the government put before Parliament – the Building Regulations are a progression from them into more specific and detailed information that is not contained within the Act. In effect, the Building Act gave power to produce regulations and any subsequent revisions.

Definitions and interpretation

The Building Act defined building work as:

▶ any temporary or permanent building or structure or part
▶ the construction or erection, or reconstruction of a building, roofing over between walls
▶ the conversion of a movable object into a building
▶ the re-erection of a building that has been pulled down, fire damaged or fallen down to leave 3 metres standing.

Material alterations

The Building Act does not cover this aspect, which is included within section 6 of the Building Regulations 2010. When a building has a material change of use (for example from a retail unit into a bar) then the full weight of the Building Regulations Approved Documents are applied to the new use. This is especially important with Part B Fire safety measures (see section B1 of this unit).

Exemptions to the regulations

The Building Act 1984 contained some exemptions (circumstances that do not have to comply with the regulations), for example:

▶ the Secretary of State (the MP who is the Secretary of State for Communities and Local Government) can direct that a building is exempt from the regulations
▶ a building belonging to the Atomic Energy Authority or Civil Aviation Authority

- a building that comes under the Transportation Act 2000
- a local authority's or county council's buildings
- an educational establishment where plans have been passed by the Secretary of State
- a Crown building or Crown Authority (i.e. a building or land owned by the reigning monarch).

The Building Regulations 2010 also contain some **exemptions** from the regulations:

- a building controlled by other legislation, for example if it contains explosives, is owned by the nuclear industry or is a monument
- buildings not frequented by people, for example a transformer substation building
- greenhouses and agricultural buildings occupied
- by livestock
- temporary buildings, for example a portable office in
- a box container
- ancillary buildings such as sheds and garages
- small single-storey detached buildings not exceeding 30 m² in floor area
- a carport, covered porch, yard, walkway or conservatory not exceeding 30 m² in floor area.

Key term

Exemption – a circumstance when regulations do not apply, as long as the proposal meets the requirements of the schedules in the legislation, e.g. Building Regulations.

Dispensation or relaxation of the Regulations

Under the Building Act 1984, the secretary of state can direct that, where the operation of the Building Regulations would be deemed unreasonable, then they can be relaxed. An application has to be made to the local authority to relax the requirements of the Building Regulations. Part of the application process involves the advertisement of the works so any member of the public can raise concerns regarding the relaxation. The regulations can be relaxed in full or partially with conditions.

Theory into practice

Find five recent or proposed planning applications, for example by searching your local authority's planning application website. Will the construction work be covered by the Building Act?

A2 Control and implementation of the Building Regulations

Application procedure

There are two main methods of applying for approval of a design or installation under the Building Regulations: the Full Plans route and the Building Notice. Full Plans needs a designer or architect to draw up the existing and proposed plans and elevations and to provide a specification for approval. The Building Notice does not require plans or a specification.

Full Plans

A Full Plans submission is used for large domestic alterations and extensions and includes a ground or first floor plan, elevations, a location plan and a full specification of each element of the work and how this complies with the Approved Documents.

The documentation is submitted to the local authority well in advance of work starting. These are checked to ensure conformity of the proposal with the Approved Documents. The result can be:

- passed plans
- passed with conditions
- rejected.

This method is better for a client because, once the Approval Notice has been received and the building is constructed in accordance with the plans and accompanying specification, then no challenge can be made against it.

Building Notice

A Building Notice can be used for small extensions and other similar works where the passing of plans is not necessary. A Building Notice is a written application to the local authority deposited 48 hours before work commences. It tends to be a single form that provides the address of where the work will be carried out, a description of the work and the name of the property owner who should be contacted. It costs the same as a Full Plans submission but saves a client the design and drawing costs.

Any work that is inspected that is found to be defective has to be removed and put right, so a strong relationship needs to be maintained between the Building Control surveyor and the main contractor. You cannot use a Building Notice for works that involve the Fire Reform Regulations, which are fire safety measures applied mainly to commercial buildings.

Theory into practice

You have taken a phone call in the office from a property owner who needs some advice on what type of application to make for their development. They want to erect a small entrance porch on the rear of their house, which is 5 m² on plan, and has a mono-pitched roof.

What advice and guidance would you give the potential applicant?

THE INFORMATION
ON THIS FORM IS IN
THE PUBLIC
DOMAIN

**BUILDING
NOTICE**
The Building Act 1984
The Building Regulations 2010

Building Regulations Ref Number:
(office use only)

BN/20   /

This form is to be completed by the person (or agent) who intends to carry out building work. If the form is unfamiliar please read the notes on the reverse side or consult the office indicated below. Please type or use block capitals.

1 Applicant's details (see note 1)
Name:
Address:
Postcode: Tel: Mob/Fax:
e-mail :

2 Agent's details (if applicable)
Name:
Address:
Postcode: Tel: Mob/Fax:
e-mail:

3 Builder's details (if applicable)
Name:
Address:

Postcode: Tel: Mob/Fax:
e-mail:

4 Location of building to which work relates
Address:
Postcode: Tel: Mob/Fax:

5 Proposed work
Description:

Number of storeys:
Is the proposed work, or any part of it subject to a current   YES/NO
LABC Type Approval (see note 6)
Date of commencement (if known, see note 7):

6 Use of building
1 If new building or extension please state proposed use:
2 If existing building state present use:

7 Fees (see note 9 and separate Guidance Note on Fees for information)
1 If Table A work please state number of dwellings and types - Total: No of types:
2 If Table B work please state internal floor area: m²
3 If Table B or C work please state the estimated cost of work excluding VAT: £
Building Notice fee: £   plus VAT (where applicable): £   Total: £

8 Additional information
Will this proposed development be a self-build project, ie DIY   YES/NO
Are there any electrical controlled works to be carried out?   YES/NO
Is the person carrying out the electrical works registered under a competent person   YES/NO
scheme?

9 Statement
This notice is given in relation to the building work as described, and is submitted in accordance with
Regulation 12(2)(a).

Name: Signature: Date:

**North Lincolnshire Council, Building Control Section, Civic Centre, Ashby Road, Scunthorpe,
North Lincolnshire, DN16 1AB Tel: 01724 297411/13/28
e-mail: buildingcontrol@northlincs.gov.uk Internet: www.northlincs.gov.uk**

Version 3   Page 1   16/10/2013

▶ An example of a Building Notice form

Notification of commencement

Full Plans submissions have to be decided within five weeks of the submission. You must give a local authority at least two days' notice of your intention to commence building work either through a Full Plans submission approval or a Building Notice. So, if you put the notice in on a Monday afternoon then you could not start the work until Thursday morning when two clear days have elapsed.

All building work must start within three years from the date of approval of Full Plans or submission of a Building Notice.

You have to give the local authority five days' notice of having completed the work (for example, within five days of occupying the building).

Supervision of works and the powers of inspectors

The Building Act 1984 Part II created the provision for supervision of submitted plans and on-site work by inspectors. Inspections can be carried out by any of the following appointed personnel:

▸ Local authority representatives can approve plans and undertake inspections on site to ensure conformance with approved plans, specifications and Approved Documents A to R.
▸ Private inspectors must be registered with the Construction Industry Council (CIC) to be able to advertise as an approved inspector for five years before being required to register again.
▸ The National House Building Council (NHBC) can undertake approval of a house design and inspection of its construction.

Testing and commissioning

Section 33 of the Building Act 1984 provides for the inclusion of testing for conformity with the Building Regulations. The types of tests can involve:

▸ tests of the soil or substructure of a site
▸ testing of any materials
▸ testing of any service, fitting or equipment
▸ the local authority instructing an owner or occupier to undertake tests to confirm conformity.

Currently the Building Regulations require that the following tests are to be carried out before approval can be given:

▸ An air tightness test to ensure conformance with Part L of the Approved Documents. This involves placing a fan into an enclosure over the front door and testing the resistance of the building to air leakage.
▸ A sound test to ensure compliance with Part E of the Building Regulations to check if the reduction measures across habitable spaces between neighbours are working correctly.
▸ Testing mechanical ventilation flow rates to ensure adequacy.
▸ Testing drainage for air or water tightness.

Inspections at certain stages of the works

Several statutory inspections have to be undertaken to ensure compliance. The Building Regulations 2010 list these as:

▸ at commencement after serving a notice to start work, the Building Control surveyor will want to meet the applicant and talk through their proposals, giving advice and guidance. The inspector will also check the site conforms to that agreed on approved plans
▸ a foundations excavation before it is concreted will check that the bearing capacity of the ground is suitable, no trees will interfere with the foundation, that the ground

Research

Investigate the CIC and find an approved inspector in your area. Take down their details and see if their certification is still valid.

Research

What is the standard method for air tightness testing and what is it called?

What are BER, DER and TER?

is not made up, and the depth of drainage will not be affected by the pressure from the foundation

▸ prior to over-site concrete, the hardcore and sand blinding is checked to ensure the integrity of the damp-proof membrane (DPM), and any flooring insulation checked
▸ any damp-proof course (DPC)
▸ drain or sewer pipework, and the drain test
▸ completion of all the work
▸ additional inspections during the occupancy of the building if required.

Certification of the works

This is a very important function that provides a certificate that all the completed works conform to the Building Regulations and their Approved Documents.

▸ If the work has been supervised by the local authority Building Control department then the certificate is called a 'completion certificate'.
▸ If the client has used an approved inspector then the certificate is called a 'final certificate'.

The certificate should be retained with the property deeds and passed on to any future purchaser of the property.

Self-certification schemes

The Building Regulations 2010 provided for the following self-certification schemes that covered the following installations. Illustrated in the table are some of the common ones used for installation covering heating, lighting, window replacement and wall insulation.

▸ **Table 8.1:** Types of self-certification scheme under the Building Regulations

Type of work	Self-certification body
Installation of a gas appliance	**Gas Safe**-certified engineer
Hot water heating systems, oil fired boilers, solid fuel heaters	A member of the Association of Plumbing and Heating Contractors or HETAS Ltd
Installation of lighting, low voltage lighting, low voltage electrical installations	A person registered with Ascertiva Group Limited, Building Engineering Services Competence Accreditation Limited, ECA Certification Limited, NAPIT Registration Limited or Stroma Certification Limited
Replacement window systems	FENSA-registered installer
Cavity wall injected insulation	A member of the Cavity Wall Self-Certification Scheme

These self-certification schemes mean that work can be undertaken without Building Regulations consent providing that the installer is a member of the applicable organisation and can provide certification on the completed installation. Schedule 2 of the Building Regulations 2010 provides more information.

Consequences of non-compliance

There are a number of options available to the inspecting authority if the regulations are breached or contravened. The building owner or developer can:

▸ be liable for a conviction and a fine for breaching the Building Regulations

▶ be fined £50 per day for each day after a conviction that the work does not conform

▶ have the work pulled down or removed by the local authority

▶ be made to alter the work so it complies with the Approved Documents.

If the building owner still refuses to correct the work so it complies, then, 28 days after serving a Section 36 notice, a local authority can undertake the work themselves and recover the costs from the building owner.

It is very rare that situations result in any dispute between Building Control and a property owner as the Building Control officers of a local authority will, when inspecting, advise clients as to the statutory standards to be achieved.

⏸ PAUSE POINT   Name a common self-certification scheme permitted under the Building Regulations.

> **Hint**   Look at the end of the Building Regulations.

> **Extend**   Is there a link between professional associations and self-certification schemes?

Assessment practice 8.1   A.P1 A.P2 A.M1 A.D1

As the graduate Building Control officer you have been tasked with writing a guidance document for clients wishing to apply for compliance with Building Regulations. This will inform them about how the regulations operate and control construction work on site. The document needs to cover the following aspects so clients are fully informed about how to adhere to the Building Regulations during their construction work:

* the requirements of the regulations outlining the processes and procedures
* the submission of a Full Plans application
* the use of a Building Notice contrasted against Full Plans submission
* how work is controlled on site
* an evaluation of how the building works can be controlled under the Building Regulations across a range of different types of work.

Plan

* How much detail do I put into the explanation of the application of the Building Regulations
* What types of work shall I evaluate in terms of controlling work?

Do

* Have I answered each part of the activity separately, so that my Pass, Merit and Distinction criteria can be assessed?

Review

* Have I given enough detail about the requirements of the Building Regulations?
* Can I explain how I would evaluate outcomes better next time?

B   Examine the requirements of the Building Regulations

B1 Approved Documents

The Building Act 1984 gave powers to issue **Approved Documents**. There are 16 current documents published by the government under the Building Regulations 2015 latest revision – these are summarised in Table 8.2. An approved document demonstrates how each part of the Building Regulations is applied to the construction of a building. They contain detailed guidance on what has to be provided in order to comply with the regulations.

▶ **Table 8.2:** Approved Documents of the Building Regulations

Approved Document	Description
A: Structure	This deals with the structural integrity of a building and its fabric. The supporting structure needs to be safe and secure and free from any movement, settlement or degradation over time.

▶ **Table 8.2:** *Continued ...*

Approved Document	Description
B: Fire safety	Important in application, especially to buildings which house or can contain a number of occupants, for example a cinema. Evacuation and protection of the structure for a minimum time period is essential in saving lives and stopping the fire spreading.
C: Site preparation and resistance to contaminates and moisture	This covers the removal of any biological vegetation from beneath a building prior to construction, and includes the provision of drainage and the passage of moisture into a building.
D: Toxic substances	This covers, for example, the provision of insulating materials giving off toxins from cavity walls.
E: Resistance to the passage of sound	Robust details must be employed to prevent the passage of sound between neighbours through walls, ceilings and floors.
F: Ventilation	The provision of natural and forced ventilation to kitchens, bathrooms and toilets.
G: Sanitation, hot water safety and water efficiency	The provision of adequate sanitation facilities, baths and showers and efficient water systems that conserve energy.
H: Drainage and waste disposal	The provision of drainage to carry away foul and surface water.
J: Combustion appliances and fuel storage systems	The safe use of combustion equipment, the discharge of combustion gases to air and all ancillary equipment.
K: Protection from falling, collision and impact	The provision of appropriate stairs, their handrails and balustrade guarding and the provision for ramped access into buildings.
L: Conservation of fuel and power	The provision of insulation to reduce heat loss and to conserve energy.
M: Access to and use of buildings	The provision of suitable and appropriate disabled access to buildings.
N: Glazing safety (withdrawn)	Part N has been withdrawn and is now incorporated into Part K.
P: Electrical safety	All electrical works on a property must be carried out by a registered electrician who can provide the correct documentation to sign off on Part P.
Q: Security in dwellings	This covers the security of new dwellings in preventing ingress (entrance) into a building by unauthorised persons by ensuring a reasonable standard of doors and windows security.
R: High speed electronic communications networks	This covers the provision of data and broadband installations and the provision of the conduits and infrastructure to enable easy installation into a building or structure.

Research

Using the Planning Portal website, locate and download each of the current Approved Documents and place these into a folder for reference during assessment.

From time to time the Approved Documents are updated and published on the **Planning Portal** (**www.planningportal.co.uk**). Approved Document 7 details materials and workmanship provisions. This lays out the provisions of Regulation 7 of the Building Regulations 2010 and the 2013 revisions.

Key term

Planning Portal – the government website that provides detailed information on all aspects of planning.

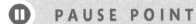

PAUSE POINT   Where can you find the current versions of Approved Documents?

Hint   Explore the Planning Portal.

Extend   How do you know they are the current versions?

On the front cover of each Approved Document are the letters, from the Building Regulations 2010 legislation corresponding to Schedule 1, Regulations 4 to 6. Always make sure that you are using the latest versions, which are all dated on the portal website.

Structure: Approved Document A

This document covers the following sections of the regulations:
▶ A1 Loading
▶ A2 Ground Movement
▶ A3 Disproportionate Collapse

Part A covers aspects of structural integrity that include:

▸ the height and thickness of walls so they remain stable in areas that are subjected to high winds

▸ sizes of structural timbers in terms of depth and width to resist deflection under load published by **TRADA**

▸ the lateral restraints that are required to hold walls together

▸ location of openings within walls

▸ foundations width and depth (see Table 8.3)

▸ wall and roof cladding

▸ resistance to disproportional collapse.

Key term

TRADA – the Timber Research and Development Association.

▸ **Table 8.3:** Minimum width of strip footings (Source: The Building Regulations 2010 Approved Document A)

Type of ground (including engineered fill)	Condition of ground	Field test applicable	Total load of load-bearing wall not more than (kN/linear metre)					
			23	30	40	50	60	70
			Minimum width of strip foundations (mm)					
I Rock	Not inferior to sandstone, limestone or firm chalk	Requires at least a pneumatic or other mechanically operated pick for excavation	In each case equal to the width of the wall					
II Gravel or sand	Medium dense	Requires pick for excavation. Wooden peg 50 mm square in cross section hard to drive beyond 150 mm	250	300	400	500	600	650
III Clay Sandy clay	Stiff Stiff	Can be indented by thumb	250	300	400	500	600	650
IV Clay Sandy clay	Firm Firm	Thumb makes impression easily	300	350	450	600	750	850
V Sand Silty sand Clayey sand	Loose Loose Loose	Can be excavated with a spade. Wooden peg 50 mm square in cross section can be easily driven	400	600	Note: Foundations on soft Type V and VI do not fall within the provisions of this section if the total lead exceeds 30 kN/m.			
VI Silt Clay Sandy clay Clay or silt	Soft Soft Soft Soft	Finger pushed in up to 10 mm	450	650				
VII Silt Clay Sandy clay Clay or silt	Very soft Very soft Very soft Very soft	Finger easily pushed in up to 25 mm	Refer to specialist advice					

Fire safety: Approved Document B

This document covers the following aspects of fire safety:
- Volume 1 covers dwelling houses that are domestic homes.
- Volume 2 covers buildings other than dwelling houses (the commercial application of fire safety).

Within both the sections the regulations apply equally and cover:
- B1 means of warning and escape
- B2 internal fire spread (linings)
- B3 internal fire spread (structure)
- B4 external fire spread
- B5 access and facilities for the fire service.

Part B covers many aspects including:
- fire detection and fire alarm systems for larger houses
- means of escape from multi floor buildings
- wall and ceiling linings resistance to fire and its spread across surfaces
- fire protection to the load-bearing elements of a structure
- **compartmentation** of areas into zones
- concealed areas such as voids and cavities
- openings through compartments
- separation of spaces between buildings
- roof coverings, roof lights and classification of coverings
- fire-fighting service equipment access.

Appendices contain performance of materials guidelines, fire doors, measurement methods, purpose groups and formal definitions.

Theory into practice

You are currently designing domestic houses with integral garages built onto the front of the homes. They have an internal door entrance into the kitchen and an automatic garage door.

What provisions need to be made within the structure and doors for fire limitation and spread?

Key term

Compartmentation – the division of a structure into separate fire-protected compartments to prevent the spread of a fire across a whole structure.

Site preparation and resistance to contaminates and moisture: Approved Document C

This document covers:
- C1 Site preparation and resistance to contaminants
- C2 Resistance to moisture.

Part C covers aspects of:
- the clearance and treatment of unsuitable materials from a site

- resistance of the structure to solid, gas and liquid contaminants from previous uses of brownfield sites and those occurring naturally such as radon gas
- subsoil drainage: maintenance of existing and installation
- provisions within suspended and solid floors to prevent damage from moisture
- provision to prevent moisture ingress into walls and driving rain measures
- resistance for roofs to moisture ingress.

Research

Using the Approved Documents list, try to answer the following research questions.
- How do you prevent moisture ingress into a solid concrete floor and its associated external cavity walls?
- What is radon gas?
- Where does it occur?
- What precautionary measures have to be provided?

Toxic substances: Approved Document D

This covers the provisions of:
- D1 Cavity insulation.

In the past, several injected foam cavity insulation materials have broken down over time and given off toxic fumes. This document specifies provisions for insulation that meets the performance required for insulation foams.

Resistance to the passage of sound: Approved Document E

Resistance of a structure to sound is essential with the proximity of our neighbours in cities and towns. The provisions of this Approved Document are:
- E1 Protection against sound from other parts of the building and adjoining buildings
- E2 Protection against sound from within a building
- E3 Sound reverberation within a building
- E4 Acoustic conditions within schools.

Part E covers the following aspects of sound control:
- performance of structures in terms of reducing airborne sound and its impact by robust insulation detailing
- requirements of a pre-completion sound test to ensure compliance has been achieved
- requirements for separating walls and floors in accordance with robust detail diagrams within the Approved Document
- treatments resulting from change of use into occupied houses and flats
- the requirements for internal walls and floors in new buildings.

Appendices include mass calculations, glossary, testing, references and details of Robust Details Ltd.

Ventilation: Approved Document F

The ventilation of a building is essential for removing humid air so it does not condense onto colder surfaces such as single-glazed windows. The requirements of F1 cover the means of ventilation that has to be provided, including:

- performance of ventilation
- provisions for new dwellings and other buildings that are not dwellings
- work on existing buildings
- standards and other reference sources.

A series of appendices provide details on how to achieve the desired performance.

Sanitation, hot water safety and water efficiency: Approved Document G

The provision within a dwelling and other buildings of hot and cold water services that are safe to use and consume is essential for a good quality of life. A modern bathroom is a priority when purchasing a family home, often along with en-suite facilities to service bedrooms.

Part G provides requirements for meeting the following parts of the Building Regulations:

- G1 Cold water supply
- G2 Efficiency in water consumption
- G3 The supply of hot water
- G4 Sanitation provisions
- G5 Bathrooms
- G6 Food preparation areas.

Approved Document G covers this by detailing:

- performance requirements of cold water supply
- the performance of water efficiency measures
- the performance of hot water systems and their design
- scale of sanitary conveniences and washing facilities
- provision of bathrooms
- food preparation scale – domestic and non-domestic.

Appendices include an efficiency calculator, potable water and references.

Drainage and waste disposal: Approved Document H

Associated with Part G is the drainage that will take away the foul and surface water from dwellings to treatment water works and sewage works. Drainage must be effective in removing waste from a building; factors include contamination of water sources and avoiding blockages by regular maintenance.

Part H provides for drainage by covering:

- H1 Foul water drainage
- H2 Waste water treatment systems and cesspools
- H3 Rainwater drainage
- H4 Building over sewers
- H5 Separate systems of drainage
- H6 Solid waste storage.

Approved Document H covers this by detailing:

- guidance on the pipework for sanitation covering sizes and falls, materials and ventilation
- foul drainage installation, its protection from loading, manholes and access provision
- waste water treatment and cesspools, maintenance of systems
- rainwater drainage, gutters and downpipes, drainage of paved areas, oil separators
- building over existing sewers
- separate systems of drainage
- provisions for solid waste storage.

Appendices provide additional guidance for large buildings, repairing drainage, adoption of sewers, and tables for drainage.

Combustion appliances and fuel storage systems: Approved Document J

Modern homes contain combustion appliances that heat our water and homes. These run on a variety of different fuel sources from heating oil, gas and liquid propane through to solid fuel. All give off toxic and harmful substances as the products of combustion.

Part J covers the following aspects for combustion:

- J1 The supply of air for combustion
- J2 Discharge of combustion products
- J3 Measures to warn about the presence of carbon monoxide

- ▶ J4 Protection of a building
- ▶ J5 Provision of information
- ▶ J6 Liquid fuel storage protection measures
- ▶ J7 Pollution protection measures.

Approved Document J deals with these aspects of the Building Regulations by making provisions for:
- ▶ the supply of air and the ventilation requirements for appliances that combust fuels
- ▶ flues to vent combustion products to air
- ▶ solid fuel, gas and oil appliances
- ▶ the external storage of liquid fuel and its supply.

Protection from falling, collision and impact: Approved Document K

This essentially covers the movement of occupants vertically and horizontally around a building. Safe movement is essential within many retail and commercial applications where glazing is used to enhance interior spaces. Stairs, ramps and access are all covered.

Part K covers the following aspects of protection from falling, collision and impact:
- ▶ K1 Provisions for stairs, ladders and ramps
- ▶ K2 Protection from falling
- ▶ K3 Barriers for vehicles and loading bays
- ▶ K4 Protection against impact with glazing
- ▶ K5 Additional provision for glazing in buildings other than dwellings
- ▶ K6 Protection against impact from and trapping by doors.

Approved Document K covers this by detailing:
- ▶ the specified gradients of stairs, landing places, loft ladders and fixed ladders, handrails and stair guarding
- ▶ construction of ramps, their width, design, handrails, landings and guarding
- ▶ protection from falling by guarding
- ▶ vehicle barriers within loading bays, heights and guarding
- ▶ provisions of toughened glazing to protect from impact damage
- ▶ protection from an open window onto a pathway or accessible area
- ▶ glazing manifestation to alert users to its presence at eye level
- ▶ safe opening and closing of windows and restrictions on height
- ▶ provisions for cleaning windows
- ▶ fixing of devices to prevent trapping in doors.

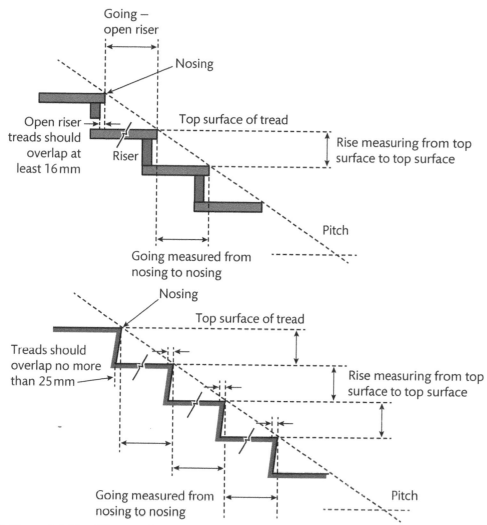

▶ **Figure 8.1:** The differences between stair construction requirements for domestic (top) and commercial (bottom) applications

Conservation of fuel and power: Approved Document L

Part L is divided into parts L1 and 2 with parts A and B for each. These are for domestic and non-domestic applications and new and existing buildings:

▶ L1A Conservation of fuel and power in new dwellings
▶ L1B Conservation of fuel and power in existing dwellings
▶ L2A Conservation of fuel and power in new buildings other than dwellings
▶ L2B Conservation of fuel and power in existing buildings other than dwellings.

Part L provides for:

▶ design standards covering target emission rates, heating, lighting, limiting standards for fabric, heat gains
▶ quality of construction and commissioning of heating and hot water
▶ providing information on energy-efficient operation
▶ model designs.

Access to and use of buildings: Approved Document M

This is a key Approved Document that covers access for people who are less mobile than others and require additional support. Two volumes deal with access to and use of buildings:

▶ Volume 1 – Dwellings

Theory into practice

You are submitting an application for a single-storey mono-pitched roof extension to the lounge of a detached building which will contain a set of roof windows and fully glazed garden bi-fold doors.

- Which Part of Part L has to be complied with?
- What provisions are needed for ventilation to the extension?
- What insulation has to be provided and in what location?
- Does the extension have to meet any u-values?

Volume 2 – Buildings other than dwellings.

Volume 1 covers:

▸ M4 Category 1: Visitable dwellings
▸ M4 Category 2: Accessible and adaptable buildings
▸ M4 Category 3: Wheelchair-user dwellings.

Volume 2 covers the following applications of the regulations to buildings other than dwellings:

▸ M1 Access and use of the building
▸ M2 Access to extensions
▸ M3 Sanitation provisions in extensions.

Approved Document M provides guidance that covers:

▸ access into a building from a level approach, to include the car parking access
▸ ramped and stepped access arrangements
▸ powered entrances and doors
▸ horizontal and vertical circulation to include lifts, corridors, doors, stairs, ramps, handrails
▸ provision of sanitation facilities.

It includes diagrams of provisions for guidance.

Research

Identify the provisions under Part M for access arrangements within a domestic home.

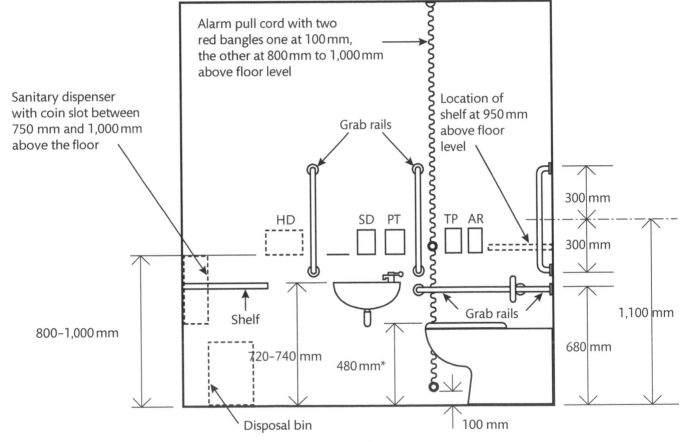

* Height subject to manufacturing tolerance of WC pan

HD: Possible position for automatic hand dryer (see also below)
SD: Soap dispenser
PT: Paper towel dispenser
AR: Alarm reset button
TP: Toilet paper dispenser

Height of drop-down rails to be the same as the other horizontal grab rails

▸ **Figure 8.2:** Building Regulations requirements for a wheelchair-accessible unisex toilet

Electrical safety: Approved Document P

In our homes, we use 240 volts at 13 amps. This is a dangerous current when it is exposed to human touch as it makes electricity flow through your body and down to earth. Part P is concerned with electrical safety devices and the installation of wiring into buildings. P1 of the regulations deals with the design and installation of electrical installations. This includes the following provisions:

▶ design and installation for new dwellings, new dwellings formed by a change of use and additions and alterations such as extensions
▶ notifiable works
▶ certification, inspection and testing requirements.

The final point is a key component as only a registered electrical contractor or a Building Control body can produce the certificates that are required for Part P to be signed off. You cannot undertake any electrical installations yourself as you will not be a certified Part P contractor. The person who undertakes the certification is deemed a 'registered competent person'. The report is termed an 'electrical installation condition report'.

Security in dwellings: Approved Document Q

This document covers the regulations through Q1, which are measures to prevent unauthorised access to a property. This is achieved by the design of:

▶ doors, door sets and their installation and fixing
▶ secure window systems and their installation and fixing.

Appendices cover bespoke timber windows.

High speed electronic communications networks: Approved Document R

With the advent of high-capacity fibre optic internet broadband connections and further advances in digital technology, provisions must be made within buildings to accommodate technology upgrades. Document R achieved this through the application of Regulation R1, covering in-building physical infrastructure:

▶ ductwork for copper and fibre optic installation
▶ satellite and wireless communication provisions.

Materials and workmanship: Approved Document 7

This is the application of Regulation 7, which covers materials and workmanship standards by:

▶ establishing the fitness of materials for inclusion into buildings
▶ short-life materials and how these are to be managed
▶ describing the use of materials which may change properties over time and use
▶ ways of establishing the adequacy of the workmanship undertaken on buildings.

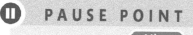 **PAUSE POINT**   Why is security in dwellings covered in the Building Regulations?

        Hint   Consider possible government priorities.

        Extend   What provisions does Part Q provide?

B2 Alternative methods of achieving compliance

The Building Regulations do make provisions for using other standards as long as they meet the standard specified within the regulations and the Approved Documents.

There are several ways to demonstrate that you have complied with the Building Regulations. These are summarised in Table 8.4.

▶ **Table 8.4:** Other methods of achieving compliance

Method of achieving standard	How this is achieved
British Standards	The British Standards Institution (BSI) publishes thousands of standards that can be used to demonstrate compliance. For example, BS 6206 is a standard for toughened safety glazing. The **kitemark** is etched into the glazing to prove it has met this standard.
European Standards	These are similar standards to BSI but are termed European Numbers (EN). They enable cross-border standards to be maintained across the whole of the European community.
NHBC Standards	The National House Building Council (NHBC) publishes its own set of standards for buildings. These can be followed along with the use of their Building Control inspections to ensure compliance with the guarantee scheme that they operate.
Competent person self-certification schemes	The Building Regulations allow for a range of different organisations to allow registered and competent members to certify various aspects of the regulations. These cover window replacement, boiler upgrades, electrical installations, air conditioning and many other services and systems installed into buildings.

Theory into practice

A client wants to install six new plug points and three new light switches in their house. They also want to install a new bath and toilet, and a new boiler. They have asked you to check whether all this work will be covered under self-certification schemes instead of referring to the Building Regulations. What would you advise?

Assessment practice 8.2

B.P3 B.P4 B.M2 B.D2

As a graduate Building Control officer with one year's experience, clients often ask you how they can achieve compliance with the Approved Documents along with other methods that could be used to demonstrate compliance with the regulations.

A property developer has approached you and asked for guidance about their existing and proposed housing developments. They are unsure of the current provisions required and are using old 2004 design details. They have asked:

- What are the outline requirements of the Approved Documents for thermal upgrading of the existing properties?
- Can you provide an analysis of any other methods that could be used to demonstrate that their architect has complied with the Building Regulations for upgrading and refurbishment works, apart from using the published Approved Documents?
- What are the advantages and disadvantages of employing these alternative approval methods for the following?
  - kitchen extensions
  - new builds
  - flat refurbishments

This must be produced in a report format that can be sent by email to the developer for comment and should contain images and diagrams so they can discuss this with the main contractor with whom they are proposing to undertake the refurbishment and extension works.

Plan
- How can I obtain copies of the relevant Approved Documents?
- What images can I use
- What other information do I need?

Do
- Am I using a report format that is presentable to a client, containing a front sheet, a contents page, a main section and conclusion?

Review
- Have I used evaluative statements that cover the Distinction criteria from the assessment table?
- What would I do differently next time?

C Undertake a Building Regulations application

C1 Types of application

The Building Act 1984 created the passing of 'Full Plans' for approval. The Building Regulations Part 3 created the additional method of the giving of a 'Building Notice', so the two methods that are available to owners and developers are:

- Full Plans – the passing of drawings with a specification that has to have been submitted for approval by the local authority and is a formal process that allows you to obtain permission before you start work.
- Building Notice – this is a quicker method and just involves submitting written notice 48 hours before you start work.

We will now examine these two types of application in detail so you will be able to undertake a full submission using all the documentation and plans produced by graphical or digital means.

C2 Preparing a Building Notice application

The Building Notice is simply a statement that you are going to undertake relevant construction work. The documentation has to be submitted prior to work starting on site and the form(s) can be found on the relevant local authority's Building Control web page.

The Building Notice

This form must include the following:

- the name, address and signature of the person who is intending to carry out the work (the contractor)
- a description of the proposed work
- the location of the proposed work
- Where the proposed work is an extension or erection then additional documentation is required:
  - a plan to a scale of 1:1250 that details the size and position of the building and its extension with adjoining related boundaries
  - the boundaries of the **curtilage** of the property
  - the size, use and position of any buildings within the curtilage of the property
  - the width and position of the adjoining streets
  - a statement specifying the number of **storeys** including any basement levels
  - the details of the drainage for the proposed building or extension
  - details of any compliance with local **enactments**.

You must sign a statement that reads:

'This notice is given in relation to the building work as described, and is submitted in accordance with Regulation 12(2)(a) and is accompanied by the appropriate fee.'

Where the work is renovation or replacement of a thermal element, a change to a building's energy status or a change of use then the local authority can request more plans within a specified time.

A Building Notice has a life of three years from the date on which it was given to the local authority unless, before the end of the three years:

- the work began
- the building changed its use or energy status.

> **Key terms**
>
> **Curtilage** – the area of the plot forming the building and its land, outbuildings and boundaries.
>
> **Storeys** – the floor-to-ceiling height of a building and the number of floors.
>
> **Enactment** – a legislative regulation.

> **Link**
>
> Section C3 of this unit gives more detail about fee schedules.

Ⅱ PAUSE POINT   Where is a Building Notice sent when it has been completed?

> **Hint**   What body is responsible for inspecting the work?
>
> **Extend**   What is the time frame before starting work?

C3 Preparing a Full Plans application

A Full Plans application will require the services of an architect or designer. This is needed to produce the existing and proposed drawings for the Building Regulations application. The same drawings can be used for both the Full Plans application and the relevant planning application.

A Full Plans application requires the following:

▶ two full sets of existing and proposed plans. Where Part B of Schedule 1 applies two further copies of plans are required to be deposited with the local authority
▶ a description of the proposed works
▶ all other associated plans required to show compliance with the Building Regulations such as:
  • details to show the size of the extension and the relationship to the adjoining boundaries
  • the boundaries of the existing curtilage, and the position and size of other buildings within the curtilage
  • location of the property in relation to the surrounding streets
  • a statement specifying the number of storeys
  • provisions made for drainage, for example for drainage connections for surface and foul water and evidence of precautions to be taken in building over a drain by providing a 'build over agreement' with the drainage authority
  • details of the works and materials
▶ a statement as to whether the work is deemed to be under the Reform Regulations (Fire Safety) Order 2005
▶ a request to issue a completion certificate by the local authority
▶ a specification, which can be by annotations to the drawings or a separate document showing how the design meets the requirements of the Approved Documents
▶ copies of calculations in support of Approved Document Part A.

You must sign a statement of application that reads:

'This notice is given in relation to the building work as described, is submitted in accordance with Regulation 12(2)(b) and is accompanied by the appropriate fee.'

Full Plans applications can now be made electronically via the websites of many local authorities. Drawings have to be submitted as PDFs along with all other documents in the file formats indicated. The local authority will:

▶ confirm receipt of the submission
▶ check that the correct fee has been submitted
▶ check the validity of the submission and issue a formal letter detailing the official receipt
▶ check the plans for conformity with the response time within the local authority service level agreement
▶ issue a passed, passed with conditions or rejected notice.

Link

See 'Fee schedules' later in this section for more details on fees.

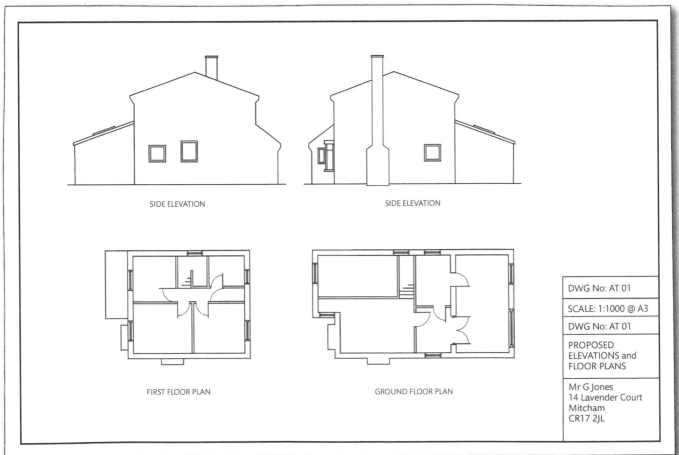

Figure 8.3: Proposed elevations and plans for extension work

Under Part 1 Section 16 of the Building Act 1984, a time period called the 'relevant period' is specified for a local authority to respond to a Full Plans submission. This period is either five weeks, or an extended period not more than two months from the deposit of the plans (if agreed in writing between both the local authority and the person depositing the plans).

Ⅱ PAUSE POINT   How many sets of drawings are required for a Full Plans application?

Hint   Look at the Local Authority Building Control website.

Extend   Do you need to send two copies of specifications?

Particular considerations

Build over agreements

Any works that cross over, under or adjacent to an existing sewer or drain have to be examined to ensure that the foundations will not place any undue pressure upon the drain, causing it to break. Where the proposed works are found to cross over a sewer or drain that is owned by the local water authority then a build over agreement has to be completed.

The initial application is made to the local authority for checking. They will ask for a formal agreement to be entered into between the property owner and the local water authority responsible for the drainage.

The measures that you are going to take must be specified and illustrated. The form illustrates how this is completed and submitted as part of the Full Plans application process.

Specifications

A specification accompanies the drawings and can be printed onto them or be a separate document. Specifications often require a high level of detail to ensure compliance with each Approved Document.

Calculations

Calculations may be required as part of the application in support of particular building elements, for example:

▶ roof trusses
▶ foundation loading
▶ structural steelwork above openings
▶ lintels
▶ traditional timber roof timber sizing
▶ raised collars for ceilings
▶ attic conversions.

An approved structural engineer should produce the calculations, as they will be covered by professional indemnity insurances and work to known standards such as the Eurocodes for structural design.

Fee schedules

The Building (Local Authority Charges) Regulations 2010 make provision authorising local authorities (LAs) in England and Wales to fix their own charges in a scheme, based on the full recovery of their costs, for carrying out their main Building Control functions relating to building regulations. This means that they are a 'not for profit' organisation and only cover the costs of administering the building regulations.

Table 8.5 is an extract taken from the Building Control fee schedule from North Lincolnshire Council (a local authority). Note that a Full Plans submission is split into two separate fees: the first is the plan fee while the second is the inspection fee for when the inspector calls on site for the 'statutory inspections'. A regularisation is the application of the Building Regulations to work that has already been completed without permission.

> **Table 8.5:** Building Control fee schedule for fees for new dwellings (no more than three storeys with a total internal floor area not exceeding 300 m²)

Number of dwellings	Full Plans submissions				Building Notices		Regularisation application
	Plan fee		Inspection fee		Building Notice fee		Regularisation fee
	Exc VAT (£)	Inc VAT (£)	Exc VAT (£)	Inc VAT (£)	Exc VAT (£)	Inc VAT (£)	Exempt VAT
1	£129.00	£154.80	£301.00	£361.20	£430.00	£516.00	£516.00
2	£198.00	£237.60	£462.00	£554.40	£660.00	£792.00	£792.00
3 and over	Consult Building Control regarding the fees						

Ⅱ PAUSE POINT   Is a Building Notice fee the same as the fee for Full Plans?

Hint   Look at the fee schedule.

Extend   What is a regularisation fee?

Assessment practice 8.3   C.P5 C.P6 C.P7 C.M3 C.D3

You are working as a trainee approved inspector for a company and have been approached by a potential client for advice and guidance. They need to apply for Building Regulations approval and are not sure which type of application to make. Produce the following for the client:

- an outline of the Building Notice application process
- a description of the Full Plans application process
- for both, describe the procedures through to completion of the work.

You have been asked to complete two sample applications for clients. These are to be used for future marketing so clients can see what is produced for each. Produce the following:

- a specimen Building Notice application with all required documents
- a specimen Full Plans application including all drawings, specification and documentation for a new dwelling
- a leaflet explaining the relationship between the documentation for both processes and how they comply with the Building Regulations.

Plan
- Do I need to do further research to describe the differences between the two types of Building Regulations application processes?

Do
- Have I used the correct forms from my local authority website?
- Can I use the work I've already done for graphical detailing and CAD in other units?

Review
- Have I filled in all the boxes within the application forms?
- Has the correct information been submitted?
- Does it all make sense?

Further reading and resources

Approved documents that accompany the Building Regulations **www.planningportal.co.uk/info/200135/approved_documents.**

Billington, M.J., Bright, K. and Waters, R.J. (2007) *The Building Regulations Explained*, 13th revised edition, Chichester: Wiley-Blackwell.

Evans, Huw M.A. (2015) *Guide to the Building Regulations*, 3rd revised edition, London: RIBA Publishing.

Legislation as passed by parliament downloadable from **www.legislation.gov.uk/uksi/2010/2214/contents/made**

Local Authority Building Control **www.labc.co.uk.**

THINK ▶FUTURE

Jatinder

Jatinder is an international graduate who has completed his final year on a Building Surveying degree programme at a UK university. The government of his native country is supporting his studies as it wishes to adopt the implementation of the UK Building Regulations in its own country. Several trainees have been sent to UK universities to become Building Control officers within their home country.

Jatinder has managed to obtain a two-year contract with a local authority as a graduate Building Control officer. He is going to spend half of his employment on the plan-checking side of applications and the other 12 months on site undertaking inspections.

Jatinder is going to use this time to gain as much commercial and domestic experience as he can to apply this when his working visa expires and he returns home. Then he will establish a Building Control system within the council districts of his country and train other students using the knowledge gained from his UK education and experience.

The establishment of a Building Control system in Jatinder's country is also going to be supported by the UK with secondments by UK Building Control officers until the system is up and running and localised for the climate and type of construction employed.

Focusing your skills

There is a wide range of skills that a Building Control officer has to use when they are checking applications for compliance and undertaking inspections on site.

- Interpretation skills – ability to examine a set of drawings and discern whether they will meet the regulations. This skill is learned by undertaking on-the-job training until you have learned the regulations and can spot non-compliance. These skills can be developed by downloading projects from a local authority planning website and investigating different projects' drawings.

- Communication skills – ability to communicate in writing with clients via approval letters and emails for additional information requests. You could acquire these skills by volunteering in the construction sector or working as an intern for the first year. Communication skills often involve

behavioural skills in responding to phone calls and face-to-face interaction.

- Time management – to react within the five-week period allowed for a Building Regulations application to be decided by being able to schedule the workload via a reporting system.

- Knowledgeable – to be able to discuss with confidence the application of the regulations on a construction site with supervisors so solutions can be agreed to problem's such as foundation depths, trees and interpretation of the regulations.

- Professional – to demonstrate a professional approach, wearing appropriate PPE on site and acting in a manner that promotes effective Building Control. This also includes your appearance where dress code and appropriate business attire presents a professional image.

Getting ready for assessment

Bella is in the first year of her diploma in Construction and the Built Environment. She works in a design office and Unit 8 is essential in ensuring that her designs meet current regulations. For her assessment, she was required to prepare a report and a presentation on:

▸ the requirements of the Building Regulations

▸ the different methods of control and demonstrating compliance with the Building Regulations.

How I got started

First I started by preparing a list of everything that I needed to cover in my presentation and report. I used this as a basis for planning my research. I decided to use PowerPoint® for my presentation so I could include photo images. I looked through my notes for this unit then started planning my presentation to cover all parts of the assignment.

For my written report, I organised the notes I had taken for this unit and separated them into different sections. I then wrote a plan for my report, mindmapping what I wanted to include in each section to make sure that I had covered everything that the assignment was asking of me.

How I brought it all together

The difficult part was preparing the evaluation for the Distinction grade. I researched what an evaluation should look like using some examples that I was able to locate online. I evidenced the main part of the LA 1 tasks using research on Building Notices and Full Plans submissions.

I then started by looking at the Planning Portal website. This is the government website that also covers the Building Regulations. I was able to find information on applying for approval and how projects are controlled in different ways. The LABC website also provides quite a lot of information on this aspect along with my local authority website.

As the Building Regulations and the methods of applying under these are a UK government initiative it was their website that I went to first. This saved me a lot of time. My local authority Building Control department also provided via their website explanations on the different types of applications that can be made, which are Building Notice and Full Plans submissions.

For learning aim B we had to outline the requirements of the Approved Documents. This I accomplished by using a small extension that included many different sub- and super-structural elements covered within the regulations.

What I have learned from the experience

I found out that planning my research first and making sure I knew what I was looking for and where is really important and doing this more could have saved me even more time. Reading information then putting it down and typing this into my own words is something that has helped me evidence the task requirements.

For my presentation, I would give myself more time to practise and plan my presentation to make sure that I was really confident with what I was saying and why. This would also help me to make sure that all the elements of my presentation were accurate and making a clear argument.

Think about it

The Building Regulations can be found on the legislation website but they are in legal terms. Find a Building Regulations guide that takes the legal terms and puts in it plain English so that the uninformed client can understand.

▸ How will you use evidence from other units for the Full Plans submissionfor the LAC merit criterion?

▸ Is this allowed and acceptable?

Management of a Construction Project 9

Getting to know your unit

The four major functions of managing a construction project are planning, organising, leading and controlling. Effective management of a construction project will ensure that it is completed successfully in terms of safety, time, cost and quality. To achieve this, the project must be properly planned, organised, led and controlled. The construction industry has a poor record in terms of completing on time and on budget. Recent reports have attempted to address these issues within the construction industry.

This unit will help you to understand management principles, techniques of the planning process and the importance of information technology in quality assurance and control of a project. You will learn about the roles of individual team members and the techniques applied by a site manager to manage a project successfully. You will understand purchasing and cost management techniques, and consider the methods used to plan and control a programme of works for a construction development.

How you will be assessed

Throughout this unit you will find useful assessment activities that will help you towards your final assignments. Completing these assessment activities will not necessarily mean that you achieve a particular grade, but they will help you to carry out relevant research or preparation that can be used towards your final assignments.

To ensure that you achieve all the tasks in your set assignments, it is important that you cover all the Pass criteria. Make sure that you check each of these before you submit your work to your tutor.

If you want to achieve a Merit or Distinction you must consider how you present the information in your assignment and make sure that you extend your responses or answers. For example, to achieve a Merit (M1) you must discuss the roles of members of the construction management team and how their individual responsibilities are applied. To achieve a Distinction (D1) you must further evaluate the different roles of the construction management team, their responsibilities and the techniques applied by a site manager to manage the project.

The assignments set by your tutor will consist of tasks designed to meet the criteria in the table on the next page.

They are likely to consist of written assignments that:

▶ explain the roles and responsibilities of each of the members within a project management team

▶ explain the cost management techniques used to monitor the profitability of construction projects

▶ produce a graphical master programme for a given construction project and explain the methods used to monitor the progress of the project.

Assessment criteria

This table shows you what you must do in order to achieve a **Pass**, **Merit** or **Distinction** grade.

Pass	Merit	Distinction
Learning aim **A** Understand the principles and application of management in construction		
A.P1 Explain the roles of the members of the construction management team and their individual responsibilities.	**A.M1** Discuss the roles of the members of the construction management team and how their individual responsibilities are applied.	**A.D1** Evaluate the different roles of the construction management team, their responsibilities and the techniques applied by a site manager to manage the project.
A.P2 Explain the techniques applied by a site manager to manage the project.	**A.M2** Discuss the techniques applied by a site manager to manage the project.	
Learning aim **B** Understand purchasing and cost management techniques		
B.P3 Explain the methods used by construction companies to facilitate the supply of appropriate materials to site.	**B.M3** Assess the methods used to facilitate the cost-effective supply of appropriate materials to site.	**B.D2** Evaluate the methods used to facilitate the ethical supply of appropriate materials to site, meeting programme requirements, and how these impact on the cost management and profitability of construction projects.
B.P4 Explain the cost management techniques used to monitor and control the cost and profitability of construction projects.	**B.M4** Analyse the cost management techniques used to effectively monitor and control the cost and profitability of construction projects.	
Learning aim **C** Develop a programme of activities for construction works		
C.P5 Produce a programme of activities with graphical representations for a given construction project.	**C.M5** Produce a detailed programme of activities, with graphical representations and appropriately detailed timings for a given construction project, and consider an appropriate method to monitor progress.	**C.D3** Produce a comprehensive programme of activities, with graphical representations and highly detailed timings that show critical and non-critical elements for a given construction project, and consider the most appropriate method to monitor progress.
C.P6 Explain the methods used to monitor the progress of construction projects.		

Getting started

The success of any construction project depends on how well the project management team plan, organise and control resources such as plant, labour and materials to ensure the project is completed safely, to the required quality, on time and on budget. Spend a few minutes researching on the internet to compile a list of major construction projects which have been completed in recent years. Now select a project of your choice and investigate whether the project was completed on time and on budget.

A Understand the principles and application of management in construction

A1 Principles of management

Management style, methods and theories

For construction projects, effective **management** makes the best possible use of the resource materials, plant, labour and finance to complete the project safely, on time and on budget.

The main management styles are:

▶ **Autocratic:** managers make all the decisions with little input from the team. This can be good in a crisis but can demotivate staff if they have no direct involvement in decisions and do not feel valued.

▶ **Laissez-faire:** managers delegate the responsibility of decision-making to the team.

▶ **Participative (democratic):** managers value the group's input but the final decision rests with them. This often boosts employee morale because they feel they are part of the decision-making process.

▶ **Transactional:** managers provide rewards or punishments to team members based on results.

▶ **Transformational:** managers motivate employees and increase output and efficiency through good communication, constant presence and regular expressions of appreciation.

Fayol's 14 principles of management

Henri Fayol was an engineer who worked in mining in France in the 19th and early 20th century. During his working life managing the mines he became aware of the importance of management principles. He published a book containing 14 principles of management. It is regarded as one of the first and most respected theories of management. His 14 principles are:

1   Division of work – workers with specialist skills become increasingly skilled and efficient, which can increase output.

2   Authority – while managers must be allowed the authority to give orders, they must be aware of the responsibility that comes with authority.

3   Discipline – there are many ways to maintain discipline in a working environment.

4   Unity of command – workers should have no more than one direct supervisor.

5   Unity of direction – work can only be well co-ordinated if teams with the same objective use a single plan and work under the direction of a single manager.

6   Subordination of individual interests to the general interest – the team as a whole is more important than the interests of one worker, including the manager.

7 Remuneration – fair financial and non-financial compensation is likely to lead to satisfied workers.

8 Centralisation – the decision-making process should be balanced in terms of worker involvement.

9 Scalar chain – workers should be clearly positioned in the business's hierarchy.

10 Order – workplaces and their facilities must be clean, tidy and safe.

11 Equity – managers should always act fairly to all their workers and maintain discipline in an even-handed way.

12 Stability of tenure of personnel – managers should try to minimise worker turnover.

13 Initiative – workers need to have the appropriate freedom to create and carry out plans of work.

14 Esprit de corps – organisations should promote team spirit, loyalty and unity.

Maslow's hierarchy of needs

Abraham Maslow was a psychologist working in the middle of the 20th century. He was primarily concerned with the human side of business and believed that all management styles are based on psychology. He wanted to determine what motivated people and, by understanding this, help businesses to motivate their staff to improve productivity. His pyramid model shows our main motivators as humans, starting from essential physiological requirements like breathing, water and food, and keeping ourselves safe. Once these aspects are in place, our motivations change to requiring love and a sense of belonging, then to elements of esteem such as confidence and mutual respect. Only when these motivations have been fulfilled can we start to consider self-actualisation, which covers creative and philosophical aspects of life. Figure 9.1 illustrates this. In the context of work, managers need to understand what might motivate their staff and how to provide these motivations.

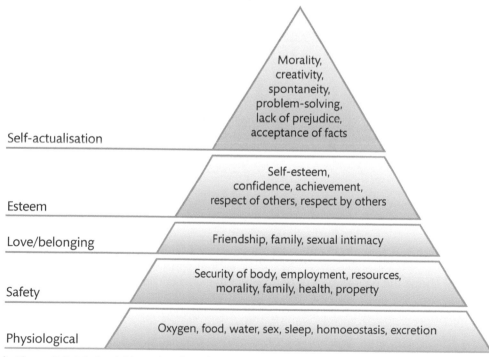

▶ **Figure 9.1:** Maslow's hierarchy of needs

McGregor's X-Y Theory

A social psychologist working in America, Douglas McGregor, proposed the X-Y Theory in his 1960 book *The Human Side of Enterprise*. He suggested that there are two main methods of managing people:

▸ Theory X – authoritarian management style – states that employees dislike work, avoid responsibility and need to be directed.

▸ Theory Y – participative management style – states that employees are motivated by job satisfaction and are keen to take responsibility.

Managers who opt for Theory X generally get poor outcomes, while managers who engage Theory Y often produce more motivated staff, resulting in better productivity and profitability.

Roles, responsibilities and interaction of a construction project management team

Construction projects involve a wide range of professions and the project team may change throughout the project. A number of key team members will be in place from concept to completion, whereas other specialist professions will only be involved for short periods of time.

It is crucial that the correct structure is put in place from the start of a project to ensure its success. Figure 9.2 shows an example of a hierarchical structure that might apply to a particular construction project.

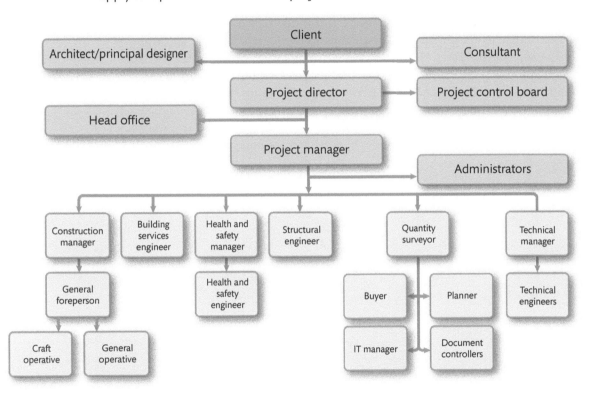

▸ **Figure 9.2:** Example of project management structure for a construction project

Table 9.1 shows the typical roles of people working on a construction project.

▸ **Table 9.1:** Typical roles on a construction site

Project team role	Description
Architect	• Usually the client's representative and may be called the principal designer for the purposes of the CDM Regulations. • Divides their time between office and site, communicating with the client and the project team. • Responsible to the client for planning, managing, co-ordinating and monitoring all the design duties during the pre-construction phase and then liaises with the principal contractor to help in the planning, managing, monitoring and co-ordination of the construction phase. • Involved at each stage of the construction process.

▶ **Table 9.1:** *Continued ...*

Project team role	Description
Quantity surveyor	• Responsible for the financial planning of the project. • Deals with all financial aspects such as payments for supplies, invoicing, claims for variations, etc. • Deals with cash flow and plans periods when payments for the work undertaken are made.
Construction/Site manager	• Responsible for all operations on site, although they may delegate to general forepersons controlling specific trade areas on the site. • Concerned with day-to-day planning, organisation and control of the site, including organisation of resources. • Ensures the project is delivered safely, on time, on budget and to required quality by carefully managing resources to maximise production and minimise wastage. • Arranges efficient disposal of waste, recycling (where possible) and arranges subcontractors.
Project manager (PM)	• Has overall responsibility for planning, organising and controlling the project from concept to completion. • Requires a wide range of skills such as identifying problems early, resolving conflicts and clearly communicating the client's objectives and vision to the team. • Must perform effectively within tight timescales and keep within strict budgets while maintaining a positive team dynamic.
Structural engineer	• Responsible for the design and physical integrity of the building to ensure safety and durability. • Structural engineers solve design challenges and help the architect achieve their vision.
Building services engineer	• Also known as mechanical and electrical engineers (M&E). • Responsible for installing and maintaining systems in buildings to make them comfortable, efficient and safe. For example: energy supply, lifts, fire safety, heating, ventilation and air conditioning (HVAC), ICT networks, plumbing, etc.
Buyer	• Responsible for purchasing and ensuring timely delivery of materials and plant to the site. • Analyses the main contract to obtain a set of delivery dates for materials. • Often places a bulk order with a supplier, who delivers specific quantities to the site when requested by the site manager.
Planner	• A technical role, usually responsible for supervising contract programmes, monitoring and reviewing progress; scheduling materials delivery, plant and labour; reporting procedures and overseeing pre-contract tender documents. • Has an overview of the project and reports on progress to the architect or project manager. • Expected to plan to make efficient use of resources, including working out how many operatives will be required; supplying buying section with materials schedules of what is required, when and how much; moving plant and machinery on and off site when required.
General foreperson	• Reports to the site manager or supervisor and is often trade specific, for example brickwork foreperson. • Assists the site supervisor with labour control, materials control and some of the trade-specific plant.
Craft operative	• Trade worker with a craft background, such as joiners, bricklayers and steel fixers. • Responsible for producing work of the correct quality. • Contributes to maintaining the construction programme. • Has a duty under health and safety legislation to work safely.
General operative	• Undertakes semi-skilled works, such as excavation of drainage trenches, working with concrete, keeping the site clean and seeing to the movement of resources.

Key terms

Variations – items that were not in the client's original budget and therefore are additional to the contract, such as changes to the design or additional quantities of material not in the original contract.

Cash flow – the amount of money flowing into the company from the client and out of the company as payments, for example salaries and payments for materials. Money flowing in should be greater than money flowing out.

Planning and forecasting a project's needs, requirements and resources

Planning the requirements for a construction project is a fundamental element in its successful management. It involves choosing the design, establishing specific work activities, estimating the resources required, predicting the length of time needed to complete each task and identifying links or overlaps with any of the activities. A good construction plan enables an accurate cost and time for completion to be established.

The plan often starts with the finished project and works backwards to initial site work. This enables the project team to identify specific activities involved in the construction of the project, and how they impact on other activities. They can identify any possible clashes or problems, so that they can accurately determine the most suitable plan and organise the required resources when necessary.

Every construction project has unique requirements or design features, but resources can generally be categorised in the same way: materials, labour, plant and finance.

On-site, short-term management for projects in progress

Table 9.2 lists the four phases of construction, and provides a brief overview of what happens and who is responsible for each phase.

▶ **Table 9.2:** Phases of a construction project

Phase	Management tasks
Pre-construction	The client appoints a principal designer to manage the pre-construction phase of a project involving more than one contractor, in line with the Construction (Design and Management) Regulations 2015.
	The principal designer must:
	• plan, manage, monitor and co-ordinate health and safety in the pre-construction phase
	• advise the client on the pre-construction information, and provide this information to the project
	• ensure everyone involved in this phase communicates and collaborates effectively
	• communicate with the principal contractor any risks to be controlled during the construction phase.
	The principal designer produces the detailed client brief for the project team, which provides the pre-construction information required under the CDM Regulations. The principal contractor must prepare the construction phase plan for the project.
Site preparation	The construction/site manager is responsible for this phase. They manage the clearing of the site, marking out of foundations, and designate the position of site office and storage of materials on site, as well as co-ordinating the human resources, plant and materials required for tasks undertaken during this phase.
Construction	During this phase, the principal contractor is responsible for the site and all the work taking place on site. Often the principal contractor will appoint specialist subcontractors to carry out specific elements of work such as foundations, steelwork, brickwork, joinery, roofing. Under the CDM Regulations the principal contractor must make sure adequate welfare facilities are provided.
Handover	The RIBA Plan of Work 2013 Stage 6 – 'Handover and Close Out' – describes handover strategy, commissioning (testing of services etc.), training of staff and other activities that are vital for the successful operation of the building. The 'as constructed' information will be updated with information from the specialist subcontractors and the principal contractor.
	This stage will also include the defects liability period (**snagging**) which may be 6 to 12 months. The final certificate will be issued at the end of this period.

PAUSE POINT   Briefly outline what is involved in the four stages: pre-construction, site preparation, construction phase and handover.

> Hint   Think about the differences between each stage.

> Extend   Explain how each stage follows on from the previous stage. Which team members are likely to have responsibilities at different stages?

Claiming interim payments

During a construction project, the contractor asks the client to make interim payments at stages or milestones throughout the building process, based on the value of work completed. A quantity surveyor carries out an interim valuation of the work that has been completed and recommends a value to the architect or designer representing the client, who issues an 'interim certificate' to the client as authorisation for payment.

Managing cash flow

For a building project to be successful it is vital there is appropriate financial planning and control to prevent the project running over budget. Cash flow management is the organisation of the money coming in and out of the company. Good cash flow management ensures there are sufficient funds to meet the needs of the company. Poor cash flow management can create a loss in profit and may even lead to the company going bankrupt. It is important to chase up payments due in good time, and to be aware of the outgoing and incoming payments due at different stages of the project. All changes to forecast costs should be recorded.

> **Research**
>
> Research construction projects that may have run over budget. Investigate the possible causes for the increase in budget and list the top five reasons.

Order and delivery of materials

Materials are the physical resources used to construct the building. Materials supply must be linked with the **construction programme** for the project to avoid delays. It is important that:

▶ the correct quantity of material is ordered (e.g. correct number of bricks or blocks)

▶ the material ordered is the required quality (concrete is the required strength, e.g. 30N/mm^2)

▶ the best value for money is achieved – balance of price and quality

▶ the materials are delivered and available at the required time.

A materials schedule is a list of the necessary materials for the job and the time they are needed to comply with the construction programme. The principal contractor completes the schedule before work starts to ensure materials are delivered on time.

The site manager needs to record and keep records of any materials delivered. They should carry out checks to ensure these meet the required specification for quantity and quality. Any damaged or incorrect materials should be returned to the supplier immediately. All records relating to the delivery of the materials should be passed to the accounts manager so they can check the quantities supplied against the invoices received.

> **Key term**
>
> **Construction programme** – the sequence in which the various tasks will be carried out during the construction project to enable it to be completed on time.

Labour requirements and training needs

> **Link**
>
> More information about subcontractors and labour requirements can be found in Unit 3: Tendering and Estimating, in section B3.
>
> More information about training labour and general site health and safety can be found in Unit 5: Health and Safety in Construction, section A6.

Labour is one of the key elements for a construction project. The labour requirements for any project must therefore be properly planned and costed well in advance of work starting and must link with the construction programme. The availability of labour depends on a number of factors:

▶ expected and reasonable pay rates

▶ government policy on spending on infrastructure and housebuilding

▶ training and upskilling of the workforce

▶ developments in technology

▶ location of the work

▶ skills needed for the project – for example, workers who know traditional building techniques or more modern methods.

The principal contractor completes a labour schedule listing the labour required throughout the project, including the number of operatives required for each activity. They should identify skilled tradespeople and specify whether they are directly employed or are subcontractors.

Maintaining skilled labour is important to ensure a high quality of work. Training is essential to maintain and upskill the workforce, and can help to motivate the workforce. A training needs analysis (TNA) will allow the company to get an accurate assessment of the type and amount of training needed.

It is essential that all the companies and employees (including subcontractors) involved in any and all stages of the project are aware of the health and safety legislation that applies and the need to adhere to it.

Plant requirements

The main purpose of using **plant** on site is to increase productivity and reduce labour costs. The plant required for the project depends on factors such as size and location of the site (for example, whether it is a restricted city-centre site or an open area), the extent of the work carried out and the cost. As with materials and labour, management of plant is critical and must be scheduled to link with the construction programme to maximise output and minimise costs. All drivers and users of plant must be fully trained and hold the appropriate qualifications and CSCS card.

Quality assurance and control

Quality assurance (QA) is an administrative process to prevent defects and failures and provide confidence to clients. Defects and failures during the construction process can result in delays and increased costs or disputes. QA aims to ensure that the building will be fit for purpose, built on time and within budget. It requires regular monitoring of the planning, design, specifications, contracts, construction and maintenance, and the links between these activities.

The main decisions relating to the quality of a construction project are made during the design phase rather than during the construction phase. At the design phase, material specifications and construction methods are planned to minimise issues that might arise during the construction phase. During the construction phase, QA involves ensuring the original design and specifications are followed, such as testing materials to ensure the quality is acceptable.

Workforce supervision

The workforce is one of the principal resources of a project and must be properly managed to ensure the project is completed to time, budget and quality. A workforce needs good leadership and supervision. It may contain employees of the company as well as staff from subcontractors, all of whom must be brought together into a productive group. It is important that any supervisor is competent and has the respect of the workforce. Good communication is an essential part of workforce supervision to ensure that all involved in the workforce understand what is required.

Decision-making

Making decisions is an integral part of a manager's job. A decision is an outcome and should be arrived at after careful consideration of all the available information. There needs to be a clear process for making decisions to avoid mistakes, accidents or failures. Everyone on the site needs to know what decisions they can make, and who to escalate problems to. This is where an organisational chart can help.

Managing unforeseen events

An unforeseen event could not have been predicted or anticipated, such as extreme weather conditions, fire, clients re-briefing the design or a subcontractor going out of business. In such circumstances, a contractor or subcontractor may demand compensation if they believe the events they encountered could not have been foreseen or were not part of the original tender or design. If the original **contract** is unclear, has insufficient detail or has not made allowance for unforeseen events, this can lead to disputes and delays which impact the project as a whole.

PAUSE POINT   What are the key resources for any construction project? Briefly outline the requirements of each resource for a construction project.

> **Hint**   Think about each resource individually.
>
> **Extend**   Explain how each resource is required at different stages of a construction project.

To manage unforeseen events, it is important the project team understands the client's exact requirements. These should be clearly and regularly communicated to the team. The project team need a process to look at claims independently while work continues. The client may need to keep a **contingency allowance**.

Handover schedule

The handover schedule is part of Stage 6 of the RIBA Plan of Works 2013. The aim of the project management team is to complete the project to the required standard so the client accepts ownership of the project. This will involve commissioning the building services of the project. This commissioning or handover period may last for 6 to 12 months and include customer training. During the handover period, other activities need to take place, such as compiling the documents and drawings for the health and safety file.

Under the CDM Regulations, the principal designer is responsible for preparing the health and safety file. The file is handed over to the client at the end of the project and must contain any information needed in relation to health and safety issues during any further work on the structure such as maintenance, conversion, adaptation or demolition (see Unit 5: Health and Safety in Construction, section B3). There will also be an agreed period of time for identifying and rectifying faults and defects.

Completion

Stage 7 of the RIBA Plan of Works 2013 is entitled 'In use'. This is the final stage and includes the completion of the activities in the handover strategy. It also includes updating the project information in response to continuous client feedback until the end of the building's life.

Managing the organisation's viable options

National and local policies, trends

Approximately two million people are employed in the construction industry, which closely follows the economic trends of the country. For example, during the economic recession in the late 2000s, the industry was hit hard, with high job losses due to a lack of investment in building. The UK is still recovering and spending on infrastructure and housing has been restricted as the government looks to reduce the country's **debt** and **deficit**.

The government has set a target of building 200,000 new homes a year. Over the last decade, the construction industry has failed to reach this target for a number of reasons such as planning issues, recession, **land banking** and a lack of available land for development. Local planning authorities have encouraged the development of brownfield sites and simplifying the planning system to try to boost the number of houses being built.

To improve the technology used in the construction industry the government has required building information modelling (BIM) to be used on large public sector projects. BIM allows all the different team members to communicate information with each other with the benefits of improved productivity, reduction in faults and reduced construction time and cost.

Labour requirements, recruitment, investment in skills and training

According to the Construction Industry Training Board's (CITB's) Construction Skills Network (CSN) forecasts for 2016–2020, the construction industry may create 230,000 construction jobs throughout the UK. During the recession, employers cut staff to a minimum and provided only limited training, and the numbers of apprentices fell drastically, all of which contributed to the lack of skilled workers available now. Employers are now finding it hard to recruit staff with the right qualifications and

---

Key terms

Contingency allowance – a percentage of the cost set aside to deal with resolving unforeseen events.

Debt – amount of money owed.

Deficit – the negative difference between spending and income.

Land banking – when developers buy up available land but do not build on it.

Discussion

As a group, discuss possible unforeseen events that may occur during a construction project. Compare these events to possible defects, which may be identified during the handover stage. Discuss who might be responsible for the cost of the remedial work and why.

Research

Search on the internet to find the government's strategy paper 'Construction 2025'. Who did the government work with to create this paper? What does the government wish to achieve? Find four key targets that are set out in the strategy paper.

Link

For more about BIM, see Unit 7: Graphical Detailing in Construction, section A2.

experience and this is set to get more difficult as demand rises. With an ageing workforce and lack of young people the industry faces a critical skills shortage.

The government, employers and training providers must work together to attract young people of the right calibre into the industry. These industry leaders must provide a clear progression pathway to allow professional development and create an attractive career.

The industry also needs to improve its recruitment of women. Currently women make up only about 10 per cent of the total workforce. The industry has much to do to improve its image and promote the positive aspects of a career in construction if it is to attract the number of young people it requires to meet the current skills shortage.

Subcontract or direct employment

It is unusual for a construction company to undertake all the tasks involved in a project using just its own staff. Table 9.3 captures the relative benefits of using direct employees and subcontractors.

▶ **Table 9.3:** Benefits of using direct employees and subcontractors

Benefits of direct employees	Benefits of using subcontractors
• Cheaper daily rates than subcontractors. • Enables more control over staff. • Ensures the quality and training of staff is suitable. • Staff should be more motivated and committed, leading to fewer disputes and discipline issues on site.	• Enables the use of specialist skills such as plumbers, electricians, fire and security alarms systems, flooring, etc. • Can be called in to carry out tasks as and when required, so a company does not have to employ them full-time, saving money. • They supply and maintain their own equipment, saving costs to the company. • The company is not responsible for the wages of the staff, holiday pay, sick pay or pension contributions.

The most successful solution is usually to have a small reliable team of directly employed staff on site to work alongside the subcontractors. This helps to ensure that the company's policies and procedures are followed and reduces the number of defects or faults encountered.

Site management structure

The management structure on a construction project will depend on the size of the project. For a project to be successful, the project team must work collaboratively to overcome design changes and unforeseen events. A company can use two main site management structures:

▶ Fully site-based – for large projects the project team may all be based on site. A principal contractor is on site at all times and they will be in regular contact with the rest of the project team overseeing the progress of the project. Regular site meetings will ensure everyone involved is clear on what work has to be carried out and able to raise any issues.

▶ Head office-based functions and support – most small to medium-sized construction projects will operate in this way. The principal contractor on site will have access to a range of company resources, which will be available at the head office. Often the company may be operating several sites and therefore it is cost-effective to centralise these resources.

Plant and equipment hire, lease or purchase

Plant and equipment are a principal resource for any construction project and must be effectively managed to maximise their use and minimise costs. For most construction projects of a short duration (16–20 weeks) it is more cost-effective to hire plant than to purchase it outright, as plant is expensive and the company may not need to use it again once the project has been completed.

Leasing plant is similar to hiring, but involves an agreement with a finance company. The contractor hires the plant for a number of years, usually three, and pays rent for it every week or month. Sometimes there are restrictions, for example the number of miles the contractor can use the equipment per year, and it is often difficult to end a lease agreement early.

Purchasing for long-term contracts or large companies is sometimes the most efficient option. The company needs to establish if it will make full use of the equipment; leaving it to sit idle for long periods is not cost-effective. Before deciding to purchase equipment, the company will need to ask some questions:

▶ How will the purchase be paid for – through a bank loan or from the company's profits?

▶ How much maintenance and servicing will the equipment require and what are the likely costs?

▶ Will an operator need to be trained to use the equipment?

▶ How much will the equipment depreciate per year?

▶ What percentage productivity a year will the equipment achieve?

> **Discussion**
>
> Discuss the advantages and disadvantages of various options available for obtaining plant and equipment for a construction project.

Organising, procuring, co-ordinating and controlling

Materials, plant and equipment delivered to site on time

The principal contractor is responsible for organising, procuring, co-ordinating and controlling materials. The availability of materials has become a problem in recent years, as suppliers no longer carry high stock levels. This is because materials standing in their yard or warehouse tie up too many financial resources. Many specialised materials are made to order and can take many months to be delivered. This needs to be considered during the planning stage and may mean the client pre-purchasing a material and including a provision in a tender document, so that when the material does arrive the contractor is paid for handling, storage and delivery.

The availability of space for storage, the rate of work, the distance materials have to travel and the reliability of the supplier are all factors to be considered when planning a delivery schedule. Good organisation can reduce the amount of waste (and therefore save money) by planning delivery dates and quantities which link with the construction plan and use available storage on site.

▶ Piggybacked site storage cabins save space

Plant and equipment need to be utilised in the most cost-effective way, for example by transporting plant and equipment from site to site so any down time is kept to a minimum. Highly trained and experienced operatives will be more productive than inexperienced ones.

Site storage facilities

The storage of plant and materials on site should be discussed at the design stage. Projects under the CDM Regulations should include arrangements for materials storage in the construction phase plan. Legislation requires sites to be kept to a good standard to prevent slips, trips and falls. Materials must be stored in clean, tidy and secure areas to prevent them becoming damaged or stolen.

A site layout plan can help identify possible storage areas. In sites with limited space, site huts and storage containers may be piggybacked (stored on top of each other). Delivery of materials will be 'as needed' due to a lack of storage space. As the project proceeds, these facilities may need to be moved to allow the construction to progress. On large sites, storage may have a permanent location.

Site distribution methods

In order to ensure that a site is operating efficiently and effectively it is vital to use a suitable store management system to issue operatives with the correct equipment and materials. Handling equipment such as cranes and forklift trucks need to be selected depending on the height and reach required for the project. The storage area will require sufficient space to allow site vehicles and delivery vehicles to access it safely. This may require a one-way system to eliminate reversing or unsafe manoeuvring.

Workforce requirements

In projects that fall under the CDM Regulations the construction work cannot start until welfare facilities have been provided. The principal contractor is responsible for the maintenance of the welfare facilities. The size and number of facilities required depends on the number of operatives working on the site. Basic welfare facilities which must be provided are toilets, washing facilities, drinking water, changing rooms and lockers for dry clothes and personal protective equipment, as well as facilities for rest.

Motivating the workforce

Labour is one of the principal resources of a project and it is vital to utilise this resource to the maximum. A content, motivated workforce is a productive workforce. Providing suitable welfare facilities such as clean toilets, a warm room to dry wet clothes and an area to eat and have warm drinks is a first step in motivating the workforce. Some other methods are shown in Table 9.4.

Table 9.4: Ways to motivate staff

Method	Description
Incentives	Bonus schemes, and/or offering incentives such as company cars and vans, can reward effort by employees. Many companies operate staff discount schemes and other benefits (e.g. vouchers employees can use for services outside work, such as private health benefit schemes for families).
Awards and rewards	Can be used to improve productivity and health and safety on site. Any scheme must be communicated fully and should be simple and easy to follow. Subcontractors can also be included in the award and reward schemes. Feedback on success should be communicated through site meetings, staff briefings, toolbox talks, site inductions and even publicising in the local press where appropriate.
Job security	When people feel secure and content in their jobs, they go above and beyond their job description to help the company, improving productivity and reducing disputes. This encourages competent staff to stay with the company, promoting staff loyalty.
Training	Employees offered regular training feel valued and appreciate the opportunity to increase knowledge and gain new skills. They tend to stay loyal to the company, and it has company benefits, including health and safety.

Ⅱ PAUSE POINT Explain why it is important to motivate staff on a construction site and identify the different approaches which can be used to motivate staff.

Hint Think about what might happen if staff are not motivated.

Extend Explain how motivating the staff can help the overall construction project.

Communication with the design and management team, the workforce and suppliers

Effective communication ensures clear understanding of the requirements on site, creates a team spirit and builds trust between the various parties involved in the project.

Part of good communication is ensuring chains of command and management structures are clear. This refers to the levels of authority within a company, and tells employees at each level who they are responsible to. These structures can vary depending on the size and nature of the company, but all employees should be clear on the structure, their responsibilities and who their manager is. If an employee does not understand the structure or ignores it, this can create problems. For example, it could lead to errors and disputes, undermine the authority of managers or create an atmosphere of uncertainty and mistrust, leading to low morale and productivity with higher rates of accidents and staff turnover.

Some of the methods a business will use for communication are shown in Table 9.5.

Table 9.5: Methods of communication on site

Method of communication	Description
Team and site meetings	Regular meetings allow team members to meet and discuss the project, communicate any necessary information and achieve a balanced opinion. Meetings can be time-consuming so it is important they are focused, meaningful and successful.
	Formal meetings are usually held at a specific time with a set agenda, allowing those attending to prepare. Minutes will record decisions, identify actions and allocate these to team members to prevent any confusion or debate later. Topics could include progress reports, reviewing planned vs actual progress, quality issues, and planning the next period of work.
Written communication	A range of written forms of communication are used during a construction project. • Letters – to request or transfer information accurately, usually formally. Should be signed and dated and can be used as reference or evidence in disputes.

▶ **Table 9.5:** *Continued ...*

Method of communication	Description
Written communication	• Report – a statement of facts and conclusions on a particular matter; often produced after an investigation by an appointed competent person. • Architect's instructions – the client's instructions must be given in writing and may include changes to the design, variations to the works, instructions to carry out tests and open up work for inspection. • Site instructions – similar to architect's instructions and given by the consultant engineer to the contractor to carry out works, purchase goods or carry out tests.
Telecommunications	On large sites, site radios can be used to communicate, such as issuing instructions to crane operators. Mobile phones are vital for quick and easy communication; however, their use on sites creates hazards and risks, for example plant drivers may be distracted by texting. It is important that the contractor has a clear policy on mobile phone use on site communicated through the site rules, site induction and toolbox talks. It may be that a safe zone for using mobile phones is established so that necessary calls can be made.
Graphical and electronic media	Electronic forms of communication, such as emails and texts, allow information to be communicated from anywhere in the world almost instantly, and is now commonplace on construction sites. Graphical communication (e.g. site drawings, building plans, etc.) are essential, as they ensure everyone can clearly see what is intended without misunderstandings. Computer-aided design (CAD) further helps communication as detailed images are produced and areas can be zoomed in on to see more detail.
Information technology	Building information modelling (BIM) creates a collaborative approach which should reduce errors and introduce more efficient ways of working. By using BIM everyone involved in a project can understand how the building is constructed by using a digital model created from information contributed from the whole team. All team members should be working to the same standards, and detailed information on the material specifications is part of the digital data contained within the model.

Ⅱ PAUSE POINT List the key methods of communicating information on a construction site. List what topics should be covered during a site meeting.

Hint Think about different forms of communication used within the construction industry.

Extend Why is good communication essential on a construction site? What are the consequences of poor communication on site?

A2 Application of construction management techniques

When working on site, you need to understand the various site management responsibilities and the techniques used to manage a site to ensure that the project is completed efficiently, from commencement to completion, on schedule and to the client's budget.

Standard planning techniques and how these are applied to control work on site

Production and use of master programmes

Programming is a planning technique which identifies the sequence and interrelationship of the activities that need to be carried out to complete the project on time. An appropriate programme is necessary for every construction project with each activity identified and given a set time for completion. Usually the programme is presented visually, such as in a chart or graph, so that the interdependence of the activities can be clearly seen. This is an essential tool to check the project's progress, by comparing work completed against the original programme.

Most construction contracts require the contractor to produce a programme of works. This can be a non-contractual reference point or part of the contract and imply obligations to deliver the works in a certain order by specific dates. Any changes in process or time may be regarded as a breach of the contract.

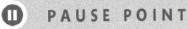

Master programmes will include:

▶ dates and time periods for each activity

▶ the sequence or order of the activities and how they may depend on other activities

▶ the resources needed for each activity such as plant, labour and materials.

The master programme identifies all the main activities throughout the project and creates an overall framework of the work and identifies the time required for each of the main activities.

Link

Drawing up delivery schedules is an important part of site management. More information about these can be found earlier in this unit in section A1.

PAUSE POINT Explain the purpose of a master programme and the difficulties which may arise during a project.

Hint Think about why difficulties may arise during a construction project.

Extend Identify the content of a master programme and explain why each aspect is required.

Three progress monitoring techniques are commonly used for construction projects, often in combination.

Gantt charts

Gantt charts are one of most common and easily understood planning methods. A Gantt chart shows a project divided into a series of activities (listed vertically) and the length of time in months or weeks (listed horizontally) required to complete each activity; see Figure 9.3.

JEP Construction

	Mar-17	Apr-17	May-17	Jun-17	Jul-17	Aug-17	Sep-17	Oct-17	Nov-17
Set up site	■								
Excavation	■								
Foundation		■							
Substructure		■	■						
Superstructure			■	■					
Roofing				■	■	■			
Services			■				■	■	
Internal finishes								■	
Landscaping									■

▶ **Figure 9.3:** Gantt chart

The length of the bar is proportional to the time required to complete the activity. Gantt charts are simple to create, easy to follow and easy to use to track progress by recording planned and actual completion dates of each activity. However, the sequence and interdependence of activities may not be completely explained. Gantt charts do not provide any information about the control of the plant, labour and materials required for each activity.

To produce a simple Gantt chart, follow this sequence:

1 Analyse the contract drawings and specification and establish the activities needed. A medium-sized construction site usually has around 20 activities.
2 Find the estimator's calculation of the length of each activity. Choose a suitable time unit to cover all the activities, such as weeks or months.
3 Record the logical sequence of activities so there is a working link between each.
4 Establish which activities are critical to the overall programme.
5 Plot each activity on a rough outline bar chart, beginning each at their earliest start point.
6 Establish the critical path through the bar chart. Some activities float within the critical activities, meaning their start or finish times can be delayed and have no overall impact on the completion date; others are critical to the overall programme and end slippage will affect the end date.
7 Adjust the non-critical activities to suit the plant, labour and material resources on site.

On the chart, time runs from left to right with the start of the programme being the first activity and handover the last. To monitor progress, place a string line across the programme at the current date, which will show the percentage of each bar that should be completed. This will identify which activities are ahead or behind schedule.

Critical path analysis (CPA)

CPA is usually used for large or complex projects. It represents the plan for a project by showing the sequence and interdependence of each activity. This allows the project team to determine the minimum time the project can be completed in. Critical and non-critical activities are identified, enabling the project management to balance the resource requirements. CPA allows the team to quickly analyse the effects of delays and to determine which activities need to be prioritised. However, they can be complex and difficult to understand unless you are competent in their use and interpretation.

Each arrow on the network represents an activity. The description of the activity is above the arrow, with the duration below it. Time runs from left to right, with the first activity representing the start of the programme. See Figure 9.4 for an example of a critical path analysis. To produce a simple network:

1 Analyse the contract drawings and specification and establish how many activities will be needed.
2 Find the length of each activity as set out by the estimator. Choose a suitable time unit to cover all the activities, such as the number of days or weeks.
3 Record the logical sequence of activities so there is a working link between each.
4 Draft the network diagram using nodes and arrows.

Add the description and durations of each activity.

5 Calculate the left-hand side of the circle's earliest start times right through the network. Where two arrows finish at one node, take the highest value calculated starting at zero.
6 Work backwards from the completion node, taking the lowest value where two arrows enter a node until arriving back at the commencement, which is zero.
7 Identify the critical path and mark it in red – this is the path where the left-hand and the right-hand figures in the circles are the same value.

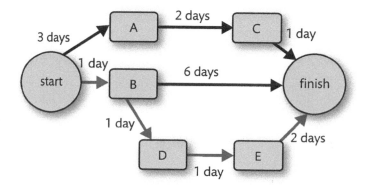

▸ **Figure 9.4:** Example of a critical path analysis (CPA)

Line of balance

Line of balance (LOB) is similar to a Gantt chart as the bars represent activities but the time is plotted on the horizontal axis with the number of units or sections requiring similar activities plotted on the left-hand vertical axis. LOB is a visual diagram representing the rate of working of repetitive activities; see Figure 9.5.

LOB is suitable for repetitive work such as constructing a large housing project or building floors in a multi-storey project.

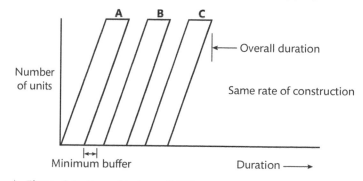

▸ **Figure 9.5:** Line of balance (LOB)

Daily activity sheets

Daily activity sheets are a record of the work which has been completed each day to track the progress made.

Production of site layout plan

Site layout plans must be prepared by the contractor before any work is carried out. On large or complex sites, a site layout plan is vital to organise the movement of plant, labour and materials effectively. The plan will be governed by the size of the site and the **footprint** of the building to be built. Though the principles for organising a site will be the same, each site will be unique. Detailed planning of the site layout can reduce the distance materials need to be moved, reduce traffic congestion and improve the health and safety on site, improving staff morale and productivity.

Figure 9.6 shows a typical site layout plan in which the contractor would construct the building. The site layout plan is drawn to scale and, once complete, passed to the site supervisor to add all the site accommodation, box containers, traffic routes, etc.

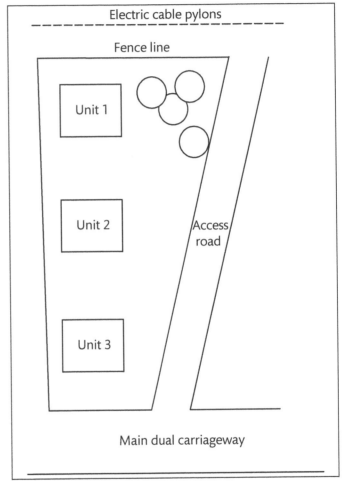

▶ **Figure 9.6:** Site layout plan

PAUSE POINT How would the location of the site affect the access and exit points?

Hint Think about the issues that might need to be considered for a site based in the city centre compared with one in the countryside.

Extend What other factors would be determined by the location of the site?

Method statements and risk assessments

Risk assessments are required under the Management of Health and Safety at Work Regulations to manage risks at work. A method statement outlines the sequence of activities, the planned resource requirement and the planned duration for the main activities. Method statements help manage the work and ensure that any precautions have been communicated to those involved. There are two types of method statement:

- A detailed description of how a task is carried out, listing the resources required, including labour, plant and materials. This helps to identify the hazards and the control measures needed.
- The estimator lists the labour, plant and machinery needed. This is used by the contracts manager to establish the construction plant required and how much it will cost.

Risk assessments and method statements are often described as RAMS. RAMS may be required by the principal contractor from subcontractors to establish that health and safety has been considered and the requirements of the CDM Regulations have been met. If they are not produced, or are not suitable, the subcontractor will not be allowed to work on the site.

The application of quality assurance and quality control requirements

QA is a quality management system designed to prevent defects, errors and accidents and ensure that agreed standards are achieved. Quality control is a method to ensure the consistent quality of the materials or work by aiming to identify defects in the materials or work produced.

There are several key processes used in QA, as shown in Table 9.6.

> **Table 9.6:** QA processes

Process	Description
Document control	A huge number of documents are used during a construction project. A system must be in place to control them. The project team may have their own quality management system or may use ISO 9001 certification to control documentation. Document management systems (DMS) automatically name, number, track changes, create versions, archive and share documents. Everyone involved must be committed to document control to prevent problems and disputes. The government requires building information modelling (BIM) to be used in public sector projects.
Drawing registers	Managed by the principal contractor to keep track of the changes made to the construction drawings as the design is amended and altered, a drawings register should include: • date and time of receipt • drawing number and revision number • drawing title • accompanying document reference, for example site instruction (SI), drawing amendment notification (DAN), variation order (VO), etc. BIM reduces the chances of out-of-date drawings being used to construct the project.
Specification use	Contract drawings show what the finished product should look like. Specifications provide essential information about the materials, finish and workmanship required. They do not lay down how the work should take place, but should confirm the requirements for the safe and successful installation of a particular component. BIM and data modelling techniques offer greater control and detail when producing specifications. Specifications are created early so that they can develop and change alongside the project. This means they should be more consistent, allow less room for errors and save time and money.
Site testing	The clerk of works reports on quality and project progress to the client and instructs on-site tests to be carried out. This monitors the quality of the materials and workmanship throughout construction. Samples are taken of various materials, sometimes by request specifically from the architect. For example, concrete is tested using a **slump test** to ensure workability when it arrives on site. Other site testing may involve load testing or crack monitoring.
Off-site testing	Some samples need to be taken to a materials testing laboratory, which should have United Kingdom Accreditation Service (UKAS) accreditation for testing and calibration of materials. Tests are carried out on a range of materials such as concrete, masonry and steel. For example, concrete strength is assessed at regular intervals by making concrete cubes and testing them to destruction at specific time intervals within a controlled environment. Tests can also be carried out on manufactured products such as windows and doors to ensure they meet specifications.
Dimensional quality control	All buildings need to be set out with absolute accuracy using modern, UKAS-accredited, calibrated equipment, to avoid errors and disruption to the project. Quality checks should be in place to identify any errors early. Surveyors carrying out checks should have the most up-to-date site plan. Building co-ordinates should be calculated correctly and uploaded onto survey equipment. Control or reference points should be located in places that will not be disturbed during construction.

Research

Research some examples of risk assessments and method statements which are used on construction projects. What do they have in common? To what extent do you think they are successful in managing health and safety risks?

Key term

Slump test – a test of concrete's workability to determine whether it is of the right quality and can be poured with the correct results.

PAUSE POINT   What is the difference between quality assurance and quality control?

     Hint   Refer to the explanations on the page.

     Extend   What methods are available for monitoring quality on a construction project?

Link

More information can be found about Building Regulations and the NHBC in Unit 8: Building Regulations in section A.

Discussion

As a group discuss the advantages of Building Control inspections. Are there any disadvantages with Building Control inspections?

Compliance with statutory liaison

A statutory liaison is a communication required by law with a group or body such as the local planning authority, Building Control, Environmental Agency, Historic England, etc.

Building Regulations Notices and inspection

The Building Regulations set out minimum standards for producing houses and commercial buildings that are safe and secure for the occupants. They ensure the safety and the quality control of buildings.

Most building work in the UK requires approval from the local authority Building Control department, gained by Full Plans approval or Building Notice approval. With a full application, Building Control inspect the drawings for any issues before building begins. With Building Notice approval, work can start immediately but, if any problems are found, Building Control can demand work is taken down and fixed. Building Notice is best suited to small domestic building projects.

The Building Regulations require inspections by Building Control officers to be carried out throughout construction to ensure all work complies with the regulations. The builder or project manager must notify Building Control that the project has started. An initial inspection will be carried out to stipulate what further inspections will be required. On completion of the work, if it meets the approved standard, a completion certificate is issued as evidence that the work meets the standards. This is required when selling the property.

National House Building Council (NHBC) inspections and standards

The NHBC is an independent body which aims to raise standards and improve quality in UK housebuilding. It is approved to carry out inspections for Building Regulations and provide other services such as energy rating of houses, and health and safety training. Most new houses built in the UK have an NHBC ten-year warranty, which is granted after a final inspection by an independent NHBC inspector, who will also conduct inspections throughout the construction process. Mortgage lenders often require an NHBC warranty to be issued before releasing any funds.

NHBC technical standards and inspections are in addition to Building Control regulations and inspections, not in place of them, and guarantee a high standard of building.

Application of on-site sampling and material testing techniques

There are many reasons for sampling and testing materials on site, mainly to ensure materials are fit for purpose, meet the quality and performance requirements of British and European standards and comply with the Building Regulations. It may also be necessary to sample and test material if there are problems or failure on site. These tests can also help resolve disputes.

Testing the quality and consistency of raw materials is vital to ensure the finished product is acceptable, for example by ensuring the quality of the concrete poured on site for walls, floors, columns and beams. Errors and defects at this stage create major problems later in the construction process. Finished products also need to be tested to ensure quality and performance, such as precast concrete products used for floor slabs or lintels.

Supervision and inspection of the quality of outcomes produced by the workforce

The quality and quantity of work produced on site must be regularly monitored and inspected to ensure the project is completed to the required standard, time and budget. If work activities fall behind schedule due to low productivity or poor workmanship, redistribution of labour resources from one activity to another can help resolve this. Progress is monitored by evaluating the level of production achieved on site against contract programme activity. Quality of work can be tested and inspected by competent staff such as the clerk of works. Output from the labourers can then be raised to meet any shortfall if an item is behind programme.

Management of direct workforce

When managing the workforce, several key issues must be addressed to ensure that the team is working safely and effectively. Some of these are shown in Table 9.7.

▶ **Table 9.7:** Factors when managing the direct workforce

Factor	Description
Recruitment and competence requirements	The company will follow a recruitment process, which may involve an application form, interview, checking qualifications and references and a possible demonstration of skills before offering a job. This ensures the candidate is both suitable and competent to do the work.
Training	Under HASAWA 1974, it is the employer's responsibility to provide information, instruction and training to ensure employees have the skills and knowledge for their role. Operatives must also have the minimum health and safety training required by the CSCS card prior to entering a construction site. CSCS cards provide evidence that operatives are competent and have the qualifications and training for their trade. The CSCS scheme is recognised throughout the construction industry and has a register of individuals working in construction and their training and qualifications. For more about CSCS cards, see Unit 5: Health and Safety in Construction, section A6.
Monitoring equal opportunities	Employers must be aware of the Equality Act 2010 and its implications. It is in place to prevent discrimination on grounds of race, religion, age, sex, disability, pregnancy or maternity, or sexual orientation. It ensures recruitment is fair and that the best candidate for the job is employed. Regular staff training ensures all staff are aware of the company's equal opportunity policy. Monitoring all applications for equal opportunity purposes is good practice and enables companies to identify potential areas where discrimination may be taking place.
Leadership skills	Strong leaders motivate and inspire the team to work productively. Good managers will be able to employ a range of skills to motivate the employees to ensure the project is completed to time, quality and budget. These may include prompt, clear communication, setting clear goals, demonstrating a positive attitude, praising workers in public and disciplining them in private, leading by example and allowing team members some responsibility and independence.

Management of subcontractors

Techniques for managing subcontractors effectively include:

▶ Communication methods – good communication is crucial as a large proportion of site work is subcontracted. Regular contact ensures co-ordination between different subcontractors. Weekly site meetings should be in place with all key subcontractors attending. Health and safety compliance, site rules, access and egress issues, use of cranage, waste disposal, and storage are just some of the items that need to be co-ordinated. Other forms of communication include, for example:

- signage – to clearly alert workers of hazards
- team meetings – to ensure a good understanding across the team
- one-to-ones – where specific points need to be raised with a particular member of the team
- site/project handbook – clearly outlining the discipline expectations, and health and safety considerations for the site.

As for the leadership skills discussed in Table 9.7, good communication requires careful use of language to instruct, motivate, persuade and discipline the workers.

▸ Checking insurance and legal requirements – The project team needs to check subcontractors are competent, with the right combination of skills, experience and knowledge to carry out work safely and correctly. The principal contractor must check the subcontractor's liability insurance to confirm they are insured to carry out any work. The principal contractor must also check the RAMS of the subcontractor to ensure they are using the correct equipment.

▸ Responsibility for compliance – The project team has responsibility to ensure the subcontractors are complying with all site rules and regulations. Clear communication and regular checks on progress will encourage this.

▸ Retention of payment – Retention is holding back a percentage of the overall amount for a specific job or activity, often between 5 and 10 per cent. This means the subcontractor will not be paid the full amount until the work is completed as required. This fee may not be released until after an agreed 'defects liability period' such as six or twelve months. Often the fee will only be released when certification of completion has been issued. If this practice is not managed correctly, it can lead to disputes over subcontractors demanding their retention fee and clients demanding aspects of the work be completed fully.

▸ Remedial work required – Defects due to inappropriate specification, detailing and design can occur in any element of the building as shown in Table 9.8. Faults and defects must be repaired at no additional cost before handover. Many defects are identified in the initial stages of the building handover and the builder will be liable to make repairs. This is referred to as the 'snagging list'.

▸ **Table 9.8:** Common defects found during construction

Section of building	Defects which may require maintenance and repair
Foundations	Settlement of building, waterproofing for leaking basements
Walls	Dampness penetration, condensation, thermal bridging
Floors	Poor noise and heat insulation
Materials	Timber (cut wrongly or split), steel (scratched or corroding)
Services – plumbing, electrics	Leaks, system not working, faulty fire alarms or wiring

Assessment practice 9.1 — A.P1 A.P2 A.M1 A.M2 A.D1

A local property developer wants to research her options in building a number of medium and large projects. She will need to engage a project management team to organise and plan the projects. The developer has mainly dealt with small projects and has asked your advice on the roles and responsibilities required for a successful project management team and the techniques required to manage key resources. She asks you to research the most suitable methods of monitoring quality and progress.

To achieve this, research and produce a brief booklet on the roles and responsibilities of each of the members of a project management team and the management of the principal resources including:

- labour and material requirements
- plant and equipment requirements
- motivating the workforce
- communication methods within the team.

The second part of the booklet should explain the techniques applied by a site manager to manage the project such as:

- programme techniques
- quality assurance and quality control techniques
- compliance with statutory requirements
- management of workforce – direct labour and subcontractors.

Plan

- How will I research the roles and responsibilities of members of the project team?
- How will I prepare brief descriptions of techniques applied by a site manager to manage the project?

Do

- Am I including as much detail as possible?
- When will I be ready to present the booklet to my peers for additional suggestions?

Review

- I can explain how I would approach the difficult elements differently.
- Once I have reflected on my own work and any feedback, I will make any necessary changes in my booklet.

B Understand purchasing and cost management techniques

B1 Application of purchasing methods

The project management team must create a system to deliver the required quantity of materials to site on time, budget and to quality. Without a robust system, the project may become delayed or incur additional costs.

List of selected suppliers

The selection of reliable suppliers for the project is crucial. By introducing a selection process, and judging suppliers against set criteria, it is possible to manage the risks associated with the supply and delivery of plant and materials. Factors to consider include looking carefully at the supplier and their:

- ▶ area of operations (where do they operate?)
- ▶ previous performance on past projects they have worked on
- ▶ operational capacity to supply what the project needs
- ▶ reputation among other clients and staff
- ▶ stock levels and the speed at which they can replace used stock
- ▶ ability to meet any changes in demand on the project.

 PAUSE POINT   What factors may influence your choice of suppliers for a project?

       Hint   List as many factors as would influence your choice of supplier.

       Extend   Looking at your list of factors, discuss which would be the most important and explain why.

Link

More information about scheduling materials or extracts from bills of quantities and the use of correct and appropriate specifications can be found in Unit 13: Measurement Techniques in Construction (section C2) and Unit 3: Tendering and Estimating (section A1).

Materials and subcontract enquiries

It is good practice to get a written quotation from at least three different subcontractors. A quote is an agreement to carry out specific work for a fixed price. It is important to be very specific about what work is required, the quality of materials to be used and the time to complete the work. The client compares the quotes to determine which offers the best value (not just the lowest price). Before accepting a quotation, the client should research the subcontractor to review their previous performance and reputation. Once a client accepts a quotation this becomes a binding agreement between the client and subcontractor.

The receipt of quotations should be acknowledged, and the quotation checked to ensure it is correct and has not omitted anything. Both the client and contractor or subcontractor should be clear on the exact work needed in terms of quality of materials and time to complete work.

Gap analysis

Gap analysis is a method of comparing actual performance against potential performance. It can identify gaps in areas of responsibility or overlaps in work processes. These gaps occur over time due to change in work processes and a lack of attention to monitoring and controlling.

Negotiating skills

As well as price, there are many criteria which should be considered when engaging a subcontractor, including quality of workmanship, reliability and their ability to work in a team. The subcontractor should have good communication skills and be able to work with the project team. There are often many changes to design and delays. It is therefore important that the subcontractor can be flexible and negotiate when resolving design changes and delays rather than causing further delays with disputes.

Planning links

Purchasing materials

Materials are one of the principal resources needed for a construction project. As projects become more complex and more expensive, it is more important to manage the purchase and delivery of materials to match the construction programme and control the potential for delays and increased costs.

Material **procurement** planning (MPP) is the process of managing purchasing materials in the right quantity, from the right supplier at the right time. Effective management of materials can result in significant savings.

Key term

Procurement – the process or act of buying or acquiring goods or materials.

Lead times

Lead time is the time taken between placing an order and the item being delivered to site. It can involve the time taken to achieve statutory approval, plant hire and production of drawings and designs.

To prevent delay it is critical to identify early any items with long lead times and their impact on the project if they were to arrive late highlighted. There may be long lead times for items such as bespoke cladding or window systems, glulam beams, steel or concrete frames, items sourced overseas and specialist plant such as cranes and tunnelling equipment.

Just-in-time deliveries

When a just-in-time delivery system is being used, a detailed tracking process of the progress of the project must be kept to ensure that materials and plant are delivered when they are needed. Although there may be additional costs in engaging suppliers to store materials longer and guarantee delivery, there are distinct advantages such as reduced waste, greater productivity and higher quality of product.

This method prevents the need for large storage areas on site. However, any supply problems can have a major impact on the cost and completion time for the project.

Discussion

Divide into two groups. One group must discuss the impact on a project of items with long lead times and how best these can be managed, while the other group discusses the advantages and disadvantages of just-in-time deliveries. As a class discuss the importance of planning deliveries and the impact poor planning may have on the construction programme.

Ethical purchasing and supply

An ethical purchasing policy aims to promote suitable working conditions and environmentally friendly standards along the supply chain, and ensure no one is exploited. Increased public awareness of poor working conditions abroad has increased pressure on companies to take responsibility for the working conditions of the workforce in their supply chain around the world. Ethical policies focus on safe working conditions, the encouragement of good health, reasonable working hours, pay meeting local legal standards and eradicating slave and child labour.

Local sourcing and minimising transportation

Local sourcing of sustainable materials ensures the materials used match other buildings in the area and blend in with the surroundings, helping to support the local community. This improves the reliability of delivery times and reduces costs.

Local sourcing also reduces CO_2 emissions because materials are transported shorter distances. Accurate planning of deliveries can reduce the number of journeys required, reducing transport costs and resulting in lower emissions. In addition, planning can reduce traffic in the area, minimising noise and emissions.

Another benefit of local sourcing is that it promotes local jobs in the supply chain.

Use of sustainable materials

The environmental costs of producing materials such as concrete and steel for the construction industry can be measured by calculating the embedded energy in the material.

As awareness for the need to protect the planet increases, so has the requirement to incorporate sustainable materials into construction project design. A sustainable material is naturally produced and can be replaced at the same speed it is used, for example timber, clay and cork. Their advantages are that they:

▶ conserve energy and save money (for example, low flow toilets and natural insulation material)
▶ minimise the carbon footprint (for example, using reclaimed or recycled materials)
▶ reduce the energy use and financial costs of transportation
▶ make a healthy, comfortable home (for example, better air quality and temperature control)
▶ can create an aesthetically pleasing finish to a building
▶ reduce the damage caused to the environment.

Many materials used in construction, such as concrete, asphalt, metals, masonry, plastic and glass, can be recycled. Recycling reduces the amount of material going to landfill and the depletion of natural resources.

Fair trade agreements

Fair trade agreements mean the producer of the product receives a fair price for their product, based on the global market price. This may make it more expensive than from other producers but it guarantees minimum standards of pay and working conditions for all the workers. It ensures workers in other countries are not exploited and helps provide them with opportunities to improve their standard of living.

Abuse of power

The Competition Act 1998 deals specifically with abuse of dominant position in business dealings. This law was created to prevent powerful companies pressurising smaller companies into agreeing unfair deals or contracts, such as making the purchase of one product conditional on the sale of another product, demanding

---

Theory into practice

You have been tasked with ordering a range of materials for a project:
- 10,000 – 7.3N solid dense block 100 mm
- 10,000 – red multi-facing bricks 65 mm
- 10 – concrete lintels 1500 mm

Calculate quotes for all the materials using prices obtained from three different online building suppliers. Was there any discount for larger quantities? Are all the materials labelled the same? List any difficulties you encountered when calculating the costs.

Link

More information on sustainability can be found in Unit 2: Construction Design (section C4).

Research

- Make a list of materials commonly used in construction projects. Research how each material may be reused or recycled.
- Find a definition of embedded energy and research the embedded energy for each material you listed.

that suppliers only supply to them, refusing to deal with other companies, charging unreasonably high prices and selling at artificially low prices with which smaller companies cannot compete.

Avoidance of corruption

Corruption is the dishonest or fraudulent behaviour of people or companies to gain advantages and often involves bribery. The construction industry has a poor record on corruption and is one of the most susceptible sectors. Corruption in the construction industry can take the form of bribery to obtain planning permission, overestimating or underestimating land values and collusion to monopolise the market.

In 2009 the Office of Fair Trading (OFT) fined 103 construction firms in the UK a total of £129.5 million as punishment for breaching competition law. The companies were found guilty of participating in anti-competitive bid-rigging in the form of **cover pricing**. As a result, a Chartered Institute of Building (CIOB) report made a number of recommendations, including:

▸ supporting the development and implementation of industry-wide anti-corruption mechanisms
▸ setting up a co-ordinated approach from the government to tackle corruption
▸ equipping industry with relevant anti-corruption training
▸ increasing awareness of corruption and measures to report it.

Social responsibility

Social responsibility is not only about delivering success for a company but also delivering benefits for the local community. Social responsibility can be beneficial for the company in that it builds trust and improves the reputation of a company. By involving the local community through employing local people on the project, investing in local projects and improving local facilities the construction project can avoid disputes with and disruption to the local community, saving money over the length of the project. The Considerate Constructors Scheme (CCS) is a national scheme set up in 1997 to improve the image of the construction industry in the UK. The CCS checklist has five major sections:

▸ Care about appearance – site, facilities and personnel.
▸ Respect the community – anyone affected by construction work.
▸ Protect the environment – protect it, enhance it.
▸ Ensure everyone's safety – minimise risks.
▸ Value their workforce – workforce health and wellbeing.

> **Key term**
>
> **Cover pricing** – an anti-competitive practice in which companies submit an artificially high bid for work with the intention of not winning the bid. This allows other companies to submit higher than necessary prices for work, with the knowledge that they will win as their competitors have submitted very high quotes. They can then divide several tenders among themselves at higher prices.

PAUSE POINT

What ethical factors should be considered before purchasing materials from a supplier?

Hint

Close this book and, in two minutes, list as many ethical factors that you can think of that should be considered before purchasing materials.

Extend

Looking at your list of factors, discuss which would be the most important and why.

Purchase orders

The process of ordering materials should follow a basic system such as:
▸ Request – create a requisition order for the required materials or items.
▸ Approve – specified staff members agree to go ahead with the purchase as part of an approval process.
▸ Purchase – create and send purchase orders to suppliers as an indication of willingness to purchase.
▸ Receive – delivery of materials or items to the specified destination (see section A1 above for more information on timing and delivery).
▸ Pay – payment made to supplier.

A purchase order is created by a buyer and details quantities, prices, payment terms and delivery dates. The information on a purchase order must be correct otherwise it can lead to disagreements, errors, delays and additional costs. The specification of the materials, and the number of items, should be clearly recorded on the purchase order and checked by a senior member of the project team before ordering, as it becomes legally binding once accepted by a supplier. The benefits of purchase orders are that they:

- can save time and money and avoid cash flow problems by allowing better control over budgets
- reduce fraud within the company by providing an audit trail and producing accurate records of expenditure
- provide easy access to records to identify trends and plan budgets
- reassure suppliers that the company is willing and able to pay.

Terms and conditions are general and specific arrangements which are part of the contract formed between the buyer and supplier as part of the purchase order. Any changes to these will not apply unless they have been agreed in writing between both parties.

The buyer can negotiate discounts for repeat and bulk orders. Materials or items that are ordered on a regular basis may be added to a supplier list with a reduced or discounted price.

Benefits and drawbacks of serial and term contracts

Annual supply contract

An annual supply contract is an agreement between a buyer and a supplier to provide materials or services over a year. They:

- protect the construction company from variations in prices, and improve financial stability in terms of prices, quality of materials and service provided
- provide both the main buyer (main contractor) and the supplier with a degree of certainty about work due and allow both to plan ahead and make arrangements
- reduce administration costs for repeat bidding, quotations and purchase orders
- enable more accurate control of spending and improvements in the cash flow to be managed
- facilitate a reliable supply of materials, which can be scheduled to meet the tight demands of the construction programme.

A drawback of an annual supply contract is that it may reduce the competitiveness of the process and lead to increased costs for the buyer and complacency from the supplier.

Multiple project contracts

Multiple contracts subdivide a large project into a series of smaller construction phases, for which contracts are awarded sequentially. For example, a contract would be awarded for the groundworks and foundations. Afterwards a further contract would be awarded for the superstructure. These contracts require more careful co-ordination and close monitoring, as several contractors may be involved in the project, and no single contractor is held responsible for the job as a whole.

Serial contracts

Serial contracts can be used for repetitive works such as housing or maintenance work. An initial contract can be agreed for one item and then be reused for a series of similar projects. For example, a developer may agree a contract with a subcontractor to build one or two houses within a housing scheme. If the houses are sold the developer can roll the contract on to build more houses but if the houses do not sell the developer is not tied into a long-term contract. Serial contracts can reduce tender costs and may encourage suppliers to keep costs low to secure regular work.

> **Research**
>
> Contact a local construction company and request examples of purchase orders. Summarise the information contained on the purchase order.

Hint   For each type of contract for purchasing materials list the advantages and drawbacks.

Extend   For each type give an example of a construction project for which it might be used.

B2 Cost management techniques

Analysis of interim claims

Preliminary items

Preliminaries are usually the first section of a bill of quantities and establish the general responsibilities of the contractor. The preliminary items tend to be time-related costs that cannot be easily priced. Typical preliminary items would be:

▸ a general outline of the site and work to be completed

▸ risk assessments and method statements

▸ requirements of statutory approvals

▸ the employer's requirements in relation to quality control and health and safety

▸ services and facilities, for example temporary water supply

▸ identification of work or materials which have been arranged by the client with specific suppliers or subcontractors

▸ requirements for insurance and product warranties

▸ details of the site waste management plan

▸ a list of the conditions of the contract.

Pricing the preliminaries in a project is one of the most difficult sections of the bill of quantities. In order to price accurately, the quantity surveyor must include all the general costs linked with managing the project.

Measured work by trade or element breakdown

This section of the bill of quantities shows quantities of work or items measured following the standard method of measurement, i.e. the new rules of measurement (NRM; see Unit 3, section B3, and Unit 13, section A2). Each item is listed individually and contains a description, unit of measurement and quantity. There is also a column for the contractor to provide a price or unit rate. Measured work by trade or element breakdown allows work that is initially difficult to quantify to be measured as the work progresses, e.g. excavations where the amount of material to be removed is uncertain.

Nominated subcontract values

A nominated subcontractor is one selected by the client to be used by the main contractor as part of the contract. A nominated subcontractor's work may be valued separately by the quantity surveyor and will be shown as a separate amount on the interim certificate. The subcontractor will be informed of this payment. If the nominated subcontractor is not paid on time by the main contractor, the client may pay them directly.

Materials on site

Generally, for construction projects, the contractor sources, orders and pays for materials and the client pays the contractor once the materials arrive. The contractor may claim payment for the materials delivered even if they have not been used in the building. In some cases, the client may pay the contractor in advance for materials to enable the contractor to purchase and deliver them on time. This improves the contractor's cash flow and reduces the chance of the contractor becoming insolvent, as material cost can be significant.

Link

More information about cost management techniques, including cash flow management and managing costs, can be found in Unit 3: Tendering and Estimating (section A5).

Discussion

As a class, discuss how the project management team can use the following to plan and manage the costs of the project: preliminary items, measured work by trade, nominated subcontract values, materials on site.

Cost value comparisons

There are several cost value comparisons that can be used to help a business make a decision. These are explained in Table 9.9.

Table 9.9: Types of cost value comparison

Comparison system	Description
Costs from management information systems (MIS)	MIS are used to gather information to generate reports and assist managers with decision-making. A contractor would use an MIS to monitor expenditure and income of the company to make it more effective and efficient. The system's accuracy will depend on the accuracy of the data collection and input. Cost estimating is incorporated into BIM.
Monthly valuations reconciled with project costs	A comparison of the monthly valuations carried out and paid through interim certificates against the value of work completed. They are usually carried out by the contractor to give an indication of the job's profitability, allowing the contractor to determine if initial cost estimates were accurate and whether these can be adjusted for future work.
Profit and loss projections	An estimate of the amount of money the contractor will be paid for work or services over a specific time or project against an estimate of the amount of money the contractor will have to spend on variable costs (e.g. materials) and fixed costs (e.g. employee wages, insurance, etc.).
	Using these, a contractor can determine how much work they need, and how much to charge, to make a profit (i.e. when the income is greater than the expenditure). If expenditure exceeds income, the contractor is making a loss and will need to make changes to the business.
Cash flow forecasting	It is important to keep track of when cash is expected to come into, and out of, the company. A cash flow profile is often called an S curve as initial payments are low but increase as the amount of work and materials used increase (see Figure 9.7).
	• For a client, cash flow will always be negative as they are continually paying out for work as it is completed. The forecast indicates when they need more finance, in this case as the project moves closer to completion.
	• For the contractor, cash flow can be positive (money received) and negative (money paid). It identifies possible deficits (more cash going out) or surpluses (more cash coming in), allowing better control of the project's budget.

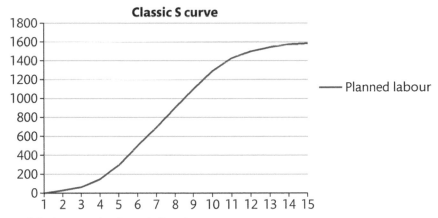

▶ **Figure 9.7:** An example of a cash flow S curve

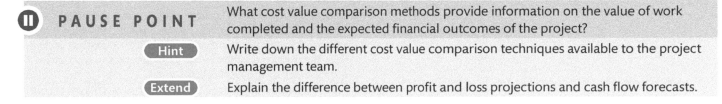

PAUSE POINT What cost value comparison methods provide information on the value of work completed and the expected financial outcomes of the project?

Hint Write down the different cost value comparison techniques available to the project management team.

Extend Explain the difference between profit and loss projections and cash flow forecasts.

Managing costs

Managing costs during a project is essential to provide financial security for the client. Factors which can affect the budget include design changes, increased material and labour costs, disputes and weather. Despite these it is essential to be alert to the build-up of costs throughout the project. Accurate cost estimates of the work must be carried out to monitor actual costs against projected costs.

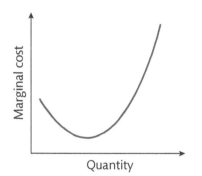

The following are financial terms used in relation to managing a project:

▶ **Estimated costs** – an approximation of the costs relating to a project using the unit method (e.g. price per bed for a hospital), superficial method (based on floor area) or elemental method (see element costing).

▶ **Variable costs** – these depend on the volume of work. Examples include the price of materials, hiring or purchasing plant and machinery, subcontractors employed, labour rates and fuel costs.

▶ **Fixed costs** – necessary costs that are independent of the volume of work, e.g. office rent, business insurance and bank loan payments.

▶ **Target costs** – these share the financial risk between client and contractor and are agreed early in the project. If the project is completed under budget, the savings are shared by contractor and client, just as additional costs from going over budget are shared.

▶ **Actual costs** – these reflect actual costs incurred on the project.

There are several techniques available to break down, itemise and control project costs, as described in Table 9.10.

▶ **Table 9.10:** Techniques for breaking down, itemising and controlling project costs

Techniques for controlling project cost	Description
Unit costing	Units of work detailed in a bill of quantities are costed individually (unit rates). Each unit rate is broken into sections such as labour, materials and plant. The quantity surveyor must determine the correct units for each item, such as weight, length, area or volume. The rates for each section are calculated and priced, based on historical data from similar projects and information from price booklets. This can be time-consuming and costly for large projects as a lot of detail and information is needed.
Element costing	The project is divided into elements. The Royal Institution of Chartered Surveyors (RICS) Building Cost Information Service (BCIS) provides current cost and price information based on a list of these elements. The list used will depend on the project. The cost for each element is calculated, allowing accurate estimates of the project cost at an early stage by following a standard system. It also allows regular comparisons against costs and work completed.
Marginal costing	Separates fixed and variable costs. Cost data is studied to understand the effect of profit changes due to volume of output. Initially costs are high but reduce as work volume increases, due to economy of scale. The larger the work volume, the greater the discount on materials and maximisation of the use of plant. This can reach a point where costs increase again as more labour and plant is required (see Figure 9.8). If the price charged is more than the marginal cost, the company makes a profit but, if the price is less than the marginal cost, the company makes a loss.
Variance analysis	Identifies the differences between planned costs and actual costs. Variance is the difference or change between a planned and actual amount. For example, a company produces 10,000 precast concrete slabs. The planned expenditure was £30,000 but actual expenditure was £35,000 – a variance of £5,000. Regular variance analysis throughout a project allows the project price to be adjusted to ensure it remains on budget.

▶ **Figure 9.8:** An example of a marginal cost U curve

(Figure axes: Marginal cost (vertical), Quantity (horizontal))

Cost savings: Labour

There are other ways to reduce labour costs than redundancies. Redundancies may come with financial costs and thus have little initial cost saving. These other methods can include:

▶ freeze on recruitment and pay rises
▶ reduce or eliminate bonus payments, overtime or temporary contracts
▶ reduced hours contracts
▶ improve management of absenteeism
▶ improve efficiency of workforce such as modernise equipment or software to speed up processes.

Cost savings: Methodologies

With increasing global competition, it is crucial for construction companies to maintain a cost saving methodology or strategy. Such a strategy may follow a system such as:

▶ creating an effective database of costs for the project
▶ identifying the major factors which influence the costs
▶ identifying areas for saving, such as improved design, change of materials, changes in processes, alternative suppliers, energy efficiency
▶ continually implementing the changes.

Cost savings: Programme acceleration or deceleration

It may be necessary to accelerate the work on a project, usually because the project has been delayed at critical points and the financial implications of falling behind are significant. Contractors and subcontractors may receive extra payments for accelerating work. Alternatively, there may be large bonuses for finishing the project early.

Programme deceleration (slowing work down) may be necessary to comply with the critical points within the programme and ensure it remains on budget. Resources may be directed to other projects until they are required.

Theory into practice

A client has bought a brownfield site on the edge of a seaside town. The town is close to a major city and has good transport links to the city. The client wants to build 200 houses on the site but is concerned about the poor reputation the construction industry has in completing projects on time and budget. He asks you to provide advice on how to identify cost savings in plant, labour, materials, site set-up and site management structure. Prepare a short report outlining the options for reducing costs in these areas.

Preparing and examining elemental and project comparison costs

The analysis of the data from preparing elemental costings and comparing them to the actual costs of the project enable the project team to monitor the project cost. This information can also be used for pricing future projects.

The elemental method is used to estimate the cost of the project and to determine the budget. Comparing this estimate against the actual cost of the project allows the management team to monitor the budget and establish the estimate's accuracy. The information gathered from one project can be taken onto future projects and help the next project to be estimated more accurately.

⏸ PAUSE POINT   What techniques can be used to obtain an accurate cost of a project?

     Hint   Think about how a project can be divided into units and elements of work.

     Extend   Why is it important to analyse the variance between planned costs and actual costs?

The local property developer was impressed by your previous research and booklet. She is now working with the project management team but would like you to prepare a short presentation to explain the methods used by construction companies to facilitate the supply of appropriate materials to the site. You should focus on:

- application of purchasing methods
- list of selected suppliers
- materials and subcontract enquiries
- planning links
- ethical purchasing
- purchase orders
- benefits and drawbacks of serial and term contracts.

The second part of the presentation should explain the cost management techniques used to monitor and control the cost and profitability of construction projects, including:

- analysis of interim claims
- cost value comparisons
- management costs
- preparing and examining elemental and project comparison costs.

Plan
- How will I research purchasing techniques to facilitate supply of materials and cost management techniques used to control costs?
- How will I create the presentation with brief descriptions of the techniques?

Do
- Have I completed the presentation in as much detail as possible?
- Can I present some additional ideas on improvements?

Review
- I can explain how I would approach difficult elements differently next time.
- Once I have reflected on my own work and any feedback from others, I will make any necessary changes to my presentation.

C Develop a programme of activities for construction works

Now you will get the chance to apply what you have learned earlier in this unit to begin planning your own programme of activities, using graphical representations of progress and detailing the timings for critical and non-critical elements of a construction project. You will need to consider and choose the most appropriate methods to monitor your progress.

C1 Production control systems

Production of programmes of activities

The project management team plans the sequence and methods (using Gantt charts, CPA, LOB, etc.) which will be used to complete the project.

The programme of activities needs to be able to be adapted to deal with any unforeseen circumstances to minimise the impact on other areas or stages of the project. It must cover the following areas (see section A2 earlier in this unit):

▶ method statements
▶ site layout
▶ site accommodation and storage
▶ Gantt charts, bar charts and linked bar charts to show and monitor progress of the project
▶ waste management – although it is no longer a legal requirement to produce a site waste management plan (SWMP) it may still be required to comply with the BRE Environmental Assessment Method (BREEAM) used to assess building sustainability. In any case, it is a good idea to prepare one before construction work starts, identifying the waste which may be produced and to estimate its quantity. This provides an action plan of how to manage each waste product (e.g. reduce, reuse, recycle or disposal).
▶ site traffic management – planning permission may require a construction traffic management plan (CTMP) to be completed before major work begins. The plan should identify suitable access roads for the delivery of plant and materials and

identify traffic routes on site to allow the safe movement of plant, reducing the likelihood of accidents. Typical recommendations are:

- separate entrance and exit points for pedestrians and vehicles, with clearly marked and lit walkways for pedestrians
- car parking away from the main work area to reduce movement of vehicles around the site
- only trained and competent personnel should drive or operate vehicles on site
- reversing on site should be kept to a minimum and be supervised. One-way systems and control measures such as mirrors, CCTV cameras and reversing alarms should be used to help reversing
- signage should be clearly displayed around the site to provide clear instructions to pedestrians and drivers of vehicles of the routes and site rules (see Figure 9.9).

Network analysis

Network analysis is a general name given to specific project management techniques such as CPA, precedence diagrams, time change diagrams and line of balance (LOB). The CPA or precedence diagram methods of planning allow project management teams to understand the interdependency of the activities and therefore to determine the effect of a delay that one activity will have on the rest of the project.

Very basic diagrams can be produced manually; however, it is more common, and more useful, to use computer-based methods (see Figure 9.10).

▶ **Figure 9.9:** This no turning sign helps to protect workers and visitors on the site

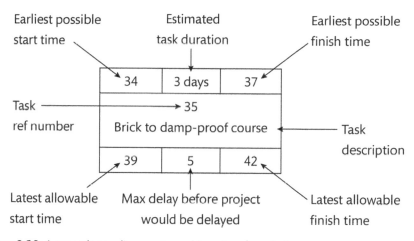

▶ **Figure 9.10:** A precedence diagram is used in network analysis

Theory into practice

Your manager has asked for a list of the welfare facilities that must be provided under the Construction Design and Management (CDM) Regulations 2015. What are they?

Link

There is more about the use of CPA and LOB earlier in this unit, in section A2.

Theory into practice

Create a programme of activities for the construction of a new house within a 12-week period. Represent the programme using a Gantt chart.

Using your programme of activities represent the activities using a critical path analysis (CPA). Show the interdependency of certain activities and identify the activities on the critical path.

PAUSE POINT Can you explain the difference between Gantt charts, critical path analysis and precedence diagrams?

Hint Look at examples of each network analysis and identify the strengths and weaknesses of each.

Extend Explain when each method would be used and why.

2001 at a cost of approximately £40 million. It eventually opened in July 2004 at an estimated cost of £430 million. The parliament building was designed by Spanish architect Enric Miralles in partnership with a local architecture firm. (Source: **www.parliament.uk**)

Check your knowledge

- What reasons might have contributed to the project being completed late and ten times over budget?
- What control measures should have been put in place to prevent the project from going so far over budget?

The new Scottish Parliament building, situated in the Holyrood area of Edinburgh, was due to be completed in

Measurement of progress

When working on site you will need to make sure that you are using an accurate system to measure your progress. The project management team monitors the actual progress of the project against planned progress on a regular basis. Generally, actual progress is measured by one representative from the project management team and one from the main contractor. These measurements are recorded and reported as they can be used for payment of the contractor and to assess the current progress of the project. These reports should:

▸ have an up-to-date Gantt chart or network analysis to show actual progress against the planned progress (see Figure 9.11 for an example). They may also include photographs to show evidence of progress

▸ highlight any delays, explain the delays and the possible impact on the completion time and cost of the project

▸ include proposals of how the lost time may be recovered.

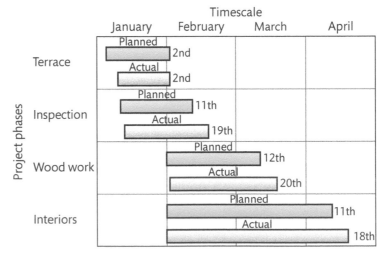

▸ **Figure 9.11:** An example of an actual versus planned Gantt chart

Overcoming the consequences of running ahead of or behind schedule

If the construction project is delayed, the project management team (with the agreement of the client) may issue the contractor with an acceleration agreement. This may incur additional costs but may be necessary when the completion time is critical. Possible options to accelerate the project to recover lost time are to:

▸ increase the plant and labour resources on critical activities or tasks

▸ review the method statements, looking for alternative safe but faster processes to complete the work

- increase working hours, such as evenings and weekends
- simplify the design specification
- consider phased completion, i.e. some sections of the project are completed at a time
- impose a plant booking system, for example on large projects where cranes may be in high demand to reduce delays due to contractors waiting to use the plant.

Network analysis identifies the critical path activities to determine if delays will impact on completion time. Delaying activities not on the critical path will not impact the overall project. If the project is ahead of schedule, it may be necessary to adjust coming activities because resources such as plant, labour and materials may be scheduled for specific arrival times. Network analysis will also show the impact of being ahead of schedule.

If it is possible to move the arrival of the resources forward, this should ensure the project remains ahead of schedule and may be completed early. If it is not possible to adjust the resources it may be necessary to redirect some of them to other projects temporarily.

Causes and effects of delays

Many issues can cause delays, such as poor planning and scheduling, lack of available resources, changes in design, adverse weather and poor site supervision. The consequences of delays can be significant to the final project:

- Going over time can increase costs for the client. Depending on the cause, the client or contractor may be entitled to compensation payments.
- Going over budget because of additional work, design changes or accelerating works needs to be paid for.
- Disputes between various parties over the cause of delays can further hinder the project. These disputes may lead to legal proceedings, which can further delay the project and increase costs significantly.
- Accidents can occur if work is being rushed.

Delays do frequently occur, so it is helpful that many computer packages allow for rescheduling. It is important to ensure specific constraints on the resources, plant, labour and materials are considered so that it is possible to see the impact on other activities of being ahead or behind the original schedule. Overtime payments to get the project back on track can be a significant cost and should be agreed by all parties in advance, to avoid disputes.

Extensions of time (EoT) applications are submitted by contractors when they feel there is a justifiable reason for a delay, such as changes in design by the client, extreme weather conditions or delays caused by a nominated subcontractor. If the contract administrator approves the application, the contractor is given a later completion date. If it is not approved it is the responsibility of the contractor to take reasonable means to ensure the contract is completed on time.

Theory into practice

Severe weather has put a construction project two weeks behind schedule. What options does the contractor have? What are the consequences of the delay to the parties involved?

Preparation of financial progress information

You will need to prepare several pieces of financial progress information to track the financial progress of the project.

▸ **Table 9.11:** Financial progress information

Financial progress information	Description
Site returns	Accurate records ensure actual costs are recorded and monitored. A daily diary could record where the resources are used each day, or goods received sheets could record each material delivered on site. Any short deliveries or discrepancies are recorded and sent to head office for processing. Supplier invoices are checked against the site record to ensure the quantity is correct. For any problems with deliveries, a credit note can be requested against the invoice. This controls costs from suppliers and checks goods received are of the right quantity and an acceptable standard.
Variations	Changes to the originally agreed contract, for example changes in design, using alternative materials, etc. These can only take place if they have been authorised in writing by the architect. However, the cost of the variations may not be agreed at the same time.
Claims	A request made by a contractor to the client for additional payments for changes or variations to the original contract. These may be due to client delays or for variations requested by the client or for extensions of time.

Link

Another piece of information you must collect includes interim valuations and payments (see Table 9.9).

Reviewing events, predicted and unforeseen

It is crucial to monitor and review the progress regularly. The impact of predicted events such as changes to design and unforeseen events such as extreme weather conditions or unexpected ground conditions should be assessed. These events may affect the completion date and the overall cost of the project. It is important for the project team to be aware of the events and quantify the impact on the project and keep the client informed.

PAUSE POINT

Why is it important to regularly monitor the physical and financial progress of a construction project?

Hint   Think about what can happen when a construction project is poorly managed.

Extend   What methods would you use to monitor the physical and financial progress of a construction project?

Assessment practice 9.3   **C.P5   C.P6   C.M5   C.D3**

The local property developer was happy with your presentation. She would now like you to produce a report which includes a comprehensive programme of activities for constructing a single detached house. This should include:

- details of the critical and non-critical activities within the programme, presented in a graphical format
- as much detail as possible in terms of timings for each activity
- an explanation of the most appropriate method for monitoring the progress of the project so it is completed on time.

Plan
- How will I find out the types of activities required to build a house, and identify which are critical and non-critical?
- How will I get an idea of realistic timings for these activities?
- Where will I research the various techniques to monitor and control the physical and financial progress of the project?

Do
- Have I completed the report in as much detail as possible?
- Is my report comprehensive enough to be used for a real project?

Review
- Can I explain how I would approach the difficult elements differently next time?
- Once I have reflected on my own work and any feedback, I will make any necessary changes in my report.
- I will be ready to present it to my peers for some additional ideas on improvements.

Further reading and resources

The Construction Industry Knowledge Base: www.designingbuildings.co.uk

Cooke, B. and Williams, P. (2009) *Construction Planning, Programming and Control*, third edition, Chichester: Wiley-Blackwell.

Crown copyright (2013) *Construction Strategy: Industrial Strategy: government and industry in partnership*, URN BIS/13/955, London: Department for Business, Innovation and Skills.

Department for Communities and Local Government (2016) *Public Land for Housing programme 2015–20 Programme Handbook*, London: Department for Communities and Local Governement.

Hore, A.V., Kehoe, J.G., McMullan, R. and Penton, M.R. (1997) *Construction 1: Management, Finance, Measurement*, Houndmills: Macmillan.

Lock, D. (2004) *Project Management in Construction*, Aldershot and Burlington, VT: Gower Publishing Limited.

THINK ▶▶FUTURE

Troy

Troy is an experienced project manager for a large design company that specialises in hospitals and education facilities. He had worked as a site manager for several years before stepping up to his current post, where he is responsible for four of the company's development sites. His role is to organise and monitor the progress of the projects, from the site investigations to the tendering and appointment of a contractor to undertake the construction work. Troy is involved in all aspects of the projects and even has to appoint architects to commission the design and production drawings.

Troy loves his job. There are so many different people and roles involved within a construction project that no two days are the same. Having to run four projects at the same time means that Troy has to manage his time efficiently to stay on top of all developments. Troy also has excellent IT skills in project management and uses contract programs to co-ordinate information and contractors, and monitor the physical and financial progress on all the sites.

Troy has targets for delivering the projects safely, to the required standard of quality, on time and on budget. This means he regularly has to make decisions that have a financial effect on a project. He has excellent skills in communication and chairs several meetings a week with different contractors and clients. Luckily, Troy has an outgoing personality and gets on with people who he interacts with; he is generally approachable and open to ideas.

Troy enjoys the feeling of completing a project that had some difficulties which have been overcome to produce a quality outcome for the client.

Focusing your skills

A project manager must possess a variety of skills when they are co-ordinating a number of different construction projects.

- Leadership – ability to lead the team as well as manage them and be able to inspire them.

- Negotiation – ability to resolve conflicts and create solutions to difficult situations.

- Time management – ability to schedule work to prevent delays or overruns.

- Cost control – ability to monitor the costs of the project to prevent it going over budget.

- Critical thinking – ability to deal with challenges and delays and find solutions.

- Health and safety management – awareness of the need for high standards of health and safety on site at all times.

- Communication skills – ability to communicate with everyone within the project team.

- Efficiency – with so many projects to manage it is essential to be organised and efficient.

Getting ready for assessment

Hayley is in the second year of her BTEC Extended Diploma in Construction and the Built Environment. For this unit she has been given an assignment that involves preparing a technical report based on the different roles and responsibilities of the project management team, the techniques used to manage the principal resources and the methods used to monitor the quality and progress of a construction project.

Hayley's report should include evidence of her own research; for example, charts showing management structures and the interaction between team members, case studies of methods and techniques to manage resources and to monitor quality and progress.

How I got started

I started this assignment by looking at the sources of information my tutor had provided on the last page of the assignment. These sources are very useful and are a great start to help you research and gather information on helpful features and items.

Then I looked through my notes and put together a list of all the different roles involved in managing a construction project. I made sure to list the responsibilities of each role. I wanted to make sure that my report was meeting all the key verbs within the assessment, so I downloaded a copy of the unit from the specification on the Pearson website which provided clear guidance on what I needed to do.

I carried out a lot of research online and in books to find out as much as I could about the different roles and responsibilities of the project management team, the techniques used to manage the principal resources and the methods used to monitor the quality and progress of a construction project. I wanted to find some interesting case studies, but this was difficult as it's such a huge topic!

How I brought it all together

To start, I wrote a very short introduction explaining the purpose of my report. I remembered my tutor telling us it was important that our introductions outlined what we were planning to say in detail, so I made sure my introduction did this.

After this, I decided to cover each role, and their responsibilities, within a project management team and how the team members interacted with each other. I then wrote a section on each of the principal resources and the techniques used to manage them. For my final section, I wrote about the methods used to monitor the quality and progress of a construction project.

Because I used lots of books and websites to support the points I was making, I put together a bibliography and references so all my points could be checked. I asked my parents to read over my report to make sure what I was writing was correct and to check for errors. I also proofread it against the assessment criteria to make sure it matched the verbs used from the grading criteria.

What I learned from the experience

I'm glad I had done some work beforehand planning my research and identifying where to look for information. It was hard to find supporting case studies and next time I would try to approach local businesses as well as looking at larger construction companies to try and get more information.

I think I spent a little too much time focused on explaining the roles and responsibilities and not enough time linking them together and evaluating the roles. Next time I would try to spend more time planning my work out, using my notes, to put together a structure for the report before I started writing it.

Think about it

▸ Type the assessment criteria into an internet search engine and see if what you find is a good starting point.

▸ Where will you find information on management of a construction project?

▸ Does the unit content help?

Building Surveying in Construction 10

Getting to know your unit

Building surveyors examine properties to establish their condition, identify any defects, suggest solutions and advise clients how best to repair, alter or extend the building.

In this unit, you will learn about carrying out building surveys, identifying issues and recording your findings so that they can be acted upon. You will find out about the impact of the methods used to construct existing buildings, and their impact on current and future maintenance requirements. Then you will explore different defects and methods of repair for low-rise residential properties and finally undertake your own measured building survey of a low-rise residential property, which will include a description of its condition, any defects, suggested remedial works, scale plans and elevations.

Some of the content in this unit links to Unit 2: Construction Design and Unit 4: Construction Technology, so it is helpful to read all three units to gain a full understanding of the topics discussed.

How you will be assessed

This unit will be assessed by a series of internally assessed tasks set by your tutor. These will cover the main theoretical knowledge required of a building surveyor followed by some practical applications of building surveys. When you work through this unit you will find a series of activities that help you practise and build up your knowledge of building surveying. The process is all about preparing you for the internal assessment that will be issued to you. To achieve the tasks in your assignment, it is important to check that you have met all the Pass grading criteria. You can do this as you work your way through the assignment.

If you are hoping to gain a Merit or Distinction, you should also make sure that you present the information in your assignment in the style that is required by the relevant assessment criteria. Merit and Distinction extend and stretch the understanding of the Pass criteria. For example, Merit criteria require you to analyse and discuss, and Distinction criteria require you to assess and evaluate.

The assignment set by your tutor will consist of several tasks designed to meet the criteria for this unit. This is likely to consist of a written assignment but may also include practical activities such as:

▶ performing a building survey on a property

▶ undertaking a measured survey

▶ producing drawings from surveys.

Assessment criteria

This table shows what you must do to achieve a **Pass**, **Merit** or **Distinction** grade, and where you can find activities to help you.

Pass	Merit	Distinction
Learning aim **A** Understand the impact of the methods used to construct existing buildings on current and future maintenance requirements		
A.P1 Describe the different styles and types of residential housing.	**A.M1** Discuss the different residential housing styles and types and how their construction methods are applied and impact on the current and future requirements for repair and remedial work.	**A.D1** Evaluate the different residential housing styles and types and how their construction methods are applied and impact on the current and future requirements for repair and remedial work.
A.P2 Describe the different methods of traditional and modern construction used for residential housing and their impact on current and future repair and remedial work.		
Learning aim **B** Explore different defects and methods of repair for low-rise residential properties		
B.P3 Describe a range of external and internal defects commonly occurring in residential properties.	**B.M2** Discuss appropriate repair and remedial measures for a range of external and internal defects for a residential property.	
B.P4 Explain different methods of repair and remediation to a range of internal and external defects of a residential property.		**BC.D2** Evaluate the repair and remedial work options for the defects identified in the building survey.
Learning aim **C** Undertake a building survey of a low-rise residential property		**BC.D3** Demonstrate individual responsibility, creativity and self-management when preparing for and undertaking the building survey and producing the survey report and drawings.
C.P5 Perform a building survey, detailing the condition and defects with required remedial works for a residential property.	**C.M3** Produce a comprehensive and detailed building survey detailing the condition, defects and remedial works required with plans and elevations for a residential property.	
C.P6 Record the findings of the survey in a survey report.		
C.P7 Perform a measured survey on a residential property.		
C.P8 Produce accurate scale plans and elevations for a residential property to standard conventions.		

Getting started

In this unit you will start to explore the role of the building surveyor, who is experienced in a wide range of construction activities for a variety of domestic and commercial clients. As a starting point, find out more about the professional association Royal Institution of Chartered Surveyors (RICS) and what a building surveyor does (www.rics.org).

Does anything about a building surveyor's job role surprise you? Why?

A Understand the impact of the methods used to construct existing buildings on current and future maintenance requirements

A1 Different styles and types of residential property

Types of property

There are many different types of properties used as residences in the UK. They are commonly classified and described as follows by UK estate agents, although there are no single definitions for each type.

- **Detached** – a property that stands alone and is not attached to another.
- **Semi-detached** – a property that is attached to another along a party wall.
- **Terraced** – a row of properties that are joined together by their party walls.
- **End terrace** – the last house in a row of terraces that only has one adjoining wall with another property.
- **Bungalow** – a single-storey house that is detached or semi-detached. Chalet bungalows have living space in the eaves of the roof.
- **Flat** – a self-contained residential property that forms part of a building that contains other flats.
- **Duplex/Maisonette** – flats containing more than one storey, each with separate entrances.
- **Cottage** – a small dwelling, often more than 100 years old and often in a village or rural location.
- **Mansion** – a large, impressive house with many bedrooms and storeys.
- **Manor house** – a large country house with substantial grounds that was probably originally built for a wealthy aristocratic owner.
- **Prefabricated** – post-war single-storey prefabricated accommodation built as quick, temporary homes for people whose houses had been destroyed in the Second World War. They were meant to last for only 20 years as new buildings were constructed to replace them but many are still habitable.

 PAUSE POINT Can you find a detached and semi-detached property in your area? What are the differences in terms of layout?

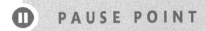

| Hint | Use a property buying website to find out more information such as floor plans. |
| Extend | Why is there a difference in property values between the two types? |

Key periods and architectural styles of residential property

A building surveyor has to survey a wide range of period buildings, many of which are listed and protected under conservation legislation. Each has been constructed using the technology that was available at the time. Alterations and adaptations therefore need a knowledge and understanding of these methods to deconstruct then adapt them sympathetically.

Table 10.1 describes some of the architectural features of different historical periods. Remember that not all houses from a particular period look the same – they reflect the taste, budgets and fashions of the owner or builder.

▶ **Table 10.1:** Key architectural styles through the ages

Period	Approx. years	Typical features
Tudor and Elizabethan	1485–1603	Steep gables, overhanging upper floors, mullioned windows, tall chimneys, timber frame and/or patterned brickwork.
Georgian	1714–1830	Local brick or stone, symmetrical with large windows each side of the main entrance doors, often terraced townhouses, smaller windows higher up, tiled hipped roof (sloping upwards from all sides) often hidden by a parapet.
Victorian	1837–1901	Bay and sash windows, solid brick construction, often laid in a pattern (Flemish brick bond), stained glass in doorways and windows, slate roofs with barge boards (decorative wooden panels), many fireplaces, porches, shallow stepped brick footings with suspended timber floor.
Pre-war	1901–39	Toilets and bathrooms often a separate room connected to mains drainage. Typical houses had three bedrooms, a lounge, dining room and kitchen with no servants' quarters. Brick built, often rendered or pebble-dashed, mock Tudor style was popular. Often cottage-style, semi-detached, two-storey homes or low, wide bungalows. New housing often built in suburbs with cavity walls, concrete strip foundations. Designed with a kitchen for the owners, not servants, to use. Generous gardens and green spaces.
Post-war	1945–60	Cheap, standard local authority builds to replace houses lost in the war, often built as part of self-contained communities (estates). Usually brick, sometimes with an additional layer of blockwork, but cheaper non-traditional methods such as precast reinforced concrete, steel frames or prefabrication also used. Garages built at same time as house. Concrete tower blocks, often prone to damp and structural issues. More space for storage and domestic appliances. Additional bathrooms, e.g. downstairs WCs. Concrete strip foundations laid with concrete floor without a damp-proof membrane. Galvanised metal single-glazed windows common.
Modern	1930s–1960s	Asymmetrical designs, often using cubic or cylindrical shapes. Demolition of characterful historical buildings, especially in city centres, in favour of modern designs. Flat roofs. Use of new materials such as reinforced concrete. Metal and glass frameworks often resulting in large windows in horizontal bands. Few ornaments or mouldings. White or cream render. Open-plan interiors.
Postmodern	1960s–1990s	Classical and literary allusions applied to contemporary buildings. Gable and sloping roofs often with concrete tiles and plastic guttering. Inconsistencies and quirks, bright colours, structural variety, variety of materials and shapes.
Contemporary	1990s onwards	Often built using modern methods of construction, low energy housing designs and low carbon technologies, may not have chimneys. Trend for high-rise accommodation in cities. Roofs have braced and restrained trusses with vapour-permeable roof underlays, eaves ventilation and deep roof insulation. Wide cavity walls with aerated block inner leaf and significant insulation. Double glazed uPVC, timber or aluminium/timber windows. Concrete strip foundations and ground bearing floor with insulation below.

Research

Find out the reasons for particular architectural choices being made after 1945. How has this affected the types of problems now found in homes built between 1945 and the late 1960s?

A2 Traditional methods of construction

Building surveyors need to know how a property was constructed so they can advise of any requirements for alteration, adaptation or extension. The age of a property will give clues to the type of construction, but it may have been extended or changed over time. We need to know how the method of construction has resulted in a latent building defect. We can then specify a method of repair or renovation for the defect.

Foundations

Strip foundations

Traditional strip **foundations** are a strip of concrete that an external wall sits upon. They are comparatively easy to construct so are commonly found in older traditional buildings but remain common today. The Building Regulations go into detail about when and how strip foundations should be used.

They are normally excavated to a metre depth so are unaffected by frost and moisture content changes within the soil. Modern strip foundations use 'trench blocks' which are wall thickness concrete blocks that raise the wall off the foundation up to ground level.

If a client wants to extend a first floor over a ground-floor kitchen, the foundations have to be checked that they can structurally support the additional load. A strip foundation must project past the supported wall the same distance as the depth of the foundation. Projection must equal depth.

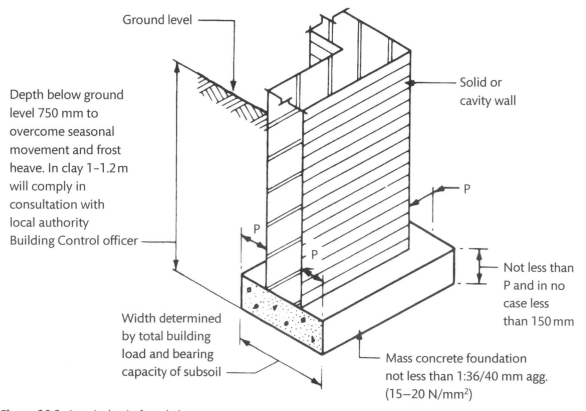

Ground level

Solid or cavity wall

Depth below ground level 750 mm to overcome seasonal movement and frost heave. In clay 1–1.2 m will comply in consultation with local authority Building Control officer

P

P

P

Not less than P and in no case less than 150 mm

Width determined by total building load and bearing capacity of subsoil

Mass concrete foundation not less than 1:36/40 mm agg. (15–20 N/mm²)

▶ **Figure 10.1:** A typical strip foundation

Raft foundations

A raft foundation is a large concrete slab designed to take the full load of a building and distribute it over a larger soil-bearing area. This was a common foundation type during the bulk-housing build after the Second World War because it was cheap, easy to install and required less excavation than strip foundations. However, they were restricted by the new Building Regulations in 1965 and this additional time and inconvenience meant they were only used in unstable ground after that.

Any additional loads due to alterations may cause one side to settle further than the other and unequal settlement develops with possible non-horizontal movement of the raft foundation.

Extending raft foundations also cause issues. The steel reinforcement within the raft foundation cannot be extended and a break is therefore created with the joining of the new foundation to the existing. Dowel bards will have to be drilled into the existing slab and grouted in to secure the new steel reinforcement to the existing structure.

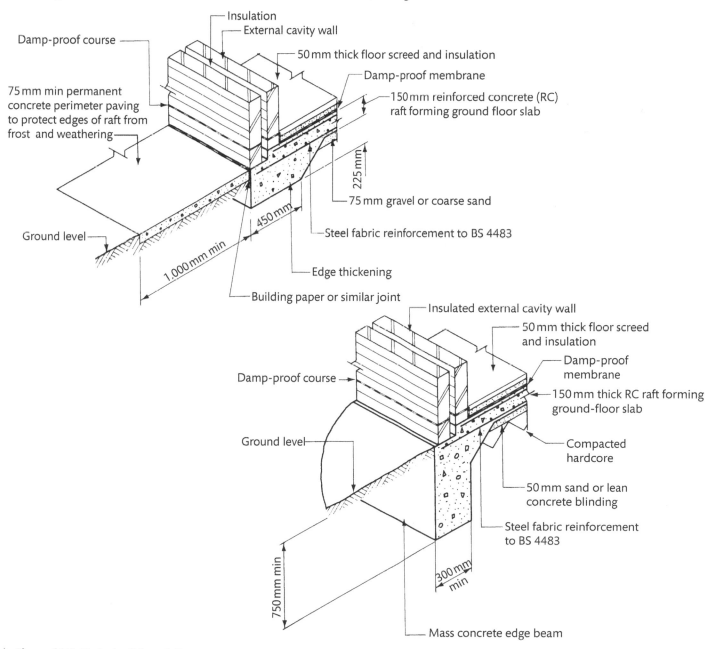

▶ **Figure 10.2:** Typical raft foundations

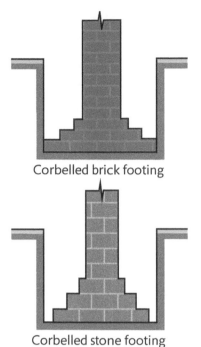

Corbelled brick footing

Corbelled stone footing

▶ **Figure 10.3:** In a corbelled footing, the masonry rests directly on the excavated foundation trench with no concrete foundation below

Spreader foundations, corbelled brickwork and masonry foundations

Until about 100 years ago, it was common for walls to be built directly onto solid ground with no excavation or reinforcement beneath. These were often constructed using corbelled brickwork with a wide course at the bottom and gradually tapering to a thickness of one brick. In theory, the load of the wall is spread over a larger area (spreader foundation). Little or no concrete fill existed under the brick foundation. Adaptation of this type of foundation may involve **underpinning** the existing brickwork to increase the bearing capacity of the foundation at lower ground levels.

Corbelled foundations were a requirement of the London Building Act of 1894 which did not mention concrete but detailed brick footings. Foundations were much shallower than their modern counterparts, which has resulted in settlement and structural issues in the long term.

Masonry foundations may also be bricks or stone laid around the perimeter of the building's footprint with the intention of spreading the weight of the building evenly across the area. Again, this method is not seen as providing adequate support.

Timber-piled foundations

Timber-piled foundations are not particularly common in the UK, and tend to be used for countryside structures like lodges and activity centres rather than for residential homes. They can, however, be a good solution in flood-prone locations as the foundations may be raised above ground level to support a building and lessen the impact of flooding. They can also be covered by a concrete capping beam that link piles together and provides a wall foundation. Timber piles driven below the water table level can last hundreds of years. The issues develop when they are exposed to air and biological degradation commences in the form of fungus attack. Concrete extensions are often added to the top section of the pile so this does not occur.

Extensions to this type of foundation can be accomplished by using bored-in-situ or precast concrete piles. The in situ method is better as no ground movement occurs as the soil is bored out and replaced with poured concrete around a reinforcement cage.

PAUSE POINT   What difference has the use of concrete made to building foundations?

Hint   How can the properties of concrete enhance foundation construction methods?

Extend   What other modern materials can be used to build more effective foundations?

Traditional wall construction

Solid

Solid walls have been used for hundreds of years. The outside of the wall was either rendered or finished in another material to make it weather proof. A single wall is only as thick as the brick, stone or other material used to build it – commonly this is 200 mm, which is not thermally efficient so loses heat easily. The solid wall can be clad externally in insulation and render or internally clad with insulation and plasterboard. With both processes a vapour control layer is used to prevent moisture ingress and damp. Single-storey solid walls must be tied at lateral points to prevent movement. Extending the solid wall foundation and building another skin inside will provide a supporting structure for a cavity wall for a two-storey extension.

Solid walls can be made from materials such as stone which is often termed a random rubble wall, or dressed stone to make it a feature wall with blocks cut into rectangular sizes. Brick walls are common, while blockwork walls require water damage prevention on their external skin.

Cavity

Cavity walls were developed in the first half of the twentieth century to prevent penetrating damp from the outside crossing the cavity. This resulted in better internal environments with the inner skin being warmer and the cavity providing an opportunity for insulation. The early developments did not have any insulation and just two skins of brickwork. Blockwork inside skins were introduced, formed of air-entrained concrete. These were lighter and had greater insulation properties. Cavity walls can be adapted if appropriate lintels are inserted above any openings and cavities are maintained so a cold bridge does not form causing a damp spot.

Cavity walls can be constructed using brick and block, brick and brick, stone and block/brick and block/brick and metal cladding.

Cavity walls were often vented to allow any moisture to escape out of the cavity via a drain.

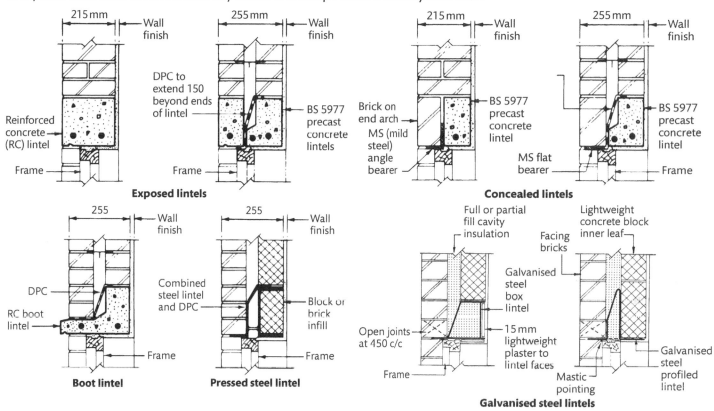

▶ **Figure 10.4:** Developments in cavity wall technology can be seen from top left to bottom right

Research

How can you tell if a house has a cavity wall? Produce a checklist. The Energy Saving Trust (**www.energysavingtrust.org.uk**) and Thinksulation (**www.thinkinsulation.co.uk**) are good starting points for your research.

Framed

A framed structure often replaces the internal skin with a frame that supports the building. For example, oak framed construction with infill panels was used for hundreds of years in domestic and commercial buildings.

▶ **Cruck frames** date back 1000 years or more and are a curved A-frame beam that ran from the **eaves** to the roof ridge in one or a joined piece. The roof structure and walls were then just hung off this frame so they were not load bearing and could be easily replaced. It was not possible to have more than one storey although the ceiling could be high. You may see them in barn conversions.

Key term

Eaves – the part of the roof that meets or overhangs the walls.

- **Box frames** were common when concrete became a common building material in the nineteenth century. They consisted of individual cells or rooms set horizontally and vertically together to create an overall frame and their regular shape enabled more storeys to be built. They had to be strong and thick enough to carry their own weight as well as weights of cells above them, which limited their use and height. Reinforced concrete and other non-load-bearing methods quickly replaced the technique.
- **Post and beam** frame techniques are thought to date back to the earliest building techniques. They utilise heavy sectioned timber posts, which are jointed with horizontal beams at eaves level. Infill panels between posts form walls. These were often brickwork or wattle and daub construction. The roof is then built off the beams, enabling an open-plan design. Cladding covers the whole structure to make it weatherproof.
- **Storey height panels** are a full storey height supporting panel in timber frames or precast concrete panel construction. The Airey house was a design of prefabricated post-war home that could be quickly erected using precast concrete columns and panels.

Roofs

The building surveyor needs to know how a traditional roof is constructed to identify any defects and advise how these could be repaired, with associated costs. A property valuation surveyors' report only needs to include the type of roof, along with any visible repairs to the roof and its finishes. Table 10.2 discusses different types of roof.

Table 10.2: Types of roofs and common issues

Type	Description	Typical issues
Pitched	Dual-pitched roof with a slope each side meeting at the ridge at the top.	Tile slippage due to failure of the nail holding them in place. Degradation of tiles due to frost. Water penetrating the roof finish because many traditional pitched roofs do not have a felt or breathable membrane below the tile battens to prevent any moisture that passes between the tiles from entering the roof space.
Mono-pitched	Mono means one and so this roof is often termed a lean-to-type roof. It has one **pitch** that is commonly used on a single-storey extension.	Ventilation (drafts) and moisture entering the existing cavity of the building that the roof abuts.
Mansard	An unusual roof that may have Dutch origins. It makes full use of the space within the roof void and makes this habitable space.	Problems occur with the sloped sides where tiles on the sloped sides are subjected to gravitational forces and nail failure means they slip. Moisture within the roof void is a common issue.
Dormer	A structure that forms a vertical window within an attic space, enabling the second-floor roof space to be used as rooms, for example in chalet bungalows.	The flat roof can collect water and decay, causing leaks.

Roof construction

Traditional

A traditional roof is normally assembled on site using sections of timber known as rafters. These are often supported on purlins (horizontal roof beams) that then transfer the weight onto the gable walls. A traditional roof has a ridge board where the rafters meet at the apex and are fixed to the wall plate at the roof eaves end. A collar ties the two sides of the roof pitch together at the ridge and a ceiling joist ties the roof laterally at wall plate level.

Discussion

Would you prefer a dormer or mansard type roof? What factors might influence your choice? Debate the merits of each.

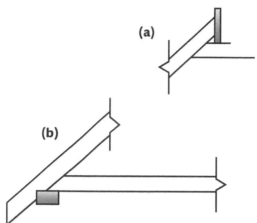

▶ **Figure 10.5:** (a) A collar and (b) a ceiling joist

Roof truss

A roof truss is manufactured in a truss plant using stress-rated timber and hydraulically pressed connection plates. It is structurally designed as a single unit that is fixed into place in accordance with the design. Bracing is then fixed for lateral restraint. There is no ridge board as with a traditional roof. These structural elements require a redesign if an attic conversion is required. They can be designed to include an attic for later conversion. A roof truss is lightweight, efficient and quickly erected.

Metal roof trusses are similar but manufactured from mild steel angles. In the past, they were often covered with an asbestos sheeting roof but now maybe covered with galvanised steel sheets for commercial units or standard roof tiles. They are quick to erect and long lasting.

Figure 10.6 shows some examples of roof trusses.

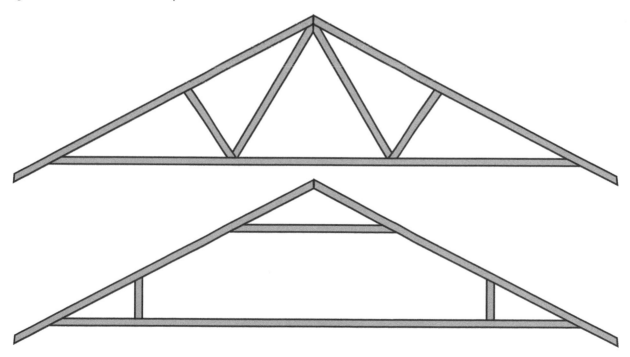

▶ **Figure 10.6:** Typical examples of manufactured roof trusses

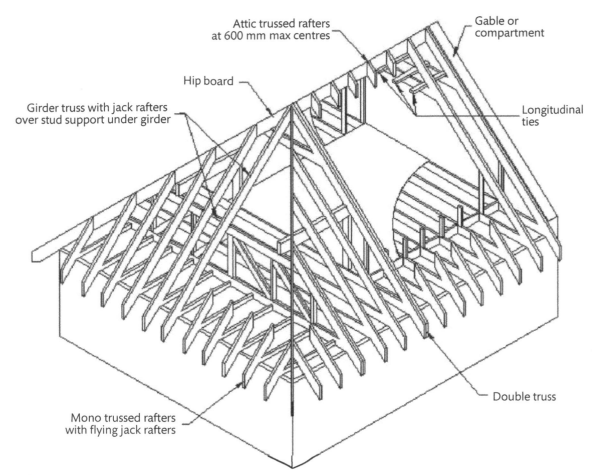

Attic trussed rafters
at 600 mm max centres

Gable or
compartment

Hip board

Girder truss with jack rafters
over stud support under girder

Longitudinal
ties

Double truss

Mono trussed rafters
with flying jack rafters

▶ **Figure 10.6:** *Continued ...*

Roof finishes

A building surveyor must inspect the roof finishes, especially if the property is within a conservation area. This is because they have to be replaced like for like under the planning designations applied to such areas. Many parts of the UK have specific localised roof coverings for traditional houses, for example slate in Yorkshire and Wales, stone in parts of Yorkshire and thatch in areas of Sussex. Industrial buildings are mostly covered with metal sheets coated with an applied paint system.

▶ **Slate** is an effective and attractive roof covering but is prone to defects such as nail failure and spalling (flaking) of the edges of the tile (see photo below). Slates can be redressed and used again or recycled. Slate roofs often do not have a felt underlay; instead, the tiles are fixed using a mortar bed and secured with nails. A slate roof can be upgraded or repaired if it is stripped and felted, re-battened and the original tiles refixed.

▶ What causes this type of damage to slate tiles?

▸ **Stone** roof tiles are large and often heavy. The tiles will wear with the action of rain upon their surface; and due to this and city pollutants they may need maintenance. New stone tiles may be sourced from local quarries to match the existing ones. The method of fixing may need to be examined to ensure it can bear the weight of replacement tiles.

▸ **Clay** roof tiles are produced as a flat tile often called a 'Rosemary' or a curved 'Pan' tile. When exposed to rain and frost they do break down over time with spalling of the tile edges and surfaces and have to be replaced as their waterproofing is weakened.

▸ **Thatch** is a specialist roof finish that naturally weathers. Thatch needs treating, and some types need replacing, after 20–30 years. All have to be reridged after 10–15 years. Bird mesh wire coverings are often employed to prevent nesting within the roof. Fire damage is a risk.

▸ **Metal sheet** in older roof sheeting tends to be galvanised steel, which deteriorates over time. The galvanised protection erodes and exposes the mild steel beneath. This, with the action of water and air, rusts the steel, and the rust can also attack the fixings holding the sheet in place. Sheets can be treated with a coating but eventually they need to be replaced with modern insulated PVC or steel roofing panels.

▸ Can you see the bird mesh that has been installed to protect the eaves of the thatch?

 PAUSE POINT   Can you explain why we often have to replace traditional finishes like for like?

> (Hint)   Look at the designation of a conservation area with a planning authority.

> (Extend)   Would you have to upgrade a roof in a conservation area so that it conforms with the Building Regulations?

Flat roofs

Table 10.3 shows the several different types of flat roof you may encounter.

▸ **Table 10.3:** Different types of flat roof

Type of roof	Description
Timber	Traditionally span the shortest distance between supporting walls. Held in place with nails or connector plates. Flat roof joists have a fillet piece fixed to the top to provide a fall so the finished roof will drain towards a gutter. The joists are boarded over with external grade plywood. This is finished traditionally with **bitumen**-bonded roof felt. Problems with this type of roof are based around moisture vapour and how this is controlled within the roof void. Traditionally it was vented to allow warm moist air to escape with no insulation within the roof void.
Metal decking	Usually made of lightweight structural steel or aluminium, enabling large spans and reducing the load on the building. The profiled design means that it must be overlaid with a timber deck or insulation and then finished with a bitumous or polymeric membrane.
Concrete	Common in blocks of flats, and pre- and post-war houses on toilet blocks or other outbuildings. They were cast in situ using supporting formwork, which relies upon the quality of the concrete as waterproof membrane was not used. Warm moist air from services within the occupied building often condensed upon the underside of the cold roof, causing mould and bacterial growth. The external surface of the concrete roof was sloped towards a gutter to collect rainwater and discharge it into a drain.
Lead	Lead was traditionally used in many significant historical buildings such as churches. Today, its high scrap value attracts vandalism and theft of lead roof sheeting. Lead also thins with age and reacts to dissimilar metals it contacts. This causes electrolytic corrosion. Lead expands and contracts with temperature variations and this must be taken into account when designing a lead roof. Lead rolls are used to accommodate this expansion and contraction.
Felt	Economical solution, traditionally uses three layers of bitumen felt bonded together with hot poured bitumen, held in place by a primer applied to the surface of plywood decking. It deteriorates over time with temperature changes, causing debonding and bubbles in the layers of felt. It can be repaired by applying additional layers. The building surveyor should report the useful service life of this type of roof to clients.
Asphalt	A more expensive product for covering a concrete roof. It is hot poured and allowed to cool into a 20 mm layer. Asphalt is hard-wearing against weather but can be damaged by machinery on its surface. Spreader plates such as concrete slabs are often used under plant installations to prevent damage to the waterproof layer. Asphalt can be laid vertically to such features as parapets.

Case study

Building survey

As the chartered surveyor for an estate agent you have been asked by a potential client to undertake a survey of the property they are interested in buying. It was built in 1950 using a cavity wall and a clay roof tile. The buyers want a report detailing the future maintenance that will be required on the roof. They need to negotiate any repairs against the asking price with the seller to see if they are willing to drop the asking price.

After viewing the property and inspecting the roof void you establish the following findings:

- There is no felt underlay to the clay roof tiles.
- Several of the tile battens are rotten.
- Some of the roof tiles are damaged with surface wear.
- There is no insulation within the roof void.

Check your knowledge

How would you advise the client?

 PAUSE POINT   What factors affect the choice of roof covering?

> **Hint**   Consider the influence of time, cost and aesthetics.
>
> **Extend**   Does a listed building's roof covering have to be repaired like for like (i.e. using the same methods and materials)?

> **Research**
>
> Investigate the construction of a traditional suspended timber floor including its detailed components and how damp is contained. Identify ways in which the floor can be modified to provide better damp protection. Present your findings to the group.

> **Link**
>
> For more on solid and suspended timber ground floors, see Unit 4, section C2.

Floors

Floors: Solid

A solid floor is just that – it does not have a void below it and has a solid dense construction.

The solid floor has been traditionally used for outbuildings such as washrooms, larders or coal sheds that serviced a larger property. The solid floor would have been excavated and backfilled with cinder hardcore compacted to form a base for an oversite concrete slab. The solid floor may not have any form of damp-proof membrane as it is covered using clay or quarry tiles. A bitumen coating may have been applied as a barrier prior to laying a floor finish. This forms a dense tile that is impervious to water and is extremely hard-wearing. Clay tiles allow a floor to breathe and for any moisture to escape. Fire-protecting membranes may also be laid under floors.

Problems with traditional solid floors include:
- damp penetration due to ineffective dpm
- mould growth against walls at floor level
- lack of insulation within the floor
- cold bridging (weak spots caused by a break in the insulation) at the edges against the inside skin.

Floors: Suspended

Traditional floors have been manufactured from timber joists and covered with floorboards. These can be used on both ground and first floors, where they are suspended by the external walls and any load-bearing internal or dwarf wall.

Timber floors are traditionally supported upon honeycomb walls. These are brick walls with holes that allow the passage of air through the floor void. The floor joists rest upon wall plates which are bedded onto the honeycomb walls. Air bricks within the external wall allow air to ventilate around the floor void.

Intermediate floors were traditionally supported by building in the ends of the joists into the inside skin of the external wall. Internal walls provided the end support for the floor joists. Herringbone strutting provided lateral support to the floor to stop it moving. This was placed down the centre of the floor. Intermediate floors span the shortest distance.

> **Link**
>
> For more on the use of joists in intermediate floors, including engineered timber joists, see Unit 4 section C2.

> **Key terms**
>
> **Intermediate floors** – those that support the first floor and floors above.
>
> **DPC** – damp-proof course that is inserted in walls to prevent the vertical movement of moisture.

Solid internal walls

A solid internal wall can be load bearing in supporting the first-floor joists. They are constructed using brick or concrete breeze blocks. There is no requirement for thermal efficiency, as the internal wall is not exposed to the same temperature range as an external wall. Internal walls tend to rest upon foundations created by thickening the floor slab to provide sufficient support for the loads. This may be the only long-term issue with settlement as the wall drops, causing floors to slope at intermediate level.

The type of finish applied depends upon the background the wall is secured to. Solid walls can be covered with a wet applied plaster, usually in two coats, which is trowelled smooth. The main issue with older lime plasters is that they debond (detach) from the background and become hollow so have to be hacked off and replastered using modern Gypsum plaster. Any damp penetration from a defective **DPC** will cause the wall finishes to develop mould which is a health hazard. This has to be repaired by removing the internal plaster finishes up to 1 m high before replastering.

Hollow internal walls

Timber stud (frame) walls have been traditionally used as they are lightweight and can quickly be erected. The timber studs are covered with wall boards which are commonly manufactured as a plasterboard. This method became popular from the 1940s. The main issue may be damp penetration into the floor causing rotting of the timber studs. Wet and dry rot can attack the timbers if the conditions encourage growth.

Traditional finishes have been plaster laths and a plaster finish. This method involves nailing strips of wood to the timber studs with a gap between each. Wet plaster is smoothed over the laths to form a finished wall. There

may be impact damage to walls which breaks the plaster and it drops off the laths. Laths can become infested with woodworm which weakens the structure.

> **Link**
>
> For more on internal walls, including metal stud walls, see Unit 4 section C1.

Doors and windows

Timber

Traditionally windows and doors have been manufactured from timber. Modern technological advances in plastics have meant that uPVC window replacements have become the norm when upgrading older properties. The removal of timber windows to accomplish this does have its problems. Often there is no lintel over the window opening and the brickwork above can drop.

Defects that occur with timber are its biological degradation and eventual rot and breakdown of the timber.

Timber doors are essentially larger versions of windows but have solid units within them for additional security and ironmongery for locking and latching. They are manufactured using hardwoods which are more durable for the UK climate and will not rot as fast as softwood. Doors have to be protected with a paint finish, which needs to be maintained to prevent moisture ingress. This finish can be a stain, which has to be applied at regular intervals to maintain the level of protection.

Metal

Although there are some spectacular examples of wrought-iron framed windows from the sixteenth century, and some cast-iron frames, surveyors are most likely to encounter metal windows manufactured from mild steel in 1960s properties. They have issues associated with single-glazing and condensation forming on the cold metal surfaces. When the paint protection is exposed to the weather then the ferrous element in the mild steel rusts. This has to be treated and the paint finish reinstated. The other main issue is their aesthetics, which clearly dates a property and may affect its value.

Metal doors tend to be created using aluminium sections or mild steel sectional doors for commercial applications such as machinery shops or roller shutter doors for vehicular access.

> **Theory into practice**
>
> Crittall windows are steel-framed windows commonly used in the early 20th century. Research how to improve the thermal efficiency of Crittall bay windows in a 1930s house. Write a step-by-step set of guidelines.

Glazing methods

Key glazing methods are shown in Table 10.4.

▶ **Table 10.4:** Key glazing methods

Type of glazing	Description
Single-glazing	A single layer of glass, bedded into the frame using linseed oil putty, initially secured with glazing nails. This dries to a hard compound, painted over to form a seal. Timber glazing beads can secure the glass but must be screwed and glued. When frames are not painted, water can ingress into the frame, causing rot. Low temperatures outside the glazing and warm moist air inside cause condensation on the inside of the glass.
Secondary glazing	An alternative solution when it is too expensive to replace single- with double-glazing is to install a secondary layer to the inside reveals of the window openings. Constructed using aluminium or plastic frames containing glazed units, which often slide for cleaning purposes. When closed they form an air gap between the secondary glazing and the single-glazed window and this trapped air insulates the air cavity.
Leaded lights	Historical, often coloured, windows formed in very small pieces held together using lead strips with channels within them. The glass is held in place by soldering joints on lead channels. The panel was cemented to seal any gaps between the lead and the glass. Failure of the lead joints can lead to a buckling of the window and movement in the wind. Repair is an expensive specialist job.

⏸ PAUSE POINT   What are the advantages of a metal door over a timber door?

       Hint   Think about the effects of moisture.

       Extend   Which is the more secure option?

A3 Modern methods of construction

Modern methods of construction (MCC) are often retrofitted to existing older structures to improve their thermal properties and to retain an existing building without demolishing and rebuilding the structure.

Link

This section has themes in common with Unit 4, which details the modern methods of construction that a building surveyor needs to be informed about so they can report it in surveys undertaken for clients. Table 10.5 indicates where these topics overlap.

▶ **Table 10.5:** Modern methods of construction (described in full in Unit 4)

Type of construction	Themes
Types of foundation (see section B4 of Unit 4)	• strip • trench fill • raft • pad • pile
Ground floors (see section C2 of Unit 4)	• solid concrete • beam and block • prestressed concrete • suspended timber
Intermediate floors (see section C2 of Unit 4)	• beam and block • prestressed concrete • timber • platform floors in timber frame construction
Internal walls and partitions (see section C1 of Unit 4)	• blockwork partitions • timber stud partitions • metal stud partitions • demountable partitions
Framed walls (see section A1 of Unit 4)	• timber frame construction • platform frames • storey height panels

Walls

The treatment and upgrading of solid walls may involve fitting some or all of the following technologies externally or internally to an existing structure. They increase thermal efficiency while allowing water vapour to breathe through the wall and also provide an attractive finish.

Thermal blocks

Also called aircrete blocks, these are made of aerated concrete with useful properties such as high compressive strength, a light weight, good thermal insulation and moisture resistance. They are often made of recycled material and are themselves recyclable. They are commonly used for external walls and sometimes paintable blocks are used internally. They may form a cavity wall, with a layer of bricks or other blocks.

Insulation

There are three main types of insulation.

▶ **External insulation** is applied to the outside face of the existing external materials or structure and is fixed using mechanical or adhesive processes to keep the insulation in place. The thickness can vary up to 100 mm. The insulation is covered with an applied coloured render for a waterproof finish. The render has stop beads at the base and angle beads around all corners to protect the edges from damage.

▶ **Rainscreen cladding** is fixed to the structure and can be used to upgrade commercial buildings. Insulation and a breather membrane layer control moisture ingress into the existing structure. Cladding rails are fixed to the structure and a rainscreen cladding finish is applied. This allows water to penetrate past the cladding and fall within what is in effect a drained cavity.

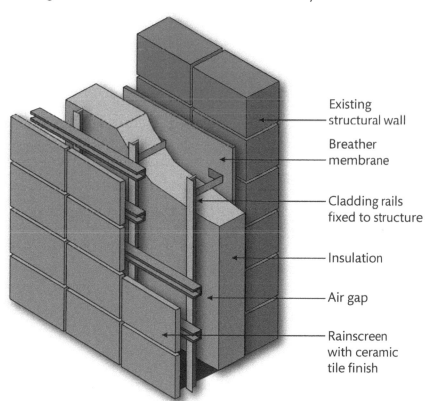

Existing structural wall

Breather membrane

Cladding rails fixed to structure

Insulation

Air gap

Rainscreen with ceramic tile finish

▶ **Figure 10.7:** An example of a wall with rainscreen cladding

A variety of finishes are available, from strips of cedar with open joints between that create a wall effect to terracotta clay tiles and ceramic tiles. Metals and wall boards can also be used for this external application.

▶ **Internal insulation** can be applied as bonded plasterboard and insulation added to the inside walls to increase the level of insulation within older buildings. The plasterboard can be covered with a skim of plaster finish.

Finishes

The type of finish chosen depends on the substrate (surface material), the age of the property (how well it fits into its surroundings and likely level of maintenance) and personal taste.

▶ **Internal finishes** may be an appropriate type of plaster for the building (for example, lime plaster for older properties), plasterboard (drylining) and paint. Board and skim is common – plastering onto a drywall for a smooth surface. Modern technologies can make even these apparently simple products advanced enough to repair and protect a building.

▶ Similarly, **external finishes** need to suit the building, whether it is rendered, painted or chemically weatherproofed. Examples of modern render systems include polymer renders (where the concrete base contains silicon for its water repellent but breathable properties or nylon for flexible strength) and acrylic renders (containing aggregates that are lighter and more porous than stone).

Structural insulated panels (SIPs)

This modern concept uses a bonded insulated panel as the structural frame for a building. A core of insulation is faced both sides with externally orientated plywood strand boards (OSB) as a very quick method of erection. They are joined using treated timber that locks into each panel and is nailed in place. An external finish of brickwork can be attached to the panel using ties to form an external weatherproof wall. These panels can also be used to construct the roof to form one composite structure.

Timber framed construction

A timber frame consists of stud framework that is faced with plywood to form a panel. Panels are then assembled on site in accordance with the timber-frame manufacturer's design details. The frame sits on a bearer at damp-proof course level and has a headbinder to tie the panels together at first-floor height.

Roofs

As we saw in section A2, roofs come in a variety of shapes and materials. Many traditional methods are still used for both the structures (such as timber and metal trusses) and for the finish (such as slate, stone or clay tiles and thatch). Here are some examples of MMCs relating to roofs:

▶ Polycarbonate is a strong, lightweight thermoplastic that can withstand significant temperature fluctuations. Although it is durable, it is better suited to small spans such as conservatories and car ports than covering a whole roof.

▶ Liquid roofing uses liquid polyurethane that dries as an elastic sheet. It is versatile, easy to apply, durable and comparatively inexpensive.

▶ Improvements to glass technology enable larger stretches to be installed overhead as roofs or ceilings. While it can be expensive and complicated to install, it can make a big impact.

Green roofs

Modern roof construction means that different forms of roof coverings can be employed for domestic and commercial applications. One such application is a green roof, a living roof where speciality plants are used as the roof covering. Soil retains the plants in a matrix which allows water to be absorbed and evaporated as the weather changes. This reduces water run off, resulting in less damp and flooding. Green roofs can also protect the roof surface material from temperature fluctuations, increasing its lifespan. Other advantages include:

▶ increasing biodiversity in the area
▶ improving thermal performance and reducing energy use
▶ providing an attractive amenity space
▶ acoustic insulation, for example lessening the noise of rain on a flat roof.

Link

For more on storey height panels, timber frames and platform frames see section A2 of this unit and Unit 4 sections A1 and C1.

Research

Find a video on how the SIPs system is erected on site to form a complete building. Identify advantages and disadvantages of the system. Discuss with your group the types of projects in which SIPs can be best used.

Ⅱ PAUSE POINT — What are the benefits of using a green roof?

Hint — Look at a green roof subcontractor's website for more information.

Extend — Are any additional structural supports required?

Sedum plants are generally used as they grow slowly and are virtually maintenance free. The increased weight of a green roof must be taken into account. Often the supporting structure is a concrete roof slab. Slight inclines can be accommodated within the design and aid drainage.

Where flat roofs are used on commercial buildings then pits can be formed within the structure to accommodate larger trees. Grassed areas will obviously grow more quickly and must be cut to maintain the appearance.

Single ply membrane roofs

Modern technology has developed roofing systems for horizontal applications with slight falls. These are single layers applied to a plywood or insulation substrate. They can be bitumen or polymer based.

The advances in glass-reinforced plastics means a roof membrane can be constructed using resins and glass fibre and finished with a top coat to form a waterproof roof. These systems have extended the life of the traditional three-layer built-up felt roof and manufacturers will now provide 25-year guarantees for their systems if authorised installers are used or instructions followed.

Doors and windows

Modern doors and windows may be constructed from the following:

▶ **Timber** is seen as a sustainable building material and is usually sourced from managed forests. The wood may be treated with coatings to prolong its durability and weathertightness.

▶ **uPVC** (or PVCu) is more durable, weatherproof and less likely to warp than timber, and can be clad to look like wood. It offers good thermal and acoustic insulation and is easy to maintain.

▶ **Thermally broken metal** doors and windows include a thermal break – a small piece of a material that does not conduct heat, such as polyurethane. This enables the doors and windows to use the benefits of a metal such as aluminium, which is strong, light, weatherproof and needs little maintenance, without having to endure its disadvantage of conducting heat and providing poor insulation.

▶ **Composite materials** are combined to take advantage of their different properties, such as PVC, wood, insulating foam and glass-reinforced plastic (GRP). They may have an insulating foam core and a strong, weatherproof outer coating.

▶ **Double- and triple-glazing** ensure a low u-value – the measurement of energy performance. The Building Regulations state that it should be no higher than 1.6. Windows and doors are weak spots with regard to insulation so really low u-values require the frame as well as the glazing to be insulated. Triple-glazing is expensive and probably is not necessary except in very cold climates, but it should increase a home's general comfort.

▶ **Thermal coatings**, such as low E (low emissivity) coatings, reflect infrared radiation, either internally to retain heat in the home, or externally to cool the home.

▶ An **inert gas**, such as argon, krypton or xenon, is sealed within the frame to minimise the conduction of heat back out of the building.

Assessment practice 10.1

You are a building surveyor who has been asked by a client with a large property portfolio to assess the styles and methods of construction. This is to be done so an estimate of its value can be established, along with the impact of future repair work.

- Produce a leaflet that describes the common styles of post-war domestic architecture.
- Describe the different methods of domestic construction that have been traditionally used.
- Contrast the traditional methods of construction with modern methods of construction for domestic applications.
- Evaluate the impact that traditional methods of construction have on future repairs and maintenance.

Plan

- How can I break down the task into manageable portions?
- Do I need to research styles and methods of construction?
- What evidence will meet the criteria targeted?

Do

- I know what level of detail is required to be evidenced.
- I can contrast between traditional and modern.

Review

- I can review evidence against targeted criteria.
- I can explain how I would evaluate outcomes better next time.

B   Explore different defects and methods of repair for low-rise residential properties

We repair our properties to maintain their value (by optimising their appearance), to prevent further damage, to improve security and to meet legal obligations (such as landlord's obligations).

Any repairs or maintenance can be managed and organised by the building surveyor, who may be working for a housing authority, a domestic client or a commercial client with a portfolio. The building surveyor can advise on repairs that will reduce the long-term maintenance costs of the properties over their life cycle. Extending a building's life by repairing and upgrading it is a sustainable strategy.

B1 Defects to the external envelope

Foundation defects

A foundation may have been designed to standards that were specified at the time the property was built. However, if a supporting foundation is not deep enough, then it can be affected by seasonal changes in the soil's moisture content and rise up and down with the increase and decrease of moisture. Clay soils are susceptible to this problem. Modern foundations are a minimum of 600 mm wide and 1 m deep so ground movement of clay soils does not affect them.

Settlement

Settlement has many causes.

- **Seasonal:** the ground dries out in summer and settles under the effects of gravity. This is due to the moisture in the pores of the soil evaporating which leaves a void; this then closes causing the ground to settle downwards, taking the foundation with it. Similarly, any water within these pores will freeze and expand during the winter, lifting the ground and causing heave.
- **Mining:** historical mining works, for example in past areas of coal extraction, cause settlement to large areas of the surface where the mine workings are beneath.
- **Compression of substrata:** subsequent layers of the substrata can compress at different rates over time. This can cause differential, uneven settlement. Too much weight or overloading on poor bearing ground will cause settlement of the foundation and the supported building.
- **Landfill:** Areas of land that have been used as rubbish tips can contain household

> **Research**
>
> How do foundation defects affect a building? Research the symptoms that a surveyor is likely to see that indicate the need for remedial works.

waste and other waste products. Landfill takes many years to break down and settle before it can be used for development. Contamination is also an issue with old tips. Similarly, **made up ground** can cause issues with settlement across sites.

▸ **Trees:** a tree will extract water from the ground as it matures, causing settlement and damage to foundations as the roots grow and expand. If a tree within the sphere of influence of the pressure from a foundation is cut down then it no longer extracts water, which increases the pore pressure within the soil and causes heave.

Sulphate attack

Soils can sometimes contain sulphates, which attack the cement within the concrete that forms the foundations. The cement bonds break down and the foundation weakens, possibly cracking and settling.

Overloading

Overloading can be caused by altering a building, internally or externally, for example adding or removing walls, enlarging doors or windows, or building an extension can all redistribute or increase the load on the subsoil. Usually, this will not significantly affect the foundations of a domestic house unless the foundations are too shallow.

Bad design/poor construction methods

The design of a foundation can cause some long-term problems. Foundation repairs are costly and difficult, so every effort should be made for them to be right first time. For example, a raft foundation is designed to rest on poor bearing soils and settle evenly over time but differential settlement can occur, leading to soil topping the foundation in places. Stepped foundations on sloping sites have to be designed to reduce the volume of excavation.

Poor quality of workmanship may cause foundation failure. For example, the depth of concrete poured should be checked by placing depth indicators within the excavated trench; if not, it may lead to thin foundations that break when the cavity wall is built upon it. Not using the correctly specified strength concrete also may affect the long-term viability of a foundation.

PAUSE POINT Can you explain a foundation defect in detail?

Hint Why does a foundation settle over time?

Extend Why does chemical degradation of a foundation occur over time?

Wall defects

Bowing/bulging

This can be caused in older walls by:
▸ failure of the cavity wall ties
▸ lack of lateral restraint at intermediate level
▸ lack of restraint at eaves level
▸ excessive lateral forces
▸ increased internal loading
▸ foundation settlement.

The outside skin of the wall can bulge as it moves if it is not attached to the internal superstructure. Similarly a solid wall can bow as it is prone to taper with height.

Research

Investigate three historical buildings in your area and see the measures that have been used to prevent and control the wall movement. Prepare a short report on this and present your findings to the rest of the group.

Poor construction detailing

The way in which water is discharged from the external envelope of a building can cause damage to the external fabric of a wall. Excessive water over time can enter through the mortar joints, and subsequent freeze and thaw action causes the spalling of facing brickwork. Lack of detailing around openings causes staining where the water takes the shortest route down the structure to the ground. Poor detailing can include:

▶ the lack of window cills and throats to drop water off the face of the wall

▶ the lack of an eaves soffit to protect the top half of an exposed wall

▶ incorrect pointing for the climatic conditions

▶ a flush roof finish with gable walls.

Wall tie failure

As we have seen, this causes the wall to bulge on a cavity. Traditionally, butterfly wall ties were used. These are painted mild steel which over time breaks down in a moist insulated cavity by rusting. The twist on the butterfly comes undone and the wall tie is rendered ineffective. The outside and inside skins are then free to move and as they are now affected by their slenderness ratio (the relationship between their width and height) the outside skin can move away from the restrained inside skin.

Lack of lateral support

The Building Regulations Approved Documents provide guidance as to where the **lateral restraint** must be placed. For example:

▶ at the intermediate floor level by using rust-resistant straps that fix to the floor joists and fix to the inside of the cavity wall

▶ at eaves level where the ceiling joists meet the wall plate

▶ at the roof level and gable walls where straps again run from the cavity across three roof joists to secure the gable walls.

Lack of lateral support results in the bulging of walls as they operate under the theory of slenderness ratios.

> **Key term**
>
> **Lateral restraint** – the horizontal restraint required by a structure to prevent sideways movement.

> **Theory into practice**
>
> Find out more about slenderness ratios and their effect on columns and walls. Prepare a single PowerPoint® slide explaining the concept in simple terms.

Overloading

Refurbishing and adapting our built environment may lead to increased loads placed upon the floors, roof and supporting walls. Increased loadings on poor foundations can cause settlement or cracking of foundations. Overloading of floors on slim joists causes deflection and cracking of the ceilings below.

Roof thrust

The pitched roof is attached to the wall plate and has a collar or ceiling joists to tie across the roof preventing the rafters thrusting outwards, causing the tops of walls to lean with the increased thrust. Where alterations have removed the collars within the attic void then issues can arise with the roof spreading under the force of the self weight and the roof finishes. The ridge indicates if this has occurred, as it is no longer straight when viewed and may contain a dip between the gable supporting walls.

Alterations and changes to original structure

This may increase the load on an existing structure either vertically or laterally as mono-pitch roofs from extensions exert lateral forces onto an existing structure. When removing walls and inserting supporting structural elements, their end bearing must be sufficient to support the structural components.

Failure of supporting elements

Supporting elements can cause a range of issues for walls if they fail.

▶ **Arches** are features over openings that support the external wall structure above. If the foundations settle on one side then differential movement will occur within an arch, which may lead to it cracking or slipping.

▶ **Lintels** in older buildings were traditionally manufactured from timber which can be subject to biological degradation and insect attack. Lintels that have been inserted with less than 150 mm end bearing may exert excessive stress onto their padstones (bases), which may crack and degrade causing settlement of the lintel.

▶ **Structural elements**, such as joists and roof timbers, can fail if they are subjected to excessive loads or have been degraded by rot or wood-boring insect damage.

▶ Expansion of embedded **steel and iron fixings** can occur on older buildings as they have no surface treatment and are subjected to moisture and develop rusting and structural degradation leading to eventual failure.

> **Theory into practice**
>
> Find an example of how a particular house or building has been affected by rot and wood-boring insects on its supporting elements. Write a report outlining the issues and consider ways in which they could be repaired.

Surface failure

The external surfaces of structures are subjected to weather (rain, wind, frost and sunlight), temperature changes and pollution. All of these age a structure and can result in the following types of surface failure:

- efflorescence – a discoloration caused by salts in the mortar and brickwork migrating to the outside and crystalling on the surface of the facing brickwork causing a white appearance. It is harmless and usually temporary but can look ugly
- freeze thaw which causes the face of the brick to spall away and wear due to continual action over a number of years
- sulphate attack
- discoloration due to ultraviolet degradation of paint pigments
- cracking in painted timber elements
- sulphate attack – failure of the mortar joint pointing, especially if incorrect pointing mixes or strong renders have been used.

Chimney defects

In traditional properties chimneys were constructed to discharge the smoke and combustion gases from coal and wood fires. Chimneys are now lined to allow new back boiler exhaust gases to use the chimney to vent the products from combustion. A traditional chimney can be prone to sulphate attack from the burning of coal. This eats into the mortar joints between the chimney bricks and expands. Expansion can also occur on the warm side of the chimney, in the lee side of the prevailing wind. This causes the chimney to lean as the expansion takes hold on the one side.

Chimney pots can also become defective if their mortar bedding and haunching (mortar at the top of the chimney) deteriorates, causing the pot to become dislodged in high winds and fall to the ground or onto the roof.

Tall, thin chimney stacks can be weak if they are not supported so often have a lateral restraint fixed to them and the main roof for additional support. If this rusts and fails then the stack can become vulnerable to lateral movement.

Roof defects

Structural failure

The main structural elements of a roof include the rafters, ceiling joists, collar ties, wall plates and valley and hip rafters.

Any failure of one of these can lead to roofing defects and water ingress into the building. Structural failure can

be caused by poor design, overloading, rot, removal by users (for example, cutting out parts of an attic). Missing flashings or lack of maintenance to flashings causes water ingress into a building and with lack of maintenance of guttering can cause the same problems. No ventilation within a roof space means moist stale air can start the process of dry rot with devastating consequences.

Roof finishes

A roof structure is only as good as the roof covering that forms the watertight envelope. Traditional roof coverings of slate and clay roof tiles are used for pitched roofs whereas built up felt roofing is used for flat roofs. The main defects associated with roof coverings are:

- poor initial construction or subsequent maintenance
- tiles slipping from defective fixings
- nail rot
- tile batten rot
- ridge and hip tiles mortar beds deteriorating
- thermal expansion on flat roofs
- damage by frost
- efflorescence
- slipping or corrosion of flashings and pointing
- wind damage lifting tiles.

Door and window defects

Doors and windows can suffer from defects such as:

- rotten timber weakens the door or window main frames
- putty failure makes any glazed panels in doors and windows rattle and is a security issue
- fixing failure – failure of screws holding door and window hinges in place leading to them dropping causing difficulty in closing
- failure of hinges and handles weakens the security of a property and may trap people inside or out of the house
- rusting to steel and iron fixings leads to hinge failure and stiffness
- swelling and poor operation of window and door jambs
- failure of decoration such as peeling and poor aesthetics.

> **Theory into practice**
>
> Collect at least five photos of roof defects, either online or by taking photographs in your local area. Discuss with your group the possible causes of each one. How do you think they can be repaired?

 **PAUSE POINT**   A wall is bulging. What might be the cause?

Hint   A wall failure often involves the cavity tie that holds the two skins of the cavity together. What wall tie commonly fails?

Extend   Are there any other types of traditional wall ties that can fail?

B2 Internal defects

Ground floors

Solid concrete floors

Traditional older properties with solid concrete floors may have defects from:

- **sulphate attack** – the chemical composition of the cement within the concrete is attacked by sulphates within the subfloor below
- **membrane failure** – the damp-proof membrane (dpm) is breached, laps are not adequately sealed or there is no dpm within a floor which is exposed to rising damp
- **damp ingress** – rising damp from external walls can be drawn into the subfloor
- **poor detailing/construction** – for example, a lack of perimeter insulation or under slab insulation causing a cold floor, poor quality compacted concrete or slabs are too thin
- **settlement** – poor compaction of the hardcore that results in the settlement and cracking of an unreinforced floor
- **heave** – movement and cracking due to water accumulation in the soil beneath, or sulphates
- **screed failure** – due to debonding of a screed from the concrete background, or from weak screed mixes resulting from site-mixed screeds
- **excessive spans for precast concrete** – resulting in bounce to a floor with excessive deflection and possible cracking of covering screed.

Suspended timber floors

These are traditionally manufactured from timber softwood floor joists that are suspended at their ends on the inside skin of external walls and intermediately using small honeycomb walls. The main related defects are:

- **poor construction and detailing**, for example badly laid, not protected from damp or overstripped when varnishing
- **rotting** of the wall plates that support the intermediate positions
- **lack of subfloor ventilation** causing fungal attack
- **damp ingress** – rising damp from external walls can be drawn into the floor
- **excessive spans** for the depth of joist causing bounce
- **insect attack** from wood-boring beetles
- **lack of strutting** across a floor leading to movement at ninety degrees to the joist span
- **ends of floor joists** touching the inside skin below damp-proof course or built into the inside skins of hollow walls causing moisture ingress and rotting the timbers.

Upper floors and ceilings

Floors

The types of defects that can occur with upper floors are:

- poor construction and detailing
- lack of support to the joist ends
- lack of support and strutting causing sagging
- alterations and service holes/notches into tops of joists
- overloading causing deflection of floors
- insect and fungal attack within the floor void
- joist ends built in with no DPC around their ends
- failure of mechanical fixings between boards and joists.

Ceilings

Ceilings, especially in older properties, can be prone to defects associated with older materials and the plaster technology that was used at the time of installation. Examples include:

- poor construction and detailing at corners, causing cracking between the external wall and ceiling
- failure of fixings to plaster laths or boards causing debonding of surface finish
- water leaks causing stains and damage to plaster lath ceiling
- overloading of attic ceiling joists causing deflection
- settlement of timber lintels causing cracking to decorative covings.

Walls

Both masonry and timber walls can be subjected to the same sorts of issues as floors, including overloading, poor detailing, construction and alteration work, insect and fungal attack and lack of support.

Rising and penetrating damp occurs, for example, where there is no DPC installed at all or the existing DPC has failed. Traditionally slate was used as a DPC layer within the external wall along with bitumen-type DPC felts. Where these fail, damp can rise a metre up a wall. This defect may be corrected by drilling in a series of holes then injecting a chemical that seals all the pores in the brickwork to prevent moisture travelling. The second method is to insert a physical DPC by removal of the external face of brickwork in short sections.

In addition, timber walls and floors may be damaged by service holes and notches, such as those drilled or sawn into the wood by plumbers laying pipes. This should be controlled under the Building Regulations but damage can still happen.

Stairs

Stairs deteriorate over time. For example:

- overloading – inadequate stairs for the traffic

- insect and fungal attack – can rot wood as it does in floors and joists
- poor detailing and design
- damage to the stair spindles that form the balustrade from impact and through movement of the handrail
- loose handrails attached to walls coming away due to defective fixings
- loose handrails on stair centre
- glue weakening on treads secured by wedges
- loose risers connected to the defective treads
- rot, if placed directly onto solid floors that contain no dpm and moisture enters the supporting stair members.

Decoration

If a building's decoration deteriorates over time, there may be secondary causes resulting from a primary failure in the structure or services within it. Causes of damage to decorations may be:

- ingress of water causing staining, mould and damp on the wall surfaces
- peeling of wallpaper as adhesives fail
- imperfections within paint due to background contaminants
- smoke or people smoking staining surfaces
- flaking of paint finishes
- fading of paint colours.

Many decoration defects can be easily repaired with some general maintenance and regular redecoration.

B3 Methods of repair and remediation

Repair

The level of a repair may range from minor surface cracking to complete replacement of an element. Buildings naturally dry out over time and will crack as a result. These minor imperfections are easily cosmetically fixed by painting, sealing or short-term filling but this does not necessarily treat the root cause.

Methods of repair for common defects

Damp

The only completely viable solution to damp is to renew the damp-proof course in the internal and external skins of the cavity or solid wall. This can be a physical or chemical DPC, which both require the inside plastering finishes to be taken off to a height of 1 m so the wall can dry out. Some damp-proof contractors use a diamond-tipped saw that cuts out the mortar joint, allowing a plastic DPC to be inserted into the correct place. This is then pinned and pointed up to make a permanent DPC. DPCs must be a minimum of 150 mm above external ground level and continue through both skins of a cavity wall.

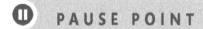 **PAUSE POINT**   What is a common cause of a ceiling cracking in the middle?

> Hint   Consider deflection.
>
> Extend   How would you correct this defect?

Case study

Damp

You have been asked to look at a semi-detached property built in the 1950s. The front of the property has one entrance door and a bathroom above this at second-floor level. The front ground-floor sitting room and the upstairs front bedroom have bay windows with an area of external tiling between them. These are concrete tiles hung on wooden battens. The ground-floor bay window wall is showing signs of damp ingress and you have been asked to make recommendations as to how this could be rectified.

You inspect the bay window and test the moisture levels which read 35 per cent. You establish the following findings:

- The damp has risen up to the bottom of the timber window board.
- The window board is soft in places.
- An area of external landscaping is bridging the damp-proof course.
- There is no wall insulation.
- The cavity has no weep holes.

Check your knowledge

What recommendations would you make for remedial works to cure the damp issues for the client?

Cracking

Minor repair: Cracks in masonry can be repaired by crack stitching across the crack using a proprietary system and then repointing the area of repair. This involves the following sequence:

1 Cut out the mortar joints that cross the area of the cracking to the depth indicated by the stitching supplier.
2 Remove and repair any cracked bricks to match those around the repair.
3 Point up the back of the joints using the adhesive supplied.
4 Insert a stainless steel rod across the cracked area within the horizontal brick joints.
5 Point up the joint to match existing brickwork.

Major repair: Underpinning of the existing foundations to prevent any further settlement would be undertaken by a specialist groundworks contractor and involves excavating small sections beneath the existing foundations and recasting them.

After the underpinning has been completed, the full areas of cracking are cut out and the new facing brickwork reinstalled and pointed to match existing brickwork. This type of repair may be subject to an insurance claim from the property's insurers. They will investigate and have the settlement surveyed by a building surveyor who will make the recommendations regarding the repairs.

Internally, with both minor and major crack repairs, the stitching can be undertaken to the inside skin and the plaster finishes reinstated to the repaired area.

Insect infestation

Insect attack involves the larva stage of wood-boring insects consuming timber. They destroy the structural strength of the timber core and evidence only appears when they leave via an exit hole. The remedy for this type of damage is to cut and replace any affected timber and to spray the surrounding area with a chemical treatment to kill off the infestation as it eats through the chemically treated surfaces. Interrupting the beetle's life cycle is the key to removing them from an infected building. The major issue is that damage is often hidden beneath inaccessible floors and areas of attic voids.

Wasps may also nest in voids such as cavity walls and eaves. While they will not consume the structural timber, they may destroy insulation. They should be poisoned with wasp powder then the damage assessed once it is clear that the nest is no longer active. It may be straightforward to replace the insulation but the area may be hard to access, so roofing materials such as tiles may need to be removed in order to make the repair.

Fungal infestation

There are two types of rot:

▶ Wet – this has an elevated level of moisture associated with it, and forms sprouting bodies which spread spores and are fungal in appearance, often brown with white edges.

▶ Dry – this still requires some moisture to thrive and spread; timber turns cuboid in structure as it takes hold and can be crumbled by hand.

Both types cause severe structural damage, which can only be treated by complete removal of the affected areas and replacing them with treated timbers. The source of any moisture needs to be stopped, for example sealing a bathroom floor perimeter or repairing a damaged water pipe joint. With dry rot, any exposed brickwork has to be treated with a proprietary chemical to stop the chance of any more spores growing and spreading.

> **Research**
>
> Obtain images of wet and dry rot and establish the visual differences between the two types.

Replace/renew

This is the most disruptive and expensive option but is often the best method to remove serious defects such as rot or heave.

Extensive repairs

Extensive repairs remove both the defects and the root cause completely. Although the job, if done properly, will involve making good the surrounding area, the repairs sometimes do not fit in with the appearance of the existing structure and will need time to weather into it.

Underpinning

If underpinning is thought to be required, first the property should be surveyed by a structural engineer as the new foundations have to be taken down to a suitable depth to support the existing loads. This might mean a trial excavation to inspect the supporting sub soil.

After the survey, the structural engineer will draw up a design indicating the size of the foundation that needs to be excavated below the existing foundation and in what sequence the excavation is undertaken. This has to be done in small sections so the rest of the building remains supported. Each section is joined to the next using steel-reinforcing dowel bars.

Various methods of underpinning can be used instead of a straight excavation of the foundation in sections. Mini piles can be inserted each side of the foundation and a beam cast between these to the underside of the existing foundation. This causes minimal disruption and is a quicker method.

Other systems available incorporate helical steel piles, which are screwed into the ground and attached to the existing foundations using a bracket system. This does away with the need for a ground beam to be inserted across the piles. Figure 10.8 shows an example of modern underpinning systems.

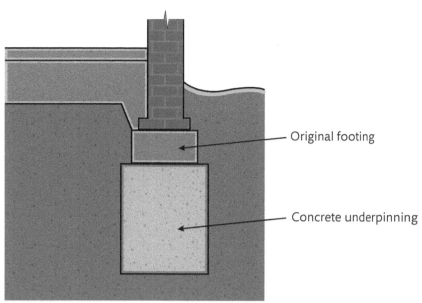

▶ **Figure 10.8:** Concrete underpinning added beneath the original foundations of a building

Taking down and rebuilding

Demolishing the affected area and rebuilding it using modern technology and materials is a disruptive and expensive undertaking but it does enable the structure to be upgraded so that it avoids a reoccurrence of the defects in the future. Some property developers favour this method as they will effectively have a new build with minimal defects that can be sold on for a profit.

⏸ **PAUSE POINT**   Why might home owners want to avoid underpinning their house?

> **Hint**   What processes might it involve?
>
> **Extend**   How might you persuade a home owner that underpinning is necessary?

Assessment practice 10.2   B.P3   B.P4   B.M2   BC.D2

You have been instructed by a property developer to review the condition of the stock of their properties. They are concerned about the long-term cost of repairs over the useful life of the buildings.

Undertake a review that describes:
- a range of common external defects that can occur over time to a domestic property
- a range of internal defects that can occur
- any related defects

Describe ways of repairing the following common defects:
- wall tie failure
- cracked facing brickwork due to settlement
- failure of an arch over a doorway
- rising damp to an internal wall
- springing on a suspended floor

Conclude the review by evaluating repair and remedial measures against factors such as cost and time.

Plan
- Research into the types of internal and external defects.
- Check that these are domestic defects.
- Ascertain a range of defects.

Do
- Have I provided enough evidence and examples?
- Are my sources fully referenced?
- Have I covered at least three defects for internal and external elements?
- Have I suggested realistic methods of repairing the defects?
- How can I use photographs, diagrams and subheadings to make my report clear and visually appealing?

Review
- Does any of the evidence relate to costs?
- Does the evidence support the higher grades?
- What would I do differently next time?

C Undertake a building survey of a low-rise residential property

C1 Types of survey

A building surveyor may be asked by clients to undertake a variety of different surveys. Table 10.6 details the different types of survey and a description of what content is covered within each.

> **Key term**
>
> **RICS** – The Royal Institution of Chartered Surveyors, a UK professional body for qualifications and standards in land, property, infrastructure and construction.

▶ **Table 10.6:** Types of building survey

Type of survey	Description
Building survey	It is a detailed survey that covers all aspects of an older property. The associated report details the structure and fabric of a property and includes: • results of a thorough inspection of all accessible areas of the building • any serious or dangerous conditions • the potential and current repairs that may be required to the property • visual defects • potential problems.
RICS Level 1 Condition Survey	Defined by RICS as an 'MOT' for a new or small building. As with a car, it checks all the important parts. The survey includes: • a description of the condition of the main elements of a building • condition ratings – colour coded • a summary of any defects found.
RICS Level 2 Homebuyers Report	There are two options for this type of survey. • Option 1 is a survey only. • Option 2 includes a market valuation for the property. The survey includes the Level 1 contents plus additional advice on any repairs or maintenance and a condition summary of ratings. The market valuation option covers the current value along with any reinstatement costs to bring the property to an acceptable standard.
RICS Level 3 Building Survey	The top standard of survey report – it a comprehensive tailored report including: • a survey of all visible and potential problems including the grounds • a defect diagnosis with advice and guidance on repair options and maintenance • full description of property condition and how each element is constructed • the energy efficiency of the building • condition ratings – colour coded • a summative assessment on the overall condition of the property including repairs required and any further recommended investigations.
Mortgage valuation	Part of a mortgage application, instructed by the lender to ensure that the property has sufficient value within it should the purchaser default and the mortgage lender recover the property. It does not report on any defects and just confirms the value and gives a brief condition of the property.
Schedule of Condition (Landlord and Tenant)	Details the condition of the property at a point in time at the start of or during a rental agreement. It covers the tenant against claims from the landlord by formally reporting the condition of the property before the tenancy agreement is signed. It includes: • the condition and construction of the property on the date it was inspected, which may be supported by photographs • any hazardous defects that are visible • any defects requiring urgent attention prior to occupancy.

▶ **Table 10.6:** *Continued ...*

Type of survey	Description
Schedule of Dilapidations (Landlord and Tenant)	Records the condition of the property at the end of the rental agreement compared with the initial condition survey and any resultant damage is charged against the tenant's deposit if they are in breach of the rental contract.
Maintenance survey	Reports upon all aspects of a property that need to be maintained, including: • guttering and drainage • fire alarm and fire-fighting measures • security alarms • window hinges and seals • ironmongery • automatic door gear. It will detail what has to be undertaken and the time interval for the maintenance.
Alteration survey	Details the existing and the subsequent alterations as drawings so the alteration is formally recorded within a client's health and safety file. This is a legal requirement under the Construction (Design and Management) Regulations 2015.
Stock condition survey	Undertaken by organisations to ensure that they meet the government's Decent Home Standard. This includes: • a property condition survey which will analyse the condition and age of the features in the property, which will then be estimated for replacement or renovation • an indication of the energy rating for the property • a general health and safety assessment, which will point out any potential risks to owners of the property.
Mortgage drawdown	You may need to finance the design and build of a property with a mortgage. Progress payments are made by the lender to you at various stages of the project. These stages have to be checked on site to ensure that the work has been completed correctly.
Access audits	Covers the Disability Discrimination Act requirements for disabled access into a building. It will cover the provision of equal movement around a building for all occupiers.
Elemental survey	Covers specific elements of a building, for example the structural timbers, ground-floor construction, roof finishes etc. It reports on the defect and the possible methods to rectify.
Insurance reinstatement survey	Follows insurable damage to a property, for example fire and smoke damage, to obtain an estimate of the costs of repairs and to prepare a schedule for contractors to price the work.
Defect analysis survey	Identifies a defect within a building or structure and analyses the causes and how it can be prevented from causing further damage, along with a detailed recommendation for the repair.
Health and safety survey	Examines all aspects of health and safety within a property, for example, the DDA access arrangements for a commercial building, fire escape routes and security arrangements.
Measured survey	A dimensional recorded survey undertaken in order to obtain dimensions for the production of drawings for existing structures and buildings.

Ⅱ PAUSE POINT   What type of survey would you advise for a client buying a domestic house built in the 1900s?

Hint   There may be structural issues – how might this affect your decision?

Extend   Would you recommend a RICS survey?

C2 Undertaking a building survey

The building surveyor will have been initially contacted by a client by telephone, face-to-face or by email or post. Depending on a client's knowledge of building surveying, the surveyor will respond and offer guidance as appropriate to the client's requests.

The following sequence is likely to be followed from this initial contact.

Pre-survey protocol

The instruction from the client must be confirmed to enable the building surveyor to act. This is effectively a contract which confirms what will be provided for the client and the cost of the work so there are no surprises for any parties with the survey work.

After this, arrangements to gain access to a property are made. This may mean arranging for premises to be unlocked, picking up keys and having temporary electrics installed for lighting if it is not otherwise available, so the surveyor can see any defects.

Remember to consider any health and safety issues with entering the building. This could include structural failure in the building, poor lighting, the presence of hazardous materials and hazards from objects in the building. It is important to carry out a risk assessment before beginning work. This should be communicated to all and reviewed and updated regularly. High risks are associated with structures that have been unoccupied for a number of years and may have structurally deteriorated. Confined spaces also present high risks.

> **Safety tip**
>
> Anyone entering a premises alone should also follow the standard personal security considerations, for example telling colleagues when and where they are going, and how long they expect to be, carrying a mobile phone, noting escape routes and leaving if they feel uncomfortable or threatened.

Property inspection requirements

The building is surveyed against the client's requirements and the type of survey that has been instructed. The survey will generally consist of:

▸ inspection of the building's main elements (walls, roof, floors, doors and windows)

▸ recording inspection findings: element condition, defects

▸ photographic records, for example of the property in general and of specific defects.

▸ measurement of defects, e.g. levels of damp, width of cracks, distance of deflection or movement.

The survey report

The survey report should be typed and we will look at the skills required for the writing of reports in section C3 of this unit. The report should be clear and concise. Photographs can be used to illustrate any defects.

The general description of the property should contain the following items: the location and address of the building as well as its type, age, current use and current services.

The details of the condition may cover all the different aspects of the building, including its structural elements, walls, doors, windows, and materials used, services and fixtures and fittings.

It must also report on any defects found against each item, perhaps on a room by room basis, along with any recommendations for repairs of the identified defects.

C3 Undertaking measured surveys

A measured survey fully records the dimensions of a structure or building so it can be replicated in a drawing and used for further design proposals. It is used to produce the as-built existing drawings as part of the planning process. Its use within structural engineering is also essential where accurate spans need to be known for floors and walls.

Measured survey requirements

Pre-survey protocol

The same survey protocol is followed as described in section C2 of this unit. The only other consideration may be access to grounds to measure the whole area and footprint of the building. Health and safety should again be observed when undertaking this survey. No area should be entered if it is a risk to health or a serious or imminent hazard that will affect the health of the surveyor.

Equipment

The main surveyor's equipment required for a measured survey is:

▸ tape measure

▸ **electronic digital measure**

▸ pencil, clipboard and paper (waterproof versions are available)

▸ digital camera

▸ any existing plans.

> **Key term**
>
> **Electronic digital measure** – a laser measure that bounces and returns a light wave to record the distance between two surfaces.

 **PAUSE POINT**   Would you use a tape measure or an electronic digital measure for your survey?

> **Hint**   What is likely to cause the least disturbance to the client?
>
> **Extend**   How do you know the digital measure is accurate?

Electronic digital measures can be useful – locate an instrument suitable for all types of measured surveys and explain why you have chosen it.

Property inspection requirements

A sketch must be made of the ground and first-floor plans and of each elevation in turn so the dimensions can be taken and recorded on each sketch. Sketches need to be in proportion to the size of the rooms surveyed. Dimensions recorded should be clear and easily read by a third party so there is no confusion.

Where survey dimensions cannot be taken then brick courses can be counted to establish lengths and heights of elevations. A full set of photographs should also be taken for reference in case questions arise during the drawing production stage.

Survey drawings

Standard scaled plans are produced at a scale of 1:100 for elevations or 1:50 for plans. Modern design drawing production is achieved using CAD software programs. These can be drawn at a scale of 1:1 and can then be printed off at any scale to suit the desired application.

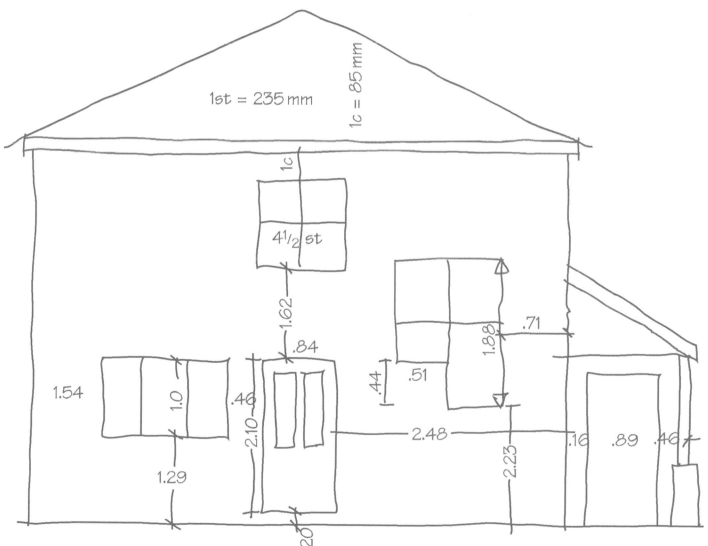

▶ **Figure 10.9:** A freehand sketch for a measured survey, showing the dimensions recorded

Link

See Unit 2 for more about scale plans and elevations.

Elevations may be produced from standard surveying methods and 2D scale drawings (see Unit 6) or, if the budget allows, a 3D scanner.

C4 Skills, knowledge and behaviours

Personal skills

The building survey must demonstrate a variety of personal and behavioural skills in order to conduct business with a range of different clients, for example:

- **Professionalism** – you are recognised by the RICS badge as a surveying professional. This means you are held to a high standard of ethics and behaviour when dealing with all types of stakeholders, for example dressing appropriately and using the correct language and terms of reference for the person you are speaking to.
- **Etiquette** – treat everyone with courtesy and respect, for example by not smoking, leaving litter or using mobile phones for personal calls, and by obeying home owners' house rules such as removing your shoes.
- **Accountability** – taking individual responsibility for your actions and decisions.
- **Evaluating outcomes** to help inform high-quality, justified recommendations and decisions. You will need to show in your report why you are making the recommendations you have made and demonstrate your reasons.
- **Accuracy in measuring** – so that surveys produce detailed drawings for secondary purposes.

Media and communication skills

Often you need to explain quite technical aspects of construction to a client who does not have the knowledge and expertise that you possess. This should be done in a way that the client will understand. Using photographs and diagrams and discussion round these is a useful tool.

Reports need to be understood by clients who are uninformed about construction technology. This means you need to develop the ability to convey your intended meaning unambiguously in different formats, including in written form (such as design documentation, recording documentation, reports, visual aids for presentation use) and verbally (for example one-to-one and group, informal and formal situations).

Remember that your choice of tone and language influence whether you make a positive and constructive impact on your audience. This may include using a positive and engaging tone and using the technical language suitable for your intended audience, whatever their level of knowledge.

Further reading and resources

Designing Buildings Wiki **www.designingbuildings.co.uk/wiki/Home.**

RICS (2015) *Homes Through the Decades: The Making of Modern Housing*, Milton Keynes: NHBC Foundation. Downloadable from **www.nhbc.co.uk**

Royal Institution of Chartered Surveyors **www.rics.org**

Assessment practice 10.3   C.P5 C.P6 C.M3 BC.D3

You have been instructed by a client to undertake a building survey on a property they are purchasing. Produce the following:

- a building survey dilapidation schedule
- a record of this survey in a report format
- a building survey for a domestic property (such as your own house) detailing the condition, defects and remedial works required, with plans and elevations
- a self-assessment explaining how you demonstrated individual responsibility, creativity and self-management when preparing for and undertaking the building survey and producing the survey report and drawings.

Plan
- Check the access to a suitable property.
- Organise a risk assessment prior to the survey.

Do
- Use photographs to enhance evidence.
- Use a RICS survey template.
- Make valid remedy recommendations.

Review
- Check the layout and content of the completed report.
- Is language and format suitable for a client?

THINK ▶FUTURE

Mohammed

Mohammed is a graduate building surveyor for a city firm of surveyors. He has been working for them for a year and is starting to develop his surveying practical skills. The company undertakes a large amount of work for an estate agency and a range of mortgage lenders. Mohammed has to take the initial phone calls from clients and confirm their instructions before ordering the surveys. Part of his role is in customer services to ensure that a client obtains the right survey to meet their needs.

Mohammed is moving onto the survey team when the new graduate starts and will then gain experience in the surveying of a range of properties and preparing reports for the different surveys. The surveys that he will undertake will be for tenants and landlords so agreements can be made for rental contracts. He will be working on mortgage valuations for lenders to ensure that a property is valued correctly with the current market value.

Mohammed took an accredited degree at university and can therefore apply to join RICS (the Royal Institution of Chartered Surveyors). Mohammed has to record two years' experience in a formal diary and undertake a professional interview in order to gain full membership of RICS.

The building surveyor often has to make recommendations regarding any defect discovered as part of a survey. These recommendations will include the methods that can be employed to rectify a defect along with the associated costs, so a client can budget for these costs. Recommendations should contain the latest technology so the repairs are futureproofed and do not deteriorate over time.

Focusing your skills

There is a wide range of skills that a building surveyor has to use when undertaking building surveys for a client:

- Communication skills – to be able to inform clients and discuss what type of survey they require, providing impartial advice and guidance through to successful completion of a purchase.

- Attention to detail – to be able to spot potential building defects, know their causes and remedies and co-ordinate repairs.

- Smartness – be appropriately dressed to provide a good first impression to clients.

- Accuracy – to produce detailed and accurate reports for clients so they have an overview of the property, its current state and recommendations.

- Co-ordination – to be able to co-ordinate a set of contractors involved in a client refurbishment, upgrade or extension.

Getting ready for assessment

Jasmine is in the second year of her Extended Diploma in Construction and the Built Environment. Her work on this unit includes a practical-based assessment in undertaking surveys on houses and properties. For her assignment, Jasmine needed to complete a survey of a low-rise residential property. She then had to complete a report detailing the condition, defects, any remedial works, plans and elevations to accompany the survey.

How I got started

We got ready for this assessment by looking through some building surveys undertaken by a local surveyor who came into lectures as a guest speaker as part of the employer engagement course element. This was very useful in providing me with guidance on how to start the surveys.

After the guest surveyor visit, I looked at a few websites for building survey examples and found a few that showed me what titles to use within each survey. The RICS website also publishes samples of the different types of surveys that are undertaken.

I wanted to look at as many examples of different surveys as possible, as I thought this would help to understand what I needed to be looking for when I did my own surveys and make sure that I was reporting on everything that would normally be covered by a survey.

How I brought it all together

I took lots of photographs when we did the visits to the properties to undertake each survey. These proved valuable as I could then look back at each aspect of the building that we had surveyed and reflect on what needed to be included within each survey heading.

I wanted to have as much evidence as possible so that I could show why I had made the decisions I made in my report and to show that I had been as thorough as possible on my survey.

I then put my report together. I decided to break it up into clear sections and to try and cover different parts of my survey in different sections. I thought this would make the report easier to read and also make it easier to see which parts of the survey related to which parts of the report.

What I have learned from the experience

This assessment has taught me about the roles and responsibilities of the building surveyor and the different types of work that they perform.

The most important thing I have learned is that attention to detail is really important in surveys. Having a thorough checklist that breaks down every single task and check that you need to carry out is really important, and is definitely something I would spend more time on if I had to do this again.

You need to make sure that in the report you have covered every aspect for a client. You do not want to miss any building defect or recommendation that may cost a client financially in the future if they went on to purchase a property. Checking and rechecking the report to make sure I had covered everything was really important.

Think about it

Professional building surveyors are highly sought after employees as they can undertake a range of different roles and responsibilities across a number of different surveys.

▸ How would you find out about the different surveys?

▸ Would it be worth asking a surveyor at a local estate agent?

▸ How would you find out about the different building defects?

Measurement
Techniques
in Construction

13

Getting to know your unit

Assessment
You will be assessed by a series of internally assessed tasks set by your tutor.

Measurement is a particular skill for a quantity surveyor. Quantity surveyors are responsible for the financial management of a contract to ensure that budgets are maintained, waste is kept to a minimum and resources are sourced, ordered and paid for within the financial constraints of the estimate.

Measurement is primarily used for the compilation of a bill of quantities – the main item used in a tender, which forms the contract sum for the project. Accuracy is therefore important, along with following any rule of measurement.

Within this unit, you will examine the rules of measurement for construction and civil engineering projects, how these differ and how they are applied. You will practise taking off quantities for a range of substructure and superstructure elements. You will also learn how to produce a bill of quantities.

This unit can help you progress into a career as a quantity surveyor, estimator or buyer in the construction sector or to Higher Nationals in Construction and degrees in construction cost control or financial management.

How you will be assessed

Throughout this unit, you will find assessment practice activity that will help you work towards your assessment. Completing these activities will not mean that you have achieved a particular grade, but you will have carried out useful research or preparation that will be relevant when it comes to your final assignment.

To achieve the tasks in your assignment, you should check that you have met all of the Pass grading criteria as you work your way through the assignment.

If you are hoping to gain a Merit or Distinction, you should also make sure that you present the information in your assignment in the style that is required by the relevant assessment criterion. For example, Merit criteria require you to analyse and discuss, and Distinction criteria require you to assess and evaluate.

The assignment set by your tutor will consist of several tasks designed to meet the criteria in the table. This is likely to consist of a written assignment but may also include activities such as the following:

▶ outlining a rule of measurement

▶ taking off quantities using drawn dimensions

▶ producing a bill of quantities for a work element.

The following table shows what you must do to achieve a Pass, Merit or Distinction grade.

Assessment criteria

This table shows what you must do to achieve a **Pass**, **Merit** or **Distinction** grade.

Pass	Merit	Distinction
Learning aim **A** Examine the measurement rules for building and civil engineering		
A.P1 Explain how approximate and accurate quantities are used for different applications by quantity surveyors.	**A.M1** Discuss the benefits of using recognised standard methods of measurement for buildings and civil engineering projects.	**A.D1** Evaluate the use of recognised standard methods of measurement to ensure consistency when tendering and estimating for buildings and civil engineering projects.
A.P2 Explain the reasons for the use of a recognised standard method of measurement.		
Learning aim **B** Undertake the production of quantities for substructure and superstructure elements		
B.P3 Perform a take off of quantities for a project using a recognised standard method of measurement and an appropriate layout of dimensions.	**B.M2** Perform an accurate take off of quantities for a project using a recognised standard method of measurement and a vocationally correct layout of dimensions.	**B.D2** Perform an accurate and comprehensive take off of quantities for a project using a recognised standard method of measurement and a vocationally correct layout of dimensions and methodology.
B.P4 Explain the difference between the production of quantities for building projects and the production of quantities for civil engineering projects.		
Learning aim **C** Undertake the production of bills of quantities		
C.P5 Produce bills of quantities for a construction project.	**C.M3** Produce accurate bills of quantities for a construction project.	**C.D3** Produce comprehensive bills of quantities for a construction project.
C.P6 Explain the different methods used to convert the take off into bills of quantities.		

Getting started

You need to understand the rules to follow to take off quantities. Examine the New Rules of Measurement, which are available from the Royal Institution of Chartered Surveyors (RICS) website, to find out what these look like, what they contain and how they are applied.

Link

Revise your knowledge and understanding of Unit 4: Construction Technology, as you will need to know how some elements are constructed in order to take off the quantities correctly.

A Examine the measurement rules for building and civil engineering

Key terms

Taking off quantities – a detailed analysis of the aspects of a construction project that can be measured and priced (such as labour and materials) so that a bill of quantities can be produced for a tender.

Bill of quantities – a document compiled by a cost consultant or quantity surveyor containing a detailed statement of the prices, units, dimensions and other details required to complete a construction project. It becomes part of the contract.

NRM – New Rules of Measurement, a standard set of measurement rules and essential guidance produced by RICS for the cost management of construction projects and maintenance works.

Estimating – the process of compiling the price for a tender or quotation.

A1 Introduction to taking off quantities

There are a number of different reasons why we **take off quantities**, relating to the life cycle of a project from its initial design through to its completion and handover to a client. In order to take off quantities you need to be able to read a drawing so that you can use the dimensions needed to calculate the quantities.

Here are some reasons why we take off quantities.

The production of bills of quantities

To produce a **bill of quantities**, detailed measurements are taken from the contract drawings and specification and, using dimension paper or a software program, the required quantities are taken off and calculated for each item. The descriptions of items in the bill of quantities should follow the rules set out within the **NRM** rule book, which covers different aspects of the construction of a project, and the measurement rules that apply to each item. The detailed measurements should contain all the necessary information for the estimator to know what they are pricing. For more about NRM, see section A2.

Tendering and estimating

During the tendering process, if a project does not have a bill of quantities, and a specification and drawing have been issued for **estimating** purposes, the contractor's estimator has to take off the quantities from the drawings in order to estimate a final price for the project. These quantities are used to obtain quotations for materials, for estimating labour costs and to judge time constraints for preliminary items.

Budgets for feasibility studies during design stages

Feasibility stages of the design are a major turning point for a project's client. This stage dictates if the project will go ahead and whether a client has the financial resources to undertake the project against the revenue it would bring. Approximate quantities are produced from the initial sketch designs against rates obtained from comparable projects. This may be as basic as measuring the floor area or a building's volume and using a comparable rate to produce an estimate of the budget.

Cost comparison of different designs

During the design stages of a project, the initial designs need to be priced so comparisons can be made between different schemes. This involves taking off approximate quantities so rates can be applied and a budget formulated so schemes can be compared. This process allows a client to make some informed decisions as to what aspects of schemes could be accommodated within their budget or what aspects would need to be changed.

DWG No: AT 01	
SCALE: 1:1000	
DWG No: AT 01	
PROPOSED ELEVATIONS and FLOOR PLANS	
Mr R Burton 14 Kings Road Notting Hill London W11 5EY	

GROUND FLOOR PLAN

FIRST FLOOR PLAN

WEST ELEVATION

SOUTH ELEVATION

NORTH ELEVATION

▶ **Figure 13.1:** A sketch does not contain any detail but basic quantities can be obtained from it

Preparation of estimates

If a client has issued the **tender documents** for a project as a specification and drawing package then the main contractor must produce the quantities themselves. These are priced and totalled to produce the net estimate for the project. Taking off quantities allows the contractor to price the materials and any plant required for their installation and to calculate labour costs using outputs per unit of materials.

Estimation of a project's value

This is the primary reason for taking off quantities where there are no bills of quantities. A project's value can be estimated using the floor area method, volume, approximate quantities or by building up elements.

Final account measurements and variations

A final account is the summation of all the variations that have occurred on a typical contract and is the final total that the client has to pay the contractor, with any previous payments they have received deducted. The final account is adjusted against the original contract sum that was agreed at the start of the project.

In compiling the final account, an architect's instruction may require carrying out measurements on site and then valuing them against the bills of quantities rates that the contractor entered within their tender. All the contract variations are worked through and the final account is prepared for agreement by the contractor and the client.

Materials ordering

Accuracy is required for this type of taking off. Estimating quantities of materials needs to be accurate to reduce wastage from:

> **Key term**
>
> **Tender documents** – the drawings, specifications, bills of quantities, letters and forms that are part of a tender.

- over-ordering
- ordering the wrong length or size
- excessive cutting.

Often orders are placed in bulk to save costs then deliveries are called off when required.

Producing a quotation for a work element

Variations often occur on projects for a variety of reasons, including:
- incorrectly drawn information
- changes by the client
- unforeseen circumstances
- incorrectly specified materials.

A quotation may therefore be requested by the client or their representative such as the quantity surveyor. This quotation may require that quantities are taken off and a price compiled using rates that have been agreed.

PAUSE POINT How would you explain the purpose of taking off quantities to a new colleague?

Hint Consider the different reasons why you might do it.

Extend What are the differences between each one?

A2 Standard methods of measurement

Measurement rules

The need for measurement rules

Standard methods of measurement rules ensure that all stakeholders take a consistent approach to the tendering and estimating process. If a client requests a tender based on an agreed set of quantities then each contractor who prices the work will follow the same quantities, so that each tender can be compared fairly. For example, the New Rules of Measurement (NRM), published by the Royal Institution of Chartered Surveyors (RICS), cover all aspects from feasibility measurement right through to the final account quantities.

Origins of measurement rules

Two sets of common rules have been adopted within the UK.

1   **The New Rules of Measurement (NRM) parts 1 to 3 published by RICS:** the NRM evolved from the Standard Method of Measurement (SMM7), which was first written in 1922 and ran until the seventh edition. Some clients still prefer it, and its contents remain valid. NRM1 was first published in 2009, NRM2 in 2012 and NRM3 in 2014.

2   **The Civil Engineering Standard Method of Measurement (CESMM):** the CESMM was first published in 1976 and is now in its fourth edition. It covers civil engineering measurement of quantities.

Measurement Initiative Steering Group

RICS originally set up the Measurement Initiative Steering Group to review the use of SMM7. It concluded that the edition required updating to reflect modern technological advances. The NRM suite was therefore proposed by the group.

Status of the RICS New Rules of Measurement (NRM)

- **NRM1 (second edition April 2012)** provides guidance on the quantification of building works to help prepare cost estimates and cost plans. It aims to support

> **Research**
>
> Find and download a copy of the NRM and note down the main points that it contains.

good cost management of construction projects by enabling more effective and accurate cost advice to be given to clients and other project team members, as well as facilitating better cost control.

▶ **NRM2 (first edition April 2012)** is written mainly to support the preparation of bills of quantities and quantified schedules of works, although it is also useful for designing and developing schedules of rates.

▶ **NRM3 (first edition March 2014)** gives guidance on the quantification and description of maintenance works to support the preparation of initial order of cost estimates. The rules also aid the procurement and cost control of maintenance works.

Status of the ICE Civil Engineering Standard Method of Measurement (CESMM)

This standard of measurement produced by the Institution of Civil Engineers (ICE) covers heavy construction and includes water engineering projects, harbours, railways and road infrastructure. CESMM4 is the latest version, published in 2012. It:

▶ covers the newest technologies so that it is completely up to date with current practices

▶ is contract neutral so can be used across a variety of contract suites

▶ retains the established structure of CESMM

▶ includes a completely updated railway work section.

(II) PAUSE POINT Can you explain why a standard method of measurement is needed?

Hint Look at the NRM on the RICS website.

Extend What other standards can be applied to measurement?

Typical considerations

A standard method of measurement contains the set of rules within a tabulated format. Table 13.1 shows the type of rules you will find in NRM2. We will examine each of the table headings in detail.

▶ **Table 13.1:** Layout of tables in the NRM

Column 1 is the title column for drawings that should accompany this type of measurement and may give minimum information that should appear on the drawings	This column details the type of drawings that might be needed, for example: • site plans • ground levels • location of boreholes or trial pits.	**Title column for any mandatory information**	This column details the type of information that might be needed, for example: • ground water levels • site contamination • date of current site survey • rock levels.	This column is for comments or a glossary, for example: Water is tidal so, high and low levels must be indicated.

Each section of measurement – for example Excavation and filling, or Ground remediation and soil stabilisation – in the NRM is accompanied by a table set out like Table 13.1.

The pages following detail the rules that must be followed for an item being measured. These are some examples:

▶ Units of measurement – in column two the units of measurement are often specified, for example nr and m^3.

▶ Deduction of voids – this is used to specify what type of void would be ignored or measured as a deduction, for example a hole in a plastered wall.

- Deemed to be included – this specifies items within the notes column that are deemed to be included, for example the screws used to fix plasterboard.
- Item description – the first column provides the main item description followed by levels 1, 2 and 3. These levels build up the item description, providing more detailed information for an estimator to price.
- Hierarchy of description – the levels provide the hierarchy which you work through in order to provide the full item description.
- Preliminaries and measured work – the front sections of a bill of quantities detail the temporary time-related charges that a main contractor will price against, such as supervision, transport, skips, tower cranes, services and scaffolding. The 'NRM' section details how these are to be specified within a bill of quantities.
- Guidance on the preparation of bills of quantities – Part 2 of NRM2 provides some guidance on the preparation of a bill of quantities, its purpose, benefits, types, preparation and composition. A bill of quantities is laid out to match the sequence of the NRM so that a quantity surveyor or estimator can easily access the sections they need.

Theory into practice

To assemble an item description for a bill of quantities, follow the starting point in the first column of the NRM then use the levels to build up the full description.

Provide an item description for the following activity on site: a trench that is 1 m deep and 0.6 m wide that will be filled with concrete to form a foundation.

Research

NRM2 contains a detailed section about preliminaries (2.7). Use this section to draw up a checklist of what would need to be considered at this stage.

The New Rules of Measurement (NRM)

Three documents comprise the New Rules of Measurement.

NRM1 – Order of cost estimating and cost planning for capital building works

This NRM is for applying a set of measurement rules to the assembly of a budget or cost estimate at the feasibility stages of a project. RICS states that the NRM1 is:

'the "cornerstone" of good cost management of construction projects – enabling more effective and accurate cost advice to be given to clients and other project team members, as well as facilitating better cost control.' (Source: *NRM 1 RICS New Rules*

of Measurement: Order of Cost Estimating and Cost Planning for Capital Building Works)

Clients do not like to have a budget formulated in the early stages of design then receive tenders which do not reflect this budget. NRM1 tries to standardise this process to enable a more accurate calculation of the cost feasibility of a project.

NRM1 covers:
- Part 1: General – purpose, use, structure, symbols and definitions.
- Part 2: Measurement rules for order of cost estimating that include:
  - the information requirements for cost estimating
  - the order of the cost estimate and the constituents
  - measurement rules for any enabling works that need to be accomplished before the main project can begin
  - measurement rules for building works
  - the elemental method and the rules for applying it
  - unit and element rates
  - measurement rules for main contractors' preliminaries
  - measurement rules for risk.
- Part 3: Measurement rules for cost planning, including:
  - constituents of the cost plan
  - stages of the cost plan
  - all the stages needed for cost estimating.

NRM2 – Detailed Measurement for Building Works

This part details the rules for taking off the dimensions of specific elements. It covers 41 work sections, covering nearly every aspect of a construction project, ranging from demolitions, alterations and repairs to masonry, carpentry and drainage.

Research

Search online for NRM2 and locate the 41 work sections covered by NRM2.

NRM3 – Order of cost estimating and cost planning for building maintenance works

These rules examine the post-completion stages of a project, which cover maintenance to keep the building's standard acceptable for its owners and users. Maintenance contracts are tendered for by infrastructure and services organisations, which manage a large portfolio of the built environment.

NRM3 contains the following parts:
- Part 1: A general section outlining the use, structure and other information
- Part 2: The rules of measurement for building maintenance works
- Part 3: Measurement rules for order of cost estimating (renewal and maintain)
- Part 4: Measurement rules for cost planning of renewal (R) and maintain (M) works

▶ Part 5: Calculation of annual costs for renewal (R) and maintain (M) works
▶ Part 6: Tabulated rules of measurement for elemental cost planning.

NRM3 has a specific application to maintenance works. It can be used for compiling feasibility cost information for maintenance budgets, annual maintenance budgets and for maintenance cost planning. Maintenance contracts engage the services of a contractor to maintain the client's portfolio of buildings so that the client does not have to employ and supervise direct labour themselves.

⏸ **PAUSE POINT**   How are concrete works measured?

     Hint   Download the NRM suite and find the concrete section of NRM1.

     Extend   How is precast concrete measured?

Civil Engineering Standard Method of Measurement (CESMM)

This standard method of measurement is published by the Institution of Civil Engineers (ICE). It covers all the aspects of heavy civil engineering, such as road construction, rail tracks, tunnelling, large concrete works, drainage, earthworks and piling. The latest edition is CESMM4.

The reason why civil engineering has its own set of standards is because a lot of the content of NRM is not applicable to civil engineering. For example, plaster finishes, woodwork and roofing are not civil engineering considerations.

Research

Investigate CESMM4 online. What does it cover?

Investigate the Institution of Civil Engineers (ICE) and identify the top five benefits it gives its members. How does your list compare with the rest of your group?

Assessment practice 13.1   `A.P1` `A.P2` `A.M1` `A.D1`

The use of standard methods of measurement is essential in obtaining an accurate competitive cost for a building project or a civil engineering project in terms of feasibility and a final cost.

In the format of a report, explain to a client the following aspects of standard methods of measurement:

- How approximate and accurate quantities are used for different stages of a project.
- Why a recognised standard method of measurement should be used.
- The benefits of using such a standard for building or civil engineering projects.

Conclude your report with an evaluation of how using recognised standards of measurement ensure consistency when tendering and estimating.

Plan
- Do I know enough about standard methods of measurement to write the report?
- If not, how do I carry out detailed research on standard methods and why they are used?

Do
- Have I referred to a standard method, using examples, and explained why its use is common for certain types of work?

Review
- Have I included a range of benefits based upon my research?
- Does the evaluation cover advantages and disadvantages in sufficient detail?
- Is there anything I would do differently next time?

B Undertake the production of quantities for substructure and superstructure elements

B1 Processes in the production of quantities

Preparation and planning, including take off lists

In section A1, we discussed establishing take off quantities. An essential part of planning the production of quantities is the take off list, a list of all the elements associated with the supplied drawings and the items that have to be measured in accordance with the method of measurement.

Take off lists are written in the correct sequence then their dimensions (or measurements) are undertaken and recorded. As each one is recorded it is good practice to strike it from the take off list.

You should plan to start and finish a section in one go, as a distraction may mean that you lose your place on the drawing and have to start again. It is good practice to mark up quantities, items and areas on drawings using a highlighter once they have been taken off. This avoids duplication of quantities that could lead to errors caused by over-measuring.

The correct format and layout of information in a dimension or direct billing paper

Dimension paper has traditionally been used for the manual taking off and recording of dimensions. Figure 13.2 is part of a full sheet to illustrate what it looks like. The columns all have functional uses:

▶ Column 1 is used for the binding where sheets are held together.
▶ Column 2 is the multiplication column where a times value can be applied to dimensions.
▶ Column 3 is where the main dimensions are written.
▶ Column 5 is the waste column where an item description is placed and any notes or rough calculations.
▶ Column 4 is the sum column for adding up quantities.
▶ Columns 6, 7, 8 and 9 repeat columns 2, 3, 4 and 5, effectively meaning that each dimension page has two pages.

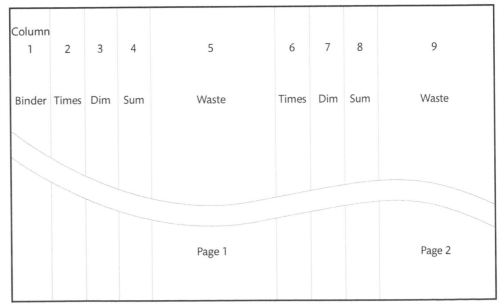

▶ **Figure 13.2:** The dimension paper layout with columns indicated

Several methods are used to take off dimensions under the rules for dimensions. Figure 13.2 demonstrates how each method is applied on the dimension paper and in what sequence the dimensions are written. Generally, dimensions are written as length (L) then width (W) followed by depth (D). This provides the data for linear measurements, areas and volumes to be dimensioned.

There is no set or recognised standard for setting out dimensions on paper. Many quantity surveying practices use a method that they have devised themselves. However, Table 13.2 shows examples of common practice in terms of how dimensions are written onto the dimension paper.

▸ **Table 13.2:** Typical types of dimensions and standard elements found in a dimension paper

Dimension	Guidance
Enumerated	A number or count of an item that has no area, linear or volume dimension, for example, a light fitting. The single item is placed in the dimension column and underlined. The number of items is then noted in the multiplication column.
Linear	Dimensions connected with items from the standard method of measurement, for example, guttering, down pipes, skirting boards, architraves and door frames.
Area	An area requires two dimensions: length and width. The measurements are placed in the dimension column then underlined to indicate an area. If there is more than one area with the same dimensions then the number of these identical areas is placed in the times column.
Volumes	A volume requires three dimensions: length, width and depth (or height). These are placed in the dimension column and multiplied in a similar way to areas if there is more than one volume with the same dimensions.
Itemised	A work element that does not have a measured value within the standard method of measurement used, for example the demolition of a building. This contains a detailed description and is listed as an item in the dimension column.
Multiplying	This is done in the times column. A number indicates how many times the dimensions should be multiplied by. Dotting on is a method of adding one or more to the multiply number.
Totalling dimensions	When all the dimensions have been written into the dimension column and are finished they are totalled and then summed. A double line underneath the final indicates the sum total.
Deductions and omission quantities	These remove elements from dimensions. For example, a window or a door in an external wall is measured then deducted from the total area.
Page numbering	Pages should be numbered to avoid confusion. There are two pages on each dimension sheet, and the page numbers are indicated on the bottom centre of the page.
Carried forward and brought forward dimension totals	This is used where a full dimension cannot be accommodated in the space left in the dimension column. The page is carried forward and then brought forward onto the next page.
Use of standard quantity surveyors abbreviations	Common abbreviations that are used by quantity surveyors include: • abd – as before described • caos – cart away off site • exc – excavation • ddt – deduct • rc – reinforced concrete • & – and • ne – not exceeding • cl – centre line.
Marking the extent of a calculation	This is done to collect together all the same calculations for a measured item. A long 'S' type bracket is used on the line of the total and waste columns.
Waste calculations	The waste column is used for: • any sketch drawings • miscellaneous calculations • centre line calculations • notes • drawing references • annotations • references to a product • location references

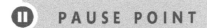

PAUSE POINT Make a sheet of dimension paper and try out some basic quantities for constructing a garden shed.

Hint Research the different ways of expressing quantities and typical dimensions and elements of a shed.

Extend How would you dimension a circle?

B2 Production of substructure quantities for a building

Elements for substructures are below ground level and include the ground floor slab. We shall now examine how each of the following elements is measured against the New Rules of Measurement 2 (NRM2).

Substructure elements

Excavations for foundations

NRM2 (page 134) defines what types of foundation the excavation covers and states that:

▶ the item is Excavation
▶ it is measured in m³
▶ it should be termed foundation excavation
▶ depths are recorded in 2 m stages
▶ details of obstructions in the ground should be recorded.

It is often necessary to calculate the centre line of a building. This is done by adding the perimeter dimension and deducting 4 x the wall thickness. In Figure 13.3 this is 288 mm or 0.288 m.

	28.80		Excavation, foundation excavation,				
	1.00		depth not exceeding 2 metres				
		28.80					
					2	10.00	
					2	5.00	
						30.00	
			&		ddt	1.20	
					cl	28.80	
			Disposal, excavated material off site,				
			to tips 5 km away				

▶ **Figure 13.3:** A typical item take off for foundation excavations

Short-bored piles

These are a method of obtaining greater load-bearing ground capacity by drilling and inserting short vertical concrete piles under a ground beam.

NRM2 (page 143) states that:

▶ the item is Piling – bored piles
▶ it is measured in metres (m) from the commencing level to the bottom w
▶ size or diameter is stated.

This would provide an adequate pile for a domestic application. A typical item take off would look like Figure 13.4.

			Piling					
20	1.50		Bored piles, in-situ concrete, 250 mm diameter					
		30.00						

▶ **Figure 13.4:** A typical item take off for short-bored piles. Note that they are measured in metres; this example shows 20 piles, each 1.5 metres long

Mass concrete foundation works

A mass concrete foundation is known as trench fill. The whole of the excavated trench is filled with concrete to form one solid foundation. Section 11 of NRM2 starting on page 151 details the rules of measurement for such a foundation. A typical take off example might look like Figure 13.5.

			In-situ concrete works					
	28.80		Mass concrete, in trench filling					
	1.00		poured against earth					
	0.60							
		17.28						

▶ **Figure 13.5:** A typical item take off for mass concrete foundation works. Note that the estimator is told that the concrete is poured against the faces of the trench and has to allow for any wastage when the quantity is priced

Formwork

This is required when detailed in-situ concrete items have to be cast in place. Formwork is the sheet materials that the concrete is poured against. Falsework is the supporting structure that holds the formwork in place. Section 11 of NRM2 on page 154 (Formwork) details the rules of measurement for in-situ concrete. A typical take off for formwork to the sides of a raft foundation is shown in Figure 13.6.

			In-situ concrete works: Framework					
	20.00		side of foundations and bases, plain					
	0.45		less than 500 mm high, 450 mm wide					
		19.007.28						

▶ **Figure 13.6:** A typical item take off for formwork. In this case, the 20 m would be the perimeter of the formwork against the face of the raft and is measured in square metres (m²)

Earthwork support

When any earthworks are excavated, a contractor has to accept the risk of the support to the sides of the excavation collapsing. There is no specific measurement for this except on page 136 of NRM2, where the risk is accepted if it is not at the discretion of the contractor.

Substructure external and internal walls to DPC level

External walls usually consist of a cavity construction. Trench blocks are often used to raise external walls off strip foundations to 150 mm below ground level. From here, engineering brickwork is used up to the damp-proof course (DPC) level. A typical take off may look like Figure 13.7.

			Masonry				
	28.80		Walls; 100 mm thick, engineering				
	0.30		brickwork				
		8.64	&				
			forming cavity				
	28.80		walls; blockwork, trench block laid				
	0.85		in cm			1.00	
		24.48			ddt	0.15	
						0.85	

▶ **Figure 13.7:** A typical item take off for walls to DPC level

Ground floor construction

When you have analysed the drawings to see how the ground floor is constructed, you produce a take off list for a ground floor that might look like the following. Notice that the sequence follows the order of floor construction, with excavation first and floor finishes last.

1 Excavation
2 Hardcore filling
3 Sand blinding
4 Damp-proof membrane
5 Insulation
6 Concrete beds
7 Finishes to concrete
8 Floor finishes

A typical dimension page for the ground floor construction is shown in Figure 13.8 for a floor with an area of 6 metres by 4.5 metres.

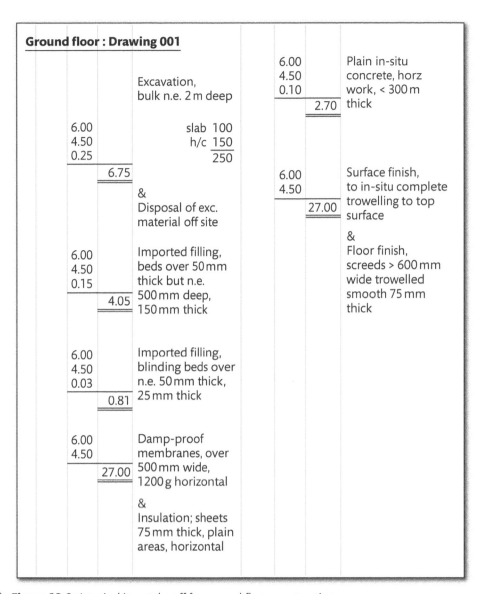

Ground floor : Drawing 001

	Excavation, bulk n.e. 2 m deep
6.00 4.50 0.25	
6.75	slab 100 h/c 150 ⎯⎯⎯ 250
	& Disposal of exc. material off site
6.00 4.50 0.15	Imported filling, beds over 50 mm thick but n.e.
4.05	500 mm deep, 150 mm thick
6.00 4.50 0.03	Imported filling, blinding beds over n.e. 50 mm thick,
0.81	25 mm thick
6.00 4.50	Damp-proof membranes, over 500 mm wide,
27.00	1200 g horizontal
	& Insulation; sheets 75 mm thick, plain areas, horizontal

6.00 4.50 0.10	Plain in-situ concrete, horz work, < 300 m
2.70	thick
6.00 4.50	Surface finish, to in-situ complete trowelling to top
27.00	surface
	& Floor finish, screeds > 600 mm wide trowelled smooth 75 mm thick

▶ **Figure 13.8:** A typical item take off for ground floor construction

Ⅱ   PAUSE POINT   Produce a take off list for an external cavity wall with openings within it for doors and windows.

Hint   You need to understand how the wall is constructed and how openings are supported.

Extend   Produce a take off list for a timber framed external wall.

B3 Production of superstructure quantities for a building

The superstructure is anything above DPC that forms the main part of a building. The following two take offs demonstrate modern timber framed construction and a traditional cavity wall construction.

External wall construction

Timber framed

The take off uses standard manufactured panels. These have been given a reference number, which can be related to the timber frame manufacturer's design drawings. A **sole plate** and **head binder** secure the panels together when the timber frame is

Key terms

Sole plate – the timber plate that the timber frame building rests upon when the first stage is installed.

Head binder – the top frame which may consist of an additional batten that is fixed across the timber frames for the floor or wall plate.

fitted on site. The damp-proof membrane covers the insulation inserted between the studs on the panels. The delivered panels are likely to have softwood stud frames and plywood coverings so these are not measured separately.

			Carpentry	Timber framing	
40/	1		Engineering or prefabrication members,		
		40	2.5 m x 2.4 m wall panels, manufacturers		
			reference A2500x		
	60.00		primary sole plate, 150 x 65 wall member		
		60.00			
			&		
			primary head binder 150 x 65 wall member		
40/	2.50		insulation, quilt, 150 mm thick, laid		
	2.40		between stud frames at 600 mm centres		
		240.00			
			&		
			damp proof membrane, 500 g, over		
			500 mm wide, vertical		

▶ **Figure 13.9:** A typical item take off for a timber frame wall. This refers to manufacturers' standard prefabricated details

Masonry

The take off for a masonry wall is as follows:

▶ damp-proof course
▶ internal skin of cavity wall in blockwork
▶ facing brickwork
▶ form cavity using stainless steel wall ties
▶ insulation
▶ lintels
▶ cavity closers.

Figure 13.10 shows a typical example.

	Facing brickwork 215 x 102.5 x 65 in cm (1.4) stretcher bond skins of walls		DDT				Form cavities 100 mm wide using stainless steel wall ties 4/m²						Blockwork 100 x 215 x 440 in cm (1.4) stretcher bond s.o.h. of wall	
1	70.15	6	6.80			3	68.75	6	6.80			5	68.75	
2	35.20	7	4.20			4	34.10	7	DDT 4.20			9	34.10	
	105.35		11.00				102.85		11.00				79.30	
DDT	11.00					DDT	11.00							
	94.35						91.85							
	94 m²						92 m²							

▶ **Figure 13.10:** A typical item take off for a masonry wall

Treatment of openings

Table 13.3 describes how different types of superstructure openings and their components are measured for take offs.

▶ **Table 13.3:** Treatment of openings for take offs

Feature	Treatment
Windows and doors	Any opening is treated as a deduction from the measured take off. You always measure straight through all openings and then add a deduction (ddt) into the dimension paper for the windows and doors. There is no take off list for these items as they are included within the brickwork, blockwork and plaster finishes.
Lintels	Measured by the metre. The type, size and description from the manufacturers is generally specified. The width of an opening plus a minimum 150 mm end bearing each side is added to obtain the dimension used on the take off.
Reveals	Normally less than 300 mm wide. These are generally measured using linear metres.
Cills and window boards	Measured in metres with any oversailing included from the drawings, type and description along with dimensioned sizes.
Thresholds	This is a linear dimension and is normally an aluminium profile which is screwed and fixed to the floor.

❚❚ PAUSE POINT   How do you record windows and doors on a take off?

Hint   Use a ddt (deduction) to help you.

Extend   How would you handle the reveals in openings?

Intermediate floors to include timber joists and precast concrete beam and block

An intermediate floor is the first floor and above. It differs from a ground floor in that it is suspended from a load-bearing wall. The method of suspension can be proprietary joist hangers or building the ends of the joists into the wall. Timber framed construction uses floor prefabricated panels which are secured to the header plate of each wall panel.

A typical timber floor take off would appear like Figure 13.11, using the rules under NRM2 (page 174).

			Carpentry	
12/	3.60		floor joists, 255 x 65 mm treated sawn	
		43.20	softwood C16 grade	
2/	12.00		4 mm welded mild steel joist hangers	
		24	for 255 x 65 joists to blockwork	
	8.00		herringbone strutting, 45 x 38 mm, treated	
		8.00	sawn wood to 225 mm deep joists	
	8.00		boarding; flooring, moisture resistant	
	3.60		over 600 mm wide, 22 mm thick, horizontal	
		28.80		

▶ **Figure 13.11:** A typical take off for an intermediate floor. Herringbone strutting stops lateral movement of the floor

Theory into practice

Obtain a simple drawing of a domestic house that is fully dimensioned. Undertake the following tasks:

- Formulate a take off list for the external cavity walls.
- Take off the external cavity walls using dimension paper.
- Include deductions for openings.

Research

Locate a roof truss manufacturer's website and analyse the specification of a roof truss. How could you use this information in a bill of quantities?

Roof construction and finishes

Trussed rafter construction are enumerated items that are specified as manufacturer's details. They are described as a finished dimension description with a manufacturer's reference if applicable.

A traditional roof is a cut roof with elements having to be assembled on site by craftspeople.

A take off list for a traditional dual pitch roof would follow the following sequence, from the installation of the supporting structure through to the roof tile finishes.

1. Wall plates
2. Wall plate straps
3. Rafters
4. Ridge board
5. Bracing
6. Roof membrane underlay
7. Treated tile battens
8. Roof tiles
9. Ridge tiles

The take off would look like Figure 13.12.

			Carpentry
			Structural timbers
2/	12.00		100 x 65 mm wall plate, bedding in cm (1:4)
		24.00	
14/	1.20		isolated metal members, 38 x 5 mm x 1.2 mm long,
		16.80	bent 100 mm, fixed using stainless steel screws
			and plugs to blockwork
14/2/	4.50		255 x 65 mm rafters, treated
		126.00	
	12.00		250 x 30 mm ridge board, treated
		12.00	
2/	12.00		Roof coverings, interlocking roof tiles ref A39
	4.50		32 degree pitch, Tyvek® breathable underlay
		108.00	fixed using 38 x 25 treated tile battens
	12.00		ridge tiles, 300 mm long, half round, horizontal
		12.00	

Figure 13.12: A typical take off for a roof

B4 Production of quantities for a civil engineering project

The taking off of quantities for a civil engineering project uses CESMM, as follows.

- **Excavation works** within the standard method excavation works are called 'earthworks' and are measured according to the rules in Class E (page 32 of CESMM4.) Excavation is measured by dredging, cuttings, foundations and general excavation. All of these employ heavy equipment. The second division (column 2) covers the materials that are excavated and column 3 specifies the depth. Further columns within the standard method cover the rules to follow, definition rules, coverage rules and any additional description rules.

▸ **Filling** is covered on page 36 of CESMM4 and the division rules cover structures, embankments, general and depth, then the type of materials used for the filling followed by the same columns as excavation - definition rules, coverage rules and any additional description rules.

▸ **Formwork** is covered under concrete ancillaries, which describes finish, vertical or horizontal direction and the range of widths of the formwork. Again, a series of rules on the opposing page help you with the descriptions against each item measured.

▸ **Reinforcement** is measured in terms of the type of reinforcing bar used, for example mild or high tensile steel bars. The nominal size is specified or, if it is a fabric, the weight per metre squared.

▸ **Concreting works** is covered under Class F and includes the design of the concrete, its strength and the aggregate size.

▸ **Drainage and manholes** are covered by Class I, which includes the pipe in the trench, and Class K which includes manholes and any pipework ancillaries. Pipes are measured by the type of material they are manufactured from, their nominal bore and the depth of the trench. Manholes are measured by the material they are made from and their depth of installation.

			Pumping Station 1
			Excavate for foundations
			depth
			98500
			topsoil 200
			98300
			rock level 94600
			3700
			Excavate for foundations, other than
9.40			topsoil, rock max depth 5–10 m
12.80			E326
3.70			
	445.18		
			and
			disposal
			E232
			depth
			94600
			bed, blinding, screed 850
			88560
			6040
9.40			Excavate rock; max depth 5–10 m
12.80			E336
6.04			and
	726.73		disposal
9.40			Prepartion of excavated surfaces
12.80			rock
	726.73		
			In-situ Concrete
			Provision of Concrete
9.40			Designed mix grade C20; sulphate
12.90			resisting cement to BS
0.70			Bases, footing and pile caps
	84.22		

▶ **Figure 13.13:** A take-off for a concrete structure in civil engineering

C Undertake the production of bills of quantities

Bills of quantities have been discussed throughout this unit. A bill of quantities is the final document that is sent out to all companies being invited to tender so that all tender prices are prepared using the same information. Now it is time to consider how you would produce one.

C1 Abstraction of quantities

Use of 'cut and shuffle' paper

This traditional method is now obsolete. Quantity surveyors used to cut up dimension paper and then collate all the quantities from one measured item and tie them with a string. It is never used today, having been replaced by digital taking-off software processes.

Use of direct billing paper

This method is employed by many main contractors who wish to take off the quantities and then apply rates on the same page without any formal abstraction process. It saves the abstraction times and writing out a bill of quantities. The whole process is sequential, following the method of construction from the ground out. Pages can be totalled and a final summary compiled quickly and efficiently. Figure 13.14 gives an example.

Times	Dims	Sum	Waste			Unit	Total	Rate	Total
	28.80		Excavation, foundation excavation,			m3	29	£6.50	£188.50
	1.00		depth not exceeding 2 metres						
		28.80			2/ 10.00				
					2/ 5.00				
					30.00				
			&	ddt	1.20				
				cl	28.80				
			disposal, excavated material off site,						
			to tip 5 km away						

▶ **Figure 13.14:** Direct billing paper is a quicker and more accurate method than 'cut and shuffle'

Use of abstract paper

Abstraction is the process of collecting all the quantities taken off into a single total for each. They are assembled onto sheets that collate all the same items from the dimension sheets. Final item quantity calculations are then performed where any deductions are collected and final quantities are rounded up with the units annotated, as shown in Figure 13.15. The final quantity is the one that is placed within the bill of quantities along with the standard method of measurement description for that item.

Facing brickwork 215×102.5×65 in cm (1:4) Stretcher bond. Skins of walls				Form cavities 100 mm wide Using stainless steel wall ties 4/m²				Brickwork 100×215×440 in cm (1:4) Stretcher bond S.O.H wall			
1	76.15	6	6.80	3	68.75	6	6.80	5	4.50		
2	29.20	7	9.20	4	34.10	7	4.20	9	74.80		
=	105.35	=	11.00	=	102.85	=	11.00	=	79.30		
–	11.00			–	11.00						
=	94.35	5 –		=	91.85						
	94 m²				92 m²						

▶ **Figure 13.15:** An example of an abstract paper. Note how each section is totalled then rounded up to a whole unit

Theory into practice

Produce an abstract to accompany the dimension sheets for the small extension that you took off in the exercise in Assessment practice 13.2.

⏸ PAUSE POINT   What are the advantages of direct billing paper?

Hint   What can it save?

Extend   Find an example of direct billing paper and identify how it has benefited the contractor and client.

C2 The production of a bill of quantities for a building or civil engineering project

Production of a bill of quantities for a building work section

As has been discussed, the bill of quantities is assembled from the dimension sheets, using one of the methods described in C1 above and the same sections, order and methods of measurement specified by NRM2.

Format and layout

The bill of quantities is usually a typed document in a tabulated format as Figure 13.16, a bill page using section 16 of NRM (Carpentry), illustrates. The main contractor would fill in the rates and then calculate the total for each item.

POTTER DRAPER PARTNERSHIP			ROOF	
ROOF		Unit rate	Total	
ROOF				
16 - CARPENTRY				
TIMBER FIRST FIXINGS				
Primary or structural timber; treated sawn softwood Grade C16				
A	75 x 125 Trimmers	574	m	
B	75 x 175 Trimmers	574	m	
	Backing and other first fix timbers, treated sawn softwood			
C	25 x 50 Ridge batten	72	m	
D	50 x 50 Tilting fillet	344	m	
E	25 x 50 Packing piece	165	m	
F	50 x 50 Packing piece	165	m	
G	50 x 50 Packing piece, shot fired to steel (approximate)	1150	m	
H	50 x 50 Packing piece, plugged and screwed	52	m	
J	50 x 125 Packing piece, shot fired to steel	52	m	
K	50 x 100 Framing	100	m	
L	75 x 75 Framing	378	m	
M	85 x 85 Framing	235	m	
N	100 x 100 Ground, once splayed	32	m	
P	100 x 130 Ground, once splayed	32	m	
Q	25 x 55 Batten	31	m	
R	50 x 50 Batten	390	m	
S	50 x 50, Shot fired to steel	455	m	
T	50 x 75 Batten	126	m	
U	50 x 100 Batten	343	m	
V	50 x 150 Batten	140	m	
W	75 x 150 Batten	279	m	
Olive Academy **Thurrock** **B7547**	3/11	*To Collection £*		

▶ **Figure 13.16:** A bill page from NRM (source: **www.thurrock.gov.uk**). The last two columns are where the unit rate is written and a subtotal

Note the following from Figure 13.16:

▶ The items are all referenced with a letter down the left-hand side (Column 1).

▶ Column 2 is the full description that follows the levels detailed in NRM2.

▶ Column 3 is the total quantity taken from the abstract.

▶ Column 4 is the unit of measurement.

▶ Column 5 is the unit rate.

▶ Column 6 is the subtotal.

Each page has a page total which is sent to a collection page, where figures are totalled and carried to a final summary. Each collection summary is held on a summary page. Finally, the main summary shows the total figure for the project.

Along with the drawn information, bill items need to contain enough detail for the contractor's estimator to price the rate for each item.

Assessment practice 13.3

The senior quantity surveyor has asked you to produce a bill of quantities for the small extension for a client that was described in Assessment practice 13.2, as they wish to go out for a competitive tender. Produce the following:

- An accurate and comprehensive bill of quantities for the extension, making assumptions, where necessary, of drawn detail.
- All finishes included in a comprehensive document.
- A description for the uninformed client of how the taking-off process operates, including the different methods that can be used to convert the take off into a bill of quantities.

Plan

- How will I ensure I will follow the sequence for the bills of quantities?
- Have I downloaded the most recent version of NRM2?

Do

- Have I covered a comprehensive range of bill items, including preliminaries?
- Have I typed and formatted the documents appropriately?

Review

- Are all the collection pages in place?
- Is a main summary included?

Further reading and resources

Cartlidge, D. (2012) *Quantity Surveyor's Pocket Book*, 2nd edition, London: Routledge.

Institution of Civil Engineers (2012) *CESMM4: Civil Engineering Standard Method of Measurement*, 4th edition, London: ICE Publishing.

Lee, S., Trench, W. and Willis, A. (2014) *Willis's Elements of Quantity Surveying*, 12th edition, Oxford: John Wiley.

Royal Institution of Chartered Surveyors (2012), *NRM 1 RICS New Rules of Measurement: Order of Cost Estimating and Cost Planning for Capital Building Works*, 2nd edition, Coventry: RICS.

Royal Institution of Chartered Surveyors (2012), *NRM 2 RICS New Rules of Measurement: Detailed Measure for Building Works*, 1st edition, Coventry: RICS.

Royal Institution of Chartered Surveyors (2014), *NRM 3 RICS New Rules of Measurement: Order of Cost Estimating and Cost Planning for Building Maintenance Works*, 1st edition, Coventry: RICS.

Website

A forum for finding and sharing construction information:
www.designingbuildings.co.uk/wiki/Home

THINK ▶▶FUTURE

Linda

Linda is a graduate quantity surveyor who has just started working for a city quantity surveying practice. The office undertakes works that include building construction and civil engineering so Linda has to know the layout and contents of the New Rules of Measurement 2 and the Civil Engineering Standard Method of Measurement.

Linda needs to understand the information that can be abstracted from drawings so that accurate quantities can be taken off. The office uses manual means but is slowly moving to digital take offs taken directly from drawings using digitisers. Linda has to have a detailed understanding of how construction projects are assembled so the take off lists she prepares are accurate and nothing is missed that may affect the cost control of a project.

The initial bill production stage is important and must be accurate so that a client receives a competitive price for their projects that will not be subject to any variations due to omissions during the take off process. Linda is developing her experience with both civil and construction projects and now understands what is contained within each method of measurement.

Focusing your skills

A quantity surveyor must possess a range of skills to accurately measure from design drawings to enable the production of budgets for tendering purposes. They must:

- be able to understand complex drawn information
- have mathematical skills to perform calculations

- show accurate attention to detail, for example when following sequences
- have digital and computer skills for recording quantities
- demonstrate patience by working methodically.

Getting ready for assessment

Jatinda has two pieces of assessment to produce for this unit, which she is completing as part of her BTEC National in Construction and the Built Environment.

▶ The first is a guidance document, which could be used as an instructional leaflet.

▶ The second is the production of a bill of quantities, using techniques involving taking off quantities using dimension paper.

How I got started

I investigated the main two standard methods of measurements for construction (NRM) and civil engineering (CESMM). I had to be able to understand and provide a summary of how they are used. This guide should explain in simple terms what a rule of measurement does and how this is achieved. Both the Royal Institution of Chartered Surveyors and the Institution of Civil Engineers publish information about their standard methods of measurements.

I looked at the Unit 13 content, specifically the A2 section, which details the standard methods of measurement. Practising the taking-off procedures for the dimension of quantities helped the final production of my bill of quantities. I used drawings from Unit 7 to support this, providing dimensions to use for take-offs.

I also referred back to the practice sessions we produced earlier in the course. This helped with applying the correct procedures to using dimension paper and the layout used. By using a drawing with foundation details, I was able to contrast the use of the two methods of measurement.

How I brought it all together

I researched all the different standard methods of measurement used for construction and civil engineering. I had to 'appraise' the documents as they are too large to reproduce, and summarise the key guidance I wanted to include in my learner document.

The second assessment is the larger of the two. This involved dimensioning quantities from construction and civil engineering related drawings. This process uses dimension paper, which is then abstracted to produce the final quantities. I looked online at some examples of the

writing of a bill item, and this helped me to understand how to achieve this for a range of quantities.

You have to be accurate when you take off. Formulating a take off list is essential as it prepares for the dimensioning of all the elements that make up a measured structure. At a Distinction level you have to demonstrate the methodology you used. When researching I found some examples that demonstrate current methods of taking off quantities. The additional guidance section in this unit's specification was very helpful in providing information on achieving higher grades against each learning objective.

What I learned from the experience

The taking-off quantities part of the assessment wasn't easy to learn. I found that practice really helped with this and to help me get used to the format required and the rules applied to taking-off.

The quantity surveying textbooks also helped a lot, particularly using the examples of the layout and dimensioning of construction elements that I was able to follow to learn how they achieved final quantities. In particular, I found practice essential in helping me to understand the calculation of a centre line, using perimeter dimensions from architect's drawings.

Think about it

▶ How can you summarise a standard method of measurement into a simple format that new learners would understand, such as a leaflet?

▶ What processes and procedures do you need to follow in the production of a simple bill of quantities for a construction and civil engineering project?

Glossary

A

Accident – an event that results in injury or ill health.

Adjudication process – a meeting held by senior management to decide what level of risk and profit needs to be placed on the net tender or estimate.

Aesthetics – the expression and perception of beauty; in construction, how attractive a building or structure looks.

Agreed surveyor – a surveyor who is appointed by the building owner or adjoining owner to resolve a party wall dispute.

Air tightness – resistance to unwanted air leakage from the building. The lower the leakage, the more efficient the building.

Approved Documents – government guidance on how to meet the Building Regulations.

Areas of Outstanding Natural Beauty (AONB) – areas of countryside designated for protection due to their landscape value, such as the Mourne Mountains in Northern Ireland.

B

BREEAM (Building Research Establishment Environmental Assessment Method) – an assessment process that measures the design, construction and operation of a building against targets based on performance benchmarks.

Backsight (BS) – first staff reading of a levelling operation or the first reading after the instrument has been moved.

Baseline – the longest line between two points on a survey framework.

Bench mark (BM) – a point of known or arbitrary height used as a reference for other height measurements

Bill of quantities – a document compiled by a cost consultant or quantity surveyor containing a detailed statement of the prices, units, dimensions and other details required to complete a construction project. It becomes part of the contract.

Bitumen – one of the oldest construction materials, a black oil-based substance that has a high melting point and is impermeable to water, so is often used to waterproof flat roofs.

Black water – water which has come into contact with faeces and may contain bacteria, which can cause diseases.

Boiling point – the temperature at which a change of state from liquid to gas will begin.

Brownfield – sites that have been previously built on and may now be disused or derelict.

Buckling – a structural failure caused when a member loaded in compression becomes unstable and deflects sideways. It is more common in long, thin structural members and can cause failure well below the material's compressive strength.

Building information modelling (BIM) – a process of creating and managing digital information about the physical appearance and properties of materials used to design and construct a building.

Building notice approval – on small projects, or when making alterations to an existing building, planning approval can be obtained by applying or building notice approval. A Building Control inspector will carry out regular inspections to approve the work as it is completed.

C

CITB – the Construction Industry Training Board, which provides support and training to the construction industry, and takes a levy from larger companies to redistribute as training grants.

CSCS card – provides proof of qualifications and training. Most sites require operatives to hold a card before allowing them to work on site.

Calibration – the process of ensuring that measuring equipment is correctly adjusted to give true readings by comparing the equipment that is going to be used with a standard version.

Carbon footprint – sum of all the CO_2 emissions used in a process (usually expressed as tonnes of CO_2 emitted).

Cash flow – the amount of money flowing into the company from the client and out of the company as payments, for example salaries and payments for materials. Money flowing in should be greater than money flowing out.

Cavity wall – two walls with a gap in between known as a cavity. The outer leaf is built with bricks and the inner leaf, which is the load-bearing wall, is built of blocks. Insulation is placed in the cavity to minimise heat loss.

Centre line – the length of a line that runs horizontally through the centre of a construction element, such as a foundation or cavity wall, when viewed from above.

Chainage or running measurements – a horizontal distance measured along a line.

Check line – a line provided to check the accuracy of surveying measurements.

Change point – a known point on the ground where a staff is held and a foresight read.

Closed traverse – a traverse that encloses a defined area, with the same start and finish point.

Coarse aggregate – gravel, crushed stone, recycled crushed concrete or blast furnace slag with a grain size greater than 5 mm.

COBie – Construction Operations Building Information Exchange – an agreed protocol to share mostly non-graphical project data to support the design and construction processes, usually in the form of simple spreadsheet.

Coefficient of linear expansion (α) – a material property that describes the amount by which a material expands upon heating with each degree rise in temperature. It is measured in inverse Kelvin (K^{-1}).

Cold bridging – where a cold spot occurs between the warm internal structure and the colder external structure of a wall.

Collateral warranty – a contract where a contractor or subcontractor warrants to a third party (such as the client) that it has complied with all the terms of its contract, possibly including using materials of an agreed quality or carrying out work in a professional manner.

Collimation – line of sight through an optical surveying instrument.

Commercial risk – when a company invests money in doing something that does not have a guaranteed profit return.

Compartmentation – the division of a structure into separate fire-protected compartments to prevent the spread from a fire across a whole structure.

Competency – the combination of training, skills, experience and knowledge that a person has and their ability to apply them to perform a task safely.

Conservation area – an area designated by the local planning authority as being of particular historical or architectural interest.

Construction programme – the sequence in which the various tasks will be carried out during the construction project to enable it to be completed on time.

Contingency – a future event or circumstance that may occur, but was not planned, which may impact the schedule or budget of a construction project.

Contingency allowance – a percentage of the cost set aside to deal with resolving unforeseen events.

Contract – a written or spoken agreement between two or more parties where there is a promise to do something (such as work) in return for a benefit (such as money).

Cover level – the reduced level of an object indicating underground services, e.g. top of an inspection chamber cover.

Cover pricing – an anti-competitive practice in which companies submit an artificially high bid for work with the intention of not winning the bid. This allows other companies to submit higher than necessary prices for work, with the knowledge that they will win as their competitors have submitted very high quotes. They can then divide several tenders among themselves at higher prices.

Curing – the process that causes workable substances, such as wet concrete, paint or glue, to permanently set and harden as a result of chemical reactions.

Curtilage – the area of the plot forming the building and its land, outbuildings and boundaries.

Cyclic loading – constantly varying dynamic loads that are applied in repeated cycles.

D

DPC – damp-proof course that is inserted in walls to prevent the vertical movement of moisture.

Dangerous occurrence – a specific, reportable adverse event, as defined in the Reporting of Injuries, Diseases and Dangerous Occurrences Regulations 2013 (RIDDOR).

Datum – a reference point or set of reference points from which measurements are made.

Debt – the amount of money owed.

Deficit – the negative difference between spending and income.

Deformation – the change in the shape of a material caused by the application of a force.

Dew point – the temperature to which humid air must be cooled to reach saturation (the maximum amount of water vapour that can be contained within the air). Further cooling forces water vapour to condense as water droplets.

E

Eaves – the part of the roof that meets or overhangs the walls.

Electronic digital measure – a laser measure that bounces and returns a light wave to record the distance between two surfaces.

Embodied energy – the total energy necessary for the extraction, processing, manufacture and delivery of building materials to the construction site. Embodied energy is used as an indicator of the overall environmental impact of building materials.

Enactment – a legislative regulation.

End bearing – the area that the lintel rests on upon the supporting structure.

Engineered timber – timber that has been engineered and reconstructed into a component.

Environmental Impact Assessment (EIA) – the process of assessing the environmental effects of a development project, implemented under the Town and Country Planning (Environmental Impact Assessment) Regulations 2011.

Estimating – the process of compiling the price for a tender or quotation.

Exemption – a circumstance when regulations do not apply, as long as the proposal meets the requirements of the schedules in the legislation e.g. Building Regulations.

Extruded – method of production where soft and malleable materials are pushed through a hole in a die plate to form long lengths of shaped material with uniform cross-section.

F

F10 – a formal document submitted to the Health and Safety Executive that advises what, where and when a construction project will start and who is undertaking the work with key contact details.

Face left – in surveying, when the vertical circle is to the left of the telescope.

Face right – in surveying, when the vertical circle is to the right of the telescope.

Feasibility – deciding whether the building is either practicable or will proceed.

Feasibility study – an assessment of the likely success of a project by looking at the advantages and disadvantages of the proposed project at an early stage. This will generally relate to the cost of the project and the benefits.

Ferrous metals – metal alloys containing iron, such as cast iron, steel and stainless steel.

Field test – a test carried out on an actual site, under real conditions.

Fine aggregate – sand with a grain size of less than 5 mm.

Footprint – the shape of the building on the ground when viewed in plan from above.

Foresight (FS) – last staff reading of a levelling operation before the level is moved.

Formation level – the reduced level of the bottom of an excavation, such as a sewer trench.

Foundation – the lowest load-bearing part of a structure, usually below ground level.

Full plans approval – the person wishing to carry out the building work submits plans and drawings to local authority Building Control showing the construction details of the project for approval before the work starts.

Functional skills – skills connected to running the business, such as marketing and commercial awareness, project management, logistic and procurement skills (receiving, storing and stock control), financial and operational skills (management of production costs and deadlines).

G

Gas Safe – the approvals organisation that licences plumbers and heating engineers to work on gas appliances through a registration and qualification service.

Going – the horizontal distance of a stair.

Greenbelt – no-build area around some towns and cities to protect the environment and prevent urban sprawl.

Greenfield – sites that have not previously been built on.

Greenhouse gases – gases such as carbon dioxide, methane and water vapour that stop heat escaping from the Earth's atmosphere. Increasing amounts of greenhouse gases can lead to global warming and climate change.

Grey water – waste water generated from water consumption that has not come into contact with black water.

Gross estimate – the net costs plus overheads and profit and risk items.

H

Hazard – anything that may cause harm.

Height line (HL) – [in construction drawings] a vertical line running parallel to the picture plane that indicates the height of construction elements.

Horizontal angle – an angle measured between three fixed points within a horizontal plane.

Hot work permit – a permit prepared by a competent person to allow any operation involving open flames or producing heat or sparks.

Hydration – a chemical reaction in which a substance combines with water.

I

Induction – an introduction to the site and the initial information that you need to enter.

Inflation – an economic situation that leads to an increase in prices and a fall in the value of money.

Initial project brief – the brief prepared by the architect and client to establish the project objectives.

Intermediate floors – those that support the first floor and floors above.

Intermediate sight – any staff reading taken between the backsight and foresight readings.

Invert level – the reduced level of the lowest part of the internal diameter of a sewer, trench, pipe or tunnel.

K

Kelvin (K) – a unit of temperature. In many scientific calculations temperature must be stated on the absolute, or kelvin (K), scale. 0° Celsius (C) corresponds to 273 K, 20°C corresponds to 293 K and minus 20°C corresponds to 253 K.

Kinetic energy – the energy possessed by an object due to its motion.

Kitemark – a quality certification mark issued by the British Standards Institution (BSI).

L

Land banking – when developers buy up available land but do not build on it.

Lateral – the horizontal restraint required by a structure to prevent sideways movement.

Legislation – a law passed by a government that sets a series of legal requirements.

Line style – types of line in CAD; typically solid, dashed or chain-linked with same representation as lines drawn manually.

Line weight – the thickness of a line drawn in CAD.

Local amenities – facilities that are available or provided for the local community, such as leisure centres, GP surgeries, waste disposal, parks, post offices and schools.

M

Made up ground – ground that has been filled with other materials to make up levels.

Management – the process of organising or controlling things or people.

Managerial skills – skills needed by senior staff, such as leadership, strategic planning, use of different management and business intelligence tools, ability to allocate resources or refining an existing strategy.

Material property – factors that are independent of the size and shape of an object that is made from a particular material. For example, density is a material property that does not depend on the size of the object. However, the mass and volume of an object depend on its size and so are not properties of the material itself.

Melting point – the temperature at which a phase or state change from solid to liquid will begin.

Method statement – documents that identify the methods used to price the work items, the plant and the labour required for each activity.

Modern methods of construction (MMC) – a range of techniques involving off-site manufacture or assembly, or more efficient onsite methods, to minimise construction time onsite.

Molecules – the smallest unit of a chemical compound, made up of groups of atoms.

N

nanometre (nm) – the unit of measurement for the tiny wavelengths of visible light, where 1nm = 0.0000000001m.

Near miss – an event not causing harm, but that has the potential to cause injury or ill health.

Necking – a phenomenon in ductile materials once they exceed their elastic limit and undergo plastic deformation as they approach tensile failure. As the material is stretched it can form a narrow neck with a reduced cross-sectional area at the point of eventual fracture.

Net estimate and net costs – the estimate for the total costs of labour, plant, materials, preliminary items and subcontractors, minus the financial benefits (profit) to be gained by the work for the contractor. They do not include overheads and costs incurred by extras or uncertainties (*see gross estimate*).

Nominated subcontractor – a subcontractor who is appointed by the client and has already tendered for the work package for the client. The main contractor is then instructed to appoint this subcontractor and the value is offset against a provisional sum placed within the main tender documents.

NRM – New Rules of Measurement, a standard set of measurement rules and essential guidance produced by RICS for the cost management of construction projects and maintenance works.

O

Open traverse – a traverse with different start and end points.

Operational resources – assets to the business that help it to function, including human resources and staff; physical resources such as machinery and industrial equipment; a distribution network; and intangible resources such as brand, licences, insurance and reputation.

Organisational capacity – the ability of the business to take on a certain amount of work, for example a small independent builder would not have the organisational capacity to build a hospital development.

Overheads – costs that need to be met in order for the company to continue to run, such as staff salaries, cost of premises, price of equipment.

Oxidation – a chemical reaction in which oxygen is gained.

P

Package of work – a group of related tasks within a project. These can be tendered for by subcontractors, and allow the main contractor to transfer some risk to other companies.

Partnering agreement – a method of procuring a contractor that enables cost sharing between client and contractor to make savings.

Party wall – a wall that is located on the land of two separate owners. It may form part of the building or be a garden wall.

Patress – container for the space behind electrical fittings, such as light switches and plug sockets.

Permeable – a material that allows water to flow through it.

Permeability – the ability of a material to transmit water. For example, a soil with many voids can allow more water to percolate through it and will therefore provide better drainage.

Perpendicular offset – a distance measured at right angles (90°) to the survey line.

Picture plane (PP) – [in construction drawings] an imaginary vertical plane where the drawing is located.

Pitch – the steepness of a roof.

Planning Portal – the government website that provides detailed information on all aspects of planning.

Plant – machinery used for a construction activity, such as cranes, concrete mixers, excavators, loaders, forklifts, bulldozers and mobile elevated work platforms (MEWPs).

Plastic deformation – permanent deformation of a material which, unlike elastic deformation, will not spring back into shape when stress is removed.

Polygonal framework – a many-sided closed shape with measured angles and plan distances.

Portal – a large gateway or doorway.

Price book – a published book that contains current prices and rates for items of work based on NRM2 (Royal Institution of Chartered Surveyors New Rules of Measurement 2).

Prime cost (PC) sums – the likely price of items difficult or impossible to price accurately at the time of estimation.

Principal contractor – under CDM 2015, this individual plans, manages and co-ordinates construction work to ensure a project can be built safely.

Principal designer – under CDM 2015, this individual plans, manages and co-ordinates the planning and design work to ensure a project can be built safely.

Procurement – the process or act of buying or acquiring goods or materials.

Procurement route – the method that a client uses to select a contractor to construct the project.

Purlin – a horizontal beam that runs the length of a dual pitch roof with gable ends.

R

Reasonably practicable – a phrase often used in health and safety legislation to mean that reasonable steps should be taken to meet the requirements.

Reconciling – settling the values of unknown costs.

Reduced level (RL) – the height of a point or feature relative to the height of a given bench mark.

Restrictive covenant – a legal agreement or clause associated with particular land or property that may limit the use of the building. It is usually a promise to do or not to do something on the land.

RICS – the Royal Institution of Chartered Surveyors, a UK professional body for qualifications and standards in land, property, infrastructure and construction.

Right of way – a privilege allowing someone to pass over land belonging to someone else.

Rise – the vertical height of a stair.

Risk – the likelihood that someone will be harmed if they are exposed to a hazard, with an indication of how serious that harm could be.

S

Safe system of work – the methods in place to minimise risks from carrying out activities in the workplace.

Safety cycle – a continuous process of developing and maintaining safety systems, for example by using a plan, do, check and act system.

Schedule of rates – a list of the hourly rates of staff, types of labour and plant hire used when a contractor is calculating prices for a tender.

Settlement – consolidation or decrease in the volume of the soil due to the weight of the building.

Shock loading – rapidly increasing dynamic loading such as that caused by an impact, for instance, a vehicle colliding with a bridge support.

Simply supported – the free ends of a beam supported at either end.

Site induction – the presentation that makes new visitors or workers on a site aware of the rules of the site and the hazards that it contains.

Sites of Special Scientific Interest (SSSI) – sites protected by law to conserve their wildlife or geology.

Skeleton frame – a rectangular framed structure made of several elements joined together.

Sketch plans – rough plans made by an architect or designer according to the client's requirements.

Slump test – a test of concrete's workability to determine whether it is of the right quality and can be poured with the correct results.

Snagging – the process of checking a new building for defects and rectifying the faults to comply with the client's requirements.

Sole plate – the timber plate that the timber-frame building rests upon when the first stage is installed.

Specification – a description of the materials and workmanship required for a specific construction project.

Spectator/observer position (S) – [in construction drawings] the point at which the observer looks at the building object. The height of the observer is the distance between the eye level and the ground line.

Steradian – a solid angle defining the size of a three-dimensional cone, used in the measurement of light.

Storeys – the floor-to-ceiling height of a building and the number of floors.

Strategic brief – initial objectives of the project are established and the client may consider different sites or choose between a refurbishment or new build.

Struck – the process of removing formwork and falsework by taking it away from the set concrete.

Stud wall – an internal wall that divides rooms, constructed as a frame with elements resting on studs or fixings.

Subsidence – the downward movement of the ground on the site that is not related to the weight of the building.

Subsoil – the soil immediately beneath the surface soil, on which a structure will be built.

Substructure – all work below the ground floor level including damp-proof membrane and foundations.

Suitable and sufficient – a phrase often used in legislation to mean that enough of something (e.g. toilets) should be provided depending on the number of people on site.

Sulphate – a water-soluble compound that reacts with the cement in concrete.

Superstructure – the part of the building above ground level.

Survey framework – a series of interlinking lines, the lengths and the angles of which uniquely fix the shape of the framework.

Survey line – any direct plan distance measured between survey stations.

Survey station – a point at the end of a survey line.

Suspended ceiling – a ceiling finish that appears to hang from the ceiling structure. Wires hold the ceiling tracks in place and are fixed to the supporting floors or roof structure.

Sustainable development – growth that meets the needs of today's society without impacting on future generations.

T

TRADA – the Timber Research and Development Association.

Taken off/taking off – a detailed analysis of the aspects of a construction project that can be measured and priced (such as labour and materials) so that a bill of quantities can be produced for a tender.

Temporary bench mark (TBM) – an arbitrary fixed point used during a survey or construction works.

Temporary dimensions – dimensions that appear on screen in a CAD program to help place elements accurately relative to a fixed reference grid or level datums.

Tender documents – the drawings, specifications, bills of quantities, letters and forms that are part of a tender.

Tendering process – a client invites companies to submit a formal, fully costed written proposal for carrying out particular work.

Thermal bridging – this is heat lost from a localised area within a building. It occurs when an area within a building has a higher thermal conductivity than the surrounding materials, which allows heat to escape from the building.

Thermal mass – the property of a material that relates to its ability to absorb and store heat energy.

Tie line – survey line connecting a point to other lines, often to check accuracy or to locate a feature. Also known as a check line.

Toolbox talk – an informal safety meeting, where work is stopped to discuss a safety aspect of a current job.

Topography – the surface features of an area.

Traverse – a series of connected lines with known lengths and directions.

Turnover – the amount of money that a business receives within a given time.

U

Underpinning – installing a system of supports or strengthening a foundation beneath an existing structure.

V

Vanishing point (VP) – [in construction drawings] a point on the horizon eye line where all parallel receding lines converge.

Variations – items that were not in the client's original budget and therefore are additional to the contract, such as changes to the design or additional quantities of material not in the original contract.

Verification – the determination of whether or not an instrument conforms to a published (normally the manufacturer's) specification.

Void – an empty space or gap in a structure.

W

Water table – the level that the water reaches within the subsoil.

Well-conditioned triangles – points on a survey that make roughly equilateral triangles with small internal angles ranging from 30° to 120°.

Index